P & S BIOMEDICAL SCIENCES SYMPOSIA Series

HENRY J. VOGEL, Editor

College of Physicians and Surgeons
Columbia University
New York, New York

Henry J. Vogel (Editor). *Nucleic Acid–Protein Recognition*, 1977

Arthur Karlin, Virginia M. Tennyson, and Henry J. Vogel (Editors). *Neuronal Information Transfer*, 1978

Benvenuto Pernis and Henry J. Vogel (Editors). *Cells of Immunoglobulin Synthesis*, 1979

Benvenuto Pernis and Henry J. Vogel (Editors). *Regulatory T Lymphocytes*, 1980

Georgiana Jagiello and Henry J. Vogel (Editors). *Bioregulators of Reproduction*, 1981

Hymie L. Nossel and Henry J. Vogel (Editors). *Pathobiology of the Endothelial Cell*, 1982

PATHOBIOLOGY OF THE ENDOTHELIAL CELL

PATHOBIOLOGY OF THE ENDOTHELIAL CELL

Edited by

HYMIE L. NOSSEL
HENRY J. VOGEL

College of Physicians and Surgeons
Columbia University
New York, New York

1982

ACADEMIC PRESS

A Subsidiary of Harcourt Brace Jovanovich, Publishers

New York London
Paris San Diego San Francisco São Paulo Sydney Tokyo Toronto

ACADEMIC PRESS, INC.
111 Fifth Avenue, New York, New York 10003

United Kingdom Edition published by
ACADEMIC PRESS, INC. (LONDON) LTD.
24/28 Oval Road, London NW1 7DX

Library of Congress Cataloging in Publication Data
Main entry under title:

Pathobiology of the endothelial cell.

(P & S biomedical sciences symposia series)
Includes bibliographical references and index.
1. Endothelium--Diseases. 2. Endothelium.
I. Nossel, Hymie L. II. Vogel, Henry James, Date.
III. Series.
RC691.4.P37 616.1'3 82-4066
ISBN 0-12-521980-6 AACR2

PRINTED IN THE UNITED STATES OF AMERICA

82 83 84 85 9 8 7 6 5 4 3 2 1

Contents

Vessel Wall Growth Control

STEPHEN M. SCHWARTZ, CORINNE M. GAJDUSEK, AND GARY K. OWENS

Angiogenesis *in Vitro:* Implications for Tumor Biology

JUDAH FOLKMAN AND CHRISTIAN HAUDENSCHILD

II COAGULATION PROTEASES

Interactions of Thrombin with Endothelium

P. CHRISTER BUSCH AND WHYTE G. OWEN

Protein C, a Regulator of Prothrombin Activation

JOHAN STENFLO, BJÖRN DAHLBÄCK, PER FERNLUND, AND KOJI SUZUKI

Leukocyte–Endothelial Interactions

RICHARD L. HOOVER AND MORRIS J. KARNOVSKY

Complement-Mediated Intravascular Leukostasis and Endothelial-Cell Injury

PHILIP R. CRADDOCK

T-Lymphocyte Activation by Antigen Presented by HLA-DR Compatible Endothelial Cells

DENIS R. BURGER, R. MARK VETTO, ANNE HAMBLIN, AND D. C. DUMONDE

A Recognition Function of Endothelial Cells: Directing Lymphocyte Traffic

EUGENE C. BUTCHER, GEORG KRAAL, SUSAN K. STEVENS, AND IRVING L. WEISSMAN

Pulmonary Endothelial Cell Injury Induced by Antibody Fragments to Angiotensin Converting Enzyme

PETER R. B. CALDWELL, H. JOACHIM WIGGER, VINCENT P. BUTLER, JR., AND IRMA GIGLI

VII VASCULAR PATHOBIOLOGY

Responses of Endothelial Cells to Injury

JOHN L. GORDON AND JEREMY D. PEARSON

Vital and Functional Activities of Endothelial Cells

UNA S. RYAN AND JAMES W. RYAN

Endothelial Cell Function in Altered Oxygen Environments

S. ALEX STALCUP, GERARD M. TURINO, AND ROBERT B. MELLINS

List of Participants

ANTHONY, ADAM, Department of Biology, Pennsylvania State University, University Park, Pennsylvania 16802

BALDWIN, WILLIAM M., University of Leiden, Leiden, The Netherlands

BERTANI, TULLIO, Department of Pathology, College of Physicians and Surgeons, Columbia University, New York, New York 10032

BIRKENBACH, MARK, Department of Pathology, College of Physicians and Surgeons, Columbia University, New York, New York 10032

BLANC, WILLIAM A., Department of Pathology, College of Physicians and Surgeons, Columbia University, New York, New York 10032

BOLHUIS, PIETER A., Department of Hematology, University Hospital, 3500 Utrecht, The Netherlands

BORNSTEIN, PAUL, Departments of Biochemistry and Medicine, University of Washington, Seattle, Washington 98195

BOYLES, JANET, Section of Cell Biology, Yale University School of Medicine, New Haven, Connecticut 06510

BRAUNHUT, SUSAN, Department of Pathology, College of Physicians and Surgeons, Columbia University, New York, New York 10032

BROEKMAN, M. J., Divisions of Hematology-Oncology, Department of Medicine, New York Veterans Administration Medical Center, New York, New York 10010

BUCHHAGEN, DOROTHY, Department of Microbiology and Immunology, Downstate Medical Center, State University of New York, New York, New York 11203

BURGER, DENIS R., Surgical Research Laboratory, Veterans Administration Hospital, Portland, Oregon 97201

BUSCH, P. CHRISTER, Department of Pathology, University of Uppsala, Uppsala, Sweden

BUTCHER, EUGENE C., Department of Pathology, Stanford University School of Medicine, Stanford, California 94305

BUTLER, VINCENT P., JR., Department of Medicine, College of Physicians and Surgeons, Columbia University, New York, New York 10032

CALDWELL, PETER R. B., Department of Medicine, College of Physicians and Surgeons, Columbia University, New York, New York 10032

CANFIELD, ROBERT E., Department of Medicine, College of Physicians and Surgeons, Columbia University, New York, New York 10032

CARON, PHILIP C., Department of Pathology, College of Physicians and Surgeons, Columbia University, New York, New York 10032

COHN, ZANVIL A., Laboratory of Cellular Physiology and Immunology, Rockefeller University, New York, New York 10021

COLLEN, DÉSIRÉ, Center for Thrombosis and Vascular Research, Department of Medical Research, Gasthuisberg, B-3000 Leuven, Belgium

COLLINS, MARGARET H., Department of Pathology, College of Physicians and Surgeons, Columbia University, New York, New York 10032

COOK, PERRY, Department of Medicine, College of Physicians and Surgeons, Columbia University, New York, New York 10032

COTRAN, RAMZI S., Department of Pathology, Harvard Medical School, Boston, Massachusetts 02115

CRADDOCK, PHILIP R., Division of Hematology–Oncology, University of Kentucky, Lexington, Kentucky 40536

DAHLBÄCK, BJÖRN, Department of Clinical Chemistry, University of Lund, Malmö General Hospital, S-21401 Malmö, Sweden

DALECKI, MARY S., Department of Medicine, College of Physicians and Surgeons, Columbia University, New York, New York 10032

DAUKAS, GAIL, Department of Biology, University of Pennsylvania, Philadelphia, Pennsylvania 19104

DEVINE, EVELYN A., Division of Medical Genetics, Mt. Sinai School of Medicine, New York, New York 10029

DOYLE, MARY JEAN, Department of Medicine, University of Wisconsin, Madison, Wisconsin 53706

DUMONDE, D. C., Department of Immunology, St. Thomas' Hospital, London, England SE1 7EH

ESMON, CHARLES T., Department of Pathology, University of Oklahoma, Oklahoma City, Oklahoma 73190

ESMON, NAOMI L., Department of Pathology, University of Oklahoma Health Sciences Center, Oklahoma City, Oklahoma 73190

FALCONE, DOMENICK J., Department of Pathology, Cornell University Medical College, New York, New York 10021

FERNLUND, PER, Department of Clinical Chemistry, University of Lund, Malmö General Hospital, S-21401 Malmö, Sweden

FOLKMAN, JUDAH, The Children's Hospital Medical Center, Boston, Massachusetts 02115

FURIE, MARTHA, Department of Cellular Immunology, Rockefeller University, New York, New York 10021

GAJDUSEK, CORINNE M., Department of Pathology, School of Medicine, University of Washington, Seattle, Washington 98195

GERRITSEN, MARY E., Department of Physiology, New York Medical College, Valhalla, New York 10595

GIANCOTTI, FRANCESCA, Department of Pathology, College of Physicians and Surgeons, Columbia University, New York, New York 10032

GIGLI, IRMA, Departments of Dermatology and Medicine, Irvington House Institute, New York University School of Medicine, New York, New York 10016

GIGUERE, L., Cancer Research Institute, University of California Medical Center, San Francisco, California 94143

GIMBRONE, MICHAEL A., JR., Vascular Physiology Laboratory, Department of Pathology, Brigham and Women's Hospital, Boston, Massachusetts 02115

GODFREY, Maurice, Department of Pathology, College of Physicians and Surgeons, Columbia University, New York, New York 10032

GODMAN, GABRIEL C., Department of Pathology, College of Physicians and Surgeons, Columbia University, New York, New York 10032

GOODMAN, DEWITT S., Department of Medicine, College of Physicians and Surgeons Columbia University, New York, New York 10032

GORDON, JOHN L., A.R.C. Institute of Animal Physiology, Babraham, Cambridge CB2 4AT, England

GOSPODAROWICZ, DENIS J., Cancer Research Institute, University of California Medical Center, San Francisco, California 94143

GRANSTRÖM, ELISABETH, Department of Physiological Chemistry, Karolinska Institutet, Stockholm 60, Sweden

GREATON, CYNTHIA J., Department of Pathology, College of Physicians and Surgeons, Columbia University, New York, New York 10032

GROSS, JANET L., Department of Cell Biology, New York University Medical Center, New York, New York 10016

GROSSMAN, BETTY A., Department of Medicine, College of Physicians and Surgeons, Columbia University, New York, New York 10032

HAJJAR, DAVID P., Department of Pathology, Cornell University Medical College, New York, New York 10021

HAMBLIN, ANNE, Department of Immunology, St. Thomas' Hospital, London, England SE1 7EH

HANDLEY, DEAN A., Department of Physiology, College of Physicians and Surgeons, Columbia University, New York, New York 10032

HASHIM, SAMI, St. Luke's Hospital, New York, New York 10025

HAUDENSCHILD, CHRISTIAN C., The Children's Hospital Medical Center, Boston, Massachusetts 02115

HEDQVIST, P., Department of Physiological Chemistry, Karolinska Institutet, S-104 01, Stockholm, Sweden

HEIDER, JOHN G., Sandoz, Inc., East Hanover, New Jersey

HEIFETZ, AARON, Department of Biochemistry, University of Texas Health Science Center at Dallas, Dallas, Texas 75230

HOAK, JOHN C., Department of Medicine, University of Iowa Hospitals, Iowa City, Iowa 52242

HOOVER, RICHARD L., Department of Pathology, Harvard Medical School, Boston, Massachusetts 02115

HOYER, LEON W., Department of Medicine, University of Connecticut Health Center, Farmington, Connecticut 06032

HURLET, ANNE, Department of Medicine, College of Physicians and Surgeons, Columbia University, New York, New York 10032

JAFFE, ERIC A., Department of Medicine, Cornell University Medical College, New York, New York 10021

JAMES, C. R. H., Moorfields Eye Hospital, City Road, London EC1, England

JOHNSON, ALICE R., Department of Pharmacology, University of Texas Health Sciences Center, Dallas, Texas 75235

KAPLAN, KAREN L., Department of Medicine, College of Physicians and Surgeons, Columbia University, New York, New York 10032

KARANIAN, JOHN W., Department of Physiology, Georgetown University School of Medicine, Washington, D. C. 20007

KARNOVSKY, MORRIS J., Department of Pathology, Harvard Medical School, Boston, Massachusetts 02115

KATZ, MICHAEL, Department of Pediatrics, College of Physicians and Surgeons, Columbia University, New York, New York 10032

KOEHN, JAMES A., Department of Medicine, College of Physicians and Surgeons, Columbia University, New York, New York 10032

KORNBLUTH, RICHARD S., Department of Pathology, College of Physicians and Surgeons, Columbia University, New York, New York 10032

KOWALIK, SHARON, Department of Pathology, College of Physicians and Surgeons, Columbia University, New York, New York 10032

KRAAL, GEORG, Department of Pathology, Stanford University School of Medicine, Stanford, California 94305

LA GAMMA, KALLIOPE S., Department of Medicine, Columbia University, New York, New York 10032

LERNER, ROBERT G., Department of Medicine, New York Medical College, Valhalla, New York 10595

LEVIN, EUGENE, Research Institute of Scripps Clinic, La Jolla, California 92037

LEVIN, RICHARD I., Department of Medicine, Cornell University Medical College, New York, New York 10021

LIU, CHUNG YUAN, Department of Medicine, College of Physicians and Surgeons, Columbia University, New York, New York 10032

LOSKUTOFF, DAVID J., Research Institute of Scripps Clinic, La Jolla, California 92037

MCCALL, ELAINE, Department of Immunology, St. Thomas' Hospital, London SE1 7EH, England

MCPHERSON, JOHN, Departments of Biochemistry and Medicine, University of Washington, Seattle, Washington 98195

MADDOX, YVONNE T., Department of Physiology, Georgetown University School of Medicine, Washington, D. C. 20007

MAGGI, FANCO, Department of Physiology, Georgetown University, Washington, D. C. 20007

MARCUS, AARON J., Divisions of Hematology-Oncology, Department of Medicine, New York Veterans Administration Medical Center, New York, New York 10010

MARONE, PALMA ANN, Department of Pathology, College of Physicians and Surgeons, Columbia University, New York, New York 10032

MARTIN, BERNICE M., Vascular Physiology Laboratory, Department of Pathology, Brigham and Women's Hospital, Boston, Massachusetts 02115

MELLINS, ROBERT B., Department of Pediatrics, College of Physicians and Surgeons, Columbia University, New York, New York 10032

MINICK, C. RICHARD, Department of Pathology, Cornell University Medical College, New York, New York 10021

MONCADA, SALVADOR, Wellcome Research Laboratories, Beckenham, Kent BR3 3BS, England

MORAN, MARY C., Sloan-Kettering Institute, New York, New York 10021

MOSCATELLI, DAVID, Department of Cell Biology, New York University Medical Center, New York, New York 10016

MOSHER, DEANE F., Department of Medicine, University of Wisconsin, Madison, Wisconsin 53706

MULLINS, DEBORRA, Department of Cell Biology, New York University Medical Center, New York, New York 10016

MUSSONI, LUCIANA, Research Institute of Scripps Clinic, La Jolla, California 92037

NACHMAN, RALPH L., New York Hospital-Cornell Medical Center, New York, New York 10021

NICOLAIDES, MARIA N., Department of Pathology, College of Physicians and Surgeons, Columbia University, New York, New York 10032

NIEMETZ, JULIAN, Department of Medicine, Mount Sinai Medical School, New York, New York 10468

NIEWIAROWSKI, STEFAN, Specialized Center for Thrombosis Research, Temple University Medical School, Philadelphia, Pennsylvania 19140

NOSSEL, HYMIE L., Department of Medicine, College of Physicians and Surgeons, Columbia University, New York, New York 10032

OTTERNESS, IVAN G., Pfizer Central Research, Groton, Connecticut 06340

OWEN, JOHN, Department of Medicine, College of Physicians and Surgeons, Columbia University, New York, New York 10032

OWEN, WHYTE G., The Cardiovascular Center and Departments of Pathology and Biochemistry, University of Iowa, Iowa City, Iowa 52242

OWENS, GARY K., Department of Pathology, School of Medicine, University of Washington, Seattle, Washington 98195

OZZELLO, LUCIANO, Department of Pathology, College of Physicians and Surgeons, Columbia University, New York, New York 10032

PALADE, GEORGE E., Section of Cell Biology, Yale University School of Medicine, New Haven, Connecticut 06510

PAWLOWSKI, NICHOLAS A., The Rockefeller University, New York, New York 10021

PEARSON, JEREMY D., A. R. C. Institute of Animal Physiology, Babraham, Cambridge CB2 4AT, England

POGGI, ANDREINA, Specialized Center for Thrombosis Research, Temple University Medical School, Philadelphia, Pennsylvania 19140

POMERANTZ, KENNETH B., Department of Medicine, College of Physicians and Surgeons, Columbia University, New York, New York 10032

PRINCIPATO, MARY ANN, Department of Pathology, College of Physicians and Surgeons, Columbia University, New York, New York 10032

PUSZKIN, ELENA G., Department of Medicine, Albert Einstein College of Medicine, Montefiore Hospital and Medical Center, New York, New York 10467

RAND, JACOB, Department of Medicine, Albert Einstein College of Medicine, Montefiore Hospital and Medical Center, New York, New York 10467

REED, ELAINE F., Department of Pathology, College of Physicians and Surgeons, Columbia University, New York, New York 10032

RIFKIN, DANIEL B., Department of Cell Biology, New York University School of Medicine, New York, New York 10016

RYAN, JAMES W., Department of Medicine, University of Miami School of Medicine, Miami, Florida 33101

RYAN, UNA S., Department of Medicine, University of Miami School of Medicine, Miami, Florida 33101

SAFIER, L. B., Divisions of Hematology-Oncology, Department of Medicine, New York Veterans Administration Medical Center, New York, New York 10010

SAGE, HELENE, Department of Biochemistry, University of Washington, Seattle, Washington 98195

SAKARIASSEN, KJELL S., Department of Hematology, University Hospital, 3500 Utrecht, The Netherlands

SAMUELS, RUTH, Department of Pathology, College of Physicians and Surgeons, Columbia University, New York, New York 10032

SAUGSTAD, JULIE, Department of Pathology, University of Oklahoma Health Sciences Center, Oklahoma City, Oklahoma 73190

SAVION, N., Cancer Research Institute, University of California Medical Center, San Francisco, California 94143

SCHULZ, WERNER W., Department of Pathology, University of Texas Health Science Center at Dallas, Dallas, Texas 75230

SCHWARTZ, STEPHEN M., Department of Pathology, School of Medicine, University of Washington, Seattle, Washington 98195

SCOTT, WILLIAM A., The Rockefeller University, New York, New York 10021

SHEPRO, DAVID, Department of Biology, Boston University, Boston, Massachusetts 02215

SILK, SUSAN T., Department of Medicine, Cornell University Medical College, New York, New York 10021

SIMIONESCU, MAIA, Institute of Cellular Biology and Pathology, University of Bucharest, Bucharest, Rumania

SIMIONESCU, NICOLAE, Institute of Cellular Biology and Pathology, University of Bucharest, Bucharest, Rumania

SIXMA, JAN J., Department of Hematology, University Hospital, 3500 Utrecht, The Netherlands

SOBEL, JOAN, Department of Medicine, College of Physicians and Surgeons, Columbia University, New York, New York 10032

SPATZ, LINDA A., Department of Pathology, College of Physicians and Surgeons, Columbia University, New York, New York 10032

SPOKAS, ERIC, Department of Pharmacology, New York Medical College, Valhalla, New York 10595

STALCUP, S. ALEXANDER, Department of Pediatrics, College of Physicians and Surgeons, Columbia University, New York, New York 10032

STENFLO, JOHAN P., Department of Clinical Chemistry, University of Lund, Malmö General Hospital, Malmö, Sweden

STERN, DAVID, Department of Medicine, College of Physicians and Surgeons, Columbia University, New York, New York 10032

STERN, KATHLEEN, Manhattan Eye, Ear and Throat Hospital, New York, New York 10021

STEVENS, SUSAN K., Department of Pathology, Stanford University School of Medicine, Stanford, California 94305

STEWART, GWENDOLYN J., Specialized Center for Thrombosis Research, Temple University Medical School, Philadelphia, Pennsylvania 19140

STUTMAN, LEONARD, Coagulation Research Laboratory, St. Vincent's Hospital, New York, New York 10011

SUSSMAN, IRA, Montefiore Hospital and Medical Center, New York, New York 10467

SUZUKI, KOJI, Department of Clinical Chemistry, University of Lund, Malmö General Hospital, S-21401 Malmö, Sweden

TAPLEY, DONALD F., Office of the Dean, College of Physicians and Surgeons, Columbia University, New York, New York 10032

TURINO, GERARD M., Department of Medicine, College of Physicians and Surgeons, Columbia University, New York, New York 10032

ULLMAN, H. L., Divisions of Hematology-Oncology, Department of Medicine, New York Veterans Administration Medical Center, New York, New York 10010

UNANUE, EMIL R., Department of Pathology, Harvard Medical School, Boston, Massachusetts 02115

VANDE STOUWE, ROBERT A., St. Luke's-Roosevelt Hospital Center, New York, New York 10019

VANE, JOHN R., Wellcome Research Laboratories, Beckenham, Kent BR3.3BS, England

VETTO, R. MARK, Surgical Research Laboratory, Veterans Administration Medical Center, Portland, Oregon 97201

VOGEL, HENRY J., Department of Pathology, College of Physicians and Surgeons, Columbia University, New York, New York 10032

VOGEL, RUTH H., Department of Pathology, College of Physicians and Surgeons, Columbia University, New York, New York 10032

VORA, SHOBHANA, Department of Pediatrics, College of Physicians and Surgeons, Columbia University, New York, New York 10032

WEISSMAN, IRVING L., Department of Pathology, Stanford University School of Medicine, Stanford, California 94305

WEITZ, JEFFREY, Department of Medicine, College of Physicians and Surgeons, Columbia University, New York, New York, 10032

WEKSLER, BABETTE B., Department of Medicine, Cornell University Medical College, New York, New York 10021

WEKSLER, MARC E., Department of Medicine, Cornell University Medical College, New York, New York 10021

WHITE, PETER, Presbyterian Medical Center, Philadelphia, Pennsylvania 19104

WIGGER, H. JOACHIM, Department of Pathology, College of Physicians and Surgeons, Columbia University, New York, New York 10032

WITTE, LARRY D., Department of Medicine, College of Physicians and Surgeons, Columbia University, New York, New York 10032

WONG, K. T. H., Department of Medicine, College of Physicians and Surgeons, Columbia University, New York, New York 10032

WONG, PATRICK Y.-K., Department of Pharmacology, New York Medical College, Valhalla, New York 10595

Preface

A harbinger of endothelial pathobiology can be seen in Lister's 1865 paper on the role of vessel wall injury in promoting blood clotting. Despite this early interest in the vessel wall, studies of hemostasis and thrombosis tended to focus on the fluid phase and on platelets which were described by Bizzozero just 100 years ago. Only rather recently have components of the vessel wall begun to attract the attention they seem to deserve in the context of hemostasis and thrombosis as well as of such areas as atherosclerosis and pulmonary function. Now, research on the endothelium and on the remarkable and versatile endothelial cell is increasing dramatically.

A symposium on the "Pathobiology of the Endothelial Cell" was held at Arden House, on the Harriman Campus of Columbia University, from June 5 through June 7, 1981. The meeting was the sixth of the P & S Biomedical Sciences Symposia. The proceedings are contained in this volume. Dr. Donald F. Tapley, Dean of the College of Physicians and Surgeons (P & S), which sponsors the symposia, welcomed the participants.

We are grateful to George E. Palade for his delivery of the Opening Address. The contributions of the session chairmen, Dr. Judah Folkman, Dr. John C. Hoak, Dr. Salvador Moncada, Dr. Ralph L. Nachman, Dr. David Shepro, and Dr. Gerard M. Turino, are acknowledged with much appreciation. A session was also chaired by one of us (H. L. N.).

Several colleagues from P & S, Dr. Vincent P. Butler, Jr., Dr. Robert E. Canfield, Dr. Shu Chien, Dr. DeWitt S. Goodman, and Dr. Turino, kindly agreed to serve as honorary hosts.

Many thanks go to Dr. Ruth H. Vogel for her contributions to the organization of the symposium and the preparation of this volume.

Hymie L. Nossel
Henry J. Vogel

Opening Address

Differentiated Microdomains in the Vascular Endothelium

George E. Palade, Maia Simionescu and Nicolae Simionescu

Yale School of Medicine and Institute of Cellular Biology and Pathology, University of Bucharest, Bucharest, Rumania

The vascular endothelium is a simple, squamous epithelium differentiated for rapid exchanges between the blood plasma and the interstitial fluid. Its permeability to water and small, hydrophilic solutes is 2 to 3 orders of magnitude higher than that of any other epithelium in a mammalian body, and its high permeability to large molecules is unique—the only possible exception being the mesothelium of the peritoneal and pleural cavities.

The structural basis of these unusual properties consists of (a) extreme attenuation to less than 300 nm in the highly functional parts of the endothelial cells, (b) simplified intercellular junctions, and (c) special, differentiated structural features such as plasmalemmal vesicles present in high volume density in all endothelia, transendothelial channels (that occur only in low volume density) in the endothelial cells of the microvasculature, and diaphragmed (or apertured) fenestrae restricted in their distribution to the capillary endothelium of glands and visceral mucosae.

STRUCTURAL SUBSTRATE OF CAPILLARY PERMEABILITY

In characterizing vascular permeability, it has been assumed that the determining parameter is simply the size of the permeant molecule, expressed as effective molecular radius (1). Evidence implicating the net charge on the molecules as an effective parameter has been limited (2) and largely disregarded. Physiological data on capillary permeability are currently explained in terms of two postulated pore systems, or water-filled channels cutting all the way across the endothelium. Small pores ($r \simeq 6$ nm, surface density $\simeq 15\ \mu m^2$ or much less in some estimates) would account for the high permeability for small- and medium-size molecules, and large pores ($r \simeq 25$ nm; surface density $\simeq 0.05\ \mu m^2$) could explain the unusual permeability of the endothelium for macromolecules (3,4).

In spite of a substantial amount of work carried out in many laboratories over the last 20 years, there is still no agreement concerning the structural equivalents of the two pore systems, primarily their possible relations with the large population of plasmalemmal vesicles involved in transcytosis or diacytosis from the blood to the tissue front of the endothelium. Our findings indicate that these vesicles, which have an average inner diameter of ~27 nm, function in the transport of water and small and large hydrophilic molecules, their contribution being particularly important in macromolecular transport. In this respect, they could be considered as an equivalent of the large pore system; they are different, however, from the original concept in the sense that they represent discontinuous transport devices rather than continuous channels. In addition, plasmalemmal vesicles generate transendothelial channels at low surface density (5). These channels could function as small pores when provided with stomatal diaphragms or structures that limit the movement of molecules of radius larger than ~10 nm or as large pores when such limiting structures are absent. An alternative location for small pores (of slit rather than cylindrical conformation) along intercellular junctions has been postulated by other investigators (6,7,14), but our results indicate that the junctions are not detectably permeable to molecules of radius >2 nm in the endothelium of the blood capillaries (the situation is different in the venular endothelium).

In capillaries provided with a fenestrated endothelium, plasmalemmal vesicles and transendothelial channels function as described above, whereas the fenestrae could act as either small or large pores according to the limits imposed by the texture of their fenestral diaphragms. For a more extensive review of the current status of structural–functional correlations in the vascular endothelium, the reader could consult references (8) and (9). It should be clear that more work is needed for achieving general agreement in this field of research.

SURFACE CHEMISTRY OF THE VASCULAR ENDOTHELIUM—DIFFERENTIATED MICRODOMAINS

In the meantime and prompted by results obtained on renal glomerular capillaries (10,11), we decided to survey the surface chemistry of the vascular endothelium in a capillary bed in which structural differentiations connected with blood–tissue exchanges, e.g., plasmalemmal vesicles, transendothelial channels, and fenestrae, occur at a high surface density. Such capillary beds are found in the pancreas and the intestinal mucosa. In these capillaries, the size and charge diffusion barrier must be located in the endothelium proper, in contradistinction with glomerular capillaries in which both barriers are in the glomerular basement membrane.[1]

Two different protocols were used. In the first, a cationic probe (cationized ferritin, pI = 88.4, r = 5.5 nm) was injected directly into the systemic circulation of mice, and specimens (pancreas and intestine) were fixed *in situ* at intervals ranging from 30 sec to 24 hr after injection. In the second protocol, the vasculature of the pancreas and intestine was perfused in sequence with a tissue culture medium (to remove the blood), with the tracer in the same medium, with the medium alone (to remove excess tracer), and finally with a buffered[2] mixture of glutaraldehyde (3%)–formaldehyde (5%), to fix *in situ* the cationic probes still attached to the endothelium. Over short intervals, the results obtained in the two types of experiments were similar, but the probes could be visualized more clearly in perfused specimens rather than in intact animal experiments, since in the former the background given by fixed plasma proteins was absent. In both cases, the first structures to bind the cationic probe were the fenestral diaphragms. They were the only structures decorated at time 0 to 30 sec. They were also the last to lose the cationic probe decoration after ~20 hr in intact animal experiments (perfusion experiments were not carried out for longer periods than 60 min). The plasmalemma proper became decorated almost as rapidly, though less heavily than the fenestral diaphragms (Figs. 1 and 2), but it lost the label by ~4–7 hr in intact animals. Hence, it can be concluded that the surface density of anionic sites as well as their affinity for the cationic probe are higher on the fenestral diaphragms than on the plasmalemma proper. In con-

[1] The fenestrae of the glomerular endothelium are large (~100 nm) and have no diaphragms. At their level, the blood plasma is in direct contact with the glomerular basement membrane.

[2] 0.1 M HCl–Na cacodylate or HCl–Na arsenate buffer, pH 7.2–7.4.

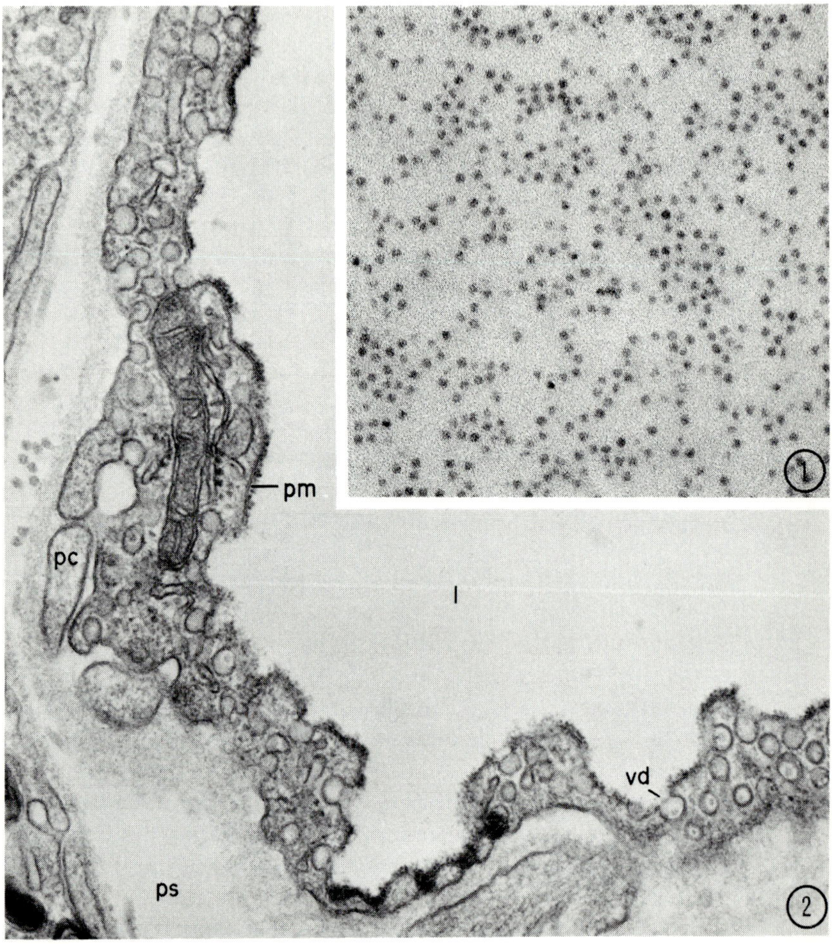

Fig. 1. Cationic ferritin molecules in the tracer solution: the molecules, in their majority, appear randomly and individually dispersed. ×140,000.

Fig. 2. Mouse pancreatic capillary, 30 sec after *in situ* perfusion with CF. The ligand decorates almost continuously the plasmalemma proper (pm); it appears in characteristically high concentration on fenestral diaphragms (f), but it does not decorate the membrane or diaphragms (vd) of plasmalemma vesicles. (e) endothelium; (l) lumen; (bm) basement membrane; (pc) pericyte; (ps) pericapillary space. ×48,000.

tradistinction to these structural elements, plasmalemmal vesicles, transendothelial channels, and (when present) their stomatal diaphragms remained unlabeled throughout the duration of the experiments (Figs. 3 and 4). In addition to plasmalemmal vesicles, the endothelial cells examined have a small population of coated vesicles. The latter bound the cationic probe shortly after the beginning (Fig. 3) of the experiments and retained it longer than the plasmalemma proper. For further details concerning the experimental protocol and the description of the results (especially after late intervals in intact animals), the reader should consult reference (13).

The findings presented so far indicate clearly that the endothelial

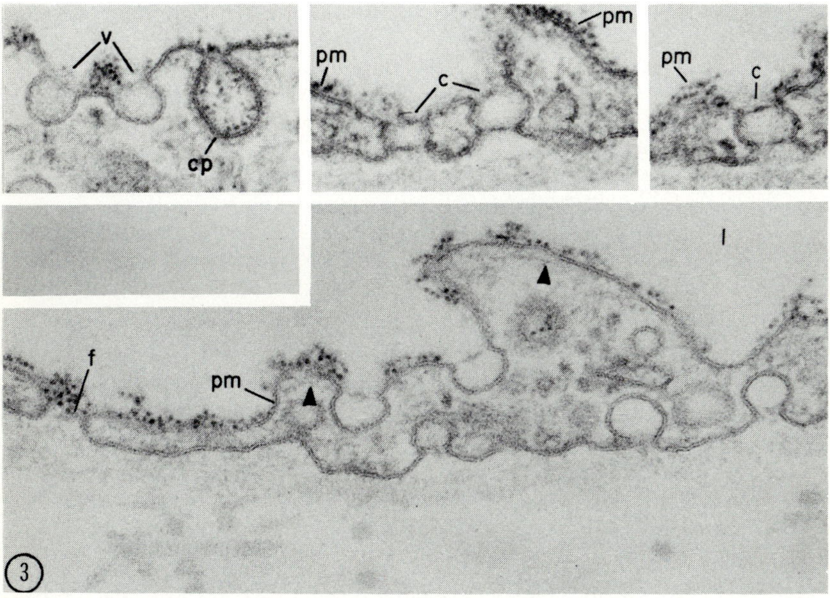

Fig. 3. Mouse pancreatic capillary 1 min after *in situ* perfusion with CF. The ligand is densely bound to an (f) and, to a lesser extent and with some discontinuities, to the pm. It is absent from the membranes of plasmalemma vesicles and their associated diaphragms (v). A grazing section through a labeled coated vesicle or pit appears at cv. Note the accumulation of fibrillar material beneath the regions of the plasmalemma to which CF is bound (arrowheads). × 65,000. *Right inset:* this micrograph illustrates the striking difference between the intense labeling by CF of a coated pit (cp) (in this case, in its course of internalization) and the absence of CF decoration from the membranes and the diaphragms of two adjacent plasmalemma vesicles (v). × 85,000. *Left insets:* details of CF decoration in a mouse pancreatic capillary 1 min after a CF injection *in vivo*. The membranes and diaphragms of transendothelial channels (c) are not labeled by CF whereas the pm is labeled. ×85,000.

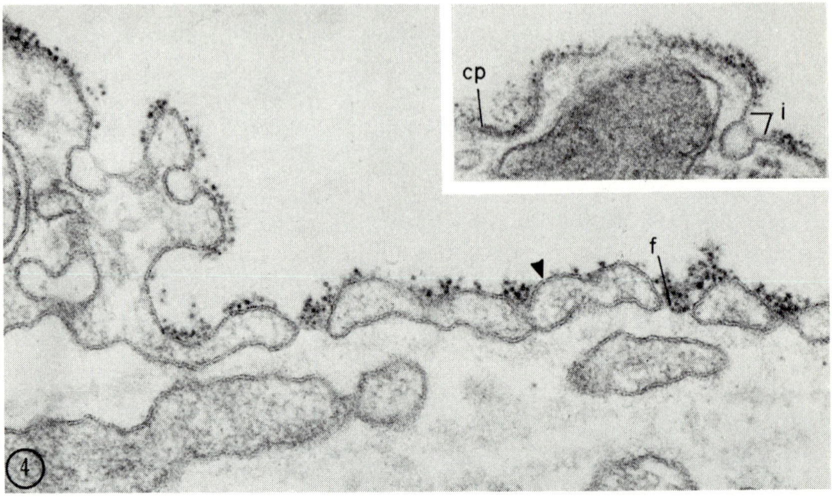

Fig. 4. Mouse pancreatic capillary 2 min after a CF injection *in vivo* (blood plasma was probably drained from the lumen during tissue processing). Whereas polycation binding to fenestral diaphragms (f) is particularly intense, the membranes and diaphragms of v are free of CF decoration. CF binding to the plasmalemma proper is moderate. In some areas, it results in a single row of CF molecules (arrow); in others it displays focal discontinuities (arrowheads). × 65,000. *Inset:* infundibula (i) leading to vesicular stomata are often also free of CF decoration. ×65,000.

luminal surface is provided with differentiated microdomains characterized both by specific structural features and by uneven distribution of anionic sites. All these domains are implicated by previous work in blood–interstitial fluid exchanges [cf. (8)].

DIGESTIVE ENZYME TESTS

To acquire more information on the nature of the anionic sites, perfusion experiments were extended to include a digestive step by appropriate enzymes (included in the perfusate). This step was intercalated between blood removal and probe perfusion. Broad specificity proteases (pronase and papain) prevent—fully or extensively—subsequent decoration by cationized ferritin (Figs. 5 and 6). We could conclude, therefore, that the anionic sites detected in our experiments were not contributed by acidic glycolipids. Trypsin weakened the structure of the fenestral diaphragms, as suggested by the occurrence of clusters of cationized ferritin displaced one-half the way or all the way across the fenestrae, but trypsin did not affect the decoration on

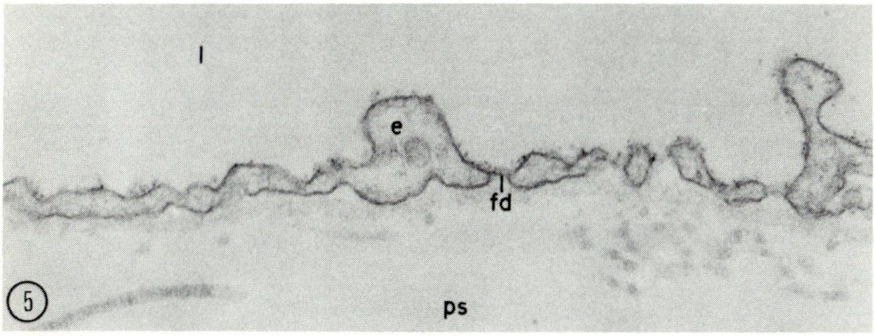

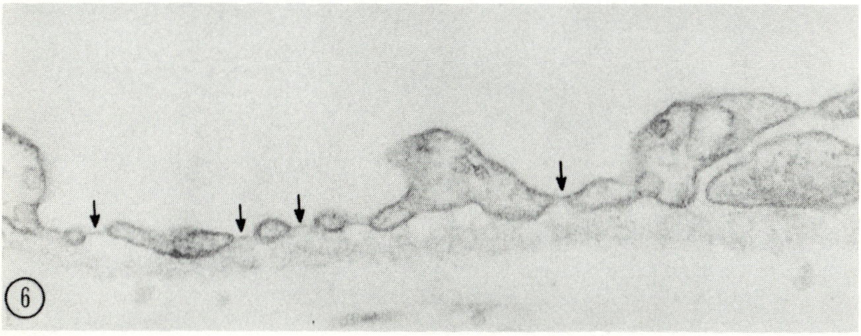

Fig. 5. Effects of papain perfusion (10 min). Practically all the anionic sites have been removed from the entire luminal surface of the endothelium. ×49,000.

Fig. 6. Pronase treatment (20 min) results in complete removal of all CF binding sites from the luminal surface of the endothelium. Note that in this case (as well as after papain treatment), fenestral diaphragms appear to be still present in their usual location (arrows). ×49,000.

the rest of the endothelium (Fig. 7). Hyaluronidase and chondroitinase ABC reduced cationized ferritin binding only partially. Heparinase, however, specifically and efficiently removed the anionic sites from the vast majority of fenestral diaphragms and thereby prevented their subsequent decoration by the cationic probe (Figs. 8 and 9). For further details concerning experimental procedures and results, the reader should consult reference (14).

A comparison of the substrate specificity of the enzymes tested indicates that the anionic sites of the fenestral diaphragms must be provided either by heparin or heparan sulfate-bearing proteoglycans. We believe that the second alternative applies in keeping with the resistance of the sites to extensive washing by perfusion (expected to remove heparin) and with the removal of the sites by broad specificity

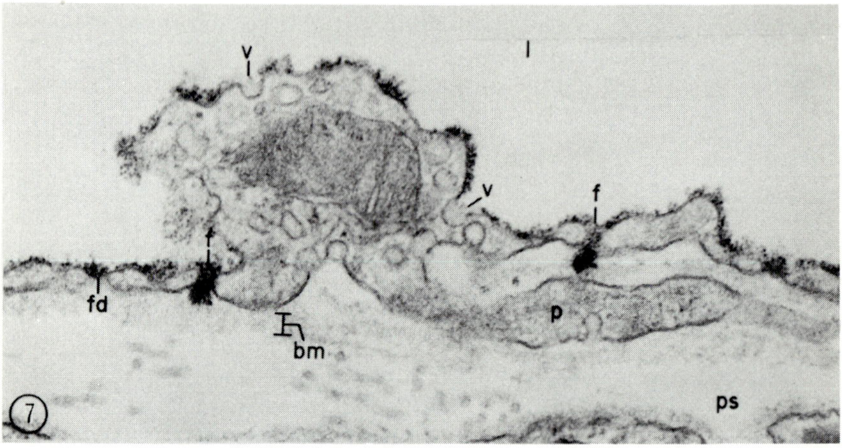

Fig. 7. Effects of trypsin perfusion (30 min). CF binding to the luminal surface of the endothelium is more uneven than in controls and is limited to the plasmalemma proper and fenestral diaphragms. Plasmalemmal vesicles and their stomatal diaphragms are not labeled. Large plugs of aggregated cationized ferritin have penetrated through two fenestrae and appear to be retained by the basement membrane (left) and by the basement membrane and a pericyte (p) (right). ×49,000.

proteases. No enzymic treatment so far tested succeeded in removing the anionic sites of the coated pits. Taken together the results indicate that the differentiated microdomains differ from one another by their morphology and by the density, affinity, and chemical nature of their anionic sites. The anionic sites on the plasmalemma proper are probably a mixture of acidic glycoproteins and proteoglycans. Our findings are of interest on two accounts: capillary permeability and maintenance of membrane specificity.

MICRODOMAINS AND CAPILLARY PERMEABILITY

The existence of differentiated microdomains indicates that the permeability properties of the endothelium must vary from domain to domain. Fenestral diaphragms are expected to retain anionic macromolecules, past a certain size yet to be determined, whereas plasmalemmal vesicles and transendothelial channels should not discriminate against such large molecules. In fact, they are known to transport native ferritin, which is anionic. The survey has been extended by using other probes (labeled lectins) on the same microvascular bed and by trying to decorate the abluminal domain of the

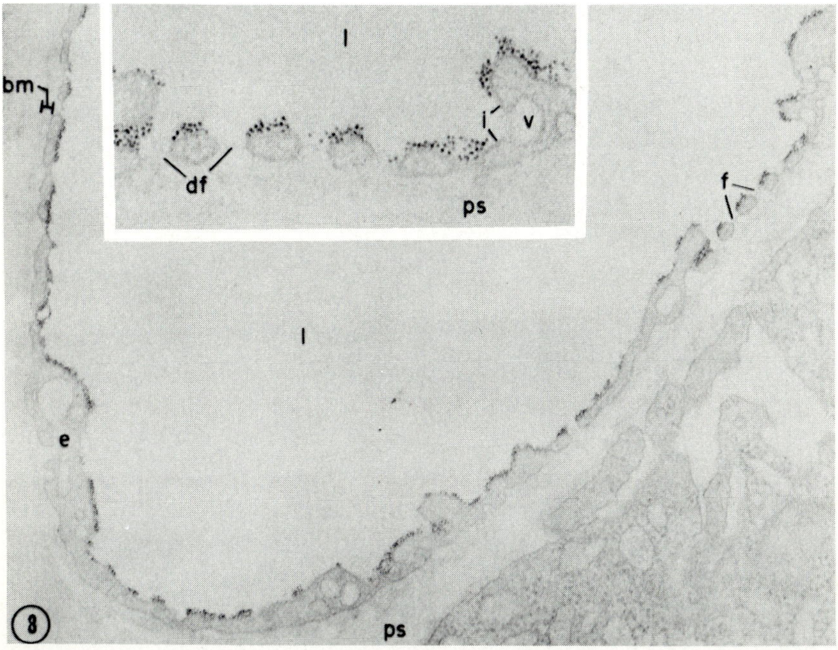

Fig. 8. General view of a blood capillary after heparinase perfusion (30 min) followed by CF decoration. Anionic sites have been effectively removed from practically all fenestral diaphragms. The decoration of the luminal surface of the endothelial plasmalemma is patchy. The decorated domains as well as alternating CF-free areas vary considerably in size. Plasmalemmal vesicles and their associated infundibula are not labeled. ×26,000. *Insert:* A series of fenestrae at higher magnification taken from the upper right corner of Fig. 8. It illustrates the extensive (but not complete) removal of CF-binding sites from fenestral diaphragms and the lack of CF binding to a plasmalemmal vesicle, its stomatal diaphragm, and its associated infundibulum. ×82,000.

endothelial surface, after removing the basement membrane by collagenase digestion *in situ*. The results obtained confirm the existence of microdomains and extend the range of the characteristics by which they differ. The same kind of inquiry is now extended to other types of capillaries. So far the literature records the presence of microdomains only on the luminal space of the sinusoids of the bone marrow (15).

Permeability must therefore be controlled in the fenestrated capillaries examined not only by the size but also by the charge of a would-be-permeant molecule. Moreover, anionic and cationic macromolecules may move through different pathways—anionic proteins via plasmalemmal vesicles, for instance, and cationic species (up to a

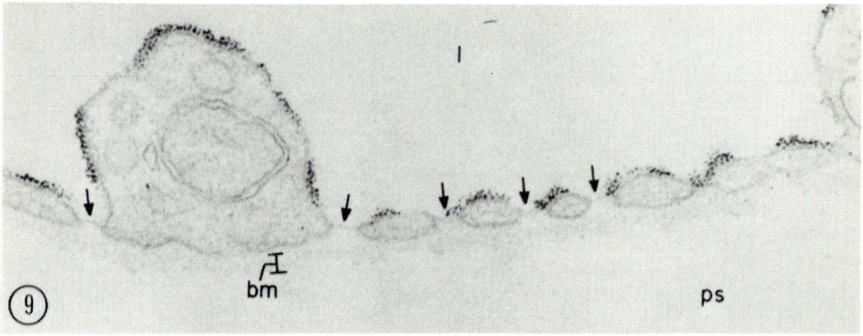

Fig. 9. Effects of heparinase perfusion (40 min). CF-binding sites have been eliminated from practically all fenestral diaphragms (arrows), but the diaphragms themselves are apparently still present in their usual location. CF decorates most of the surface of the plasmalemma proper. ×63,000.

certain size) via diaphragmed fenestrae. In this respect, it is worth while pointing out that, in their vast majority, plasma proteins are anionic. Diaphragmed fenestrae appear to be devices that allow extensive exchanges of small molecules without extensive leakage of plasma proteins to the interstitial fluid in the pericapillary spaces.

MICRODOMAINS AND MEMBRANE SPECIFICITY

The existence of differentiated microdomains implies that locally specific chemistry can be achieved and maintained in a membrane provided with a continuous lipid bilayer known to be fluid (and, hence, to promote randomization) at the temperature of the body. Such domains can be generated and maintained by interactions among integral membrane proteins (in our case acidic glycoproteins and proteoglycans) and a stabilizing infrastructure built up of peripheral proteins. In the case of coated pits and coated vesicles, this stabilizing infrastructure is the geodetic cage formed by clathrin and associated protein molecules. In the case of the plasmalemmal vesicles, the evidence obtained so far suggests that endothelial cells stabilize (by a fibrillar infrastructure) the plasmalemma proper rather than the plasmalemmal vesicles. Such an infrastructure can define ports of entry and sites of departure of transporting vesicles, in addition to maintaining the distinct chemistry of their membranes (by preventing lateral diffusion from and into the plasmalemma proper).

REFERENCES

1. Landis, E. M., and Pappenheimer, J. R. (1963) Exchange of substance through the capillary walls. *In* "Handbook of Physiology" (W. F. Hammilton and P. Dow, eds.), vol. 2 (2), p. 961. Amer. Physiol. Soc., Washington, D.C.
2. Garby, L., and Areekul, S. (1970) Reflection coefficients of neutral and sulphate-substituted dextran molecules in the capillaries of the isolated rabbit ear. *In* "Alfred Benzon Symposium. II. Capillary permeability." (C. Crone and N. A. Lassen, eds.) pp. 560–563. Munksgaard, Copenhagen.
3. Renkin, E. M. (1979) Relation of capillary morphology to transport of fluid and large molecules. A review. *Acta Physiol. Scand., Suppl.* **463,** 81.
4. Crone, Ch. (1979) Permeability of single capillaries compared with results from whole-organ studies. *Acta Physiol. Scand., Suppl.* **463,** 75.
5. Simionescu, N., Simionescu, M., and Palade, G. E. (1975) Permeability of muscle capillaries to small hemepeptides. Evidence for the existence of patent transendothelial channels. *J. Cell Biol.* **64,** 586.
6. Karnovsky, M. J. (1970) Morphology of capillaries with special reference to muscle capillaries. *In* "Capillary Permeability." Alfred Benzon Symposium II (Ch. Crone and N. A. Lassen, eds.), p. 341. Academic Press, New York.
7. Wissig, S. L., and Williams, M. C. (1978) Permeability of muscle capillaries to microperoxidase. *J. Cell Biol.* **76,** 341.
8. Palade, G. E., Simionescu, M., and Simionescu, N. (1979) Structural aspects of the permeability of the microvascular endothelium. *Acta Physiol. Scand., Suppl.* **463,** 11.
9. Renkin, E. M. (1977) Multiple pathways of capillary permeability. *Circ. Res.* **41,** 735.
10. Kanwar, Y. S., and Farquhar, M. G. (1979) Presence of heparan sulfate in the glomerular basement membrane. *Proc. Natl. Acad. Sci. U.S.A.* **76,** 1303.
11. Brenner, B. M., Hostettler, T. H., and Humes, H. D. (1978) Molecular basis of proteinuria of glomerular origin. *N. Engl. J. Med.* **298,** 826.
12. Farquhar, M. G. (1978) Structure and function in glomerular capillaries. Role of the basement membrane in glomerular filtration. *In* "Biology and Chemistry of Basement Membranes." (N. Kefalides, ed.) p. 43. Academic Press, New York.
13. Simionescu, N., Simionescu, M., and Palade, G. E. (1981) Differentiated microdomains on the luminal surface of the capillary endothelium. I. Preferential distribution of anionic sites. *J. Cell Biol.* **90,** 605.
14. Simionescu, M., Simionescu, N., Silbert, J. E., and Palade, G. E. (1981) Differentiated microdomains on the luminal surface of the capillary endothelium. II. Partial characterization of their anionic sites. *J. Cell Biol.* **90,** 614.
15. DeBruyn, P. P. H., and Michelson, S. (1979) Changes in the random distribution of sialic acid at the surface of the myeloid sinusoidal endothelium resulting from the presence of diaphragmed fenestrae. *J. Cell Biol.* **82,** 708.

PATHOBIOLOGY OF THE ENDOTHELIAL CELL

PART I

ENDOTHELIAL GROWTH REGULATION

Stimulation of Vascular Cell Growth by Macrophage Products

MICHAEL A. GIMBRONE, JR., BERNICE M. MARTIN, WILLIAM M. BALDWIN, EMIL R. UNANUE, AND RAMZI S. COTRAN

Vascular Pathophysiology Laboratory, Department of Pathology
Brigham and Women's Hospital
and the Department of Pathology
Harvard Medical School
Boston, Massachusetts

INTRODUCTION

Macrophages, the tissue derivatives of circulating blood monocytes, play a vital role in host defense mechanisms, including inflammation, wound healing, and the immune response. In addition to their well-recognized capacity for scavenging foreign material (pathogenic organisms, tissue debris, etc.), mononuclear phagocytes secrete a broad spectrum of biologically active molecules, including proteolytic enzymes, complement components, prostaglandins, cyclic nucleotides, chemotactic factors, pyrogens, and cytotoxins (1–3). Certain of these products appear to act as modulators of the growth, differentiation, and functional status of lymphoid cells. Recent studies also suggest that macrophages produce substances that can influence the growth or regeneration of other types of mesenchymal cells. For example, the proliferation of fibroblasts in healing skin wounds is retarded in monocytopenic guinea pigs (4), and the conditioned medium from macrophage cultures has been found to contain a growth-promoting activity for cultured guinea pig wound fibroblasts (5). However, although the phenomenon of macrophage-induced cell proliferation has been described (5–10), characterization of the active factor(s) involved and the stimuli responsible for their production is largely lacking.

This chapter briefly reviews a series of experiments conducted in our laboratory (11–15) on the effects of macrophage-derived factors on

PATHOBIOLOGY OF THE ENDOTHELIAL CELL

ISBN 0-12-521980-6

the growth of certain nonlymphoid mesenchymal cells involved in chronic inflammation, wound healing, and atherosclerosis. Our results indicate that mononuclear phagocytes can be stimulated to produce a potent growth factor (macrophage-derived growth factor, MDGF) that induces neovascularization *in vivo* and the proliferation of vascular endothelium, smooth muscle, and fibroblasts *in vitro.* The conditions that influence secretion of MDGF, properties that distinguish it from other macrophage products, and its potential role in certain pathophysiological processes are discussed.

MACROPHAGE-INDUCED NEOVASCULARIZATION: *IN VIVO* STUDIES

Neovascularization ("angiogenesis") refers to the formation of new blood vessels, in the adult, usually as capillary-like outgrowths from an existing microvascular bed. Macrophages appear capable of stimulating neovascularization in the context of both immunological and nonimmunological inflammatory reactions and wound healing.

The association of endothelial proliferation with various immunological phenomena has been suggested by a number of studies. For example, Sidky and Auerbach (16) observed increased capillary density in the skin during graft-versus-host reactions. Herman and associates (17) and Anderson *et al.* (18) found an increase in capillary density and extensive proliferation of postcapillary venular endothelium in lymph nodes undergoing strong immunological reactions. Dvorak *et al.* (19) described ultrastructural changes suggestive of endothelial regeneration in delayed-type hypersensitivity reactions in human skin. Subsequently, in a quantitative morphological study, Polverini, Cotran, and Sholley (20) documented that the endothelial proliferation associated with delayed hypersensitivity reactions coincides in time, distribution, and magnitude with mononuclear cell infiltration of the lesions.

To investigate the mechanisms of this endothelial growth stimulation and to establish more clearly the role of mononuclear phagocytes in the process, Polverini and co-workers (11) utilized the corneal neovascularization model. This bioassay procedure had been developed previously in this laboratory to evaluate the angiogenic capacity of malignant tumor cells and their products (tumor-angiogenesis factors) (21). Neovascularization is assessed in the normally avascular cornea of the guinea pig (or rabbit) eye by introducing either cells or cell products into the stroma and observing the directional outgrowth of vascular sprouts from the peripheral limbal plexus toward the implantation site (Fig. 1). This procedure allows direct visualization and daily measurements of new blood vessel growth in the live animal; furthermore, the absence of preexisting vessels in the normal corneal stroma simplifies the interpretation of the neovascular response.

Macrophages were obtained from the peritoneal cavities of mice and guinea pigs by simple lavage (resident, "nonactivated" macrophages), or following induction of a sterile inflammatory exudate by paraffin oil, thioglycolate, or peptone injection ("activated" macrophages). The latter cell populations are processed by standard separative techniques to yield a preparation consisting of 85–90% viable macrophages. In some experiments, peritoneal exudate macrophages also were adherence-separated and cultured in plastic petri dishes to obtain a 24-hr (serum-free) conditioned medium. Implantation of live, *in vivo*-activated, peritoneal macrophages induced a neovascular response in more than 75% of the corneas tested (Fig. 1A), whereas resident, nonactivated macrophages showed activity in less than 10% of corneas (Fig. 1B). Similar results were obtained with activated macrophages from guinea pigs of an outbred (allogeneic) strain and a highly inbred (syngeneic) strain, and even from the same individual guinea pig (autologous), thus indicating that the neovascular response could not be due solely to an immunological reaction. Brief *in vitro* exposure of resident peritoneal macrophages to a phagocytic challenge (latex particles), before implantation in corneas, appeared to augment their neovascular activity.

For comparative study, neutrophils were harvested from sterile peritoneal exudates and lymphocytes were separated from the spleens and lymph nodes of normal or PPD-sensitized donors. Normal and immunologically primed lymphocytes injected immediately after isolation or following incubation with and without PPD showed no evidence of neovascularization in 58 corneas tested. Implantation of comparable numbers of neutrophils resulted in a positive response in only 1 out of 15 corneas tested (P. Polverini, unpublished observations).

Samples of cell-free, dialyzed, concentrated media from 24-hr cultures of activated peritoneal macrophages also induced corneal neovascularization when implanted in a noninflammatory vehicle (Hydron polymer) (Fig. 1C). This response showed the same temporal and spatial pattern as that elicited by viable macrophages and was not associated with an acute inflammatory response in the corneal stroma, as seen clinically or histologically. The angiogenic activity of macrophage-conditioned media was enhanced following exposure of the cultured macrophages to gram-negative bacterial endotoxin (*Escherichia coli* lipopolysaccharide), latex particles, viable microor-

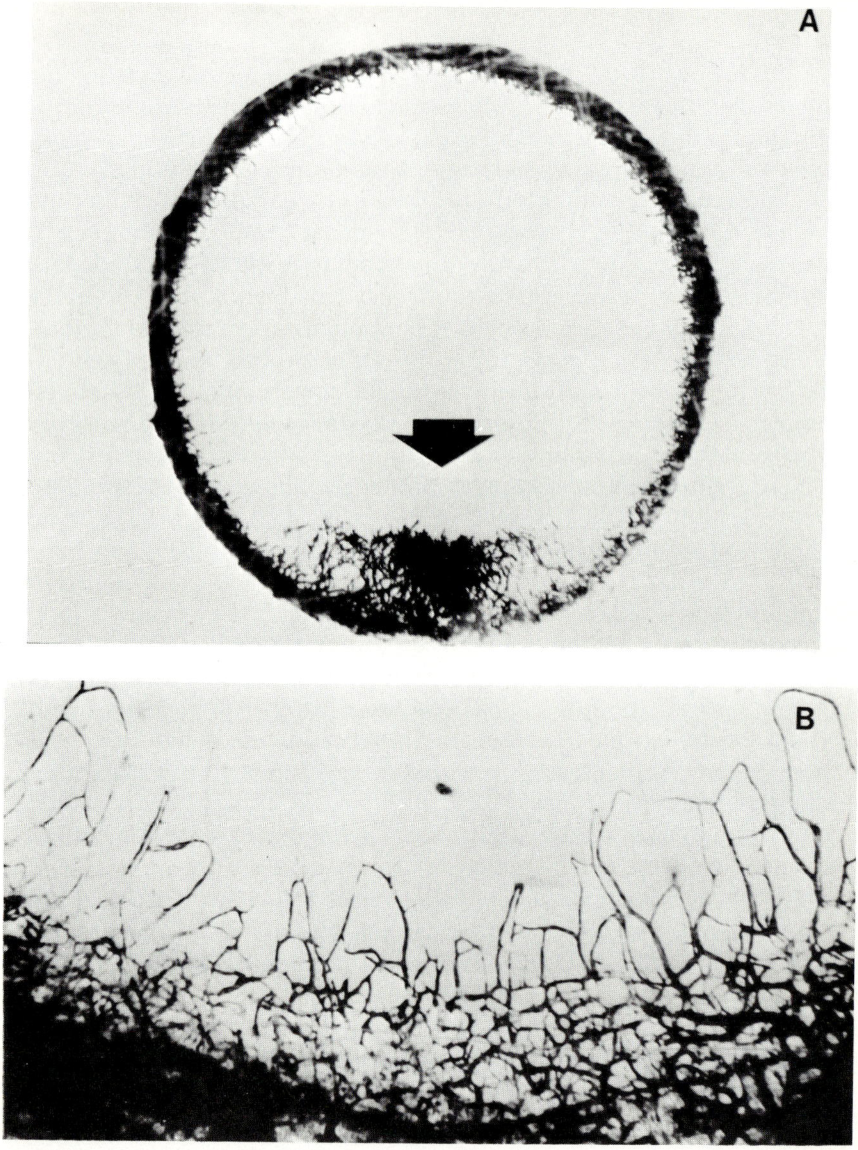

Fig. 1. Neovascularization induced in Hartley albino guinea pig corneas by live macrophages or macrophage-conditioned culture medium. (A) Positive neovascular response stimulated by injection of 5×10^5 homologous (*in vivo*-activated, paraffin oil exudate), peritoneal macrophages. Note directional outgrowth of small vessels from limbal plexus adjacent to the injection site (arrow) and the normal vascular pattern around the remainder of the corneal circumference (carbon-perfused, whole-mount preparation). (B) Higher-power view of negative response following injection of 5×10^5 homologous, resident (nonactivated) peritoneal macrophages. Limbal vessels show normal arcade pattern without evidence of capillary sprouting.

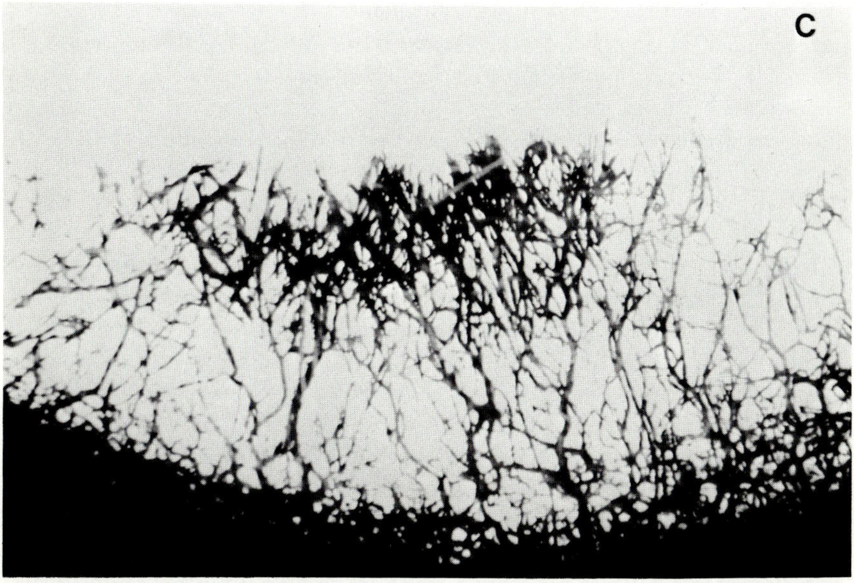

(C) Neovascular response, similar to that seen in (A), elicited by Hydron implant of cell-free culture medium conditioned by activated peritoneal macrophages. A meshwork of newly formed capillary sprouts has penetrated the normally avascular corneal stroma in several planes. Modified from Polverini *et al.* (11).

ganisms (*Listeria monocytogenes*), or phorbol myristate acetate (the active component of croton oil) (W. Baldwin, unpublished observations).

Taken together, these results indicate that activated peritoneal macrophages can stimulate microvascular proliferation *in vivo* through the action of an inducible, secreted factor. Similar angiogenic activity also has been reported for wound macrophages (22,23). Thus, activated mononuclear phagocytes appear to have the potential to induce both fibroplasia (4,5) and neovascularization, essential components of the basic wound-healing response.

STIMULATION OF CELL GROWTH *IN VITRO* BY MACROPHAGE PRODUCTS

To study more directly the mechanisms of action of macrophage-derived growth factor(s) (MDGF), the conditioned media from macrophage cultures was assayed for growth-promoting activity on various cell types *in vitro*. This approach permitted studies of conditions influencing secretion of MDGF, its spectrum of target cell reactivity, and its preliminary biochemical characterization.

Activated macrophages were obtained by adherence separation of mouse peritoneal exudate cells as previously described (12). More than 95% of these adherent cells were typical macrophages, as judged by their morphology and expression of surface markers (e.g., ability to bind red cells opsonized with IgG or C_3 and by the presence of macrophage differentiation antigens) (24). For routine preparation of macrophage-conditioned media, 2–3× 10^6 macrophages per petri dish (2.5–3 × 10^5 cells/cm^2) were incubated for 24–72 hr in 1 ml of Dulbecco's Modified Eagle's Medium (DME), containing either 1% calf (whole blood) serum, 1% platelet-factor-deficient serum (25), 0.6 mg/ml bovine serum albumin, or, in some cases, no added protein. Conditioned media were centrifuged at 500 *g* for 10 min at 4°C to remove cellular debris. Aliquots were assayed for the presence of lactate dehydrogenase to monitor macrophage viability during culture. In all experiments, samples of media (1 ml/35-mm dish) also were incubated in parallel in the absence of macrophages (control-conditioned media). Both types of conditioned media were dialyzed against a 10- to 20-fold excess of Hanks balanced salt solution twice and against DME once for 20–24 hr at 4 °C to remove low-molecular-weight inhibitors of DNA synthesis (1) and replenish nutrients.

To evaluate growth-promoting activity, macrophage-conditioned media and control-conditioned media were assayed for stimulation of cell division in sparsely plated cultures (5 × 10^3 cells/cm^2) of Balb/c 3T3 cells and other test cells. Figure 2 (upper panel) shows that the addition of conditioned media from macrophages cultured for 72 hr in 1% whole blood serum markedly enhanced the growth of Balb/c 3T3 cells (left side) and bovine aortic smooth muscle cells (right side). The proliferation of smooth muscle cells in response to the macrophage-conditioned media was comparable to that observed with 5% whole blood serum. In the case of 3T3 cells, the macrophage-conditioned media stimulated twice as much growth as 5% serum. Neither control-conditioned media (incubated for 72 hr) nor fresh 1% calf serum stimulated significant growth of 3T3 cells or smooth muscle.

The cationic platelet-derived growth factor (PDGF), present in whole blood serum, is a potent growth promoter for various cell types, including 3T3 and smooth muscle cells (25). To examine its influence on MDGF production by cultured macrophages, PDGF was removed from calf serum by ion exchange chromatography using CM-Sephadex (25). Macrophages then were incubated in this platelet-factor-deficient serum (PDS) for 72 hr. As shown in Fig. 2 (lower panel), after preincubation in 5% PDS, both 3T3 cells and smooth muscle cells remained quiescent in fresh 1% PDS and were only minimally stimulated by the addition of fresh 5% PDS. In contrast, macrophage-conditioned media

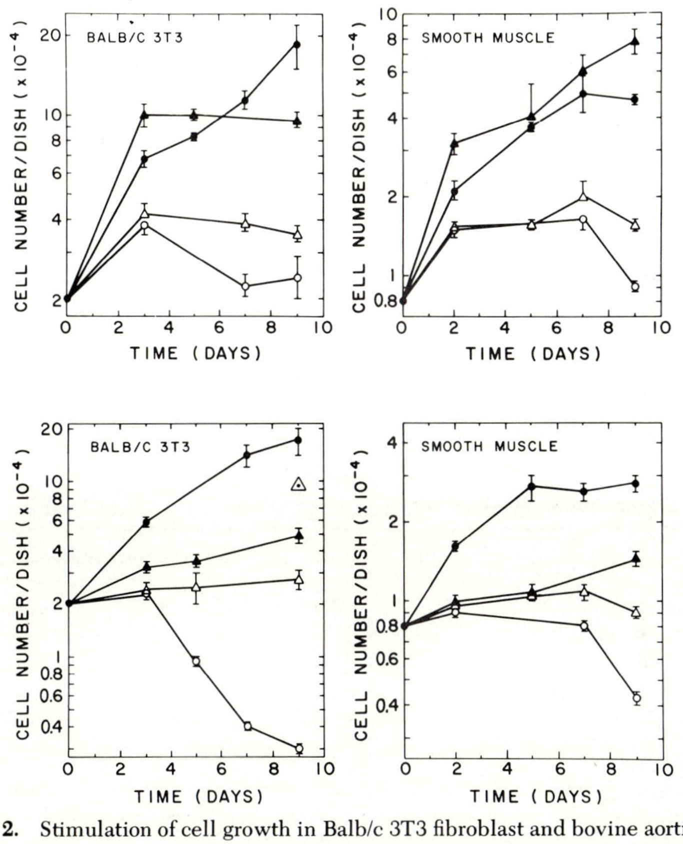

Fig. 2. Stimulation of cell growth in Balb/c 3T3 fibroblast and bovine aortic smooth muscle cultures by macrophage-conditioned media containing either whole blood serum (upper panels) or platelet factor-deficient serum (lower panels). In both sets of experiments, Balb/c 3T3 cells (left side) and bovine aortic smooth muscle cells (right side) were replicate plated in Dulbecco's Modified Eagle's Medium (DME) with 5% whole blood serum and then maintained in DME with 5% PDS for 3 days prior to initiating the growth assay. (Upper panels): At zero time and at 48-hr intervals thereafter, macrophage-conditioned media (●) or control-conditioned media (○) containing 1% whole blood serum, or fresh DME containing 1% (△) or 5% whole blood serum (▲) were added. (Lower panels): At zero time and at 48-hr intervals thereafter, macrophage-conditioned media (●) or control-conditioned media (○) containing 1% platelet factor deficient-serum (PDS), or fresh DME containing 1% (△) or 5% (▲) PDS were added. For reference the 3T3 cell growth elicited by fresh 5% whole blood serum at day 9 is indicated (◬). Each point represents cell number (mean ± S.D.) measured in triplicate. Reproduced from Martin *et al.* (12), with permission of the publisher.

containing 1% PDS stimulated the growth of 3T3 and smooth muscle cells to a similar extent as macrophage-conditioned media generated in 1% whole blood serum. Furthermore, in the case of 3T3 cells (lower panel, left side) the macrophage-conditioned media in 1% PDS exhibited more growth-promoting activity than fresh 5% whole blood serum. The effect of the macrophage-conditioned media was especially striking when compared with control-conditioned media, which did not maintain 3T3 or smooth muscle cell number.

We also examined the ability of macrophage-conditioned media to stimulate DNA synthesis in nonconfluent cultures that were rendered quiescent by serum limitation (Table I). Macrophage-conditioned media, generated in either 1% whole blood serum or 1% PDS, stimulated two to three times more DNA synthesis in smooth muscle and 3T3 cells than control-conditioned media. In addition, these macrophage-conditioned media also stimulated DNA synthesis in bovine aortic endothelial cells.

These results indicate that cultured macrophages produce a growth promoter for fibroblasts, smooth muscle, and endothelial cells. Furthermore, the production of this activity, as well as its action on these

TABLE I
Stimulation of DNA Synthesis by Macrophage-Conditioned Media[a]

	^{3}HTdR stimulation index		
Macrophage culture conditions	3T3	Smooth muscle	Endothelium
Group A			
1% whole blood serum	2.96(17)	2.08(4)	2.01(4)
1% platelet-factor-deficient serum	3.01(12)	2.39(1)	3.53(3)
Group B			
0.6 mg/ml BSA	14.38	3.71	N.D.
Protein-free media	15.23	3.08	N.D.

[a] Macrophage-conditioned media were obtained from 72-hr cultures of activated mouse peritoneal macrophages maintained in DME containing 1% whole blood serum, 1% platelet-factor-deficient serum, 0.6 mg/ml bovine albumin (BSA), or in protein-free DME. Following dialysis, macrophage-conditioned media or control-conditioned media were added to test cells plated in triplicate, along with 0.5 μCi [^{3}H]methylthymidine (^{3}HTdR), and ^{3}HTdR incorporation into DNA was determined 20 hr later. Data are expressed as the stimulation index: the ratio of ^{3}HTdR incorporation in the presence of macrophage-conditioned media to that in the presence of control-conditioned media. The total number of experiments performed with each cell type in group A is indicated in parentheses. In the representative experiments in group B, stimulation indices appear much higher because of very low levels of DNA synthesis obtained with the control-conditioned media, which did not contain any serum-derived growth factors. Adapted from Martin *et al.* (12,13).

various target cells, does not appear to be dependent on platelet-derived growth factor.

Since macrophages were cultured in serum-containing media in the preceding experiments, it is conceivable that some serum component was processed or modified by the macrophage to produce the observed growth-promoting activity. Therefore, further experiments were conducted to establish whether the growth promoter was initially derived from serum, as suggested by Leibovich (6), or whether it was primarily of macrophage origin. As seen in Table I (group B), when macrophages were incubated in serum-free media, supplemented only with 0.6 mg/ml bovine albumin (the protein equivalent of 1% whole blood serum), significant amounts of growth-promoting activity were obtained. Furthermore, macrophages cultured in the *total* absence of exogenous protein produced comparable amounts of growth stimulation as those cultured in 0.6 mg/ml BSA. These results indicate that none of the substances present in serum are required for production of macrophage-derived growth factor(s).

MODULATION OF MACROPHAGE-DERIVED GROWTH FACTOR PRODUCTION

Cultured macrophages can be "activated" by a variety of stimuli, including exposure to a phagocytic challenge, such as latex particles; bacterial products, such as gram-negative bacterial endotoxin; or certain chemicals, such as phorbol esters. Activated macrophages characteristically exhibit increased secretory, phagocytic, and antimicrobial capacity (1,3). As was observed for macrophage-induced angiogenesis *in vivo*, the production of growth factor(s) for nonlymphoid mesenchymal cells *in vitro* is enhanced by such activating stimuli. Thus treatment of cultured mouse peritoneal macrophages with latex particles, bacterial lipopolysaccharide, or phorbol myristate acetate results in severalfold increases in growth factor production (12). Furthermore, culture of resident mouse peritoneal macrophages (harvested by saline lavage) in bovine serum albumin also causes morphological changes characteristic of the activated state and results in growth factor production comparable to that of peritoneal exudate macrophages, induced by mineral oil or peptone injection. Recent studies with human monocytes, isolated directly from peripheral blood, indicate that cold-insoluble globulin (plasma fibronectin) also may augment growth factor production by adherent mononuclear cells (14,15). This monocyte culture system may permit further analysis of pathophysiological stimuli that can trigger MDGF production as circulating blood monocytes evolve into tissue macrophages.

SOURCES OF MACROPHAGE-DERIVED GROWTH FACTOR

The capacity to secrete growth-promoting activity for nonlymphoid mesenchymal cells, such as fibroblasts, vascular smooth muscle, and vascular endothelium, appears to be a property of a variety of mononuclear phagocytes obtained from different tissue sources and animal species. Thus in addition to activated murine macrophages isolated from sterile peritoneal exudates, we have observed MDGF production by cultured murine resident peritoneal macrophages, cultured murine bone marrow macrophages, and cultured human peripheral blood monocytes. Other laboratories also have reported on growth factor production by macrophages cultured from guinea pig peritoneal exudates (5), human peritoneal dialysis washings (7), and rabbit wound fluids (23), as well as cultured human peripheral blood monocytes (9,26). It is noteworthy that MDGF activity appears to act across species lines; i.e., human MDGF can stimulate murine target cells and vice versa. This lack of species specificity is consistent with the known action of other growth factors, such as platelet-derived growth factor, fibroblast growth factor, and epidermal growth factor and may aid in bioassay during purification.

PROPERTIES OF MACROPHAGE-DERIVED GROWTH FACTOR

The effects of various physical and chemical treatments on the growth-promoting activity generated by cultured mouse peritoneal macrophages are shown in Table II. These data indicate that MDGF is a protein, which appears to contain at least one essential disulfide bond but no essential ribonucleic acid. The failure of soybean trypsin inhibitor, or phenylmethanesulfonyl fluoride, a serine esterase inhibitor, to neutralize MDGF activity suggests that it is not related to various of the proteolytic enzymes known to be released or secreted by activated macrophages, such as plasminogen activator or elastase (1,3). Furthermore, since its activity is destroyed by boiling but it is not adsorbed by CM-Sephadex under hypotonic conditions, MDGF appears to be biochemically distinct from the cationic platelet-derived growth factor (25). The factor is nondialyzable and can be concentrated

TABLE II
Macrophage-Derived Growth Factor: Biochemical Properties

Treatment of conditioned media[a]	Initial activity (%)
Heat: 100°C, 2 min	15
56°C, 30 min	>95
Freeze/thaw (4 cycles)	>95
Trypsin: 4×10^{-6} *M*, 30 min	5
4×10^{-7} *M*, 30 min	85
Ribonuclease A: 120 Kunitz units, 30 min	>95
Dithiothreitol: 20 m*M*, 30 min	17
Phenylmethanesulfonyl fluoride: 2 m*M*, 30 min	>95
Soybean trypsin inhibitor: 1 m*M*, 30 min	>95
CM-Sephadex C-50	>95
DEAE-Sephadex A-50	35
Phenyl-Sepharose CL-4B	< 5

[a] Macrophage-conditioned media and control-conditioned media obtained from 72-hr cultures of peritoneal macrophages in DME containing either 1% whole blood serum or 0.6 mg/ml BSA were exposed to the treatments indicated. Net stimulation of DNA synthesis in 3T3 or bovine aortic smooth muscle cells was determined for each pair of samples before and after treatment. Adapted from Martin *et al.* (12,13).

by ultrafiltration on 10,000-dalton exclusion membranes; activity is stable to freezing and thawing but it is adsorbed by both anion exchange and hydrophobic resins.

Preliminary gel filtration chromatography of macrophage-conditioned media on Sephadex G-100, under both hypertonic (2 *M* NaCl) and isotonic conditions (Fig. 3), indicates that MDGF activity fractionates in a high-molecular-weight range (>80,000 daltons). Although this result may reflect aggregation of MDGF, rather than its separation in monomeric form, it is noteworthy that the activity in fraction I (void volume) stimulates DNA synthesis in all three cell types tested *in vitro,* Balb/c-3T3 fibroblasts, bovine aortic smooth muscle, and bovine aortic endothelium, and also induces capillary proliferation in the corneal neovascularization assay *in vivo* (unpublished data). This growth-promoting effect of MDGF on vascular endothelium is another property that distinguishes it from the platelet-derived growth factor (25). In addition, MDGF activity is separable from the previously described T-lymphocyte mitogen or "lymphocyte-activating factor" secreted by macrophages (1), which is recovered in fraction IV in this system.

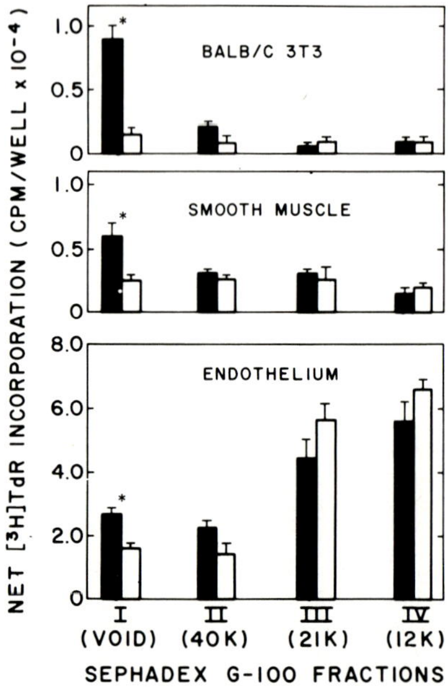

Fig. 3. Gel filtration chromatography of macrophage-conditioned media. Macrophage-conditioned media (solid bars) and control-conditioned media (open bars) containing 1% whole blood serum were chromatographed on Sephadex G-100 under isotonic conditions. Samples in phosphate-buffered saline were added to serum-free DME (1 : 1) and ^{3}HTdR incorporation (mean ± S.D.) was determined in triplicate for each cell type. Statistical significance of differences between the activities of macrophage- and control-conditioned media were determined by the *t*-test; asterisk indicates $p < .01$. Reproduced from Martin *et al.* (12), with permission of the publisher.

BIOLOGICAL SIGNIFICANCE OF MACROPHAGE-DERIVED GROWTH FACTOR

The capacity of the monocyte/macrophage to secrete growth factor(s) for a broad spectrum of nonlymphoid mesenchymal cells implies a more active involvement of mononuclear phagocytes in certain pathophysiological processes than previously realized. For example, in addition to participating as a phagocytic scavenger in the debridement of damaged tissues, the macrophage appears to be capable of directly stimulating fibroplasia during wound repair. Activated macrophages also can mediate the proliferation of small blood vessels, presumably augmenting the delivery of nutrients and the removal of catabolites from a healing area. However, in contrast to these beneficial roles, there are potential pathological manifestations of macrophage-induced mesenchymal cell proliferation. Excessive stimulation of fibroblast activity may result in certain disease processes, such as pulmonary fibrosis (27), and the invasive growth of an inflamed, vascularized connective tissue pannus can result in the destruction of normal tissues, as in rheumatoid arthritis. Recent studies have shown that monocytes infiltrate the renal glomerulus during the development of glomerulonephritis, and *in vitro* experiments indicate that a macrophage-derived growth factor can stimulate mesangial cell proliferation (28).

Stimulation of vascular smooth muscle proliferation by macrophage products may be especially relevant to the problem of atherosclerosis. As several studies have shown (29,30), circulating blood monocytes infiltrate the aortic intima during the development of atherosclerotic lesions. Smooth muscle proliferation is considered to be a central event in the development of the atheromatous plaque (31), and as our data indicate, activated macrophages can secrete a potent growth-promoting activity for this cell type. In contrast to the boluslike release of stored alpha-granule contents by platelets at sites of vascular injury (25,31), activated macrophages presumably could continue to synthesize and secrete growth factor(s) locally over a prolonged time course. However, any speculation regarding the *in vivo* effects of MDGF must take into account possible interactions with other factors (e.g., the cationic platelet-derived growth factor, low-density lipoproteins, etc.) that also can influence vascular smooth muscle growth (31).

The significance of the broad spectrum of cell types stimulated by MDGF is not yet clear. Conceivably, several growth factors may be secreted by the activated macrophage, each with its own target cell specificity, or different subpopulations of monocyte/macrophages may be recruited under different circumstances (24,32,33). Also, the stimulus for secretion of MDGF may vary with the particular pathophysiological setting. Isolation and purification of putative monocyte/macrophage-derived growth factor(s) and further studies of their secretion and cellular effects will be required to resolve these questions.

SUMMARY

Mononuclear phagocytes secrete a growth-promoting activity that acts on several different nonlymphoid mesenchymal cell types, including vascular endothelium, vascular smooth muscle, and fibroblasts.

This monocyte–macrophage-derived growth factor (MDGF) induces the proliferation and directional growth of capillaries (neovascularization) *in vivo* and stimulates DNA synthesis and cell replication *in vitro*. Production of MDGF by cultured mononuclear phagocytes is augmented by phagocytic and biochemical stimuli but does not require the presence of platelet-derived growth factor or other serum constituents. Preliminary studies indicate that MDGF is a nondialyzable protein with properties that differentiate it from certain other macrophage products and growth factors. Secretion of MDGF may be an important aspect of mononuclear phagocyte function in several pathophysiological processes, including the neovascularization and fibroplasia of wound healing and the intimal smooth muscle hyperplasia of atherosclerosis.

ACKNOWLEDGMENTS

This research was supported by Grants HL-22602, HL-08251, NCI-1473 from the National Institutes of Health, a grant from the Council for Tobacco Research and NIH Biomedical Research Support Funds. Dr. Gimbrone is an Established Investigator of the American Heart Association.

REFERENCES

1. Unanue, E. R. (1976) *Am. J. Pathol.* **83,** 396–417.
2. Page, R. C., Davies, P., and Allison, A. C. (1978) *Int. Rev. Cytol.* **52,** 119–157.
3. Nathan, C. F., Murray, H. W., and Cohn, Z. A. (1980) *N. Engl. J. Med.* **303,** 622–626.
4. Leibovich, S. J., and Ross, R. (1975) *Am. J. Pathol.* **78,** 71–100.
5. Leibovich, S. J., and Ross, R. (1976) *Am. J. Pathol.* **84**(3), 501–513.
6. Leibovich, S. J. (1978) *Exp. Cell Res.* **113,** 47–56.
7. Wall, R. T., Harker, L. A., Quadracci, L. J., and Striker, G. E. (1978) *J. Cell. Physiol.* **96,** 203–213.
8. Greenburg, G. B., and Hunt, T. K. (1978) *J. Cell. Physiol.* **97,** 353–360.
9. Ziats, N. P., and Robertson, A. L., Jr. (1981) *Atherosclerosis* **38,** 401–410.
10. DeLustro, F., Sherer, G. K., and LeRoy, E. C. (1980) *J. Reticuloendothel. Soc.* **28**(6), 519–532.
11. Polverini, P. J., Cotran, R. S., Gimbrone, M. A., Jr., and Unanue, E. R. (1977) *Nature (London)* **269,** 804–806.
12. Martin, B. M., Gimbrone, M. A., Jr., Unanue, E. R., and Cotran, R. S. (1981) *J. Immunol.* **126**(4), 1510–1515.
13. Martin, B. M., Gimbrone, M. A., Jr., Unanue, E. R., and Cotran, R. S. (1981) *In* "Plasma and Cellular Modulatory Proteins" (D. H. Bing and R. A. Rosenbaum, eds.), pp. 83–94. Center for Blood Research, Inc., Boston.
14. Martin, B. M., Gimbrone, M. A., Jr., Unanue, E. R., and Cotran, R. S. (1981) *Fed. Proc., Fed. Am. Soc. Exp. Biol.* **40**(3), 335.

15. Martin, B. M., Gimbrone, M. A., Jr., Majeau, G. R., Unanue, E. R., and Cotran, R. S. (1981) *Circulation* **64,** IV–214.
16. Sidky, Y. A., and Auerbach, R. (1975) *J. Exp. Med.* **141,** 1084–1100.
17. Herman, P. G., Yamamoto, I., and Mellins, H. Z. (1972) *J. Exp. Med.* **136,** 697–714.
18. Anderson, N. D., Anderson, A. O., and Wyllie, R. G. (1975) *Am. J. Pathol.* **81,** 131–153.
19. Dvorak, A. M., Mihm, M. C., Jr., and Dvorak, H. F. (1976) *Lab. Invest.* **34**(2), 179–191.
20. Polverini, P. J., Cotran, R. S., and Sholley, M. M. (1977) *J. Immunol.* **118**(2), 529–532.
21. Gimbrone, M. A., Jr., Cotran, R. S., Leapman, S. B., and Folkman, J. (1974) *J. Natl. Canc. Inst.* **52,** 413–427.
22. Clark, R. A., Stone, R. D., Leung, D. Y. K., Silver, I., Hohn, D. C., and Hunt, T. K. (1976) *Surg. Forum* **27,** 16–18.
23. Thakral, K. K., Goodson, W. H., III, and Hunt, T. K. (1979) *J. Surg. Res.* **26,** 430–436.
24. Beller, D. I., Kiely, J.-M., and Unanue, E. R. (1980) *J. Immunol.* **124**(3), 1426–1432.
25. Ross, R., and Vogel, A. (1978) *Cell* **14,** 203–210.
26. Glenn, K. C., and Ross, R. (1981) *Cell* **25,** 603–615.
27. Allison, A. C. (1978) *Int. Rev. Exp. Pathol.* **18,** 303–346.
28. Cotran, R. S. (1978) *J. Lab. Clin. Med.* **92**(6), 837–840.
29. Gerrity, R. G. (1981) *Am. J. Pathol.* **103**(2), 181–190.
30. Schaffner, T., Taylor, K., Bartucci, E. J., Fischer-Dzoga, K., Beeson, J. H., Glagov, S., and Wissler, R. W. (1980) *Am. J. Pathol.* **100**(1), 57–73.
31. Ross, R., and Glomset, J. A. (1976) *N. Engl. J. Med.* **295,** 369–377.
32. Arenson, E. B., Jr., Epstein, M. B., and Seeger, R. C. (1980) *J. Clin. Invest.* **65,** 613–618.
33. Norris, D. A., Morris, R. M., Sanderson, R. J., and Kohler, P. F. (1979) *J. Immunol.* **123**(1), 166–172.

The Control of Proliferation and Differentiation of Endothelial Cells

D. GOSPODAROWICZ, N. SAVION, AND L. GIGUERE
Cancer Research Institute and the Department of Medicine
University of California, Medical Center
San Francisco, California

INTRODUCTION

The ultrastructural, developmental, and physiological functions of the endothelial cells that line various organs are well understood. Not yet fully understood, however, are the factors and control mechanisms that govern and mediate their proliferation during early morphogenesis or during the repair process subsequent to a wound. The mechanisms by which various endothelia perform their physiological functions have also yet to be fully explored. The recent development in tissue culture of endothelial cells that retain their ability to proliferate actively and that, once confluent, can adopt the morphological configuration and phenotypic expression of the endothelial territories to which they once belonged has provided *in vitro* models for the study of their growth control and biological functions. In the present review, we describe the use of bovine corneal endothelial cells for studies of the differentiated functions of that tissue, as well as the use of the biomatrix produced by such cells for the development in culture of other cell types. Such studies could lead to a better understanding of the factors and mechanisms involved in cell proliferation and differentiation.

THE CORNEA: STRUCTURE AND DEVELOPMENT

Although deceptively simple at first glance, the cornea is a complex organ composed of three different tissues separated from each other by

PATHOBIOLOGY OF THE ENDOTHELIAL CELL

ISBN 0-12-521980-6

acellulars layers that have the biochemical characteristics of basement membranes. The organization and interrelationship of each tissue with the others are essential for the proper embryonic development as well as for the proper function of the cornea during the life of the individual.

The embryonic development of the cornea is complex and results from a series of inductive interactions between germ layers [for a review, see (1)]. After the detachment of the lens from the overlying ectoderm, the presumptive corneal epithelium is at first two cells thick. Under the inductive influence of the lens, the basal layer hypertrophies and assumes the morphology of a secretory tissue producing an extracellular corneal matrix: the primary stroma. Following the formation of the primary stroma, the endothelium of the cornea migrates into spaces between the primary stroma and the lens. The corneal endothelium is derived from the neural crest. It is closely associated with the developing blood vessels in the vicinity of the lip of the optic cup and remains in functional continuity with the adjacent vascular endothelium. The migrating endothelial cells flatten out and migrate on top of the inner surface of the primary stroma. Dividing repeatedly until they become a monolayer, they then adopt the configuration of a cuboidal endothelium with a free surface facing the lens. The space between the lens and the endothelium enlarges and gives rise to the anterior chamber. The endothelium then produces a basement membrane on its basal surface that thickens and will become the Descemet's membrane.

During this process the primary stroma has begun to swell. It is likely that the edema of the stroma triggers the subsequent fibroblast invasion. This invasion proceeds rapidly, and mesenchymal cells derived from the secondary mesenchyme penetrate between the lamellae of the primary stroma but not into the area of primary stroma, which is immediately adjacent to the epithelium. This zone is later destined to become the definitive acellular Bowman's membrane.

Development of the cornea is henceforth dominated by the activities of the newly differentiated fibroblasts. As a result of their secretory activity, numerous collagen fibrils are added to the corneal stroma in a highly organized fashion that follows the blueprint represented by the lamellar organization of the inner part of the primary stroma. This results in a three-dimensional, orthogonal organization of the collagenous lamellae. Although opaque at the outset, the cornea starts to lose water rapidly. This could result from a loss of ability of the stroma to bind water and the newly acquired ability of the corneal endothelium to perform its pump function.

The developing cornea provides several examples of tissue interactions that might be more easily approached experimentally, especially with respect to the role of extracellular materials in cell migration and cell proliferation. After an epithelial–epithelial interaction represented by the interaction between the basal surface of the lens and the presumptive corneal epithelium, there occur in the cornea two interactions in which mesenchymal cells seem to respond to the products of the epithelia. In the first of these, the vascular mesenchyme at the limbus of the eye inserts itself between the optic cup and the ectoderm. Endothelial cells derived from the neural crest migrate along the posterior surface of the primary corneal stroma. In the second interaction, changes consisting principally of disruption of collagen fibrils and swelling of the extracellular interstices may be the result of actions taken by the endothelium and epithelium and seem to be involved in bringing about the invasion of the stroma by fibroblasts (1).

The different embryological origins of the various tissues composing the cornea make possible a study of the ability of various growth factors to stimulate the proliferation of cells originating from embryological territories as diverse as the ectoderm, the primary and secondary mesenchymes, and the neural crest. It also allows one to study how the extracellular material produced during the differentiation of the various cell layers could later affect the proliferative ability of the cells, as well as their response to various growth factors. Finally, the various tissues that compose the cornea have widely varying abilities to regenerate after being wounded. While the ability of the epithelium is the greatest, that of the endothelium is the lowest. The endothelium therefore offers an excellent model for testing whether growth factors stimulate the proliferation of tissues that otherwise have a low potential *in vivo* for regeneration.

EFFECTS OF GROWTH FACTORS ON THE PROLIFERATION AND DIFFERENTIATION OF CORNEAL ENDOTHELIAL CELLS *IN VITRO*

PROLIFERATION OF LOW-DENSITY CORNEAL ENDOTHELIAL CELL CULTURES EXPOSED OR NOT TO GROWTH FACTORS

Significant limitations on the culture of corneal endothelial cells have been imposed in the past by the slow doubling time of these cultures, which can be passaged only at a high cell density if precocious senescence is to be avoided (2–4). Even in this case, though, most, if not all, corneal endothelial cell cultures thus far established have shown a short *in vitro* life span and exhibit early signs of dedifferentiation (2–4). The use of mitogens such as fibroblast growth factor (FGF) or epidermal growth factor (EGF), which are mitogenic *in vitro* for a wide variety of mesenchyme-derived cells, and in particular for corneal endothelial cells, could greatly facilitate the culture of that cell type (5–10), since low-density cell populations have been shown to proliferate actively when exposed to either mitogen (5).

Although neither EGF nor FGF is required to establish primary cultures of bovine corneal endothelial cells when cells are plated at a high enough cell density, the development of a monolayer will depend on the presence of FGF or EGF in the culture medium if cultures are started at clonal cell density (three cells per square centimeter). In 10% calf serum alone, small colonies develop from cell aggregates during the first few days but the cells look unhealthy, proliferate slowly, and quickly become senescent. However, if as little as 10 ng/ml of FGF or EGF is added to the culture, the population doubling time is reduced to 20 hr, and the culture upon becoming confluent exhibits the morphology characteristic of that tissue *in vivo* (6).

When cell proliferation is analyzed as a function of the mitogen concentrations to which the cells are exposed, EGF is effective at concentrations as low as 0.01 ng/ml (1.5×10^{-12} *M*), with a half-maximal response at 1.5×10^{-10} *M*. FGF has a slight mitogenic effect at 0.1 ng/ml (7×10^{-12} *M*) and is optimally effective at 100 ng/ml (7×10^{-9} *M*). The half-maximal response is at 6×10^{-10} *M*. Such cultures maintained and propagated in the presence of saturating concentrations of FGF or EGF divide with an average doubling time of 18 hr when seeded at either a high (up to 1 : 1000) or low split ratio and, on reaching confluence, the cells adopt a morphological configuration similar to that of the confluent culture from which they originated (6). The response of the cells to both mitogen is a strict function of the plasma or serum concentrations to which they are exposed. This indicates that plasma or serum factors are also involved in the proliferative response of this cell type.

THE EFFECTS OF FGF AND EGF ON THE LIFE SPAN OF BOVINE CORNEAL ENDOTHELIAL CELLS

The response of corneal endothelial cell cultures to mitogens has been shown to be a function of the age of the culture, of the substratum on which they are maintained, and of the initial cell density to which cultures are seeded (8–10).

The Response to FGF as a Function of the Age of the Culture and Substrata. The mitogenic effect of FGF depended on the age of the cultures. Although FGF had a minimal effect in primary culture (threefold increase in cell number over control), cells in their second passage, when maintained in the absence of FGF, already had begun to lose their potential for proliferation (9). This was reflected by an increased average doubling time (from 18 to 48–72 hr) and by cell enlargement (two- to threefold) (8,9). The addition of FGF to the cultures allowed cells to proliferate with a doubling time of 20 hr and prevented cell enlargement. The resulting cell monolayer is therefore composed of small, nonoverlapping hexagonal cells with a cell density of 1500–2000 cells/mm^2 similar in morphology to that seen *in vivo* (8,9).

The mitogenic effect of FGF further depended on the substrate on which the cells were maintained and could also be potentiated by the addition of dextran (40,000 MW) to the medium. While cells maintained on gelatinized dishes responded better to FGF than did cells maintained on plastic, the addition of 5% dextran alone to the medium greatly accelerated the proliferation of cultured cells maintained in the absence of growth factor. It also resulted in a higher final cell density when added to cultures maintained in the presence of FGF (9). The addition of dextran affected not only the rate of proliferation of the cell population but also the synthetic ability of the cells and their morphological appearance. Cultured cells maintained in the presence of dextran and FGF produced an extensive basement membrane that appeared earlier and was far more developed than that seen in cultures maintained in the absence of dextran (9,11,12). The morphological appearance of late-passage cultures (200 generations) maintained in the presence of FGF and dextran was that of a highly contact-inhibited monolayer with a nonthrombogenic surface exposed to the medium and similar in appearance to that observed *in vivo* (8). In contrast, late-passage cultures maintained in the presence of FGF but without dextran exhibited cell overlapping in some areas that disappeared when the cultures were maintained at confluence for over a week (8).

The Response to FGF as a Function of Cell Density. Cultures maintained in the presence of FGF proliferate actively, regardless of the split ratio to which they are passaged. In contrast, seeding of the same cells in the absence of FGF results, even at a low (1 : 10) split ratio, in a much longer doubling time (60–78 hr) and in a strikingly different morphology (8). When seeded at a high split ratio (1 : 128) and in the absence of FGF, the cells proliferated poorly and became senescent. The alterations in growth behavior and morphological appearance were best demonstrated after three to four passages (15–20 generations) in the absence of FGF (8). The cells, by then 5- to 10-fold larger in size, failed to adopt the configuration of a nonoverlapping cell monolayer even after being split at a 1 : 10 ratio. Instead, at sparse density they were flattened and highly spread (8) and at confluence grew on top of each other, leaving intercellular spaces. These cells exhibited a short life span, as reflected by vacuolization and cell degeneration after 30 generations in the absence of FGF (8).

Endothelial cells maintained for three passages in the absence of FGF and beginning to show degenerative changes were still capable of responding to the mitogen, since its addition resulted in a greatly increased growth rate of the cultures (8), which at confluence adopted the characteristic endothelial configuration of a cell monolayer composed of small, highly flattened, and closely apposed cuboidal cells, rather than their prior appearance as overlapping and large cells (8).

A similar effect of EGF could also be observed in earlier passage cultures, however, and unlike FGF, EGF could not prevent cell senescence beyond six to eight passages when the cultures were split weekly at a high split ratio (1 : 100).

MORPHOLOGICAL DIFFERENTIATION OF LONG-TERM CONFLUENT CULTURES GROWN IN THE PRESENCE OR ABSENCE OF GROWTH FACTORS

Confluent cultures of corneal endothelial cells grown in the presence of FGF exhibited every one of the attributes of endothelial cells *in vivo*, including the ability to synthesize a basal lamina with a given polarity (11,12). An extracellular matrix typical of basement membrane *in vivo* is seen underneath the cell monolayer. Pinocytotic vesicles with an electron-dense coat are present near the base of these cells; it is presumed that these vesicles are involved in the secretion of the amorphous material found underneath the basal cell membrane. Furthermore, cultured bovine corneal endothelial cells *in vitro* show a distinct polarity in the localization of their organelles. Bundles of microfilaments run parallel to the apical plasma membrane and delimit a zone free of organelles. A junctional complex is present at the apex of the cells. Microplicae and villi characterize the apical cell membrane as well as a marginal fold found adjacent to the junctional complex (11,12). The cell mitochondria exhibit the unique arrangement of the cristae mitochondriales that is characteristic of corneal endothelial cells (11,12).

Confluent cultures of corneal endothelial cells exhibit, as *in vivo*, an asymmetry of cell surfaces. While the apical cell surface is a nonthrombogenic surface to which platelets do not bind (11,12), the basal cell surface is involved in the synthesis of a highly thrombogenic basal lamina (8), which, when examined by immunofluorescence, is composed of collagen type III and, to a lesser extent, of collagen type IV (13,14). Chemical analysis of the collagen types synthesized by bovine corneal endothelial cells has led to the conclusion that type III collagen is the major component both deposited in the basal lamina and secreted into the media (15). The basement membrane collagens, types IV and V, are also found in each compartment, though the latter is associated preferentially with the cell matrix. The ratios of types I : III : IV + V collagens synthesized by corneal endothelial cells and present in the basal lamina is 3 : 16 : 1 (15). Associated with the basal lamina collagens are proteoglycans and glycosaminoglycans, composed mostly of heparan sulfate and Dermatan sulfate (16). Also present in large quantities are two glycoproteins, fibronectin and laminin (13,14), which have been held responsible for cell attachment to the basal lamina (17–20).

The presence of both laminin and collagen types IV and V, as well as that of fibronectin, in the basal lamina produced by cultured corneal endothelial cells has led to the conclusion that this matrix has the *in vivo* characteristics of a basement membrane (13,14). Indeed, previous studies have demonstrated that laminin is an antigen present within the lamina lucida portion of basement membranes produced by various epithelia *in vivo*, whereas collagen type IV and fibronectin are localized to the lamina densa (17,18).

STRUCTURAL CHANGES IN THE CORNEAL ENDOTHELIAL CELL SURFACES ASSOCIATED WITH THE FORMATION OF A CONFLUENT CELL MONOLAYER

Among the factors that could be responsible for cell attachment to the basal lamina are fibronectin and laminin (19,20). The presence of both fibronectin and laminin in such matrices is well documented (17,18,21–24). We have therefore looked at the ability of corneal endothelial cells to produce laminin versus fibronectin and have analyzed the localization of such antigens within the cell surface and basal lamina produced by the cells as a function of cell growth (sparse, actively dividing versus confluent, resting cultures) and organization.

Laminin, when observed by indirect immunofluorescence following staining with affinity purified antilaminin antibodies, could not be detected on the apical cell surface of sparse corneal endothelial cell cultures. Instead, it was found localized within the basal cell surface, where it was present as a characteristic "food pad" composed of irradiating spikes localized at the periphery of the cell (14). The unique distribution of that antigen is consistent with a primary role in cell migration rather than in cell adhesion (14). In contrast, in subconfluent cultures laminin was distributed in a fibrillar pattern forming an extensive reseau in the basal cell surface and in the region of cell–cell contact (14). As in sparse cultures, no laminin could be detected on the apical cell surface. When cultures became confluent and were still suborganized, laminin was present in foci concentrated in the region of cell–cell contact and none was expressed on the apical cell surface. Likewise, in 2-week-old confluent and fully organized cultures (14) all of the laminin present on the cell surface was totally localized underneath the cells and was therefore inaccessible to the antibodies, unless one permeabilized the cell layer with acetone treatment or removed it by detergent treatment (14).

In contrast to laminin, the appearance of fibronectin was density-dependent. Although it is absent in isolated cells, when cells come into contact, fibronectin appears on both their apical and basal cell surfaces, as well as in the area of cell–cell contact (14). This could imply a role for fibronectin in cell attachment but not in cell migration. When cultures became confluent, fibronectin disappeared from the apical cell surface (14) and could be found only beneath the cells. Therefore, in contrast to laminin, in subconfluent cultures fibronectin can be found associated with both the apical and basal cell surfaces.

In order to determine whether laminin, like fibronectin, could be present primarily in the basal lamina located beneath the cells, its possible association with this matrix has been analyzed. The association of laminin with the basal lamina produced by corneal endothelial cells was very apparent when cells of a confluent monolayer were removed following exposure to 0.5% Triton X-100 and the material left on the dish was stained by immunofluorescence (13,14). When the immunofluorescent pattern observed with laminin was compared to that of collagen type IV or that of fibronectin (13,14), it was found to be similar. These results therefore demonstrate that, like fibronectin, laminin in confluent cultures can be found associated primarily with the basal lamina (which is in turn closely associated with the basal cell surface), but not with the apical cell surface, and that within the basal lamina it codistributes with type IV collagen (13,14).

When the amount of laminin released into the incubation medium was analyzed (following metabolic labeling and specific immunoprecipitation with specific antibodies) as a function of cell density, it was observed that it decreased drastically (from 0.3 to 0.01%) once cultures became confluent (14). Since it is at that time that cells start to produce an extensive basal lamina, this decreased release of laminin into the media may reflect the preferential accumulation of laminin within the basal lamina (14). This would also be consistent with the polarity of cell surfaces shown by endothelial cells. Indeed, previous studies have shown that the basal cell surface is where the secretion and accumulation of the basal lamina take place (7,11). The apical cell surface, which is nonthrombogenic, has been shown to be devoid of a basal lamina (7,11,25–27). The polarity of corneal endothelial cell surfaces may therefore be reflected by the preferential accumulation of laminin in the basal cell surface. In contrast to laminin, the amount of fibronectin released into the incubation medium was not affected by cell density or growth stage, since it increased from 2.7% in sparse, actively growing cultures to 5.8% once cultures became confluent (14).

STRUCTURAL AND MORPHOLOGICAL ALTERATIONS OCCURRING WHEN CELLS ARE GROWN IN ABSENCE OF FGF

In contrast to cultures grown in the presence of FGF, cultures maintained in its absence lose within three passages their ability to form at confluence a monolayer of closely apposed and flattened cells. Instead, the cultures were composed of large and overlapping cells that are no longer contact inhibited (8). Parallel to these changes in cell morphology, the apical cell surface becomes thrombogenic, as reflected by an increase in platelet binding capacity (8). Likewise, marked changes in the distribution and appearance of cell surface proteins such as fibronectin and collagen can be observed. Fibronectin and collagen, which in confluent and highly organized cultures grown in the presence of FGF is detected only in the basal, and not in the apical cell surface, now appear in both basal and apical cell surfaces (8). Release of fibronectin into the medium, which in normal differentiated cultures amounted to 5–6% of the total amount of protein released, increased to 40%. These morphological and biosynthetic alterations are similar to those previously observed with vascular endothelial cells maintained and propagated in the absence of FGF (25–28).

These observations suggest that corneal endothelial cells maintained in the absence of FGF exhibit, in addition to a much slower growth rate, morphological as well as structural alterations that mostly involve changes in the composition and distribution of the basal lamina. This raises the possibility that the basal lamina produced by these cells could either directly or indirectly affect the cells' ability to proliferate and to express their normal phenotype once confluent (13).

SYNTHESIS OF CELL PROTEINS IN SPARSE VERSUS CONFLUENT CULTURES AND THE EFFECTS OF FGF AND EGF ON THE SYNTHESIS OF CELLULAR PROTEINS

Metabolic changes associated with the formation of a confluent, contact-inhibited endothelial cell monolayer have been analyzed by studying the metabolic activity of the cultures after various periods of time in culture. The parameters followed were DNA and protein synthesis, which were studied by measuring [^{3}H]thymidine and [^{35}S]methionine incorporation, respectively. A comparison of the respective rates of incorporation of [^{35}S]methionine and of [^{3}H]thymidine into the cultures demonstrated that corneal endothelial cells are capable of reaching a truly resting state at confluence. This is reflected by a 50-fold decrease in the rate of [^{3}H]thymidine incorporation into DNA once cultures became confluent. Consistent with the drop in DNA synthesis, corneal endothelial cell cultures, in the process of forming a confluent and resting monolayer, showed more than a 14-fold reduction in the incorporation of [^{35}S]methionine (8,29). One could therefore argue that the formation of a contact-inhibited endothelial cell monolayer should, in addition to having the morphological appearance of highly flattened and closely apposed cells, also meet two additional criteria: (1) an almost complete inhibition of DNA synthesis, and (2) a 10- to 20-fold decrease in the rate of protein synthesis (8,29).

The effects of FGF and EGF on corneal endothelial cell growth have also been studied at the level of newly synthesized protein patterns. Sparse and confluent cultures of corneal endothelial cells maintained in the presence and absence of FGF or EGF were incubated for 1 hr in the presence of [^{35}S]methionine, and the cellular proteins were subjected to two-dimensional gel electrophoresis (8). The effect of FGF on sparse, actively growing cultures was a preferential reduction in the production of several proteins [(peptides 1 (fibronectin), 10,22,30)] and the induction of production of two other proteins (peptides 3 and 32) (8). EGF, which induced cell proliferation as well as FGF but, unlike FGF, could not prevent cell senescence beyond six to eight passages when the cultures were split weekly at high split ratio, was found to induce only some of the changes in the newly synthesized protein patterns induced by FGF, since it induced preferential reduction only in the production of peptides nos. 1 and 10 while inducing the production of peptide no. 32. Peptide no. 10 was found to be induced by the absence of growth factors in all cases studied thus far (bovine corneal endothelial cells, adult bovine aortic endothelial cells, and fetal bovine heart endothelial cells). The appearance of this protein correlates with a loss of cell surface polarity when cultures are maintained in the absence of growth factors (8,25–28). In confluent and organized cultures maintained in absence of growth factors the rate of [^{35}S]methionine incorporation declined with the culturing time (8). The pattern of newly synthesized proteins remained the same, however, as indicated by two-dimensional gel electrophoresis. The effect of FGF in confluent, resting, contact-inhibited cultures was confined to a preferential reduction in the production of peptides nos. 1, 10, 22, 36, and 41 (8). It did not induce the synthesis of a new protein. The reduction in the number of proteins synthesized by confluent cultures in the presence of EGF could reflect the differentiation process that occurs in these cultures. As observed with sparse cultures, EGF also has a partial effect in confluent cultures and reduced the appearance of peptides nos. 10, 36, 41, and 3 but did not reduce the appearance of peptide no. 1 (8). In addition, EGF also induced the synthesis of two new peptides (nos. 33 and 40), which did not appear in corneal endothelial cells under any other conditions (8).

THE SYNTHESIS AND DISTRIBUTION OF CYTOSKELETAL ELEMENTS IN CORNEAL ENDOTHELIAL CELLS AS A FUNCTION OF CELL GROWTH AND ORGANIZATION

The studies outlined above indicate that once corneal endothelial cell cultures become confluent, their reorganization into a single cell monolayer composed of closely apposed and nonoverlapping cells is reflected in a redistribution of cell surface proteins. Since cell surface proteins can be connected (directly or indirectly) to cytoskeletal elements, their redistribution in confluent endothelial cell cultures could reflect a reorganization of the cytoskeleton. In view of the role of cytoskeletal elements in cell–cell interaction, as well as in cell shape, we have investigated whether the transition from sparse, actively growing to confluent and resting endothelial cell cultures could also be reflected in changes in the synthesis and cellular distribution of cytoskeletal elements (29).

The distribution of newly synthesized cytoskeletal elements in endothelial cell cultures has been analyzed and compared as a function of cell density. Sparse (5-day-old) and confluent (21-day-old) cultures maintained in the presence of FGF were incubated for 1 hr in the presence or absence of [^{35}S]methionine. Cellular proteins were then extracted and subjected to two-dimensional gel electrophoresis. The pattern of newly synthesized proteins was found to be almost the same when either sparse or confluent cultures were studied, and the cells did not synthesize greater amounts of the 58 K molecular weight cytoskeleton protein (vimentin) on reaching confluence, as had been observed with vascular endothelial cultures (25,29). A Coomassie blue staining pattern of the slab gel revealed, however, that the relative amount of the 58 K molecular weight protein (vimentin) increased dramatically in confluent cultures as compared with that present in sparse cultures (29). This, together with the previous data on metabolic labeling of the cells with [^{35}S]methionine, indicates that the rate of turnover of vimentin is slower in confluent than in sparse BCE cell cultures. This results in its preferential accumulation in confluent cells (for reviews, see 30,31). In contrast to vimentin, when the synthesis of tubulin and actin was analyzed in sparse versus confluent cultures, no density-dependent changes in the rate of actin or tubulin synthesis were observed (29,32).

INVOLVEMENT OF THE BASAL LAMINA PRODUCED BY CORNEAL ENDOTHELIAL CELLS IN THE CONTROL OF CELL PROLIFERATION

BASAL LAMINA AND CELL ATTACHMENT, MIGRATION, AND PROLIFERATION

Cell migration and growth *in vivo* are the result of a complex balance between cell–cell and cell–substrate interactions. Those forces that combine to modulate the cell shape may either permit or prevent cell proliferation and differentiation (33–36). Following its original proposal by Grobstein (37), a role for cell–substrate interactions in the control of cell proliferation and morphogenesis has been demonstrated (for reviews see 38–43). In the case of epithelial tissues with a high rate of cell turnover, such as the epidermis or the corneal epithelium, active cell proliferation is restricted to their basal layer, composed of tall and columnar cells. These cells are in close contact with a basal lamina. In contrast, cells in the upper layers, which have lost their ability to proliferate and gradually adopt a flattened configuration, are no longer in contact with the basal lamina. Thus contact between the cells and their substrate rather than contact between cells could have a permissive influence *in vivo* on cell proliferation. Likewise, mammalian cells maintained under tissue culture conditions require, in order to proliferate and to express their normal phenotype, not only nutrients and growth factors (44,45), but also an appropriate physical substratum upon which they can attach and spread (for reviews, see 13,46). Recent investigations have suggested that, *in vitro*, cell–substratum interactions may play other roles in addition to cell anchorage (47–49). The substrate upon which a cell rests can dictate its shape and may play a role in its response to serum factors (50) and a differential response to various growth factors (42,50). Components of the basal lamina produced by cells have also been shown to be involved in cell differentiation and the control of gene expression (51–54), as well as in cell attachment (47–49). Yaoi and Kanaseki (55), as well as Weiss *et al.* (56), using microexudate carpets from chick or mouse embryo cells, have demonstrated a role for this material in cell proliferation. The substrate upon which cells rest *in vitro* could therefore be a decisive element in their proliferative response to various factors.

BASAL LAMINA AND THE CONTROL OF CORNEAL ENDOTHELIAL CELL PROLIFERATION

Identification of the components present within the basal lamina that could be involved in controlling cell proliferation either *in vivo* or *in vitro* has been made difficult mostly by its intricate nature. Because the *in vitro* reconstruction of the basal lamina from its isolated native components into the correct highly ordered structure that it represents would be a formidable task, advantage was taken of the fact that cultured corneal endothelial cells have the ability to produce basal lamina *in vitro* that closely resembles the basement membrane upon which cells rest *in vivo* (11,12). This membrane can easily be denuded from the cell monolayer that covered it by exposing confluent cultures to either detergents (0.5% Triton X-100 in PBS) or weak alkali (20 m*M* NH_4OH in water) and subsequent washing (57–60).

When the ability of the basal lamina to support cell proliferation was analyzed using either rabbit or bovine corneal endothelial cells seeded at low cell density in the presence of high (10%) concentrations of serum, cells maintained on basal lamina-coated dishes proliferated even faster than cells maintained on plastic and exposed to FGF (8,61). It can therefore be concluded that when the proliferation of corneal endothelial cells from two different species is compared, cells maintained on plastic proliferate poorly and FGF is needed in order for the cultures to become confluent. In contrast, when similar cultures are maintained on basal lamina, they proliferate actively and no longer require FGF in order to become confluent. Similar results were observed when lens epithelial cells were maintained on plastic versus basal lamina-coated dishes. Although cells maintained on plastic hardly proliferated, yielding within a few days a population composed of large and binucleated cells, cells plated on basal lamina proliferated actively, with an average doubling time for lens epithelial cells of 15 hr during their logarithmic growth phase (8,61). One could therefore conclude that although FGF greatly increases the growth rate of cultures maintained on plastic and exposed to either plasma or serum, it does not affect the growth rate of cultures maintained on a basal lamina. This is to be expected, since cells on basal lamina had a proliferative rate that is nearly optimal (15 hr doubling time).

In order to decide whether the basal lamina itself was mitogenic for cells seeded on it or whether it had a permissive effect, the final cell density of cultures maintained on plastic versus basal lamina-coated dishes was examined as a function of increasing concentrations of serum or plasma to which cultures were exposed (61). If the basal lamina has a permissive effect, the proliferation of the cells would be dependent on serum or plasma factors. One would therefore expect that the final cell density of the cultures would be directly proportional to the plasma or serum concentration to which they are exposed. If the basal lamina itself is mitogenic, one would then expect little difference in the final cell density of cultures exposed to increasing concentrations of serum or plasma. When such experiments were done, the results clearly showed that the rate of proliferation and final cell density of cultures maintained on a basal lamina was a direct function of the serum or plasma concentration to which the cells were exposed (61). This indicates that the basal lamina has a permissive effect on cell growth. Furthermore, since the growth rate and final cell density of cultures maintained on basal lamina and exposed to increasing concentrations of serum or plasma were roughly similar, it is likely that in both cases the cells respond to the same set of factors. One can therefore conclude that the simple change of substrate from plastic to basal lamina will restore the sensitivity of these cells to agents present in plasma.

The permissive effect of the basal lamina produced by corneal endothelial cells on cell proliferation has also been observed in the case of granulosa cells (59,62), adrenal cortex cells (59), vascular smooth muscle cells (57,60), and vascular endothelial cells (58,63). Although these various cell types when maintained on plastic require serum and/or FGF in order to proliferate, they no longer require FGF when maintained on a basal lamina, and the addition of plasma alone is sufficient to make them proliferate at an optimal rate. The permissive effect of the basal lamina produced by corneal endothelial cells is therefore not restricted to the corneal endothelium but also extends to cell types of unrelated origin.

These observations raise the possibility that although the final effect of FGF is that of a mitogen (10,64,65), its action could be indirect. FGF either could replace the cellular requirement for a substratum such as the basal lamina or could induce the synthesis and secretion of the basal lamina produced by the cells. This could in turn make the cells sensitive to factors present in plasma. That the latter alternative could occur finds support in previous observations that FGF can control the production by vascular or corneal endothelial cells of extracellular and cell surface components such as fibronectin and various collagen types (8,25–27). Since sparse cultures of bovine corneal endothelial cells proliferate poorly when maintained on plastic but not when maintained on a basal lamina, it may be that, at very low cell density, cultures maintained on plastic are unable to produce enough extracellular material to support further growth. The mitogenic effect of FGF on these cells could therefore be the result of an increased or decreased potential of the cells for synthesizing specific components of their basal lamina that could be beneficial or harmful, respectively, to the support of cell growth (13,66). Therefore, one can drastically modify the proliferative response of a given cell type to serum or plasma factors in accordance with the substrate on which cells are maintained. It is possible that the lack of response of different cell types maintained under tissue culture conditions to agents responsible *in vivo* for their proliferation and differentiation could be attributed to the artificial substrate, whether plastic or glass, upon which the cells rest and which limits their ability to produce a basal lamina either of the correct composition or in adequate amounts (67).

The ways in which the basal lamina exerts its permissive effect on cell proliferation can only be the object of speculation. One possible effect is to modify the cell shape in order to make it responsive to factor(s) to which the cells do not respond unless they adopt an appropriate shape. Recently, Folkman and Moscona (50), using vascular endothelial cells maintained on tissue culture dishes coated with an agent that modifies the adhesiveness of the cells to the dish, were able to control precisely the cellular shape in morphologies ranging from highly flattened to almost spheroidal. When the extent of cell spreading was correlated with DNA synthesis or cell growth, it was found to be highly coupled. Whereas highly flattened cells responded to serum factors, spheroidal cells no longer responded and intermediate degrees of response could be observed, depending on how flattened the cells were. Likewise, with corneal epithelial cells, changes in cell shape that depend on the substrate on which the cells are maintained correspond to drastically altered sensitivities of the cells to EGF versus FGF (7,36,68). An attractive hypothesis proposed by Yaoi and Kanaseki (55) is that the basal lamina could play a key role in mitosis and facilitate cytokinesis. This hypothesis was based on their observation that, while both high- and low-density cultures maintained on microexudate exhibit a high rate of DNA synthesis and a high mitotic index, only high-density cultures maintained on plastic have both a high rate of DNA synthesis and a high mitotic index. In contrast, low-density cultures maintained on plastic, although they have a high rate of DNA synthesis, have a low mitotic index, thus suggesting that cells do not enter into mitosis. It is therefore likely that, although plastic provides a foreign substrate on which cells can attach tenaciously and spread in a vain attempt to phagocytose it, the basal lamina provides a natural substrate that the cells recognize and upon which they can undergo their characteristic changes in morphology (rounding up) occurring at mitosis (cytokinesis). These morphological changes probably reflect the rearrangement of the cellular cytoskeleton, so that cells can go through the cleavage steps, giving rise to two progeny cells instead of undergoing endomitosis.

PLASMA FACTORS INVOLVED IN THE CONTROL OF PROLIFERATION OF BOVINE CORNEAL ENDOTHELIAL CELLS MAINTAINED ON A BASAL LAMINA

In order to investigate the factors present in plasma that could affect the proliferation of bovine corneal endothelial cells, the innovative approach of Gordon Sato and his colleagues, who have shown that the serum or plasma requirement for growth of a number of cell lines can be satisfied by the addition of specific hormones and growth factors to synthetic media (69–71), was followed. Further incentive to follow this approach was provided by previous observations indicating that bovine corneal endothelial cells maintained on basal lamina-coated dishes had a much lower requirement for either serum or plasma in order to proliferate actively than when they were maintained on plastic (61). It is therefore possible that these cells, when maintained on basal lamina-coated dishes, could survive and still be responsive to plasma factors even when maintained in a well-defined synthetic medium unsupplemented with either plasma or serum. Among the factors that are likely to promote the growth of bovine corneal endothelial cells in serum-free conditions are FGF and EGF, which have been shown to be mitogenic for cultures exposed to either plasma or serum (6). Insulin and transferrin are also likely candidates, since these two agents are required for the growth of most cell types cultured in serum-free medium (69–71). Other less well-studied factors are represented by high- (HDL) and low- (LDL) density lipoproteins, both of which have been shown to be mitogenic for several cell types when cultures are exposed to lipoprotein-deficient serum (72–74) or to serum-free medium (75,76).

When the effects of these various factors were compared, it was observed that bovine corneal endothelial cells seeded on basal lamina-coated dishes in the total absence of serum can actively proliferate when exposed to a synthetic medium supplemented with transferrin and HDL. In contrast, LDL had little if any effect at low concentration and at high concentration was found to be cytotoxic. The requirement for transferrin was absolute, since in its absence cells did not proliferate. However, the addition of HDL and transferrin alone did not ensure a growth rate or a final cell density comparable to that achieved by cultures exposed to plasma alone (77). This suggests that agents other than HDL are required in order to promote optimal endothelial cell growth. When insulin, EGF, and FGF were tested, although they had little mitogenic effect by themselves, they were strongly synergistic when added to cultures exposed to HDL and transferrin, and low-density cultures exposed to transferrin (10 μg/ml), HDL (250 μg protein/ml), insulin (2.5 μg/ml), and FGF (100 ng/ml) or EGF (50 ng/ml) had a growth rate and a final cell density similar to that of cultures exposed to optimal (10%) plasma concentration. Since transferrin, HDL, insulin, and EGF are present in plasma (FGF alone being absent), they may be part of the plasma constituents involved in the control of the proliferation of corneal endothelial cells when such cells are maintained on basal lamina-coated dishes and exposed to plasma. Since similar cultures maintained on plastic did not proliferate when exposed to transferrin, HDL, insulin, and FGF or EGF, it is demonstrated that, as with other cell types, the substrate on which cells are maintained is crucial for a response to these various agents (77).

Exposure of bovine corneal endothelial cell cultures to a synthetic medium supplemented with HDL, transferrin, insulin, and FGF or EGF not only ensures their active proliferation but also allows them to be subcultured repeatedly in the total absence of serum and to undergo 90 generations before senescing (77). In these conditions, the longevity of the cultures is clearly dependent on the use of basal lamina as substrate, since cultures maintained on plastic and exposed to similar conditions cannot be passaged even once. The effect of basal lamina in extending the life span of bovine corneal endothelial cell cultures is even more impressive in the case of cells exposed to a synthetic medium supplemented with optimal concentration of serum (77). While cultures maintained on plastic and exposed to serum can be maintained at best for 20 generations, cultures maintained on basal lamina can be passaged for 160 generations. It may therefore be concluded that the basal lamina on which cells were maintained not only made them sensitive to factors present in serum but also delayed to a considerable extent the ultimate senescence of the cells when they were exposed to a synthetic medium supplemented with either well-defined factors or with serum. Furthermore, since the life span of cultures exposed to serum is longer than that of cultures exposed to a combination of transferrin, HDL, insulin, and FGF or EGF, it is likely that there are other factor(s) present in serum that could further prevent their senescence (77).

The ability to maintain in a synthetic medium supplemented with defined factors actively proliferating bovine corneal endothelial cells which at confluence can still exhibit the morphological characteristics of the corneal endothelium *in vivo* or that of confluent cultures maintained in the presence of serum and FGF-supplemented media could help to shed some light on the *in vivo* growth requirements of this cell type. It could also result in a cell model that more closely mimics the *in vivo* situation, since *in vivo* the corneal endothelium is exposed to the aqueous humor, which, depending on the species considered, has a protein concentration only 0.1 to 1% that of plasma.

THE USE OF CORNEAL ENDOTHELIAL CELL CULTURES IN STUDIES OF THEIR DIFFERENTIATED PROPERTIES

The development in culture of corneal endothelial cells has led to the study of some of its properties which would otherwise have been difficult to study *in vivo*. Among these are its roles in hemostasis and as a selective permeability barrier. Corneal endothelial cells can also be used as a model for studying the interaction of mitogens such as EGF with the cell surface and its processing by the cells. Finally, the functionality of these cultured cells which can be tested in transplantation studies can also have implications in the field of surgery.

THE CORNEAL ENDOTHELIUM AND HEMOSTASIS

Hemostasis, or the cessation of blood flow, is the result of complex interactions between multiple factors leading to blood coagulation. Among the key factor(s) involved in blood clotting is the protease thrombin, which serves as an initiator of the biochemical and functional changes seen in platelets early in hemostasis and in the conversion of fibrinogen into fibrin. One of the principal factors involved in the disappearance of blood clots or fibrinolysis is the protease plasmin, which appears as a result of the activation of plasminogen. This factor has a great affinity for fibrin and adsorbs to it during blood clot formation. The activation of plasminogen into plasmin can be controlled by various tissue kinases, one of which is the plasminogen activator present in the vascular endothelium. Clotting in blood vessels following various types of injury often has a beneficial effect, since it interrupts blood flow, thereby preventing hemorrhage. In contrast, when it occurs in the anterior chamber of the eye, it has only undesirable effects. In addition to preventing proper vision, adhesion of a fibrin clot to the endothelium could cause cell death. Since this tissue has a limited ability to regenerate *in vivo*, this would result in a permanent denudation, leading to stromal edema and progressive opacification of the cornea.

Hemostasis properties of the corneal endothelium and mechanism through which the cells could prevent clot formation or accelerate the fibrinolysis process is difficult if not impossible to study *in vivo*. Cultured corneal endothelial cells have therefore been used to study how this endothelium could prevent blood clot formation. Studies have focused on the interaction of thrombin with the endothelium (78–80) as well as on the generation of plasmin from plasminogen catalyzed by corneal endothelial cells (81).

Studies on the interaction of thrombin with the corneal endothelial cell surfaces have demonstrated the presence of specific receptor sites to which thrombin can be linked covalently (78,79). Following binding to the cell surface, rapid internalization occurs which proceeds at a rate of 4 ng thrombin $\times 10^6$ cells/hr. Following internalization, the thrombin receptor complexes are degraded within the lysosomal system and degradation produces are released into the medium. In addition to being able to link thrombin covalently to specific cell surface binding sites, corneal endothelial cells also release thrombin binding sites into the medium (79). Similar thrombin binding sites can also be detected *in vivo* in the aqueous humor (unpublished results). These soluble sites are capable of binding thrombin. This leads to the formation of a 77,000 MW complex capable of binding to the cells and of being internalized and degraded. Even more impressive is the ability of corneal endothelial cells, once thrombin has bound to their cell surface, to up-regulate their thrombin–receptor binding sites present on the cell surface and to increase greatly their rate of release into the medium (79). Therefore, once cells have sensed the presence of thrombin in their medium, this up-regulation and increased release of thrombin–receptor binding sites into the medium leads to an increased ability on the part of the cells to inactivate whatever thrombin can appear in the medium. The interaction of thrombin with corneal endothelial cells is therefore unique in four respects: (1) the ability of corneal endothelial cells' thrombin receptor binding sites to bind covalently with thrombin; (2) the ability of the cells to internalize and degrade such complexes rapidly; (3) the ability of the cells to release free thrombin receptor binding sites into the medium, where *in vivo* they could at all times bind thrombin and control its concentration in either the medium or aqueous humor; and (4) the ability to up-regulate that release once they have sensed the presence of thrombin in their environment. These four characteristics of the interaction of thrombin with the corneal endothelium can result in the rapid inactivation of thrombin generated within the aqueous humor, thereby limiting the risk of blood clot formation (32,78–80).

Equally remarkable is the ability of corneal endothelial cells to lyze fibrin clots rapidly. This is due to the direct activation of plasminogen by these cells through the mediation of plasminogen activator (81). This kinase is far more active in corneal endothelial cells than in vascular endothelial cells and is capable of rapidly generating high concentrations of plasmin, which can then dissolve the fibrin clot. Also unlike the membrane-bound plasminogen activator present in vascular cells, that of corneal endothelial cells is present in the cytosol and upon cell death could be released in a soluble form into either the medium or the aqueous humor, where it could be diffused into various parts of the eye such as the trabecular meshwork complex in order further to dissolve fibrin clots located far away from the endothelium (81).

Finally, corneal endothelial cells, like vascular cells, can synthesize prostacyclins, which are potent inhibitors of platelet aggregation (13,28,63).

One could therefore conclude that corneal endothelial cells are uniquely prepared to defend themselves against fibrin clot formation and can prevent it in at least three different ways: by neutralizing thrombin generated during the clot formation process, by preventing platelet aggregation, by rapidly dissolving the fibrin clot.

THE CORNEAL ENDOTHELIUM AND ITS SELECTIVE PERMEABILITY BARRIER FUNCTION

The formation of a highly confluent and fully organized corneal endothelial cell monolayer is associated with changes in the cell surface membrane that impede the process of adsorptive endocytosis and allow cells to form an efficient block against the internalization of various macromolecules such as low-density lipoproteins (82). The barrier formed by the confluent endothelium is correlated with a massive accumulation, mainly in the pericellular matrix, of disulfide cross-linked fibronectin, which might stiffen the cell membrane and, as observed at confluence, restrict the lateral mobility of various cell surface receptors (11,12). Confluent endothelial cell monolayers will therefore no longer internalize LDL receptor complexes. From these studies the notion emerged that the organization of a tissue, such as the strict nonoverlapping and highly flattened morphology of the endothelium, which leads to the adoption of a two-dimensional cell monolayer configuration, could in turn dictate its physiological corneal properties, one of these properties being to act as a selective barrier between the aqueous humor and the corneal stroma (82).

THE CORNEAL ENDOTHELIUM AS A MODEL SYSTEM FOR STUDYING THE INTRACELLULAR PROCESSING OF MITOGENS

Epidermal growth factor (EGF) is a mitogen for epidermal cells *in vivo* (83,84), as well as for a wide variety of cells in culture (85,86), including bovine corneal endothelial cells (87). EGF binds to specific cell surface receptor sites and induces phosphorylation of a specific cell surface protein (88). The receptor–EGF complexes are then internalized and degraded by the lysosomal system (89,90). Although the fate of the EGF within the target cells is thought to be well defined, the mechanism by which EGF can drive the cell through the G_1 phase and commit it to enter the S phase of the cell cycle is still unclear. Various theories regarding the primary mechanisms through which EGF exerts its growth-promoting effect have been proposed (91,92).

The recent observation (93) that degradation of EGF has no role in its mitogenic response in bovine granulosa cells led us to focus on the fate of the intact cell-associated ^{125}I-EGF within the cell. We have been looking for a possible site of action for the intact EGF molecule. The use of the lysosomal inhibitors chloroquine and leupeptin, which results in a greatly increased intracellular accumulation of intact EGF, has allowed us to follow the fate of the cell-associated ^{125}I-EGF within the endothelial cells. It was found that in inhibitor-treated cells the cell-surface-bound ^{125}I-EGF was translocated to the nucleus and that about 15% of the total amount of EGF that had accumulated in the cytoplasm of inhibitor-treated cells became associated with the nuclei. The observation that two different inhibitors acting through different mechanisms (chloroquine by raising the lysosomal pH and leupeptin by specifically inhibiting the lysosomal protease cathapsin B), increasing the cellular accumulation to different extents but can both increasing the nuclear association of ^{125}I-EGF, indicates that the observed nuclear accumulation is a true phenomenon and not just an artifact induced by a nonspecific action of the inhibitors. In untreated cells, because of a rapid degradation rate of the internalized EGF, only a small percentage (0.7%) of the internalized growth factor can be detected in the nucleus, and because of high background it is difficult to study this nuclear accumulation phenomenon. In the inhibitor-treated cells, on the other hand, the EGF degradation was greatly diminished, thus allowing a cytoplasmic accumulation of EGF and making a greater proportion of the internalized, intact EGF available for association and accumulation with the nuclei (93).

The cell-surface-bound EGF was previously shown to be internalized and degraded by the lysosomal system (90,91). The nuclear association of EGF that has been observed with corneal endothelial cells therefore represents a new stage in the fate of the cell-surface-bound EGF.

Linsley *et al.* (94) have observed that a small portion of radiolabeled EGF that binds specifically to 3T3 cells becomes directly and irreversibly linked to a polypeptide of molecular weight 185,000. The biological role for direct linkage of hormones to their specific receptors has not yet been defined. Schecter *et al.* (95) have proposed that a small fraction of "very high affinity binding sites," possibly identical to the covalent EGF–receptor complex, may play an important role in mediating EGF action. We have studied this irreversible linkage of EGF to its receptor in bovine corneal endothelial cells and have demonstrated the appearance of such a ^{125}I-EGF–receptor complex, which accounts for 0.6% of the total ^{125}I-EGF accumulated in chloroquine-treated cells. Interestingly, 50% of it was found associated with the nuclei in the chloroquine-treated cells, representing a preferential accumulation of the directly linked EGF–receptor complexes within that organelle (93). The possible role of the directly linked ^{125}I-EGF–receptor complex in inducing the mitogenic response has been discussed previously (94,95). The preferential accumulation of this complex in the endothelial cell nuclei may therefore suggest that the nuclear association of ^{125}I-EGF is involved in the mitogenic activity of EGF.

The mechanism of translocation of ^{125}I-EGF from the cell surface to the nuclei is still unclear. It has been shown that both EGF (96) and insulin (97) can be internalized by endocytotic vesicles. Cytoplasmic vesicles containing fluorescein-labeled EGF were observed to form a perinuclear ring in a time- and temperature-dependent fashion (96). The internalized ^{125}I-EGF was found to be translocated to the lysosomal system and to be degraded there. One may wonder whether a small fraction of the ^{125}I-EGF internalized by the endocytotic vesicles does not become associated with the nucleus, while the major amount of the internalized ^{125}I-EGF is translocated into the lysosomal system, where it is degraded. Use of chloroquine or leupeptin to inhibit the lysosomal degradation could result in an accumulation of endocytotic vesicles containing ^{125}I-EGF, thereby making it possible to detect the translocation pathway of EGF to the nuclei which exists in untreated cells but is slightly or not at all detectable due to a rapid lysosomal degradation.

THE TRANSPLANTATION *IN VIVO* OF CULTURED CORNEAL ENDOTHELIAL CELLS

Of all cell types that are currently maintained in tissue culture, one of the most attractive, for purposes of transplantation, is the corneal endothelial cell. Studies on the *in vivo* transplantation of cultured cells had been hindered in the past by two factors. One was that, unless one uses animals, such as the "nude mouse," having a nonexistent immunological system, the transplanted cells are ordinarily rejected by the host within a short period of time. The second factor was that it is extremely difficult to evaluate the performance of transplanted cells *in vivo* as far as their differentiation and integration into the host is concerned. In most cases, to evaluate such functions one has to kill the host, thereby ending the experiment. In the case of the transplantation of cultured corneal endothelial cells back *in vivo*, however, both these factors could be obviated.

The corneal endothelium *in vivo* is to be found at the interface between the corneal stroma and the aqueous humor, where it forms the inner lining of the anterior chamber and of the cornea. This location is among the most privileged immunological sites that exist *in vivo* and thereby severely limits, when compared with other sites of transplantation, the possibility of graft rejection. The second factor could also be obviated if one takes into account the physiological function of the corneal endothelium. This tissue, which functions as a transport pump, maintains the corneal stroma in a permanent state of deturgescence. It thereby prevents corneal edema and ensures the transparency of the corneal stroma. If the endothelium is no longer functional because of increased retention of water in the stroma, corneal thickness increases and corneal edema will develop. One therefore had a simple test to evaluate the physiological performance of the transplanted endothelium. Should it perform its normal function, the cornea of the host would stay clear, corneal thickness would be normal, and no edema would develop. Should the transplanted endothelium fail to perform its normal function, the cornea of the host would become edematous and increase in thickness within a matter of days.

Studies on corneal endothelial cell transplantation, in addition to their theoretical implications, may also have considerable potential for further clinical applications, since if corneal endothelial cells grown in tissue culture for a considerable number of generations were to remain functional, they could then be used to replace *in vivo* either a malfunctioning endothelium or areas with extensive endothelial denudation resulting from an injury. This would be of particular importance in species such as primates whose corneal endothelium has a limited regenerative capacity.

Studies done by our group have demonstrated that such transplantation is feasible. Using cultured bovine corneal endothelial cells, we have reported the successful transplantation of bovine corneal endothelial cells into heterologous recipients (rabbits and cats). In both cases, transplanted cells resumed their normal functions and were capable of maintaining the corneal stroma of the recipient in a permanent stage of deturgescence (98–101). Identification of the transplanted cells after various periods of time in the eye of the recipient was done either by putting the cells back into culture and karyotyping them (12,98–101) or by autoradiography of *en face* preparations of transplanted endothelium. In the latter case, cells were labeled with [^{3}H]thymidine before transplantation (98–101).

One of the main implications resulting from such studies has been the demonstration that cultured cells maintained *in vitro* and passaged repeatedly for long periods of time can stay fully differentiated and resume their normal function when put back in an *in vivo* environment. Our ability to transplant heterologous endothelium in various species could also increase our understanding of the factors that control their proliferation *in vivo*. Such factors may be intrinsic to the cells, which may lose their ability to proliferate early in their life, or they could be found in the limiting environment (aqueous humor) to which cells are exposed and/or in the substrate upon which cells rest and which may or may not favor their proliferation. This could be explored by transplanting cultured cells into various species and analyzing how they react to a wound situation.

THE USE OF THE BASAL LAMINA PRODUCED BY CULTURED CORNEAL ENDOTHELIAL CELLS FOR STUDIES OF THE CONTROL OF PROLIFERATION OF MALIGNANT VERSUS NORMAL CELLS

BASAL LAMINA AS A BIOMATRIX FOR THE STUDY OF THE INTERACTION OF TUMOR CELLS WITH A NATURAL SUBSTRATE

A central issue in tumor biology is the understanding of the interactions between tumor cells and their environment. It is of interest in this regard to study the tumor cells' interaction with extracellular matrices and basal laminae, since this could throw light on the need for stromal and fibroblastic support for tumor cell growth (102,103); on the ability of tumor cells to reorganize their local environment in order to grow and invade (104,105); and on the mechanisms through which various artificial substrates that are introduced *in vivo* into the animal result in the production of malignant mesothelioma and fibrosarcoma (106). We have therefore explored the use of the basal lamina produced by cultured corneal endothelial cells as a substrate on which to culture human cells originating from solid tumors (colon carcinoma and Ewing's sarcoma).

Tumor Cell Attachment, Migration, Morphology, and Organization When Maintained on Plastic versus Basal Lamina–Coated Dishes. In the case of either colon carcinoma cells or Ewing cells, as many as 50–80% of the cells seeded on basal lamina-coated dishes attached firmly within 30 min, and 70–90% by 1 hr, as compared to only about 5–10% of the cells on plastic. Even after 24 hr, no more than 5% of the seeded cells attached firmly to plastic (107). The rapid attachment of colon carcinoma cells and Ewing's sarcoma cells was shown in subsequent studies to be mediated by different adhesion factors present in the basal lamina (108). While colon carcinoma cells attached to the basal lamina through their interaction with laminin, the Ewing's cells interacted with fibronectin (108). The presence of specific adhesion factors in basal lamina, as well as the ability or inability of tumor cells to produce and/or interact with such factors, may therefore determine the localization of given tumor cell types to specific matrices. Such specific cell–matrix interaction may then direct the growth, the migration, as well as the invasiveness and metastasis of tumor cells.

Tumor cell migration was studied qualitatively by seeding cell aggregates rather than single cells on plastic or basal lamina-coated dishes and observing the extent of cell migration out of the aggregates. When seeded onto plastic dishes, Ewing cells remained mostly in the form of floating and loosely packed cell aggregates, whereas colon carcinoma cells remained as tightly packed, ball-like aggregates. In neither case, and even after 10 days in culture, did cells migrate out of the cell aggregates (107). In contrast, seeding of the same cell aggregates on basal lamina–coated dishes was associated with spreading and migration of cells that could be observed within 5–15 min after seeding. Since cell migration is observed prior to cell proliferation, it can be regarded as a specific cellular response to the mere contact between cells and the basal lamina. This may reflect the preferential adhesion of the cells to the basal lamina components such as laminin or fibronectin (108), as well as to other as yet unidentified components. These interactions of cells with basal lamina adhesion factors can be stronger than the interaction between cells and are obviously much stronger than the interaction between cells and plastic.

Both colon carcinoma and Ewing sarcoma cells adopted a flattened morphology when seeded on basal lamina and at confluence both cell types develop as a continuous monolayer of organized, nonoverlapping cells (107). The morphological configuration of confluent cultures of the Ewing cells was similar to that of a confluent vascular endothelial cell monolayer, thereby corroborating the initial impression of Ewing (109) concerning the cellular source of the disease, which he regarded as endothelial.

Effect of the Basal Lamina on Tumor Cell Growth. Normal human epithelial cells are difficult to grow using conventional culture techniques. This applies even to malignant cells of epithelial origin (110), some of which, when maintained in culture, have an extremely long doubling time (10–20 days). Since cell shape has been shown to be a major factor in regulating cell growth (34,35), we compared the growth rate of cells plated on plastic (spherical configuration) to that of cells plated on basal lamina which adopt a flattened morphology. Although differences in growth rate were observed, both the Ewing and colon carcinoma cells proliferated in either configuration. This observation therefore demonstrated that adhesion of the cells to the basal lamina and their subsequent flattening did not impose a restriction on their proliferation. On the contrary, when cells flattened out a stimulation of cell growth that depended on the cell type and culture conditions was observed. This stimulation was particularly evident with the colon carcinoma cells, which, when seeded on plastic, showed a lag period of 3–4 days before resuming a logarithmic growth rate (16 hr doubling time). In contrast, seeding of the same cells on basal lamina resulted in an active cell migration concomitant with an immediate resumption of proliferation, so that after 3–4 days the cell density of cultures maintained on basal lamina was 4- to 10-fold higher than that of cultures maintained on plastic (107). An increased growth rate of Ewing cells maintained on basal lamina was also observed. Since these cells showed a very loose attachment to plastic, both attached and floating cells had to be counted in order to reach a meaningful conclusion regarding an effect on the total number of cells. When cultures were maintained on basal lamina, the number of firmly attached and floating cells exceeded by 2- to 4-fold that of cultures maintained on plastic. However, if only the number of firmly attached cells on both substrata was considered and floating cells were ignored, active proliferation of attached cells was observed only with basal lamina–coated dishes, resulting in an infinite increase in the final density of attached cells in cultures maintained on a basal lamina versus those attached on plastic. Similar studies with human choroidal melanoma cells have also demonstrated that cells maintained on basal lamina proliferate rapidly (107). This contrasts with a very slow rate of proliferation of these cells (12–18 days average doubling time) when maintained on plastic. Effects similar to those of the basal lamina as far as cell attachment and proliferation are concerned were not observed when tumor cells were seeded on plastic dishes coated with purified preparations of collagen types I, III, or IV (107).

The preceding results therefore demonstrate that cultured tumor cells can proliferate both when they are highly flattened and firmly attached to the substrate and when they are loosely attached or in suspension. Adhesion of cells to the basal lamina, however, results in a stimulation of cell growth, which is best observed when the cell fractions grow in the anchorage-dependent configuration, but not in the floating configuration, thereby indicating the need for actual contact.

EFFECT OF BASAL LAMINA ON THE ATTACHMENT, MIGRATION, PROLIFERATION, AND DIFFERENTIATION OF NERVE CELLS

Nerve cells do not have the ability to produce a basal lamina *in vivo* and therefore rely on that produced by other tissues during the early steps of embryogenesis for both their migration and neurite outgrowth. This is shown by the very firm attachment of neurites and of their growth cones to the basal lamina *in vivo*. The interaction of nerve cells with a proper substrate could therefore be important insofar as nerve cell migration and differentiation are concerned. Likewise, a close contact with the basal lamina seems to be important in the control of nerve cell proliferation. In both the cortex and spinal cord, dividing nerve cells can only be found in close contact with the basal lamina which forms the inner lining of the cortex or which is located along the central cavity of the spinal cord. Following division, nerve cells can migrate to the outer part of the cortex or of the spinal cord and differentiate.

To study the interaction of nerve cells with the basal lamina, we first used as a model the PC12 cell line, which is derived from a pheochromocytoma and has been shown to respond to nerve growth factor (NGF) by extending neurites on collagen-coated dishes. When the percentage of firmly attached PC12 cells was compared as a function of the substrates upon which they were seeded, it was found to be 5% for plastic, 20% for collagen-coated dishes, and 90–95% for basal lamina-coated dishes (111,112). Following their attachment to the basal lamina, the cells, which on either plastic or collagen substrata remained rounded, became flattened and began to emit fine neurites characterized by branching, varicosities, and growth cones. This outgrowth of neurites is only temporary and by day 3 they start to regress unless NGF is present. In its presence, however, extensive thin neurite outgrowth can be seen, and apparent fasciculation of some neurites is observed (111,112). In contrast, no neurites are extended by NGF-treated cells on plastic, and those cells on collagen-coated dishes extend thick, straight neurites poorly attached to the substrata at only a few points and with little branching or varicosities. Thus, basal lamina may be permissive for neurite outgrowth, and the trophic effects of NGF are only necessary for long-term differentiation (111,112).

The ability of the basal lamina to support extensive cell differentiation is not its only biological effect, since it can affect nerve cell migration and proliferation. While on plastic or collagen-coated dishes single cells will eventually aggregate and give rise to ball-like clumps of cells, single cells on basal lamina will grow as a monolayer of flattened cells. Only when cultures are kept for extended periods of time at confluence will cells eventually overgrow each other. Moreover, whereas the rate of proliferation of cells maintained on plastic is very slow, that of cultures maintained on basal lamina is faster and can be improved by the addition of EGF to the medium (111,112).

THE BASAL LAMINA AS A BIOMATRIX FOR GROWING NORMAL CELLS IN ABSENCE OF SERUM AND FOR IDENTIFYING THE PLASMA FACTORS RESPONSIBLE FOR THEIR PROLIFERATION

Vascular Endothelial Cells. Previous studies (10, 25–28) have shown that vascular endothelial cell cultures maintained on plastic and propagated in the presence of fibroblast growth factor (FGF) divide with an average doubling time of 18 hr when seeded at either a high (up to 1 : 1000) or low split ratio. Upon reaching confluence, the cells adopt a morphological configuration similar to that of the confluent culture from which they originated. In contrast, seeding of the same cells at a high split ratio in the absence of FGF results in a much longer doubling time (60–78 hr), and within a few passages cultures maintained in the absence of FGF exhibit, in addition to a much slower growth rate, morphological as well as structural alterations that mostly involve changes in the composition and distribution of the basal lamina (25,27). This raises the possibility that the basal lamina produced by these cells could have an effect on their ability to proliferate and to express their normal phenotype once confluent.

This is what is in fact observed when the growth rates of bovine vascular endothelial cells seeded at low density on plastic versus basal lamina are compared. Regardless of the initial cell density (sparse culture, 10 cells/mm^2, or clonal density, 1 cell/cm^2) at which cells were seeded, they divided extremely rapidly when maintained on basal lamina–coated dishes (58). Addition of FGF to such cultures did not decrease their mean doubling time, which is already at a minimum (18–20 hr), nor did it result in a higher final cell density, which is already at a maximum (900–1000 cells/mm^2). One can therefore conclude that, while low-density cell cultures maintained on plastic proliferate poorly and therefore require FGF in order to become confluent within a few days, when the cultures are maintained on basal lamina, they proliferate actively and no longer require FGF in order to become confluent (58). However, in either case (either maintained on plastic and exposed to FGF or maintained on a basal lamina), the rate of proliferation is a direct function of the serum or plasma concentration to which cultures are exposed. It is therefore likely that, as with corneal endothelial cells, the effect of the basal lamina is more a permissive than a direct mitogenic effect, since cells still required serum or plasma in order to proliferate (13,58).

Among the plasma factors that could be held directly or indirectly responsible for the active proliferation of vascular endothelial cells are the high-density lipoproteins (HDL), as well as the low-density lipoproteins (LDL). Earlier studies have shown that both LDL and HDL can interact specifically with vascular endothelial cells (113–116), and others have shown that LDL could be mitogenic for vascular smooth muscle cells and dermal fibroblasts when added to lipoprotein-deficient serum (LPDS) (72,117) or to serum from abetalipoproteinemic subjects (118). Likewise, in the case of cells that have a limited ability to make cholesterol *de novo,* or in the case of cells maintained in the presence of compounds such as compactin that totally inhibit their ability to make cholesterol, addition of LDL to the medium leads to resumption of cell proliferation (119). In that case, LDL could act by providing an exogenous source of cholesterol to the cells, thereby obviating the block in cholesterol synthesis resulting from the presence of the inhibitor in the medium (119). We have therefore compared the respective mitogenic activities of HDL and LDL on vascular endothelial cell cultures exposed to LPDS or to serum-free medium.

Vascular endothelial cells maintained in the presence of medium supplemented with LPDS grow poorly (75). Such cultures therefore require the presence of lipoproteins in order to proliferate optimally. Of the two classes of lipoproteins (HDL and LDL) that have been studied, HDL seems to be the major factor involved in the proliferation of vascular endothelial cells. This is due primarily to its lack of toxicity when added at high concentration, as well as to its lack of dependence of LPDS in order to exhibit its mitogenic properties (75,76).

LDL, unlike HDL, had a biphasic effect. Although mitogenic for vascular endothelial cells when added at low concentration, once physiological concentrations are reached it becomes toxic for the cells (75). Moreover, and in contrast with HDL, the mitogenic effect of LDL was found to be a function of the LPDS concentration to which cultures were exposed. LDL at a concentration of 200 μg protein/ml did not stimulate cells to proliferate at an optimal growth rate unless cultures were maintained in high (5–10%) LPDS concentration. This mitogenic effect on the part of HDL, as opposed to the cytotoxic effect of LDL, was observed regardless of the density at which cultures were seeded (clonal or high-density cultures). Therefore, HDL at physiological concentrations can replace serum or plasma. This is best exemplified by our observation that cells maintained in serum-free medium on basal lamina will proliferate at an optimal rate, provided that HDL (250 μg protein/ml) is added together with transferrin (10 μg/ml) to the medium. In contrast, in the absence of LPDS, LDL concentrations as low as 80 μg protein/ml resulted in cell death and at 30 μg protein/ml had only a small mitogenic effect in comparison to that of HDL (75,76).

The substrate upon which cultures were maintained was found to be of crucial importance if a mitogenic effect on the part of either HDL or LDL is to be observed. When maintained on plastic, cells exposed to LPDS did not survive and therefore could not respond to either lipoprotein (75). In contrast, when maintained on basal lamina they survived quite well, thereby making it possible to observe the mitogenic effect of either HDL or LDL. This suggests that, *in vivo*, the integrity of the basal lamina upon which endothelial cells rest and migrate is an important factor in determining the cells' response to lipoproteins present in plasma (75,76).

Vascular Smooth Muscle Cells. Previous studies (37) have indicated that vascular smooth muscle cells maintained on basal lamina-coated dishes have a much lower requirement for either serum or plasma in order to proliferate actively than when they are maintained on plastic. It is therefore possible that vascular smooth muscle cells, when maintained on basal lamina-coated dishes, could survive and still be responsive to plasma factors even when maintained in a well-defined synthetic medium unsupplemented with either plasma or serum (60). Preliminary studies demonstrated that, as with corneal endothelial cells, transferrin, which is the main iron-carrying protein in the bloodstream, must be present if any mitogenic response to plasma factors on the part of vascular smooth muscle cells seeded and maintained in total absence of serum is to be observed. It is likely that this absolute requirement for transferrin reflects either its role in delivering iron to the cells or its ability to detoxify the medium by removing toxic traces of metals (71).

When low-density vascular smooth muscle cell cultures were maintained on basal lamina-coated dishes and exposed to a synthetic medium supplemented with transferrin (10 μg/ml), HDL (250 μg protein/ml), insulin (2 μg protein/ml), and FGF (100 ng/ml), cells proliferated as actively as when they were exposed to optimal serum concentration. The single omission of HDL, insulin, or FGF resulted in a lower growth rate of the cultures, as well as in a lower final cell density (60). This indicates that all of these factors have an additive effect upon one another and must be present simultaneously in order to induce optimal cell growth. Neither FGF nor insulin, either singly or in combination, had a significant effect on cell growth. When the ability of EGF (50 ng/ml) to substitute for FGF was tested, it was found to be as potent as FGF. Likewise, somatomedin C at low concentration (10 ng/ml) can fully substitute for insulin (60).

Since HDL, as well as insulin and EGF, are all normally present in plasma, these factors may reflect the plasma constituents involved in the control of the proliferation of vascular smooth muscle cells when such cells are maintained on basal lamina-coated dishes and exposed to plasma. The concentrations at which insulin was mitogenic are clearly pharmacological. However, since insulin can be replaced by somatomedin C, and since it is known to have a low affinity for somatomedin binding sites, it may be that the high concentrations of insulin required are due to its weak interaction with the somatomedin binding sites. If this is the case, the mitogenic activity of insulin on vascular smooth muscle cells is not directly mediated through its interaction with high-affinity insulin binding sites but rather through its weak interaction with somatomedin C binding sites. In contrast, the effect of EGF, which was further documented by the identification on the cell surface of EGF receptor sites, is probably due to a direct interaction of EGF with the cells (60).

The use of a synthetic medium supplemented with transferrin HDL, insulin, and EGF has allowed vascular smooth muscle cells not only to proliferate actively in the total absence of serum but also to be passaged repeatedly for 60 generations. This therefore makes realization of the goal of serial passage in a totally defined medium possible.

The substrate upon which vascular smooth muscle cells are maintained is an important factor in their response to the various factors to which they are exposed. When cells are maintained on plastic and exposed to a synthetic medium, they will not proliferate. The addition of FGF, which in previous studies has been shown to replace the requirement for a basal lamina, will allow the cells to proliferate and to respond to either HDL or LDL. Yet the life spans of the cultures are far from impressive, since even when exposed to HDL and FGF, cells stop dividing after undergoing 23 generations. In contrast, when cells are maintained on basal lamina-coated dishes, even when exposed to a synthetic medium unsupplemented with any factor, cells have a life span of 15 generations and, while the addition of FGF to the medium no longer has any effect on the life span of the cultures, that of HDL, insulin, and FGF or EGF will allow the cells to undergo 46 generations (60). The effect of the basal lamina in delaying cell senescence is even more impressive if one considers the case of cells exposed to synthetic medium supplemented with optimal concentration of serum. While cultures maintained on plastic and exposed to serum can be maintained at best for 16 generations, cultures maintained on basal lamina can undergo 90 generations before losing their ability to proliferate (60). It may therefore be concluded that the basal lamina upon which vascular smooth muscle cells are maintained not only makes cells sensitive to factors present in plasma, but also delays to a considerable extent the ultimate senescence of the cells when they are exposed to a synthetic medium supplemented either with well-defined factor(s) or with serum.

Granulosa and Adrenal Cortex Cells. Both of these cell types *in vivo* have the ability to produce steroids. Earlier studies done on the control of proliferation of bovine granulosa cells have shown that this cell type, when maintained on plastic and exposed to optimal serum concentration, proliferates slowly or not at all (120–122). The addition of either EGF or FGF to the media could trigger substantial cell proliferation, since low-density cell cultures exposed to either mitogen can have an average doubling time as low as 18 hr (120–122). Addition of either mitogen also delays the ultimate senescence of the cultures (123). With adrenal cortex cells, on the other hand, although they do divide extremely slowly and rapidly senesce when maintained on plastic and exposed to serum, the addition of FGF, but not that of EGF, results in both a shortening of the average population doubling time and in the prevention of their precocious senescence in culture (124). In contrast, maintaining either granulosa or adrenal cortex cells on an basal lamina renders the requirement for growth factors obsolete, and the addition of plasma alone is enough to ensure an optimal growth rate of the cultures (59). However, the plasma factor(s) to which granulosa and adrenal cortex cells respond when maintained on basal lamina were found to be quite different from those to which endothelial or smooth muscle cells responded. In particular, HDL, which was shown to be mitogenic for endothelial or vascular smooth muscle cells over a wide range of concentrations, was cytotoxic for granulosa cells when present at concentrations above 50 μg protein/ml (125). With adrenal cortex cells, cytotoxicity was observed at a concentration above 30 μg protein/ml (126). Insulin was observed to be the main mitogen for both granulosa and adrenal cortex cells, and both cell types responded to concentrations as low as 30 ng/ml by proliferating actively, reaching a final cell density that was 60% that of cultures exposed to optimal serum or plasma concentrations. Addition of either EGF or FGF helped the cultures to reach a final cell density similar to that observed when cells were maintained in the presence of either plasma or serum. As with other cell types, the importance of the substrate in supporting the proliferative response to insulin of both granulosa and adrenal cortex cells was demonstrated by the fact that such response could be observed only with cultures maintained on basal lamina-coated dishes but not with those maintained on plastic (125,126).

Both of these cell types when cultured in the absence of serum and in the presence of insulin, HDL, and FGF were still capable of producing steroids in response to dibutyryl cyclic AMP. In the case of granulosa cells, exposure to FSH improved the ability of the cells to produce steroids, and in the case of adrenal cells exposure to ACTH also stimulated steroid production above its normal level. Addition of either LDL or HDL to the culture medium of confluent cultures grown in the absence of serum improved the ability of the cells to produce steroids (127–129).

Therefore, not only can one grow both adrenal cortex and granulosa cell cultures in a defined medium, but the differentiated properties of these cells are also preserved, as shown by their ability to respond to stimuli provided by their trophic hormones.

THE NATURE OF THE FACTORS PRESENT IN BASAL LAMINA THAT ARE INVOLVED IN CELL ATTACHMENT VERSUS CELL PROLIFERATION

The intricate nature of the basal lamina is exemplified by the difficulties encountered in identifying its composition and by the complex interaction of its known components (collagen, glycoproteins, proteoglycans, and glycosaminoglycans), which form a highly stable scaffolding upon which cells rest *in vivo*. Because the reconstitution of this unique structure from its known constituents would be a formidable task, we have examined the role that these components play in cell proliferation by indirect methods that rely on selective inactivation by chemical, enzymatic, or heat treatment of intact basal lamina produced by cultured corneal endothelial cells. Although all the treatments we have used are known to affect glycoproteins, and specifically proteoglycans, only three were effective in inactivating the basal lamina's ability to support cell proliferation. Exposure of basal lamina-coated dishes to 14 *M* NH_4OH (pH 13.8), which results in the cleavage of proteoglycoproteins or glycopeptides at the *O*-glycosidic bond between the protein and carbohydrate moieties (130–132), inhibited the rate of proliferation of vascular endothelial cells by 97%. Likewise, treatment of basal lamina-coated dishes with 4 *M* guanidine-HCl, which extracts up to 80% of the glycosaminoglycans (GAGs), caused a 92% reduction in cell growth. Treatment of basal lamina-coated dishes with nitrous acid (HNO_2), which results in the degradation of heparin or heparin sulfate into sulfated disaccharides and nonsulfated oligosaccharides (133–135), affected cell proliferation by 90% and led to the release of $^{35}SO_4$-labeled macromolecules, 50% of which were disaccharides. Although the effect of HNO_2 on intact basal lamina is not known, its specific degradation of isolated heparan and heparan sulfate could indicate that it has a similar effect on basal lamina, and this may correlate with its adverse effect on the ability of the basal lamina to support cell growth. Chemical treatment of the basal lamina with a reducing agent such as dithiothreitol (from 0.1 to 100 m*M*), which has been reported to destroy the biological activity of platelet-derived growth factor completely, did not affect its ability to support growth, nor did treatment with sodium dodecyl sulfate (2%) or urea (2 or 8*M*). Specific enzymatic treatment of the basal lamina with collagenase, pepsin, trypsin, chymotrypsin, or neuraminidase did not inhibit cell proliferation. It is therefore doubtful that any of the substrates of these enzymes plays a direct role in cell proliferation. The lack of effect of either hyaluronidase or chondroitinase ABC, as well as the presence of chondroitinase-resistant material in extracts of basal lamina that affect cell proliferation, would tend to rule out a role for hyaluronic acid or chondroitin sulfates in the permissive effect of the basal lamina on cell proliferation (136).

The importance of cell shape in proliferation suggests that attachment factors present within the basal lamina which, because of their direct contact with the cell membrane, could modify cell shape, could also play a dual role in controlling cell proliferation.

To investigate the relationship between growth and attachment, we have compared the ability of vascular endothelial cells to attach to basal lamina following treatment by alkaline pH and heat. In the absence of serum, only 10% of the cells attach to plastic after 1 hr, and maximal cell attachment (40%) is not reached before 18 hr. In contrast, cells seeded on basal lamina-coated dishes rapidly attach and spread and most (90%) have attached by 1 hr. No significant difference is seen in the rate of cell attachment after treatment of the basal lamina at pH 13.8, which destroys 97% of the ability of the basal lamina to support cell proliferation. In fact, initial cell attachment (1 hr) was slightly enhanced by alkaline treatment. It therefore appears unlikely that the component of the basal lamina that is removed or destroyed by high pH treatment is involved in cell attachment (136).

In contrast to the results obtained after treatment of the basal lamina with alkaline pH, thermal treatment of the collagen bed, which results in its denaturation, had the opposite results. Treatment of the basal lamina at 70°C, which has no effect on cell proliferation, greatly reduces the rate of cell attachment. The rate of cell attachment is reduced even further after treatment of the basal lamina at 90°C, although cell growth is reduced by only 60%. This indicates that the basal lamina component(s) implicated in cell growth is probably distinct from those involved in cell attachment. Thermal disruptions of the basal lamina also revealed the importance of the spatial geometry of the basal lamina component responsible for cell proliferation (136). The present study provides evidence regarding the nature of the components of the basal lamina that are responsible for conveying the permissive effect of the basal lamina on cell proliferation. It is likely that the active component is a sulfated glycoprotein or proteoglycan that is susceptible to extraction by 4 *M* guanidine-HCl, is degraded by nitrous acid, and contains an *O*-glycosidic bond and glucosamine and/or galactosamine. Its resistance to hyaluronidase and chondroitinase ABC suggests that neither hyaluronic acid nor chondroitin sulfates are involved (136). Identification of the active component, however, must await a detailed analysis of the basal lamina and of the material extracted from the basal lamina under conditions that inhibit cell proliferation. Although cell attachment and proliferation appear to be separate processes mediated through different mechanisms, the role of the component of the basal lamina responsible for cell proliferation and cell differentiation has yet to be examined. It is well known that cell proliferation precedes many differentiated steps during embryogenesis, and numerous basal lamina components have been shown to be important in differentiation. It will be of interest to compare the component of the basal lamina that is active in cell proliferation to the alkaline-labile component of demineralized bone matrix described by Reddi and Huggins (137,138) as inducing differentiation of fibroblasts. It may very well be that it will turn out to be calcium, which, in the form of microcrystalline precipitate, has been shown to be as good a commitment factor as FGF or PDGF (139). In that regard it is of interest to note that, *in vivo*, basement membranes in all types of organs are rich in calcium deposits which are bound to the numerous cationic groups present within the GAGs and proteoglycans, particularly in heparan sulfate. The ability of basement membrane to accumulate calcium is reflected by its ability to stain with alizarin red and may represent an evolutionary mechanism through which the commitment factors are automatically incorporated into the substrate upon which cells will migrate, proliferate, and differentiate.

CONCLUSION

Growth factors have been shown to affect, at least *in vitro*, the proliferation of all cell types derived from the various tissues that compose the cornea (32). They have proved to be useful tools in developing corneal cells in culture, particularly corneal endothelial cells. Although studies of the mode of action of growth factors are still in their infancy, they have provided insight into the role and importance of the various basement membranes upon which corneal cells rest in supporting their proliferation and eventual differentiation.

The number of biological problems that can be studied using only cultured corneal endothelial cells is truly astonishing. The following is a list of only some of the questions that can now be explored using cultured corneal endothelial cells: the cause(s) and mechanisms leading to the acquisition of a cell surface polarity; the biochemical characteristics of nonthrombogenic surfaces; the mechanisms underlying contact inhibition of cell proliferation, so well exemplified by the corneal endothelium *in vivo*; the involvement of various cytoskeletal elements in controlling corneal endothelial cell shape and the reorganization of those elements when cells pass from an active proliferative stage to a resting stage, as reflected in the formation of a confluent monolayer composed of tightly packed and nonoverlapping cells; the cell surface proteins involved in active cell migration versus cell attachment; the relationship between cell shape and a permeability barrier function.

The ability of cultured corneal endothelial cells to produce a basal lamina with a given polarity and in unparalleled amounts provides us with an opportunity to investigate the biological properties of such a "biomatrix." Among the features that can be studied is the composition of the basal lamina produced by actively dividing cell populations versus that of confluent, resting cultures, as well as the modulation of synthesis of the various components of the biomatrix by growth factors and the subsequent permissive effects of those factors in cell proliferation. One can also analyze the effects of the biomatrix on the synthesis of the basal lamina produced by cells freshly seeded on it and the role of various substrata in modulating the synthesis of the various elements of the basal lamina.

The biomatrix can itself be used as a tool for the development of new cell types in serum-free medium. It can also be used to study the differentiation of nerve cells, as exemplified by its ability to support neurite outgrowth. In the field of tumor cell biology, such biomatrices can be used to study the interaction of tumor cells with a normal substrate and the effect it has on tumor cell attachment, migration, and proliferation. In the field of aging, one can study the role of such biomatrices in cell senescence and how they can delay it.

In the field of pharmacology, cultured corneal endothelial cells can be used to test various drugs used to combat ocular disease and to test their side effects. In the field of surgery, one can foresee the day when a diseased and incapacitated endothelium could be replaced by cells grown in tissue culture. Because of the progress that will, in all likelihood, be made in the next few years in the culture of corneal endothelial cells, it is likely that the long-term culture of human corneal endothelial cells will be established. Provided that adequate controls of the karyotypes of these cells are made, and provided that their preneoplastic transformation in tissue culture is carefully assessed, one could use cultured endothelial cells as an endless supply to replace this tissue *in vivo* when it is no longer functional. This is even more likely in view of the foreseen shortage of corneas available for transplantation. This predicted shortage will be due primarily to the increased longevity of the population, which will result in both a diminished supply of young cornea donors and in an increased demand for cornea transplants on the part of older individuals. Transplantation of cultured corneal endothelial cells, for experimental purposes, can be considered not only on a homologous basis but on a heterologous basis as well. This is mostly due to the unique location of the corneal endothelium, which protects it from immediate rejection.

REFERENCES

1. Hay, E. D., and Revel, J. P. (1969) *In* "Fine Structure of the Developing Avian Cornea," Monogr. Dev. Biol., pp. 4–15. Karger, Basel.
2. Mannagh, J., and Irving, A. R., Jr. (1965) *Arch. Ophthalmol. (Chicago)* **74,** 847.
3. Perlman, M., and Baum, J. L. (1974) *Arch. Ophthalmol. (Chicago)* **92,** 235–239.
4. Perlman, M., Baum, J. L., and Kaye, G. I. (1974) *J. Cell Biol.* **63,** 306–308.
5. Gospodarowicz, D., and Zetter, B. (1977) *Dev. Biol. Stand.* **37,** 109–130.
6. Gospodarowicz, D., Mescher, A. L., and Birdwell, C. R. (1977) *Exp. Eye Res.* **25,** 75–89.
7. Gospodarowicz, D., Greenburg, G., and Birdwell, C. R. (1978) *Cancer Res.* **38,** 4155–4171.
8. Gospodarowicz, D., Vlodavsky, I., and Savion, N. (1981) *Vision Res.* **21,** 87–103.

9. Gospodarowicz, D., and Greenburg, G. (1979) *Exp. Eye Res.* **28,** 249–265.
10. Gospodarowicz, D., Greenburg, G., Bialecki, H., and Zetter, B. (1978) *In Vitro* **14,** 85–118.
11. Gospodarowicz, D., Greenburg, G., Vlodavsky, I., Alvarado, J., and Johnson, L. K. (1979) *Exp. Eye Res.* **29,** 485–509.
12. Gospodarowicz, D., Vlodavsky, I., Greenburg, G., Alvarado, J., and Johnson, L. K. (1979) *Recent Prog. Horm. Res.* **35,** 375–448.
13. Gospodarowicz, D., and Tauber, J.-P. (1980) *Endocr. Rev.* **1,** 201–227.
14. Gospodarowicz, D., Greenburg, G., Foidart, J.-M., and Savion, N. (1981) *J. Cell. Physiol.* **107,** 173–183.
15. Tseng, S. C., Savion, N., Gospodarowicz, D., and Stern, R. (1981) *J. Biol. Chem.* **256,** 3361–3365.
16. Gospodarowicz, D., and Fujii, D. K. (1981) *Miami Winter Symp.* **18,** 113–135.
17. Foidart, J. M., and Reddi, A. H. (1980 *Dev. Biol.* **75,** 130–136.
18. Foidart, J. M., Bere, E. W., Yaar, M., Rennard, W. I., Gullino, M., Martin, G. R., and Katz, S. I. (1980) *Lab. Invest.* **42,** 336–342.
19. Yamada, K. M., and Olden, K. (1978) *Nature (London)* **275,** 179–184.
20. Timpl, R., Rohde, H., Robey, P. G., Rennard, S. I., Foidart, J.-M., and Martin, G. R. (1979) *J. Biol. Chem.* **254,** 9933–9937.
21. Rohde, H., Wick, G., and Timpl, R. (1979) *Eur. J. Biochem.* **102,** 195–201.
22. Vaheri, A., Kurkinen, M., Lehto, V.-P., Linder, E., and Timpl, R. (1978) *Proc. Natl. Acad. Sci. U.S.A.* **75,** 4944–4948.
23. Stenman, S., and Vaheri, A. (1978) *J. Exp. Med.* **147,** 1054–1064.
24. Alitalo, K., Kurkinen, M., Vaheri, A., Krieg, T., and Timpl, R. (1980) *Cell* **19,** 1053–1062.
25. Vlodavsky, I., Johnson, L. K., Greenburg, G., and Gospodarowicz, D. (1979) *J. Cell Biol.* **83,** 468–486.
26. Vlodavsky, I., and Gospodarowicz, D. (1979) *J. Supramol. Struct.* **12,** 73–114.
27. Greenburg, G., Vlodavsky, I., Foidart, J.-M., and Gospodarowicz, D. (1980) *J. Cell. Physiol.* **103,** 333–347.
28. Gospodarowicz, D., Vlodavsky, I., Savion, N., and Tauber, J.-P. (1980) *Soc. Gen. Physiol. Ser.* **35,** 1–38.
29. Savion, N., Vlodavsky, I., Greenburg, G., and Gospodarowicz, D. (1982) *J. Cell. Physiol.* (in press).
30. Goldberg, A. L., and Dice, J. F. (1974) *Annu. Rev. Biochem.* **43,** 835–869.
31. Goldberg, A. L., and St. John, A. C. (1976) *Annu. Rev. Biochem.* **45,** 747–803.
32. Gospodarowicz, D., and Giguere, L. (1982) *In* "Cellular Aspects of the Eye" (D. S. McDevitt, ed.). Academic Press, New York.
33. Thompson, D. (1961) *In* "On Growth and Form" (J. T. Bonner, ed.), pp. 1–64. Cambridge Univ. Press, London and New York.
34. Maroudas, N. G. (1973) *Nature (London)* **244,** 353–354.
35. Folkman, J. (1977) *Recent Adv. Cancer Res.: Cell Biol. Mol. Biol. Tumor Virol.* **1,** 119–130.
36. Gospodarowicz, D., Vlodavsky, I., Greenburg, G., and Johnson, L. K. (1979) *Cold Spring Harbor Conf. Cell Proliferation* **6,** 561–592.
37. Grobstein, C. J. (1955) *J. Exp. Zool.* **130,** 319–340.
38. Grobstein, C. (1967) *Natl. Cancer Inst. Monogr.* **26,** 279–299.
39. Grobstein, C. (1975) *In* "Extracellular Matrix Influences on Gene Expression" (H. C. Slavkin and R. C. Greulich, eds.), pp. 9–16, 804–814. Academic Press, New York.

40. Kratochwil, K. (1972) *In* "General Properties in Tissue Interaction in Carcinogenesis" (D. Tarin, ed.), pp. 1–48. Academic Press, New York.
41. Bernfield, M. R. (1981) *In* "Morphogenesis and Pattern Formation (T. G. Connally, L. L. Brinkley, and B. M. Carlson, eds.), pp. 139–162. Raven, New York.
42. Wessels, N. K. (1964) *Proc. Natl. Acad. Sci. U.S.A.* **52,** 252–259.
43. Wessels, N. K. (1977) *In* "Tissue Interaction and Development," pp. 213–229. Benjamin, New York.
44. Holley, R. W., and Kiernan, J. A. (1968) *Proc. Natl. Acad. Sci. U.S.A.* **60,** 300–304.
45. Temin, H. M., Pierson, R. W., and Dulak, N. C. (1972) *In* "Growth, Nutrition and Metabolism of Cells in Culture" (G. H. Rothblat and J. Cristofalo, eds.), pp. 49–81. Academic Press, New York.
46. Bornstein, P., and Sage, H. (1980) *Annu. Rev. Biochem.* **49,** 957–1003.
47. Murray, J. C., Stingl, G., Kleinman, H. K., Martin, G. R., and Kidwell, W. R. (1979) *J. Cell Biol.* **80,** 197–202.
48. Rollins, B. J., and Culp, L. A. (1979) *Biochemistry* **18,** 5621–5629.
49. Wicha, M. S., Liotta, L. A., Garbisa, and Kidwell, W. R. (1979) *Exp. Cell Res.* **124,** 181–190.
50. Folkman, J., and Moscona, A. (1978) *Nature (London)* **273,** 345–349.
51. Hay, E. D. (1977) *In* "Cell and Tissue Interaction" (J. W. Lash and M. M. Burger, eds.), pp. 115–137. Raven, New York.
52. Bernfield, M. R. (1978) *Int. Congr. Ser.—Excerpta Med.* **432,** 111–125.
53. Sun, T. T., and Green, H. (1977) *Nature (London)* **269,** 489–492.
54. Yang, J., Richards, J., Bowman, P., Guzman, R., Eham, J., McCormick, K., Hamamoto, S., Pitelka, D., and Nandi, S. (1979) *Proc. Natl. Acad. Sci. U.S.A.* **76,** 3401–3405.
55. Yaoi, Y., and Kanaseki, T. (1972) *Nature (London)* **237,** 283–285.
56. Weiss, L., Poste, G., Mackearnin, A., and Willett, K. (1975) *J. Cell Biol.* **64,** 135–145.
57. Gospodarowicz, D., and Ill, C. R. (1980) *Proc. Natl. Acad. Sci. U.S.A.* **77,** 2726–2730.
58. Gospodarowicz, D., and Ill, C. R. (1980) *J. Clin. Invest.* **65,** 1351–1364.
59. Gospodarowicz, D., Delgado, D., and Vlodavsky, I. (1980) *Proc. Natl. Acad. Sci. U.S.A.* **77,** 4094–4098.
60. Gospodarowicz, D., Hirabayashi, K., Giguere, L., and Tauber, J.-P. (1981) *J. Cell Biol.* **85,** 568–578.
61. Gospodarowicz, D., and Ill, C. R. (1980) *Exp. Eye Res.* **31,** 181–199.
62. Savion, N., Lui, G.-M., Laherty, R., and Gospodarowicz, D. (1981) *Endocrinology* (in press).
63. Gospodarowicz, D., Savion, N., and Vlodavsky, I. (1980) *J. Supramol. Struct.* **13,** 339–372.
64. Gospodarowicz, D., Mescher, A. L., and Birdwell, C. R. (1978) *Natl. Cancer Inst. Monogr.* **48,** 109–130.
65. Gospodarowicz, D., Mescher, A. L., and Moran, J. (1978) *Symp. Soc. Dev. Biol.* **35,** 33–61.
66. Gospodarowicz, D., and Greenburg, G. (1981) *In* "The Biology of Normal Human Growth," pp 1–21. Raven, New York.
67. Gospodarowicz, D. (1980) *Adv. Biochem. Psychopharmacol.* **28,** 243–262.
68. Gospodarowicz, D., Mescher, A. L., Brown, K., and Birdwell, C. R. (1977) *Exp. Eye Res.* **25,** 631–649.

69. Sato, G., and Reid, L. (1978) *Int. Rev. Biochem.* **20,** 219–251.
70. Bottenstein, J., Hayashi, I., Hutchings, S., Masui, H., Mather, J., McClure, D., Ohasa, S., Rizzino, A., Sato, G., Serrero, G., Wolfe, R., and Wu, R. (1979) *In* "Methods in Enzymology" (W. B. Jakoby and I. H. Pastan, eds.), Vol. 58, pp. 94–109. Academic Press, New York.
71. Barnes, D., and Sato, G. (1980) *Cell* **22,** 649–655.
72. Ross, R., and Glomset, J. A. (1973) *Science* **180,** 1332–1339.
73. Brown, G., Mahley, R., and Assmann, G. (1976) *Circ. Res.* **39,** 415–424.
74. Fischer-Dzoga, K., Fraser, R., and Wissler, R. W. (1976) *Exp. Mol. Pathol.* **24,** 346–359.
75. Tauber, J.-P., Cheng, J., and Gospodarowicz, D. (1980) *J. Clin. Invest.* **66,** 696–708.
76. Tauber, J.-P., Cheng, J., Massoglia, S., and Gospodarowicz, D. (1981) *In Vitro* **17,** 519–530.
77. Giguere, L., Cheng, J., and Gospodarowicz, D. (1982) *J. Cell. Physiol.* (in press).
78. Isaacs, J. D., Savion, N., Gospodarowicz, D., Fenton, J. W., and Shuman, M. A. (1981) *Biochemistry* **20,** 398–403.
79. Savion, N., Isaacs, J. D., Shuman, M. A., and Gospodarowicz, D. (1981) *J. Biol. Chem.* **256,** 4514-4519.
80. Shuman, M. A., Isaacs, J. D., Maerowitz, T., Savion, N., and Gospodarowicz, D. (1981) *Ann. N.Y. Acad. Sci.* **370,** 57–66.
81. Fehrenbacher, L., Gospodarowicz, D., and Shuman, M. A. (1979) *Exp. Eye Res.* **29,** 219–228.
82. Goldminz, D., Vlodavsky, I., Johnson, L. K., and Gospodarowicz, D. (1979) *Exp. Eye Res.* **29,** 331–351.
83. Birnbaum, J. E., Sapp, T. M., and Moore, J. B. (1976) *J. Invest. Dermatol.* **66,** 313–318.
84. Savage, C. R., and Cohen, S. (1973) *Exp. Eye Res.* **15,** 361–366.
85. Rheinwald, J. G., and Green, H. (1977) *Nature (London)* **265,** 421–424.
86. Savage, C. R., and Cohen, S. (1972) *J. Biol. Chem.* **247,** 7609–7611.
87. Gospodarowicz, D., Mescher, A. L., and Birdwell, C. R. (1977) *Exp. Eye Res.* **25,** 70–89.
88. Carpenter, G., King, L., and Cohen, S. (1978) *Nature (London)* **276,** 409–420.
89. Carpenter, G., and Cohen, S. (1976) *J. Cell Biol.* **71,** 159–171.
90. Vlodavsky, I., Brown, K. D., and Gospodarowicz, D. (1978) *J. Biol. Chem.* **253,** 3744–3750.
91. Carpenter, G., and Cohen, S. (1979) *Annu. Rev. Biochem.* **48,** 193–226.
92. Das, M., and Fox, F. C. (1978) *Proc. Natl. Acad. Sci. U.S.A.* **75,** 2644–2648.
93. Savion, N., Vlodavsky, I., and Gospodarowicz, D. (1980) *Proc. Natl. Acad. Sci. U.S.A.* **77,** 1466–1470.
94. Linsley, P. S., Blifeld, C., Wrann, M., and Fox, C. F. (1979) *Nature (London)* **278,** 745–748.
95. Schechter, Y., Herneus, L., and Cuatrecasas, P. (1978) *Proc. Natl. Acad. Sci. U.S.A.* **75,** 5788–5791.
96. Haigler, H., Ash, J. F., Singer, S. J., and Cohen, S. (1978) *Proc. Natl. Acad. Sci. U.S.A.* **75,** 3317–3321.
97. Schlessinger, J., Schechter, Y., Willingham, M. C., and Pastan, I. (1978) *Proc. Natl. Acad. Sci. U.S.A.* **75,** 2659–2663.
98. Gospodarowicz, D., and Greenburg, G. (1979) *Exp. Eye Res.* **28,** 147–157.
99. Gospodarowicz, D., Greenburg, G., and Alvarado, J. (1979) *Proc. Natl. Acad. Sci. U.S.A.* **76,** 464–468.

100. Gospodarowicz, D., Greenburg, G., and Alvarado, J. (1979) *Arch. Ophthalmol. (Chicago)* **97,** 2163–2169.
101. Gospodarowicz, D. (1979) *Mol. Cell. Biochem.* **25,** 79–110.
102. Aaronson, S. A., Todaro, G. J., and Freeman, A. E. (1970) *Exp. Cell Res.* **61,** 1–5.
103. Leighton, J. (1957) *Cancer Res.* **17,** 929–935.
104. Liotta, L. A., Abe, S., Robey, P. G., and Martin, G. R. (1979) *Proc. Natl. Acad. Sci. U.S.A.* **76,** 2268–2272.
105. Rifkin, D. B., Loeb, J. N., Moore, G., and Reich, E. (1974) *J. Exp. Med.* **139,** 1317–1328.
106. Thomassen, M. J., Buoen, L. C., and Brand, K. G. (1975) *JNC1, J. Natl. Cancer Inst.* **54,** 203–207.
107. Vlodavsky, I., Lui, G.-M., and Gospodarowicz, D. (1980) *Cell* **19,** 607–616.
108. Vlodavsky, I., and Gospodarowicz, D. (1980) *Nature (London)* **289,** 304–306.
109. Ewing, J. (1921) *Proc. N.Y. Pathol. Soc.* **21,** 17–24.
110. Rafferty, K. A. (1975) *Adv. Cancer Res.* **21,** 249–292.
111. Fujii, D. K., Massoglia, S., Savion, N., and Gospodarowicz, D. (1982) *J. Neurosci.* (submitted for publication).
112. Gospodarowicz, D., Fujii, D. K., Giguere, L., Savion, N., Tauber, J.-P., and Vlodavsky, I. (1981) *In* "The Prostatic Cell, Structure and Function." Progress in Clinical and Biological Research, Vol 759, pp. 95–132. A. R. Less, Inc. New York. (in press).
113. Fielding, P. E., Vlodavsky, I., Gospodarowicz, D., and Fielding, C. J. (1979) *J. Biol. Chem.* **254,** 749–755.
114. Reckless, J. P. D., Weinstein, D. B., and Steinberg, D. (1978) *Biochim. Biophys. Acta* **529,** 475–487.
115. Stein, O., and Stein, Y. (1976) *Biochim. Biophys. Acta* **23,** 563–568.
116. Tauber, J.-P., Goldminz, D., Vlodavsky, I., and Gospodarowicz, D. (1981) *Eur. J. Biochem.* **119,** 317–325.
117. Brown, G., Mahley, R., and Assmann, G. (1976) *Circ. Res.* **39,** 415–424.
118. Layman, D. L., Jelen, B. J., and Illingworth, D. R. (1980) *Proc. Natl. Acad. Sci. U.S.A.* **77,** 1511–1515.
119. Goldstein, J. L., Helgeson, J. A. S., and Brown, M. S. (1979) *J. Biol. Chem.* **254,** 5403–5409.
120. Gospodarowicz, D., Ill, C. R., and Birdwell, C. R. (1977) *Endocrinology* **100,** 1108–1120.
121. Gospodarowicz, D., Vlodavsky, I., Bialecki, H., and Brown, K. (1978) *In* "Fifth Brook Lodge Meeting on Novel Aspects of Reproductive Biology" (J. Wilkes, ed.), pp. 107–178. Plenum, New York.
122. Gospodarowicz, D., Savion, N., Vlodavsky, I., and Johnson, L. K. (1980) *In* "Control Mechanisms in Animal Cells" (L. Jimenez de Asua, R. Levi-Montalcini, R. Shields, and S. Iacobelli, eds.), pp. 61–83. Raven, New York.
123. Gospodarowicz, D., and Bialecki, H. (1978) *Endocrinology* **103,** 854–858.
124. Gospodarowicz, D., Ill, C. R., Hornsby, P. J., and Gill, G. N. (1977) *Endocrinology* **100,** 1080–1089.
125. Savion, N., Lui, G.-M., Laherty, R., and Gospodarowicz, D. (1981) *Endocrinology* **109,** 409–421.
126. Gospodarowicz, D., and Ill, C. R. (1982) *Endocrinology* (in press).
127. Savion, N., and Gospodarowicz, D. (1981) *In* "Functional Correlates of Hormone Receptors in Reproduction" (V. B. Mahesh and Muldoon, eds.), pp. 437–461. Elsevier/North-Holland, New York.

128. Savion, N., Laherty, R., Cohen, D., Lui, G.-M., and Gospodarowicz, D. (1982) *Endocrinology* **110,** 13–24.
129. Savion, N., Laherty, R., Lui, G.-M., and Gospodarowicz, D. (1981) *J. Biol. Chem.* **256,** 12817–12822.
130. Anderson, B., Hoffman, P., and Meyer, K. (1965) *J. Biol. Chem.* **240,** 156–167.
131. Marshall, R. D., and Neuberger, A. (1972) *In* "Glycoproteins" (A. Gottschalk, ed.), 2nd ed., Vol. 5, Part B, pp. 322–336. Elsevier, Amsterdam.
132. Castenalli, A. A., Balduinia, D., and Brovelli, A. (1970) *In* "Chemistry and Molecular Biology of the Intercellular Matrix" (E. A. Balazs, ed.), Vol. 2, pp. 945–957. Academic Press, New York.
133. Cifonelli, J. A. (1968) *Carbohydr. Res.* **8,** 233–242.
134. Kosher, R. A., and Searls, R. L. (1973) *Dev. Biol.* **32,** 50–68.
135. Lindahl, V., and Roden, L. (1972) *In* "Glycoproteins" v. 45, (A. Gottschalk, ed.), 2nd ed., Vol. 5, Part B, pp. 491–515. Elsevier, Amsterdam.
136. Greenburg, G., and Gospodarowicz, D. (1982) *Exp. Cell Res.* (in press).
137. Reddi, A. H., and Huggins, C. B. (1974) *Proc. Soc. Exp. Biol. Med.* **145,** 475–486.
138. Reddi, A. H. (1976) *In* "Biochemistry of Collagen" (G. N. Ramachadran and A. H. Reddi, eds.), pp. 449–477. Plenum, New York.
139. Stiles, C. D., Capne, G. T., Scher, C. D., Antoniades, H. M., Van Wyk, J. J., and Pledger, W. J. (1979) *Proc. Natl. Acad. Sci. U.S.A.* **76,** 1279–1283.

Vessel Wall Growth Control

STEPHEN M. SCHWARTZ, CORINNE M. GAJDUSEK, AND GARY K. OWENS
Department of Pathology
School of Medicine
University of Washington
Seattle, Washington

INTRODUCTION

Growth control has emerged as a key issue in our understanding of atherosclerotic lesions. This resulted from four events during the 1970s:

1. Several investigators demonstrated that smooth muscle proliferation was a major feature of the lesions (1–3).
2. Smooth muscle cells and endothelial cells became available in cell culture (4,5).
3. Ross and his colleagues demonstrated that platelet thrombosis was able to stimulate smooth muscle cell growth *in vitro* (5,6).
4. Benditt demonstrated that smooth muscle cells in human lesions have a mono- or oligoclonal origin (7).

These discoveries and methodological developments led to quite different views of the role of growth control in the two cell types that make up the artery wall. For the smooth muscle cell, the critical issue is abnormal cell growth. If the Ross hypothesis is correct (6), this is basically a problem of growth factor stimulation of G_0–G_1 transition, analogous to the studies of action of growth factors on other cell types. If the Benditt hypothesis is correct, the issue is still G_0–G_1 transition, but now the mechanism becomes intrinsic to the cells themselves and the model systems come from neoplastic, rather than normal growth (7). The issues are different for the endothelium. The critical event in the Ross hypothesis, release of platelet-derived growth factor, depends on loss of endothelial integrity. In contrast, the Benditt hypothesis does not propose endothelial injury as a critical, initiating event. In

PATHOBIOLOGY OF THE ENDOTHELIAL CELL

ISBN 0-12-521980-6

either scheme, however, there is no doubt that thrombosis, and therefore loss of endothelial continuity, occurs as lesions progress. In summary, the loss of endothelial integrity plays a role in either the initiation or progression of atherosclerotic lesions. Thus in the smooth muscle cell we are interested in the etiology of abnormal cell growth. For the endothelium we are concerned with growth as a repair process that is necessary to replace endothelial cells lost through normal turnover. This view of the vessel wall raises several questions that are addressed in this review:

1. What is the initial resting state of these two cell types and what are the stimuli that initiate growth?
2. Does failure of the endothelium to regenerate lead to exposure of subendothelium with resulting platelet release and smooth muscle mitogenesis?
3. What are the interrelations between growth of the two cell types?

INITIAL GROWTH STATE OF SMOOTH MUSCLE CELLS

A recent study from our laboratory puts the problem of initiation of growth of the smooth muscle cell into a new light (8). These studies grew out of our interest in hypertension. Classical studies by Bevan (9) and Crane and Dutta (10) showed that smooth muscle cell replication, as measured by ^{3}HTdR autoradiography, is increased during the early phases of hypertension. This period of DNA synthesis was followed by an increased DNA content of the vessel wall (11), implying an increase in cell number. This conclusion, however, came into question with the development in 1978 of a relatively simple morphometric method to assay the frequency of cells in the vessel wall. Olivetti *et al.* (12) failed to find any increase in cell number. Instead they found only an increase in cell size. The apparent paradox was resolved by Dr. Gary Owens (8). Dr. Owens began by isolating smooth muscle cells from rat aortas in order to measure cell size, independent of the uncertainties associated with morphometry. The data, as summarized in Fig. 1, demonstrate that hypertensive smooth muscle cells, from both genetically hypertensive strains and animals with renal artery constriction, have an unusually high mean DNA content. The mean cell mass, including actin/cell, is increased in proportion to the change in DNA content. When the vessel wall DNA content is corrected for the change in DNA/cell, there is no increase in cell number/unit length. We concluded from this that the change in DNA was not due to an increase in

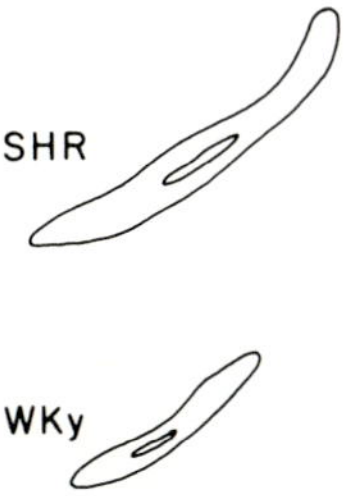

	Aver. protein / cell	Aver. DNA / cell	Protein / DNA	% Actin
SHR	*560±60	*10.3±1.0	54.7±4.3	41.4±2.5
WKy	320±50	7.1±0.6	46.3±5.1	44.4±3.3

Fig. 1. DNA, protein, and actin content of isolated aortic smooth muscle cells. SHR, spontaneously hypertensive rat; WKy, Wistar Kyoto Rat (nonhypertensive control). *, significantly greater than WKy ($p < .05$, t-test).

cell number. Flow cytometry and Feulgen microdensitometry were done to determine the location of this excess DNA and showed (Fig. 2) that the increase was due to an alteration in ploidy. Normal animals showed about 10% tetraploid cells. Hypertensive animals, depending on severity and etiology, showed values ranging from 20 to 40%. These tetraploid cells do not represent cycling cells, since the frequency of cells in S as measured by ^{3}HTdR autoradiography of dispersed cells, was approximately 10^{-3} or less.

Finally, in collaboration with Drs. Thomas Barret and Earl Benditt, we have demonstrated that normal adult human arterial smooth muscle cells also have tetraploid cells (unpublished data). The principal importance of these data is for the etiology of hypertension. A number of forms of hypertension have as their common end point a "fixed" increase in the structural mass of the vessel wall in small arteries. As discussed by Folkow (13), this fixed change may be the structural cause of elevated pressure.

Beyond hypertension, these data also have direct significance for atherosclerosis. The proposition stated above, that the issue for smooth muscle cell growth is transition from G_0 to S is now seen to be simplistic, since this cell type is able to endoreplicate. Furthermore, ^{3}HTdR index data, whether applied to hypertension or atherosclerosis, do not constitute a satisfactory measure of changes in smooth muscle cell number (14–16). This may have implications for existing cell kinetic studies of cell proliferation in lipid and traumatic models of atherosclerosis. The observation of tetraploidy, however, is important at a more fundamental level. Until now the homogeneity of the smooth muscle cell population of mammalian aorta has been a major argument against monoclonality. It was hard to see how a proliferative event

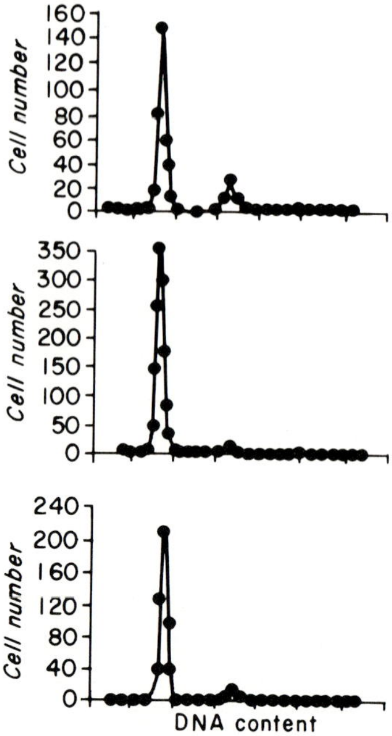

Fig. 2. Flow microfluorimetry of DNA content of smooth muscle cells isolated from hypertensive (SHR) and normotensive (WKy, SD) rats. The large peak is the diploid peak. The small peak to the left is the tetraploid or G_2 peak. The negligible number of cells with DNA content between diploid and tetraploid (i.e., S phase cells) assures us that the G_2 peak represents either true tetraploidy or cells arrested in G_2.

would produce a mono- or oligoclonal proliferation if the cells of vessel walls have random proliferative properties. If the vessel wall is heterogeneous, the analysis of proliferation may become quite different. For example, there may well be subpopulations of cells in quite different states of growth arrest. Furthermore, very little is known about the properties of tetraploid cells in any system. There is, for example, evidence from cardiac muscle cells that endoreplication may occur as a function of age or hypertension (17,18). The properties of these altered cells are unknown, although hypertensive heart itself has several abnormal properties. Finally, Pero *et al.* (19) reported an abnormal rate of unscheduled DNA synthesis (i.e., "repair") in nuclei of lymphocytes from hypertensive patients. It is interesting to speculate whether instabilities of the genome of tetraploid cells might not predispose to the sort of mutational change in vascular smooth muscle proposed by Benditt to be the key event in atherosclerotic lesion formation.

INITIAL GROWTH STATE OF ENDOTHELIAL CELLS

The initial status of the endothelium is quite different. Like the smooth muscle cell, resting endothelial cells have only a few cells per thousand in the S phase of the cell cycle. Endothelial cells have only rare cells in G_2, by Feulgen microspectrophotometry of endothelium *in vivo* or flow cytometry of cultured cells (Fig. 3). The problems of initiation of growth are also different. For the smooth muscle cell, as for fibroblasts and 3T3 cells, cell density is a function of the concentration of growth factors in the medium. For example, as shown by Vogel and Ross, growth is stimulated by the availability of platelet-derived growth factor, and saturation density is dependent on concentration of this mitogen (20). In contrast, once quiescence is reached, growth factors or serum concentration have no effect on cell density or growth of endothelium (21). Furthermore, if the endothelial cells are stimulated to grow by *in vitro* wounding, there is nearly maximal initiation of growth even in serum-free conditions and there is very little effect of added growth factors.

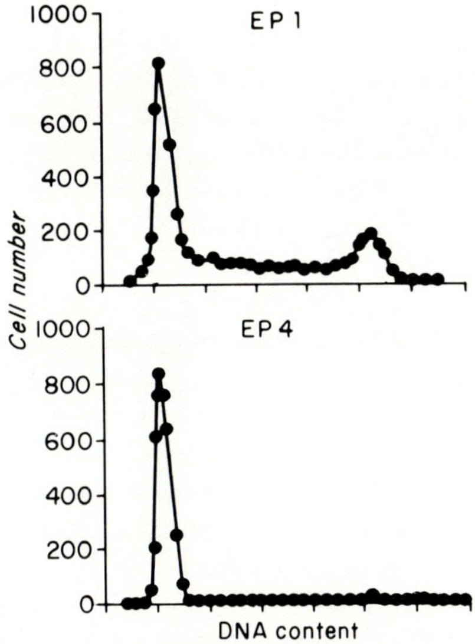

Fig. 3. Flow microfluorimetry of DNA content of bovine aortic endothelial cells in culture. The lower figure shows a normal stationary density culture with very few cells in the cell cycle. The small "G_2" peak probably represents an artifact. The upper figure shows the response to wounding with increases in the number of cells between the two peaks; in S, and in the small peak, which is in G_2 + M.

This wounding model has been quite important in studies both *in vivo* and *in vitro*. Endothelial regeneration *in vitro* is, at least formally, comparable to regeneration following a wound *in vivo*. In both cases the apparent stimulus is loss of a portion of the cell layer. Also in both cases, growth occurs without addition of exogenous growth factors to the overlying medium during growth stimulation. We have presented data on a fairly extensive comparison of growth responses of endothelium *in vivo* and *in vitro*. The only apparent differences are morphological variations in the polarity of cell movement and the cell density of responding cells (22). These may be a function of differences in substrate or differences in flow of the overlying medium. Finally, in both cases the cells giving rise to the proliferative response start from saturation density. This is also the initial growth state for the other known cases of growth stimulation of the endothelium: angiogenesis and vascular embryology (23–27). In both of these, new endothelium only appears as new vessels branching or "sprouting" off from existing vessels. There is no recruitment of undifferentiated cells from the surrounding mesenchyme, except in the earliest phases of differentiation of the primitive hematopoietic islands and vascular sacs. Thus there is no evidence that endothelium *in vivo* ever grows from sparse density. This presents problems for the interpretation of studies of growth requirements of endothelium plated at sparse density. Several laboratories, including our own, have reported requirements for various growth factors for endothelial cell growth (28–30). There is, however, no evidence that these "requirements" for growth from sparse density *in vitro* reflect a physiological stimulus for growth. Indeed, reports of Gospodarowicz and his colleagues, as well as from this laboratory, indicate that endothelial cells are able to "condition" their own medium and may have no requirement of exogenous factors when grown at any density on their own extracellular matrix (28,31). These observations may explain the absence of growth factor requirements under culture conditions that mimic cell density *in vivo*.

It seems reasonable then to consider the possible requirements for growth of endothelium from a high-density stationary state. We have a limited amount of data of this sort. We know, for example, that growth can be initiated at a high rate even in serum-free medium, and defined growth factors have no additive effect when added to medium containing plasma-derived serum (21). The ability of cells to move into a wound area is inhibited by the cytochalasins (32,33). Inhibitors of protein synthesis, of course, will also inhibit cells from entering the cell cycle (28). What was unexpected, was our finding that the cytochalasins also selectively inhibit DNA synthesis at the wound edge (33). This suggests that endothelial replication is somehow related to stimulation of endothelial cell movement or shape change, a suggestion supported by the studies of Folkman and Moscona (34) on cell shape at wound edges and cell shape on different surfaces. Our study of the kinetics of entry into S after arrest with cytochalasin D suggests that the event effected by this drug must be one quite close to the time of wounding itself, since the time course of response after arrest for 20 hr was essentially identical to the response after a fresh wound (28).

Two other observations are consistent with the hypothesis that endothelial growth depends on cell shape, cell movement, or cell contact. Although conventional mitogens have not been reported as being about to restimulate growth in stationary density endothelium, we have found three chemicals that will do this. These agents are colchicine, vinblastine, and polymerized collagen (32,35) (Fig. 4). In the case of the microtubule drugs, the effective doses cause a marked change in cell shape with partial retraction of the cells from each other. The cells enter S with kinetics similar to those seen after an *in vitro* wound. If vinblastine is removed, cells progress through mitosis. Daughter cells, however, are lost so that the culture maintains normal density. In the case of collagen, cells also retract and enter S when soluble collagen is allowed to gel over the cells or when a preformed gel is added to the cells. In this case a new, higher cell density is achieved. Thus in a somewhat bizarre sense, collagen or vinblastine are "growth factors" for confluent endothelium.

The significance of these phenomena *in vivo* is difficult to identify. It may be worth while to include some speculation. Most of our attempts to look at control of cell growth have concentrated on soluble factors. This is convenient for biochemistry but may not reflect the reality of mechanisms *in vivo*. A good example of this is the recent report by Gospodarowicz and his colleagues of the ability of cells to grow in the absence of exogenous growth factors if the cells were plated on triton-extracted cell layers called "extracellular matrix" (31). It is obviously difficult to know what component of this solid phase growth support is responsible for this effect, particularly since the available literature suggests that material prepared in this way is likely to include components of the cytoskeleton and cell surface (36–38), and it is difficult to exclude the possibility that growth factors are adsorbed from the culture medium. Despite these difficulties, the concept that substrate controls growth remains attractive. For example, it has been known for some time that basement membranes are able to survive cell necrosis and provide a scaffold for regeneration of surrounding tissues, includ-

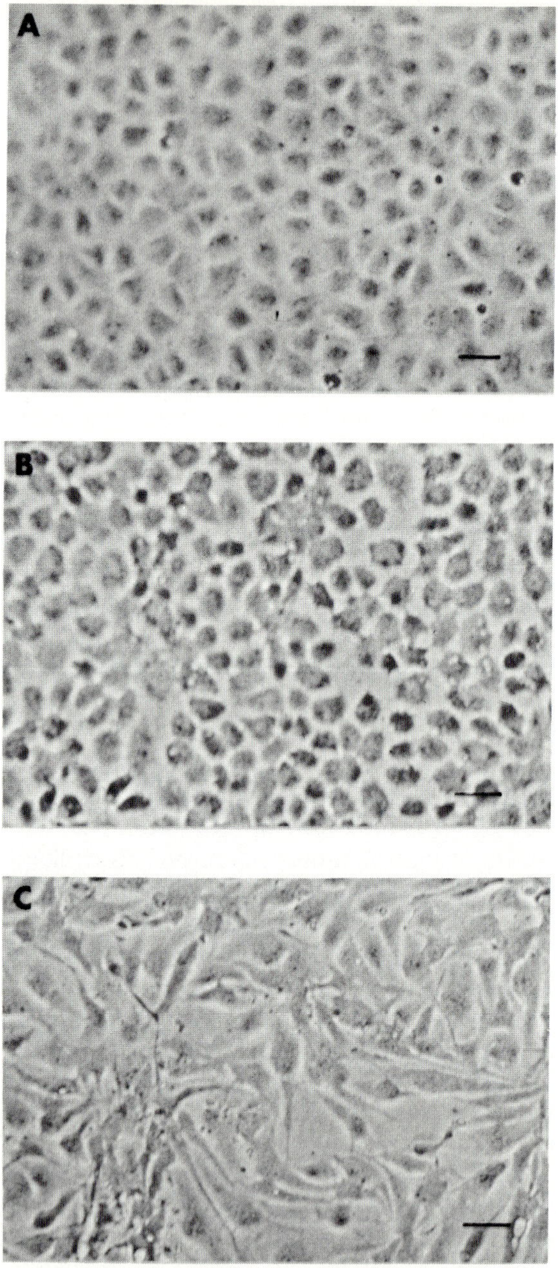

Fig. 4. Stimulation of growth by disruption of the monolayer. These three micrographs show endothelial cells: (A) in the normal state, (B) after addition of vinblastine, or (C) after addition of collagen. These disrupted monolayers show initiation of the cell cycle. Marker = 10 μm.

ing blood vessels (39). The idea of substrate control of growth is also attractive, since stimulation of endothelial cell mitosis is obviously only one part of the series of events that must occur when a new vessel is formed. Migration of endothelial cells into the surrounding tissues must occur at the same time and clearly requires some morphogenic event that releases these cells from their normal, tightly contact-inhibited pattern. We have data that may support this concept. As discussed in the next section, endothelial cells make a growth factor that is mitogenic for other cells, but not for endothelial cells themselves (40,41). This factor, endothelial-cell-derived growth factor (EDGF), has a molecular weight of approximately 30,000 and can be separated from low-molecular-weight components of endothelial cell conditioned medium by dialysis (41). When this is done, the retained fraction is mitogenic for smooth muscle cells but has no effect on endothelial cells at sparse or confluent densities. This same material, however, applied to a chick chorioallantoic membrane is markedly angiogenic (42). Thus it seems that angiogenesis may be a separable event from endothelial mitogenesis. We have proposed that the target cell of this sort of angiogenesis is the smooth muscle cell and that movement and division of the underlying smooth muscle cell may disrupt endothelial continuity, resulting in endothelial cells leaving their contact-inhibited cell layer.

INTERRELATIONSHIPS OF ENDOTHELIAL AND SMOOTH MUSCLE GROWTH

Perhaps the most obvious connection is the effect of loss of the endothelium. As shown by Ross and his colleagues *in vitro* (6) and by Friedman and his colleagues *in vivo* (43), loss of endothelial integrity leads to stimulation of smooth muscle growth as a result of release of the platelet-derived growth factor. The role of an intact endothelium in this process, however, is less clear. Gajdusek *et al.* (41), from this laboratory, and Fass *et al.* (44) have reported that endothelial cells synthesize a polypeptide mitogen, endothelial-cell-derived growth factor (EDGF). In collaboration with Paul DiCorleto and Russell Ross, we have presented evidence that EDGF is a unique peptide, unrelated to PDGF, FGF, EGF, or the somatomedins (45). In addition to the macrophage-derived growth factor (46,47), EDGF represents one of the few growth factors known to be produced by a nontransformed, nonestablished cell. Since monocytes do not grow in culture, EDGF may represent a unique opportunity to study growth factor synthesis.

As already noted, EDGF does stimulate smooth muscle cells, but has no apparent effect on endothelial cells themselves (40,41).

In contrast to the mitogenic effect of EDGF, Castellot *et al.* (48) described a growth inhibitor for smooth muscle cells in conditioned medium from endothelial cells. These apparently conflicting data are simply resolved. Castellot's report deals with growth inhibition where endothelial cells condition medium already contained growth factors for smooth muscle cells, that is, whole blood serum. Medium containing whole blood serum shows a decrease in activity after being on endothelial cells. In our own case, an increase in activity is seen in medium containing plasma-derived serum or in serum-free medium, without the platelet-released material required for smooth muscle growth. It is important to point out that Castellot's effect is not simply the depletion of essential components of the medium. The data suggest that the cells actively synthesize a heparan sulfate-like proteoglycan responsible for this effect. Thus cultured endothelial cells are able to synthesize both inhibitors and stimulators of smooth muscle growth. The nature of the interrelationship between these forces remains totally unknown.

At this point we can return to the hypothesis that endothelial injury is the initial event in atherosclerosis. Surprisingly little is known about actual loss of endothelium either spontaneously or in response to atherosclerosis risk factors. Studies from several laboratories, including our own, have shown that the basal rate of replication in normal adult animals is extremely low but that there are areas of the endothelium that show an increased replication (49). These areas appear to be more frequently associated with branches (50), although similar areas can be seen at some distance from branches (49). Endothelial replication is increased in response to hypertension (51) or hyperlipemia (52). It is important to point out, however, that increased turnover does not necessarily imply that areas of the endothelium become denuded. Although a number of morphologic changes have been described in the endothelium, evidence for spontaneous denudation is lacking, either in normal animals or in animals exposed to hypertension or hyperlipemia. Recent studies by scanning electron microscopy in hyperlipemic animals have found evidence of denuded areas in the early stages of hyperlipemia only after lesion formation was advanced. This lack of convincing evidence for spontaneous denudation does not prove that subendothelium cannot be exposed for brief, repeated episodes. It does, however, make it worth while to consider the possible role of nondenuding injuries in atherosclerosis (Fig. 5).

Fig. 5. Denuding versus nondenuding injuries. Note that these categories are not mutually exclusive.

One possible nondenuding injury might be alterations in the normal nonthrombogenicity of endothelial cells. Endothelial cells are able to secrete quantities of PGI_2 sufficient to inhibit thrombosis (53,54). It is important to point out, however, that nonthrombogenicity is not dependent on prostaglandin metabolism alone. Endothelium remains nonthrombogenic even when prostaglandin synthesis is inhibited (55). The only example of "thrombogenic" endothelium is when the cells have somehow undergone neoplastic transformation or are grown under conditions that produce totally abnormal growth patterns (53,55,56). Under these conditions, thrombogenicity seems to correlate with the loss of normal endothelial polarity, an extremely abnormal growth pattern, and the appearance of exoskeletal proteins on the luminal surface. Phenomena of this sort *in vivo* occur only in neoplasia. Thus for now we can only relate thrombosis to actual denudation of the vessel surface.

There are at least three other ways in which alterations of a number of endothelial cell functions might lead to smooth muscle proliferation: altered transport, nonthrombotic platelet interactions, and altered control of growth factor production. Altered bulk phase or high-affinity transport has been shown in regenerating endothelial cells in culture by Davies, in this laboratory (57), and by Vlodavsky and his co-workers (58). Accelerated uptake does occur across replicating cells and this could account for localization of lesions at sites of increased cell turnover. Altered transport presumably would not include the platelet-derived growth factor, since that requires platelet release. Other mitogens including epidermal growth factor, however, are available in plasma at concentrations close to mitogenic levels (59). Furthermore, there is evidence that lipoproteins may contribute to cell proliferation as well as to lipid accumulation (60). It is also conceivable that endothelial cells, while maintaining integrity, may become altered in ways that allow altered platelet interaction without adherence. These could include contact with platelets followed by release, uptake, and transport of the platelet factor. In this regard it is interesting to note that endothelial cells have receptors for other platelet components, including β-thromboglobulin (61), platelet factor IV (62), and thrombin (60,63). We have already discussed the possible role of endothelial-cell-derived growth factor and heparan sulfate.

CRITICAL LESION SIZE

If we assume that injury does occur via denudation, it is possible to make some statements about the probable duration of exposure of the thrombogenic subendothelium and about the interval of denudation required for stimulation of smooth muscle proliferation. Our *in vivo* studies, in collaboration with Christian Haudenschild and Michael Reidy (22), have examined the response of rat aorta to denudation. As in other animals, denudation by the balloon technique leads to a characteristic intimal accumulation of smooth muscle cells. At least in the rat, we found smooth muscle lesions are formed only in those areas requiring more than seven days for endothelial regeneration. This suggests that there is a "critical lesion size," i.e., a critical amount of endothelium that must be removed before smooth muscle lesion formation will be stimulated (22). Work by Hirsch *et al.* and by Michael Reidy shows that small areas of endothelial denudation 2–20 cells wide do not result in smooth muscle proliferation (64,65). Thus it appears that brief exposures to platelet releasate are not sufficient to stimulate smooth muscle proliferation. Moreover, recent reports by Goldberg *et al.* (66) and Groves *et al.* (67), based on use of the balloon catheter in rabbits, indicate that most of the platelet adhesion occurs in the first 24 hr following experimental denudation. Release of granules, which contain platelet-derived growth factor, into the vessel wall can only be detected in the first 4 hr after denudation. It is interesting to speculate whether the nature of the platelet interactions in the days after initial exposure may play a special role in stimulation of smooth muscle proliferation or whether an accumulation of platelet-released materials may be required over a longer time period.

We attempted to make a more reasonable model for spontaneous cell loss by removing strips of endothelium only 1–2 cells wide. These wounds are completely closed within 6–8 hr, without stimulation of DNA synthesis in the endothelium (65). The rapidity of this response suggests the possibility that regeneration might be coordinated with endothelial desquamation. This would explain the paradox of cell replication studies that show focal areas of high cell turnover while scanning electron microscopy studies continue to show a continuous endothelium. Recent work from our laboratory supports that hypothesis. Endotoxin has been known to produce endothelial cell desquamation. We repeated the endotoxin experiments using high-resolution scanning electron microscopy. Although we could detect a fivefold increase in endothelial cell replication in response to endotoxin, there was no evidence of denudation (68). These data imply that the endothelium may be able to regenerate rapidly enough to obviate any exposure of the subendothelium. If this is true, endothelial cell replacement may occur without release of platelet-derived growth factor.

SUMMARY

Growth control of endothelial cells and smooth muscle cells differs both in respect to growth state under resting conditions and in respect to the nature of the stimuli required for release from that state. Endothelial cells rest in a diploid state and do not respond to conventional polypeptide mitogens. Smooth muscle cells rest in the diploid state, but also show a large number of tetraploid cells. The nature of these cells is not known. Unlike endothelial cell growth, smooth muscle cell growth does appear to be dependent on exogenous growth factors. Furthermore, it appears that factors made by endothelial cells are able either to stimulate or to inhibit smooth muscle growth.

This leaves us with a picture of endothelial growth control by mechanisms intrinsic to the relationships of cells to one another or to the cell substrate. The simplest example of this comes from studies of the response to wounding. Less obvious examples come from studying effects of agents that affect cell shape or the cytoskeleton.

Somehow, one would like to bring this concept of endothelial growth control together with the abnormal growth of smooth muscle cells. As of now, however, we simply do not know a great deal about the interrelationships of growth of these two cell types. When large amounts of endothelium are removed *in vivo,* smooth muscle cell proliferation results. This does not, however, occur when the extent of the wound is very small. Our data suggest that most spontaneously occurring desquamation does not produce denudation. Furthermore, denudation, unless repeated or extensive, is not a sufficient stimulus for smooth muscle proliferation.

ACKNOWLEDGMENTS

This work was supported by NIH grants HL-03174, HL-18645, HLC-7312, and HL-26405. Dr. Schwartz is an Established Investigator of the American Heart Association.

REFERENCES

1. Geer, J. C. (1965) *Am. J. Pathol.* **47,** 241–269.
2. Haust, M. D., More, R. H., and Movat, H. Z. (1960) *Am. J. Pathol.* **37,** 377–389.
3. Parker, F. P., and Odland, G. F. (1966) *Am. J. Pathol.* **48,** 197–239.
4. Gimbrone, M. A., Jr. (1977) *In* "International Cell Biology" (B. R. Brinkley and K. R. Porter, eds.), pp. 649–658. Rockefeller Univ. Press, New York.
5. Ross, R. (1971) *J. Cell Biol.* **50,** 172–186.
6. Ross, R., and Harker, L. (1976) *Science* **76,** 1094–1100.
7. Benditt, E. P. (1977) *Am. J. Pathol.* **86,** 693–702.
8. Owens, G. K., Cornhill, J. F., and Schwartz, S. M. (1981) *Fed. Proc., Fed. Am. Soc. Exp. Biol.* **40,** 614.
9. Bevan, R. D. (1976) *Blood Vessels* **13,** 100–128.
10. Crane, W. A. J., and Dutta, L. P. (1963) *J. Pathol.* **86,** 83–97.
11. Wolinsky, H. (1972) *Circ. Res.* **30,** 301–309.
12. Olivetti, G., Anversa, P., and Melissari, M. (1980) *Lab. Invest.* **42,** 559–565.
13. Folkow, B. (1971) *Clin. Sci.* **41,** 1–12.
14. Hassler, O. (1970) *Lab. Invest.* **22,** 286–293.
15. Thomas, W. A., Reiner, J. M., Florentin, R. A., and Scott, R. F. (1979) **31,** 124–144.
16. Webster, W. S., Bishop, S. P., and Geer, J. C. (1974) *Am. J. Pathol.* **76,** 245–284.
17. Eisenstein, R., and Wied, G. L. (1970) *Proc. Soc. Exp. Biol. Med.* **133,** 176–179.
18. Sandritter, W., and Scomazzoni, G. (1964) *Nature (London)* **202,** 100–101.
19. Pero, R. W., Bryngelsson, C., Mitelman, F., Thalin, T., and Norden, A. (1976) *Proc. Natl. Acad. Sci. U.S.A.* **73,** 2496–2500.
20. Vogel, A., Ross, R., and Raines, E. (1980) *J. Cell Biol.* **85,** 377–385.
21. Schwartz, S. M., Selden, S. C., III, and Bowman, P. (1979) *Cold Spring Harbor Conf. Cell Proliferation* **6,** 593–610.
22. Schwartz, S. M., Gajdusek, C. M., Reidy, M. A., Selden, S. C., III, and Haudenschild, C. C. (1980) *Fed. Proc., Fed. Am. Soc. Exp. Biol.* **39,** 2618–2625.
23. Auerbach, R., Kubia, L., and Sidky, Y. (1976) *Cancer Res.* **36,** 3435–3440.
24. Ausprunk, D. H., and Folkman, J. (1977) *Microvasc. Res.* **14,** 53–65.
25. Clark, E. (1915) *Anat. Res.* **9,** 67–68.
26. Hughes, A. F. (1942) *J. Anat.* **77,** 266–286.
27. Schoefl, G. I. (1963) *Virchows Arch. A: Pathol. Anat.* **337,** 7–141.
28. Schwartz, S. M., Gajdusek, C. M., and Selden, S. C., III (1981) *Arteriosclerosis (Dallas)* **1,** 107–126.
29. Gospodarowicz, D., Brown, K. D., Birdwell, C. R., and Zetter, B. R. (1978) *J. Cell Biol.* **77,** 774–788.
30. Gospodarowicz, D., Moran, J. S., and Braun, D. (1977) *J. Cell Physiol.* **91,** 377–385.
31. Gospodarowicz, D., and Ill, C. (1980) *J. Clin. Invest.* **65,** 1351–1364.
32. Selden, S. C., III, Rabinovitch, P. S., and Schwartz, S. M. (1981) *J. Cell. Physiol.* (in press).
33. Selden, S. C., III, and Schwartz, S. M. (1979) *J. Cell Biol.* **81,** 348–354.

34. Folkman, J., and Moscona, A. (1978) *Nature (London)* **273,** 345–349.
35. Delvos, U., Sage, H., Gajdusek, C. M., Harker, L. A., and Schwartz, S. M. (1981) *Fed. Proc., Fed. Am. Soc. Exp. Biol.* **40,** 624.
36. Brown, S., Levinso,, W., and Spudich, J. A. (1976) *J. Supramol. Struct.* **5,** 119–130.
37. Trotter, J. A., Foerder, B. A., and Keller, J. M. (1978) *J. Cell Sci.* **31,** 369–392.
38. Webster, R. E., Osborn, M., and Weber, K. (1978) *Exp. Cell Res.* **117,** 47–61.
39. Vracko, R., and Benditt, E. P. (1970) *J. Cell Biol.* **47,** 281–285.
40. Gajdusek, C., DiCorleto, P., Ross, R., and Schwartz, S. (1980) *J. Cell Biol.* **85,** 467–472.
41. Gajdusek, C. M., and Schwartz, S. M. (1982) *J. Cell. Physiol.* (in press).
42. Harris, S. A., Gajdusek, C. M., Schwartz, S. M., and Wight, T. N. (1979) *J. Cell Biol.* **83,** CU444.
43. Friedman, R. J., Stemerman, M. B., Wenz, B., Moore, S., Gauldie, J., Gent, R., Tiell, M. L., and Spaet, T. H. (1977) *J. Clin. Invest.* **60,** 1191–1201.
44. Fass, D. N., Downing, M. R., Myers, P., Bowie, E. J. W., and Witte, L. D. (1978) *Blood* **52,** 181.
45. DiCorleto, P. E., Gajdusek, C., Schwartz, S. M., and Ross R. (1980) *J. Supramol. Struct., Suppl.* **4,** 205.
46. Glenn, K. C., and Ross, R. (1982) *Cell* (in press).
47. Martin, B. M., Gimbrone, M. A., Jr., Unanue, E. R., and Cotran, R. S. (1981) *J. Immunol.* (in press).
48. Castellot, J. J., Jr., Addonizio, M. L., Rosenberg, R., and Karnovsky, M. J. (1980) *J. Cell Biol.* **87,** CI424.
49. Schwartz, S. M., and Benditt, E. P. (1976) *Proc. Natl. Acad. Sci. U.S.A.* **73,** 651–653.
50. Schwartz, C. J., Gerrity, R. G., and Lewis, L. J. (1978) *Atheroscler. Rev.* **3,** 109–124.
51. Schwartz, S. M., and Benditt, E. P. (1977) *Circ. Res.* **41,** 248–255.
52. Florentin, R. A., Nam, S. C., Lee, K. T., and Thomas, W. A. (1969) *Exp. Mol. Pathol.* **10,** 250–255.
53. Fry, G. L., Czervionke, R. L., Hoak, J. C., Smith, J. B., and Haycraft, D. L. (1980) *Blood* **55,** 271–275.
54. Jaffe, E. A., and Weksler, B. B. (1979) *J. Clin. Invest.* **63,** 532–535.
55. Curwen, K. D., Gimbrone, M. A., and Handin, R. I. (1980) *Lab. Invest.* **42,** 366–374.
56. Czervionke, R. L., Smith, J. B., Fry, G. L., Hoak, J. C., and Haycraft, D. L. (1979) *J. Clin. Invest.* **63,** 1089–1092.
57. Davies, P. F., Selden, S. C., III, and Schwartz, S. M. (1980) *J. Cell. Physiol.* **102,** 119–127.
58. Vlodavsky, I., Fielding, P. E., Fielding, C. J., and Gospodarowicz, D. (1978) *Proc. Natl. Acad. Sci. U.S.A.* **75,** 356–360.
59. Dailey, G. E., Kraus, J. W., and Orta, D. N. (1978) *J. Clin. Endocrinol. Metab.* **46,** 929–936.
60. Fischer-Dzoga, K., Chen, R., and Wissler, R. W. (1974) *In* "Arterial Mesenchyme and Arteriosclerosis" (W. D. Wagner and T. B. Clarkson, eds.) pp. 299–311. Plenum, New York.
61. Hope, W., Martin, T. J., Chesterman, C. N., and Morgan, F. J. (1979) *Nature (London)* **282,** 210–212.
62. Busch, C., Dawes, D. S., Wasteson, P., and Wasteson, A. (1980) *Thromb. Res.* **19,** 129–137.
63. Awbrey, B. J., Hoak, J. C., and Owen, W. G. (1979) *J. Biol. Chem.* **254,** 4092–4095.

64. Hirsch, E. Z., and Robertson, A. L. (1977) *Atherosclerosis* (*Shannon, Irel.*) **28**, 271–287.
65. Reidy, M. A., and Schwartz, S. M. (1981) *Lab. Invest.* **44,** 301–308.
66. Goldberg, I., Stemerman, M. B., and Handin, R. I. (1980) *Science* **209,** 610–612.
67. Groves, H. M., Kinlough-Rathbone, R. L., Richardson, M., Moore, S., and Mustard, F. (1979) *Lab. Invest.* **40,** 194–206.
68. Reidy, M. A., and Schwartz, S. M. (1980) *Fed. Proc., Fed. Am. Soc. Exp. Biol.* **39,** 1109.

Angiogenesis *in Vitro*: Implications for Tumor Biology

JUDAH FOLKMAN AND CHRISTIAN HAUDENSCHILD
Children's Hospital Medical Center
Boston, Massachusetts

INTRODUCTION

Angiogenesis describes the growth of new capillary blood vessels. It is becoming generally recognized that most, if not all, solid tumors have angiogenesis capacity and that progressive tumor growth depends on continuous induction of new capillaries (1–3). Experimental studies on inhibition of angiogenesis have provided further evidence that tumors are angiogenesis-dependent (4). This has led to the notion that angiogenesis is a control point in neoplastic growth. Furthermore, angiogenesis plays an important role in other pathological states such as diabetic retinopathy (5), psoriasis (6), and arthritis (7).

For these reasons there has been increased interest in the study of angiogenesis. However, until recently, angiogenesis could only be studied *in vivo,* as, for example, in the rabbit ear chamber (8), the hamster cheek pouch, or the rabbit cornea (9). Previous attempts to grow capillaries *in vitro* from tissue explants were made by Lewis (10) and by Woodard and Pomerat (11). However, there was no proof of the identity of endothelium in these cultures, and there was rapid overgrowth of fibroblasts. The first successful culture of identifiable endothelial cells was the report by Jaffe *et al.* (12) of the isolation of umbilical vein endothelial cells. Gimbrone *et al.* (13) also cultured umbilical vein endothelium, and subsequently Buonassisi (14) reported the culture of aortic endothelial cells.

However, attempts to grow capillary endothelium failed. It was found that capillary endothelium did not grow under conditions that were appropriate for the culture of endothelium from large vessels. For example, capillaries were isolated from retina (15), from the marginal vein of the rabbit ear (16), and from the epididymal fat pad of the rat (17). However, none of these "endothelial cells" survived subculture.

PATHOBIOLOGY OF THE ENDOTHELIAL CELL

ISBN 0-12-521980-6

In 1978 Del Vecchio reported the culture of capillaries from rat adrenal cortex (18). Endothelial cells spread out from the capillary segments, but there was minimal growth and the cells could not be subcultured.

We used Del Vecchio's method of nylon filtration, but we used bovine adrenal cortex instead of rat adrenal and grew the cells on gelatin-coated plates in the presence of tumor-conditioned medium (19). Capillary cells grew and could be subpassaged at least 12 times. They were cloned from a single cell using feeder layers of 3T3 fibroblasts. For convenience we use the term "capillary" endothelial cell when endothelial cells are derived from a single capillary segment of not more than three cells and usually containing a red cell in the lumen. The term "microvascular" endothelium is used if the source of endothelium is a pool of cells or vascular fragments that have passed through nylon meshes of 55 to 110 μm.

METHOD OF ISOLATION OF CAPILLARY ENDOTHELIAL CELLS

The method of isolating capillary endothelial cells has been previously described in detail (19, 20). Briefly, tissue is submerged for a few minutes in Betadine solution and washed with sterile Ringer's solution. Using sterile technique the tissue is cut into pieces about 2 mm in diameter and these are incubated in 0.75% collagenase (Worthington type II) with 0.5% bovine serum albumin for at least one-half hour at room temperature with occasional shaking. This releases capillary segments. The suspension is then passed through a nylon mesh of 110-μm pore size. Capillary segments smaller than 110 μm pass through the filter. These segments are centrifuged at 650 rpm and the collagenase is discarded. The pellet is then suspended in Dulbecco's modified Eagle's medium (DMEM) supplemented with 10% calf serum and washed three times with this medium. The final pellet is suspended in tumor-conditioned medium and dispersed among two to four gelatin-coated plastic culture dishes (35-mm-diameter Falcon 3001). An alternative method is to pass the 110-μm filtrate through a 55-μm nylon screen, wash with culture medium, and collect the material that is retained on the second screen by inverting it and flushing from the back with a stream of culture fluid. This second technique tends to eliminate any larger venules or arterioles. However, it requires a large amount of starting tissue.

Capillary segments can be identified by the next day. They usually contain two or three endothelial cells surrounding a lumen. There is often a red cell in the lumen [Fig. 1]. These capillary segments are encircled with a marking pencil on the bottom of the culture dish. After two or three more days when endothelial cell proliferation is evident, the nonendothelial cells close to the capillary segment are weeded from the plate with the blunt end of a Pasteur pipette. The remaining cells in the dish are swept away by a rubber policeman. Capillary endothelial cells are allowed to proliferate until they can be subpassaged.

The evidence that these cells are capillary endothelium is as follows:

1. They arise from single capillary segments in culture.
2. They contain factor VIII antigens and angiotensin converting enzyme.
3. Their ultrastructure is similar to the capillary endothelial cells of their organ of origin.
4. Often these cells form "rings" when they are in sparse cultures. (These rings are rarely observed in cultures of other cell types when such cells are incubated in tumor-conditioned medium.)
5. The capillary endothelial cells have a requirement for conditioned medium and grow best in tumor-conditioned medium. In contrast, aortic or umbilical vein endothelium do not require conditioned medium for growth.
6. Capillary endothelial cells increase their rate of migration in a dose-dependent manner with increasing concentrations of tumor-conditioned medium or tumor-derived factors (21). In contrast, the migration of aortic endothelium is not affected by tumor-conditioned medium.
7. Capillary endothelial cells have a special matrix requirement, such as gelatin, for their survival. Such a matrix is not required for vascular endothelium from larger vessels. Recently, Davison and Karasek have shown that fibronectin-coated plates will also support the growth of microvascular endothelial cells (22).
8. Human capillary endothelial cells contain Weibel–Palade bodies. Bovine cells do not.
9. When capillary endothelial cells are confluent they typically show a large amount of cytoplasm compared to the nucleus. Their cytoplasmic-to-nuclear ratio is much larger than that of aortic endothelium.
10. Finally, tubular structures are formed by confluent capillary endothelial cells, i.e., "angiogenesis *in vitro*."

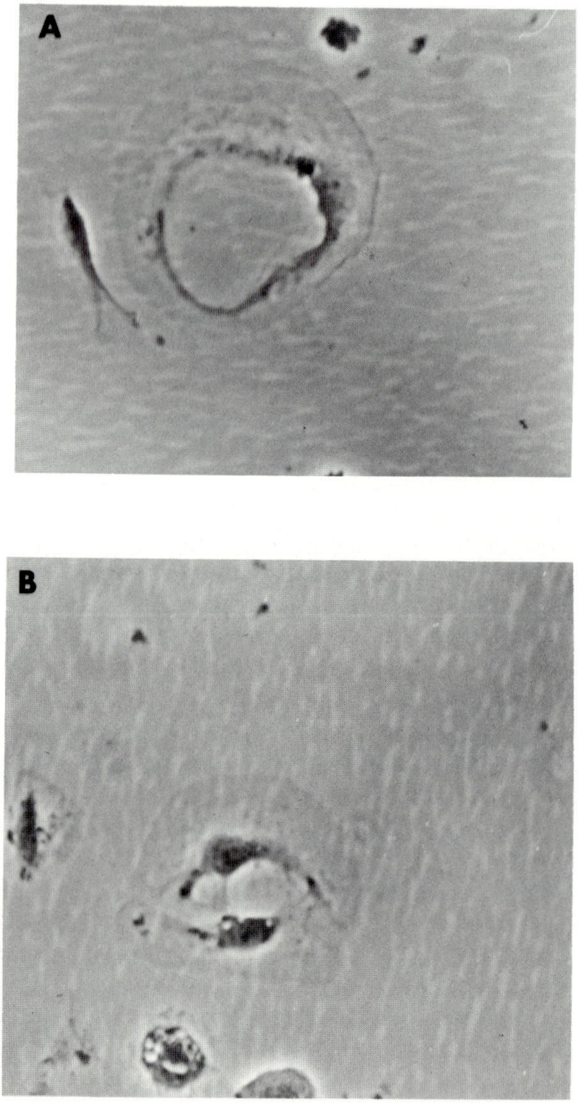

Fig. 1. Capillary segment with (A) one endothelial cell, and (B) 2 endothelial cells after 24 hours in culture. Source was bovine adrenal cortex. ×460.

ANGIOGENESIS *IN VITRO*

When capillary endothelial cells are allowed to remain confluent, they form "tubes" that resemble capillaries *in vivo* (20). In primary cultures of both human and bovine capillary endothelial cells the tubes first appear after about 3 weeks. However, once capillary tubes are formed in the primary culture they appear earlier in subsequent passages. By the fourth passage, tubes have developed as early as 5 to 10 days.

The first evidence of tube formation is the appearance by phase contrast of a cylindrical vacuole within an endothelial cell. By 48 hr contiguous cells develop similar vacuoles that form a long tube connecting one cell to another (Fig. 2). Branches begin to appear after four or five cells become connected and within 5 to 10 days the entire network of tubes replaces the dense regions of the culture. It is not necessary for the entire plate to be confluent. Small colonies of a few hundred cells often form tubes in the center of the colony. Branch formation can be observed to begin within one cell (Fig. 3). An occasional branch will be directed out of the plane of the substratum. Then

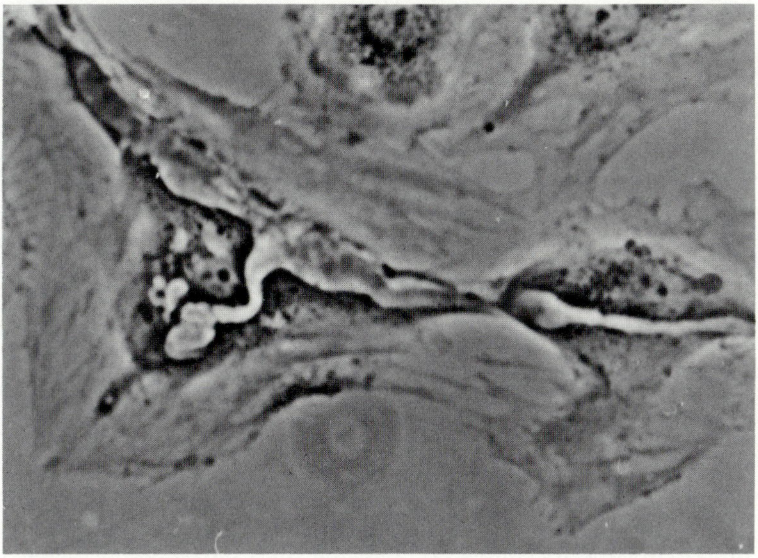

Fig. 2. Vacuoles appearing within human foreskin capillary endothelial cells. Cells were isolated 108 days prior to this photograph. They began to form vacuoles and "tubes" approximately 30 days before this photograph. The vacuole is branched and will eventually become the site of a "branch" in the capillary network. Phase contrast ×460.

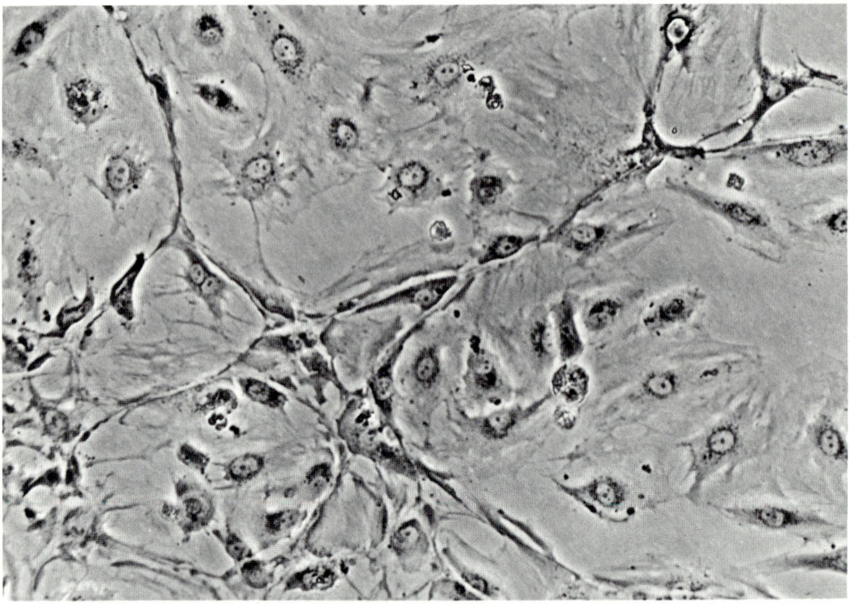

Fig. 3. Tubes formed by capillary endothelial cells obtained from human rhabdosarcoma. Photograph taken at 30 days after isolation of the cells. Tubes began to form at 15 days. Phase contrast ×230.

a second layer of capillaries organizes on the dorsal surface of the cell layer (Fig. 4). This three-dimensional growth results in a capillary network of multiple layers.

Electron micrographs show that the tubes represent continuous lumina bordered by cell cytoplasm and membranes (Fig. 5). When tube formation is still confined to a single capillary cell, the lumen appears to be a true vacuole and is surrounded by one uninterupted membrane that contains fragments of fibrillar, membranous, and amorphous material of undetermined origin. Later the cytoplasm from capillary cells surrounding the tubes forms an extremely thin wall. The tubular wall of one cell will connect to the tubular wall of another cell or sometimes to itself. At these connections junctional complexes form.

These experiments suggest that all the information necessary to construct a capillary tube, form branches, and build an entire capillary network *in vitro* can be expressed by a single cell type, the capillary endothelial cell. One implication of this finding is that angiogenesis factors may function to turn on the endothelial cell program for production of a capillary network. It is not known how new capillary tubes

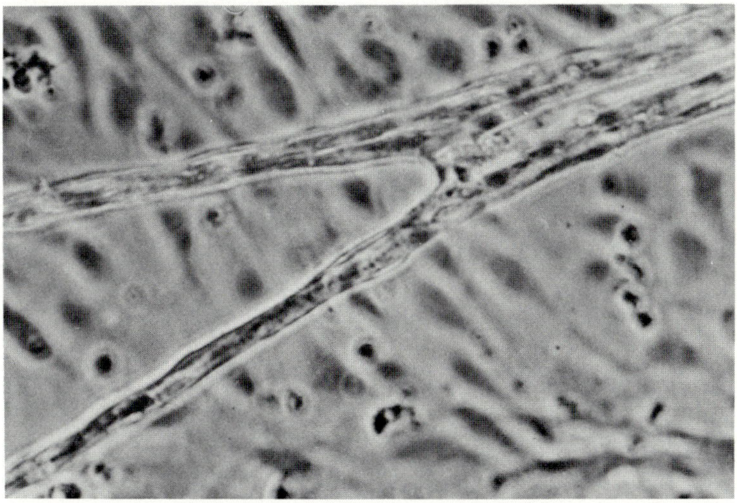

Fig. 4. A network of tubes formed in the original monolayer of these bovine capillary endothelial cells. Several branches were directed out of the plane of the substratum and formed a second layer of tubes. (Approx. ×400). Reprinted from "Cold Spring Harbor Symposium on Growth of Cells in Hormonally Defined Medium" (Gordon Sato and Arthur Pardee, eds.) 1981, in press.

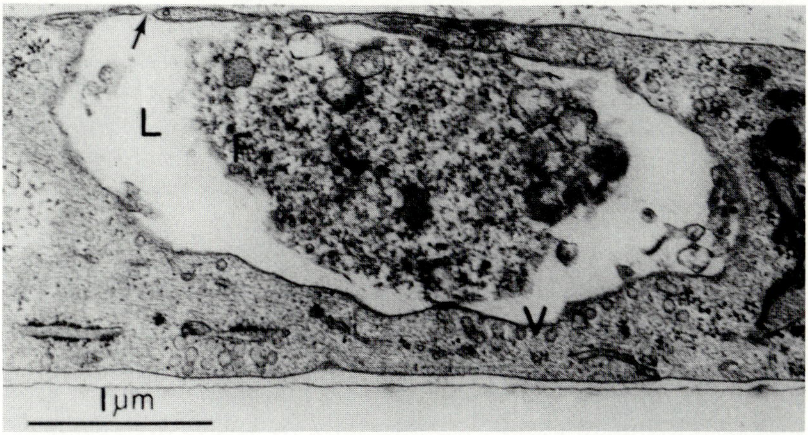

Fig. 5. Transmission electron micrograph of capillary formation. Earliest phase of tube formation within one cultured capillary endothelial cell derived from bovine adrenal cortex. The micrograph is taken in a plane perpendicular to the culture dish and demonstrates the thin cytoplasmic plate which covers the lumen (L) and which shows structures resembling fenestrations (arrow). Note the intralumenal material and the pinocytotic vesicles (V). Reprinted from *Nature* **288**, 551, 1980 with permission of the publisher.

then form venules and arterioles and obtain a coating of smooth muscle. The present experiments did not address this problem because the cultures contained only capillary endothelial cells. Furthermore, these experiments do not exclude the possibility that other types of endothelium from venules or from fetal vessels may also form tubes *in vitro*. Maciag (23) has recently shown that umbilical vein endothelial cells will form tubes *in vitro* under appropriate conditions. Also, Nicosia and Leighton (24) have shown that capillary tubes with true vascular endothelium can grow from segments of rat aorta implanted in plasmafibrin gels.

TIME-LAPSE STUDIES

Time-lapse movies of capillary tubes being formed *in vitro* show extensive remodeling. A branch may appear from the side of a long tube over a period of 6 to 8 hr. The branch may remain in place for a day and then it may shorten or retract completely. Occasionally a long tube lined by multiple endothelial cells may contract and pull apart and the endothelial cells may form a monolayer. In many areas of the culture the tube that is formed is surrounded by capillary endothelial cells which continue to migrate along the tube and will even follow its branches. In some areas of the culture, the tubes appear to be "inside out." Capillary endothelial cells encircling a tube are never seen to divide, although they do migrate. In contrast, endothelial cells out in the monolayer away from the tube divide and migrate. The tubes do not undergo contraction. However, many tubes appear to be filled with fluid. Others contain the amorphous fibrillar material to which we previously referred.

OTHER FACTORS AFFECTING GROWTH OF CAPILLARY ENDOTHELIAL CELLS

Bovine capillary endothelial cells that are obtained from primary culture or early passage will not grow when seeded at low density in Dulbecco's modified Eagle's medium with 10% calf serum. However, if the cells are taken from later passages (i.e., 5–10), they will grow slowly when sparse in only DMEM with 10% calf serum and they have a population doubling time of 67 hr. When the same cells are grown in tumor-conditioned medium (1 : 1 with fresh DMEM containing 10% calf serum), cells grow rapidly with a doubling time of 28 hr

TABLE I
Capillary Endothelial Cell Proliferation Effect of Growth Factors[a,b]

Growth factor	Optimal concentration	Cell number (day 7)
Cartilage growth factor	1 μg/ml	13.0×10^4
Fibroblast growth factor	100 ng/ml	10.2×10^4
Sarcoma–conditioned medium	50%	7.6×10^4
Bovine colostrum	2.5%	2.6×10^4
Epidermal growth factor	10 ng/ml	2.5×10^4
Insulin	1 μg/ml	1.6×10^4
Thrombin	1 μg/ml	1.4×10^4
Platelet-derived growth factor	10 ng/ml	1.3×10^4
None (10% calf serum)	—	1.1×10^4

[a] 1×10^4 Bovine capillary endothelial cells were plated in each well of a 24-well Falcon plate coated with gelatin.
[b] From Zetter *et al.* (25).

(19). Although we routinely use cells derived from mouse sarcoma-180 to prepare tumor-conditioned medium, capillary endothelial cells will also proliferate rapidly in other types of tumor medium, including B-16 mouse melanoma and an established line of human bladder carcinoma. A variety of growth factors have also been tested for their ability to stimulate capillary endothelial cell division when added to nonconditioned medium containing 10% calf serum (25). In these experiments, after endothelial cells were plated, the growth factors were added in log increments from 1 ng/ml to 1 mg/ml. Fresh growth factors were added every second day and the medium was changed on the fourth day. Of the factors tested only cartilage-derived growth factor (CDGF) (26) and fibroblast growth factor (FGF) (27) stimulated growth. Other factors were inactive, including epidermal growth factor, the platelet-derived growth factor, thrombin, human milk (28), and insulin. (See Table I.) Each of these factors was assayed in the presence of 10% calf serum.

MIGRATION OF CAPILLARY ENDOTHELIAL CELLS

While tube formation and proliferation are essential requirements of new capillary formation *in vivo,* a key component of neovascularization is the ability of endothelial cells to *migrate* from a preexisting venule toward an angiogenic source. Ausprunk and Folkman (29) showed with combined autoradiography and electron microscopy that the leading cells of an advancing capillary tip neither synthesize DNA nor mitose but only migrate toward the stimulus. Only the trailing cells further back in the capillary tube undergo mitosis. It is conceivable that the migratory stimulus may dominate the other components of angiogenesis in such a way that inhibition of migration of capillary endothelial cells might inhibit angiogenesis.

Therefore, it has become important to be able to study capillary endothelial motility *in vitro*. Zetter has accomplished this by employing the phagokinetic track assay devised by Albrecht Buehler (30). In this assay cells are plated onto glass cover slips that have been previously coated with colloidal gold particles (21). The cells ingest the gold and, as they move, leave bare areas or tracks that serve as a record of their movement. Time-lapse films showed that the size of the phagokinetic tracks was proportional to the rate of endothelial cell migration. When the capillary endothelial cells were plated on the gold-coated cover slips and incubated for 18 hr in standard Dulbecco's modified Eagle's medium with 10% calf serum, there was little phagokinetic movement [Fig. 6a]. Yet when the assay was performed in medium conditioned by mouse sarcoma-180 cells, phagokinetic tracks were significantly larger (Fig. 6b). By using a digital image

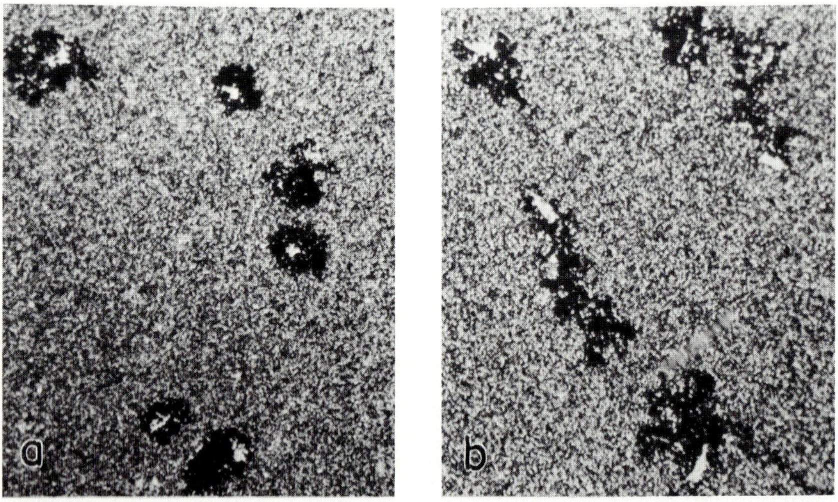

Fig. 6. Phagokinetic tracks made by bovine capillary endothelial cells plated on colloidal gold. (a) Dulbecco's modified Eagle's medium (DMEM) with 10% calf serum, or (b) DMEM + 10% calf serum 1 : 1 with tumor-conditioned medium from sarcoma-180 cells. Reprinted from *Nature* **285**, 41–43, 1980 with permission of the publisher.

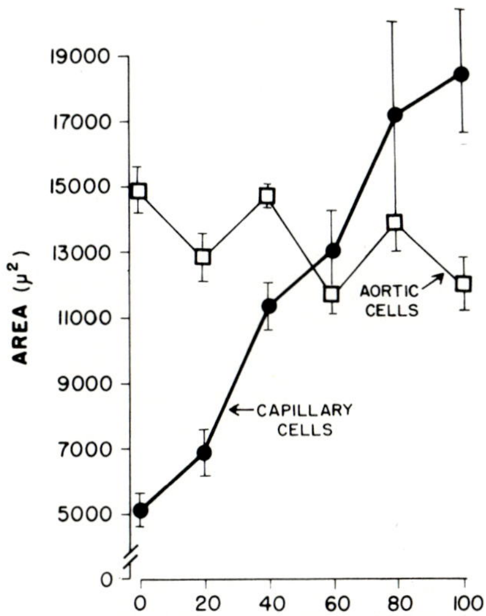

Fig. 7. Migration of bovine capillary endothelial cells (closed circle) compared to bovine aortic endothelial cells (open square) in increasing concentrations of tumor-conditioned medium. Reprinted from *Nature* **285,** 41–43, 1980 with permission of the publisher.

analyzer to measure the area of tracks projected from an inverted microscope to a television screen, it was found that the mean area of the tracks was directly proportional to the concentration of tumor-conditioned medium (Fig. 7). Furthermore, aortic endothelial cells moved rapidly whether or not tumor-conditioned medium was present. Medium that was conditioned by smooth muscle cells or aortic endothelial cells was completely inactive. It did not stimulate capillary endothelial migration. Medium conditioned by the capillary endothelial cells themselves induced a 20% increase in the size of the phagokinetic track compared to a 90% increase with the sarcoma cell-conditioned medium.

THE EFFECT OF MAST CELLS ON MIGRATION OF CAPILLARY ENDOTHELIAL CELLS

Ever since Ehrlich first described mast cells, they have been found to be associated with neovascularization. However, their role, if any, in neovascularization was unknown. Kessler *et al.* (31) reported that an early response to a tumor challenge was the appearance of mast cells at the tumor site. The number of mast cells at the tumor site increased 40-fold compared to normal tissue levels of mast cells prior to the tumor implantation. This increase occurred within 24 hr after tumor inoculation and the appearance of mast cells preceded the appearance of new blood vessels. Furthermore, it was found that mast cells could not initiate angiogenesis themselves. The finding suggested that mast cells may play an important role in the development of new capillaries although not in the initiation of angiogenesis.

Subsequently, we found that medium conditioned by mast cells caused a maximal stimulation of migration of capillary endothelial cells (32). The stimulation of capillary endothelial cell migration by mast cell-conditioned medium was dose dependent (Fig. 8). However, proliferation of capillary endothelial cells was not. A variety of mast cell products was tested on capillary endothelial migration. These included histamine; heparin; eosinophilic chemotactic factor-A; chondroitin sulfate (mixed isomeres); 5-hydroxytryptamine; chondroitin sulfates A, B, and C; trypsin; and chymotrypsin. These compounds were tested in log increments over a concentration of 10 ng/ml to 1 mg/ml. Only heparin significantly stimulated capillary endothelial cell migration, and only heparin could completely substitute for mast cell-conditioned medium (Fig. 9). Furthermore, specific inhibitors of

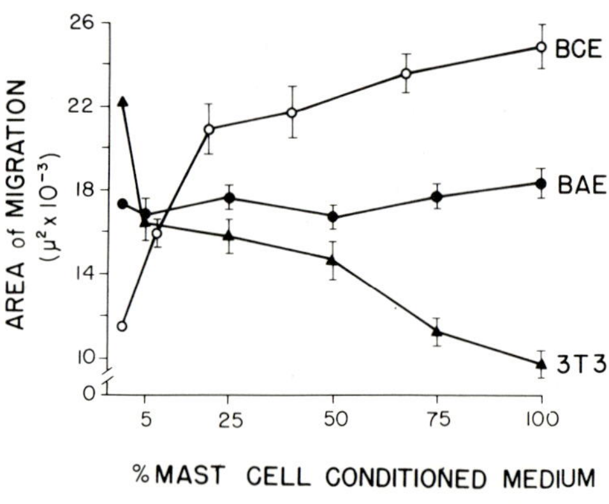

Fig. 8. The effect of mast cell conditioned medium on the migration of bovine capillary endothelial cells (BCE), bovine aortic endothelial cells (BAE), and Swiss 3T3 cells. Increasing concentrations of mast cell conditioned medium from 0–100% were incubated with BCE, BAE, and 3T3 cells. Each point on the graph represents the mean area of 100 phagokinetic tracks ± SEM. Reprinted from the *Journal of Experimental Medicine* **152**, 931, 1980 with permission of the publisher.

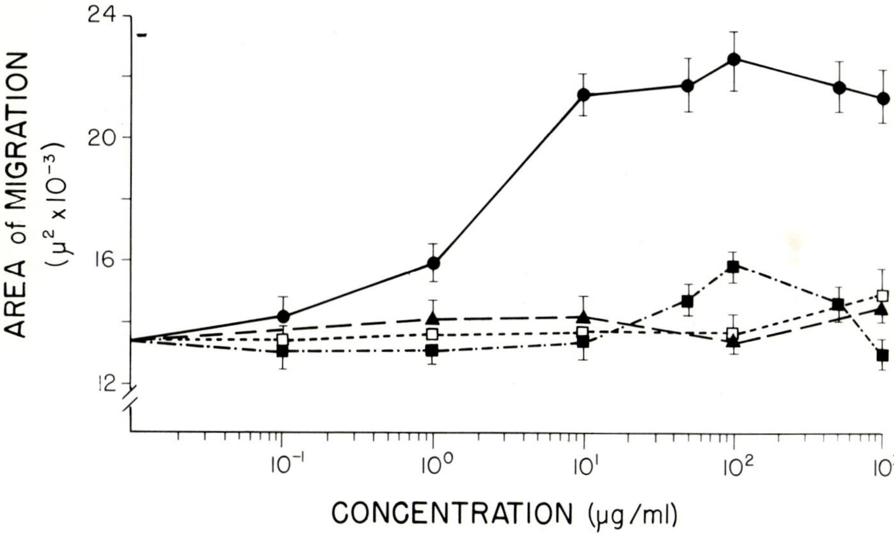

Fig. 9. The effect of isolated mast cell products on migration of bovine capillary endothelial cells. The mast cell products were solubilized in DMEM 10% calf serum and tested at log-dose intervals from 10 ngm to 1 mg/ml. The mast cell products tested included heparin (closed circle), chondroitin sulfate A, B, C (closed square), ECF-A (open square) and histamine phosphate (closed triangle). Each point represents the mean area of 100 phagokinetic tracks ± SEM. Reprinted from the *Journal of Experimental Medicine* **15,** 931, 1980 with permission of the publisher.

heparin such as protamine and heparinase completely blocked the stimulatory effect of heparin on capillary endothelial cells.

Tumor-conditioned medium stimulates proliferation as well as migration, but mast cell-conditioned medium stimulates migration only. Thus endothelial cell migration and proliferation can be uncoupled *in vitro*. These results cannot predict whether separate factors will regulate migration and proliferation *in vivo*. However, they suggest that these two events are controlled independently. Thus the migration component of angiogenesis may be modulated by the action of mast cell products. One possibility is that mast cell heparin could amplify the *rate of capillary growth* once new capillaries had been induced to grow toward an angiogenesis stimulus such as a tumor focus. Thus a present working hypothesis is that mast cells may be attracted to an angiogenesis source such as a tumor by a separate diffusable signal and that they in turn may facilitate the directional growth of the capillary tubes through tissue.

SUMMARY

The ability to clone capillary endothelial cells and to carry them in a prolonged culture has allowed the mechanism of capillary growth to be studied *in vitro*. It is now becoming clear that the growth of a capillary involves an ordered sequence of events that include lysis of the basement membrane of the parent venule, migration of capillary endothelial cells, lumen formation, branch formation, and anastomosis of the tip of one tube with another to form a loop. In addition, mast cells appear to act in a helper role for endothelial cells, in some way speeding up their rate of directional migration.

This new field is still in an early phase of development. A major problem is the tedious work required to establish capillary endothelial cells, although once established in culture they are easily subcultured. The development of more expeditious methods for isolating capillary endothelial cells from any tissue should enhance progress in the field of angiogenesis.

REFERENCES

1. Folkman, J., Merler, E., Abernathy, C., and Williams, G. (1971) *J. Exp. Med.* **133**, 275–288.
2. Gimbrone, M. A., Jr., Leapman, S. B., Cotran, R. S., and Folkman, J. (1972) *J. Exp. Med.* **136**, 261.
3. Folkman, J., and Cotran, R. S. (1976) *Int. Rev. Exp. Pathol.* **16**, 207.
4. Folkman, J. (1972) *Ann. Surg.* **175**, 409–416.
5. Patz, A. (1978) *N. Engl. J. Med.* **298**, 1451–1454.
6. Folkman, J. (1972) *J. Invest. Dermatol.* **59**, 40–43.
7. Folkman, J., Ausprunk, D., and Langer, R. (1980). *In* "Textbook of Rheumatology" (W. M. Kelley, E. D. Harris, S. Ruddy, and C. B. Sledge, eds.), Saunders, Philadelphia, Pennsylvania.
8. Wood, S. (1958) *Arch. Pathol.* **66**, 550–568.
9. Gimbrone, M. A., Jr., Cotran, R. S., Leapman, S. B., and Folkman, J. (1972) *JNCI, J. Nat. Cancer Inst.* **52**, 413–427.
10. Lewis, W. H. (1931) *Bull. Johns Hopkins Hosp.* **48**, 242–253.
11. Woodard, W. C., and Pomerat, C. M. (1953) *Anat. Rec.* **117**, 663.
12. Jaffe, E. A., Hoyer, L. W., and Nachman, R. L. (1973) *J. Clin. Invest.* **52**, 2757–2764.
13. Gimbrone, M. A., Jr., Cotran, R. S., and Folkman, J. (1974). *J. Cell Biol.* **60**, 673–684.
14. Buonasissi, V. (1973) *Exp. Cell Res.* **76**, 363–368.
15. Tripathi, B., Knight, G., and Ashton, N. (1974) *Exp. Eye Res.* **19**, 449–475.
16. Karasek, M., and Charlton, M. (1974) *J. Invest. Dermatol.* **62**, 542.
17. Wagner, R. C., and Matthews, M. A. (1975) *Microvasc. Res.* **10**, 286–297.
18. Del Vecchio, P. J., Ryan, U. S., and Ryan, J. W. (1977) *J. Cell Biol.* **75**, 73a (abstr.).
19. Folkman, J., Haudenschild, C. C., and Zetter, B. R. (1979) *Proc. Nat. Acad. Sci. U.S.A.* **76**, 5217–5221.

20. Folkman, J., and Haudenschild, C. C. (1980) *Nature* (*London*) **288**, 551–556.
21. Zetter, B. R. (1980) *Nature* (*London*) **285**, 41–43.
22. Davison, P., and Karasek, M. (1978) *Clin. Res.* **26**, 208A.
23. Maciag, T. (1981) *Cold Spring Harbor Symp. Cell Proliferation* (in press).
24. Nicosia, R. F., and Leighton, J. (1981) *In Vitro* **17**, 204 (abstr.).
25. Zetter, B. R., Azizkhan, R. G., Azizkhan, J. C., Brouty-Boye, D., Folkman, J., Haudenschild, C. C., Klagsbrun, M., Potash, R., and Scheiner, C. J. (1981). *In* "Plasma and Cellular Modulatory Proteins" (D. H. Bing and R. A. Rosenbaum, eds.), pp. 59–73. Center for Blood Research, Inc., Boston, Massachusetts.
26. Klagsbrun, M., and Smith, S. (1980) *J. Biol. Chem.* **255**, 10859–10866.
27. Gospodarowicz, D. (1974) *Nature* (*London*) **249**, 123–127.
28. Klagsbrun, M. (1978) *Proc. Natl. Acad. Sci. U.S.A.* **75**, 5057–5061.
29. Ausprunk, D., and Folkman, J. (1977) *Microvasc. Res.* **14**, 53–65.
30. Albrecht-Buehler, G. (1977) *Cell* **11**, 395–404.
31. Kessler, D. A., Langer, R. S., Pless, N. A., and Folkman, J. (1976) *Int. J. Cancer* **18**, 703.
32. Azizkhan, R. G., Azizkhan, J. C., Zetter, B. R., and Folkman, J. (1980) *J. Exp. Med.* **152**, 931–944.

PART II

COAGULATION PROTEASES

Interactions of Thrombin with Endothelium

P. CHRISTER BUSCH* AND WHYTE G. OWEN
The Cardiovascular Center
and the Departments of Pathology and Biochemistry
University of Iowa
Iowa City, Iowa

The hemostasis system, and coagulation in particular, presents a unique and difficult regulatory dilemma. On one hand, the system must be poised, with a sensitive trigger, to respond to every breach in the vasculature. On the other hand, the growth of the hemostatic plug must be limited strictly to the site of hemostasis, and, in addition, any procoagulant that escapes into the systemic circulation must be managed. This becomes problematic when the clotting system is regarded as a cascading sequence of irreversible reactions that, if left uncontrolled, would proceed to completion at the first initiation. Shed blood indeed clots rapidly with virtually complete consumption of several of its clotting factors, whereas *in vivo*, even marginal consumption of fibrinogen requires a drastic stimulus. Thus we have looked to the primary difference between shed and circulating blood, the circulatory system, for a major regulatory role in hemostasis.

Because circulating blood, except for a fraction perfusing the liver, is in contact almost exclusively with vascular endothelium, we turned to cultured endothelium as a model of the vessel lumen. We and others have observed several interactions of thrombin with endothelium *in vitro* (Table I).

Although the biological significance of none of the reactions is known, thrombin has been useful as a reagent for investigating the cells. Furthermore, the first reaction in Table I has directed our attention to two regulatory roles of endothelium *in vivo:* the activation of

* Present address: Department of Pathology, University of Uppsala, Uppsala, Sweden.

PATHOBIOLOGY OF THE ENDOTHELIAL CELL

ISBN 0-12-521980-6

TABLE I
Reactions of Thrombin with Endothelium

Primary interaction	Consequence	Reference
Active site independent[a] binding to cultured endothelium (K_D = 3 nM)	Unknown	1,2
Unknown (active site dependent)	Release of arachidonate (prostacyclin), nucleotides	3–6
Unknown (active site dependent)	Mitogenesis	7–10
Unknown (active site dependent)	Inhibition of plasminogen activator	11
Active site dependent binding of thrombin *in vitro*	Catabolism of thrombin	2,12,13

[a] Active site dependency denotes the ability of the cells to distinguish inactive, diisopropylphosphothrombin from active thrombin.

protein C (14,15) (to be discussed elsewhere in this volume) and the inactivation of circulating thrombin.

When high-affinity binding of thrombin to endothelium *in vivo* was regarded in light of the law of mass action and the geometry of cylinders, it became clear that in capillaries the concentration of binding sites should exceed the dissociation constant, thus driving equilibrium toward the bound state. By studying the pharmacokinetics of both inactive and active thrombin (16) we found (a) that most of the thrombin injected into the circulation interacts with endothelium whenever it reaches a capillary, (b) that thrombin reacts with antithrombin III much faster *in vivo* than *in vitro,* and (c) that inhibition of the interaction of thrombin with endothelium *in vivo* inhibits the rate of inhibition by antithrombin III. From these findings we proposed that a cell surface factor in the microcirculation catalyzes the inactivation of thrombin as diagrammed in Fig. 1. Thus this factor has the specifity ascribed to heparin: that of an antithrombin III cofactor.

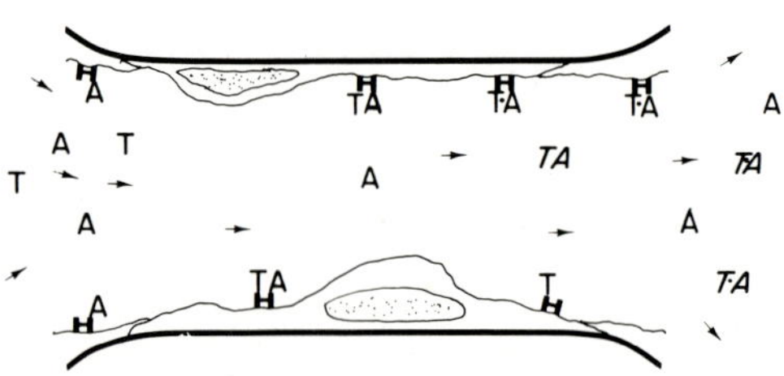

Fig. 1. Schematic representation of the inactivation of thrombin (T) by antithrombin III (A) in the microcirculation; the reaction is catalyzed by a vessel wall cofactor (H).

The study of thrombin inhibition *in vivo* is cumbersome, consumes large quantities of reagents, and, because the reaction cannot be interrupted readily, does not lend itself to detailed functional analysis. As an initial approach to this problem, we have used the Langendorff heart preparation (17) in which the myocardial capillary bed serves as a large endothelial surface.

Perfusion of hearts with the thrombin–antithrombin III mixture resulted in about 50% inhibition of the thrombin during the transit through the heart; individual hearts promoted a constant degree of inhibition that could be maintained for more than 1 hr of perfusion, whereas inhibitions among hearts varied between 40 and 60%. Addition of a 50-fold excess of diisopropylphosphothrombin reduced the capacity of the heart to promote inhibition.

A second approach was provided by endothelium cultured on microcarrier beads. Columns packed with cell-covered beads were used in place of the isolated heart and promoted similar inhibition of thrombin.

Thus in a functional sense the cultured cells and the isolated heart both can function in a manner analogous to that expected of a bed of stationary-phase heparin (or heparan-SO_4). Indeed, we have found that small (50-μl) columns of heparin-agarose, when used in similar perfusion experiments, gave comparable results. Of the glycosaminoglycans known to enhance the activity of antithrombin III, heparan sulfate has been found associated with endothelium (19) and thus is a candidate for the capillary antithrombin III cofactor. However, we have no structural data that can clarify this issue, nor do we know whether the cofactor is integral with the membrane, adsorbed to it, or associated with the intercellular matrix. With the finding that isolated organs and cultured cells can express the activity, an approach to these questions should now be feasible.

Regardless of the identity of the cofactor, its location provides a mechanism that insures a very brief lifetime for circulating thrombin and perhaps other procoagulants that escape sites of hemostasis, while not influencing the lifetime of these same enzymes in a wound, where circulation is interrupted. When the velocity of the circulation is considered together with its geometry (Fig. 2), we envisage a molecule of thrombin (or antithrombin III) experiencing a dramatic but brief "flash" of cofactor activity every 5–6 sec, the transport time from any point in the system to a capillary. In effect, circulating blood behaves as if it were heparinized, while noncirculating blood, either *in vitro* or *in vivo,* may clot without impediment.

The capillary wall antithrombin III cofactor provides another exam-

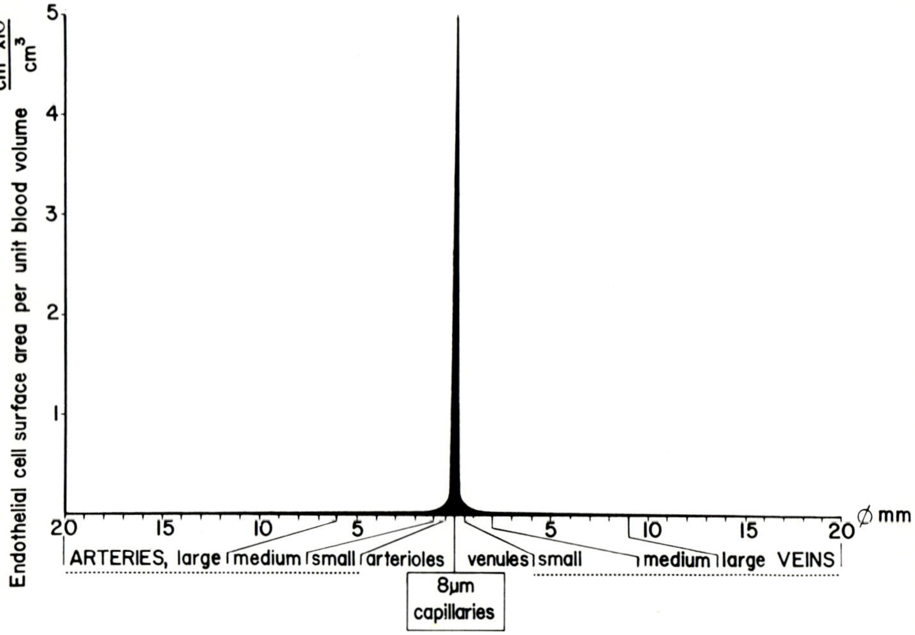

Fig. 2. Variation of cell surface "concentration" within the circulatory system.

ple of a recurring theme in hemostasis: regulation by cofactors that enhance the activities of relatively ineffective enzymes, or, in this case, an inhibitor. Whether on platelets or the vessel wall, localization of these cofactors to a stationary phase ensures that control is appropriately restricted. With the finding that the vessel wall controls a second anticoagulant regulatory system, protein C activation (14,15), plus the localization of prostacyclin biosynthesis to endothelium, the microcirculation with its vast surface area now looms as the seat of systemic control of hemostasis.

ACKNOWLEDGMENTS

Susan M. Hale and Christina Hallberg provided expert technical assistance. This work was supported by grants HL22478-03, HL14230-10 (SCOR), and Career Development Award HL00348(WGO) from the National Heart, Lung, and Blood Institute, and Grant 12x-5191-4 from the Swedish Medical Research Council.

REFERENCES

1. Awbrey, B. J., Hoak, J. C., and Owen, W. G. (1979) *J. Biol. Chem.* **254,** 4092–4095.
2. Lollar, P., and Owen, W. G. (1980) *J. Biol. Chem.* **255,** 10279–10283.
3. Weksler, B. B., Ley, C. W., and Jaffe, E. A. (1978) *J. Clin. Invest.* **62,** 923–930.
4. Lollar, P., and Owen, W. G. (1980) *J. Biol. Chem.* **255,** 8031–8034.
5. Pearson, J. D., and Gordon, J. L. (1979) *Nature (London)* **281,** 384–386.
6. Lollar, P., and Owen, W. G. (1981) *Ann. N.Y. Acad. Sci.* **370,** 51–56.
7. Glenn, D. H., and Cunningham, D. D. (1978) *Cell* **15,** 1341–1349.
8. Zetter, B. R., Chen, L. B., and Buchanan, J. M. (1976) *Cell* **7,** 407–412.
9. Blumberg, P. M., and Robbins, P. W. (1975) *Cell* **6,** 137–147.
10. Gospodarowicz, D., Brown, K. D., Birdwell, C. R., and Zetter, B. (1978) *J. Cell. Biol.* **77,** 774–788.
11. Loskutoff, D. J. (1979) *J. Clin. Invest.* **64,** 329–332.
12. Baker, J. B., Low, D. A., Simmer, R. L., and Cunningham, D. D. (1980) *Cell* **21,** 37.
13. Isaacs, J. D., Savion, N., Fenton, J. W., and Shuman, M. A. (1979) *Blood* **54,** Suppl. 1, 284.
14. Esmon, C. T., and Owen, W. G. (1981) *Proc. Natl. Acad. Sci. U.S.A.* **78,** 2249–2252.
15. Owen, W. G., and Esmon, C. T. (1981) *J. Biol. Chem.* **256,** 5532–5535.
16. Lollar, P., and Owen, W. G. (1979) *J. Clin. Invest.* **66,** 1222–1230.
17. Langendorff, O. (1897) *Pfluegers Arch. Gesamte Physiol. Menschen Tiere* **66,** 355–400.
18. DeBault, L. E., Kahn, L. E., Frommes, S. P., and Cancilla, P. A. (1979) *In Vitro* **15,** 473–487.
19. Busch, P. C. (1981) *In* "The Biology of Endothelial Cells" (E. A. Jaffe, ed.) (Review). Nijhoff, The Hague (in press).

Protein C, a Regulator of Prothrombin Activation

JOHAN STENFLO, BJÖRN DAHLBÄCK, PER FERNLUND, AND KOJI SUZUKI
Department of Clinical Chemistry
University of Lund
Malmö General Hospital
Malmö, Sweden

INTRODUCTION

In the early 1960s Seegers and co-workers found an activity in the so-called prothrombin complex that appeared to be an inhibitor of factor X_a-mediated activation of prothrombin (1). The inhibitor termed *autoprothrombin II-A* was thought to derive from the prothrombin molecule and it was found in the prothrombin complex after incubation with purified thrombin. Besides acting as an inhibitor of factor X_a, autoprothrombin II-A could induce fibrinolysis both *in vivo* and *in vitro* (2). The molecular nature of the inhibitor or the nature of its precursor was not disclosed at that time, although in 1972 Marciniak (3) provided evidence suggesting that autoprothrombin II-A was not a derivative of prothrombin, but was a separate protein.

In 1974 γ-carboxyglutamic acid was identified as the vitamin K-dependent structure required for Ca^{2+} binding and biological activity of bovine prothrombin (4–6). Soon thereafter the newly discovered amino acid was found to be present in all hitherto known vitamin K-dependent proteins, i.e., prothrombin, factor VII, factor IX, and factor X (7,8). The search for γ-carboxyglutamic acid in proteins soon provided a powerful tool in the search for previously unidentified vitamin K-dependent proteins in blood plasma and in tissues. As a result of such efforts a previously unknown γ-carboxyglutamic acid containing protein was purified from barium citrate eluates of bovine

PATHOBIOLOGY OF THE ENDOTHELIAL CELL

ISBN 0-12-521980-6

plasma (9). The protein was distinct from the classical vitamin K-dependent clotting factors, and it is now referred to as protein C. It was found to have two polypeptide chains linked by at least one disulfide bridge and an approximate molecular weight of 56,000. Although protein C was shown to be a zymogen of a serine esterase (10), no function was initially associated to the protein. Seegers and coworkers, using immunological techniques, soon demonstrated that protein C and autoprothombin II-A were related, i.e., that autoprothrombin II-A was derived from protein C by limited proteolysis (11). The activated form of protein C is now commonly referred to as protein C_a. Kisiel *et al.* (12) showed that protein C could be activated to a serine protease not only by thrombin, but also by the purified factor X activator from Russel's viper venom and that the amino acid sequence around the active site serine residue in protein C was homologous to the corresponding sequences in other serine proteases. Active protein C had anticoagulant properties and it was first shown that factor V was susceptible to proteolysis by protein C_a and later that the activity of factor VIII was inhibited by protein C_a (13–15). In this context it should be mentioned that, unlike the other serine proteases participating in blood coagulation protein, C_a is not inactivated by antithrombin III, not even in the presence of heparin. Recently, Marlar and Griffin (16) identified an inhibitor of protein C_a that is distinct from the previously described plasma protease inhibitors. This inhibitory activity was missing from the plasma of patients with a very rare bleeding disorder due to combined factor V and factor VIII deficiency. The two substrates for protein C_a are similar in that they are both so-called accessory factors in blood coagulation; i.e., they have no known enzyme activity of their own, but function in conjunction with the active enzymes and thereby enhance the rates of factor IX_a- and factor X_a-mediated activations of factor X and prothrombin, respectively (17). Factor V–V_a and factor VIII–VIII_a are the only known substrates for protein C_a. The potential importance of protein C as a regulator of the rate of prothrombin activation *in vivo* is illustrated by the contribution of factor V to the prothrombinase complex (factor X_a, factor V, Ca^{2+} ions, and phospholipid). The role of factor V in the complex is demonstrated by decreases in the rate of prothrombin activation by up to 1000-fold in its absence.

Protein C and factor X, as found in plasma, have two polypeptide chains, whereas prothrombin, factor VII, and factor IX have a single polypeptide chain (17). The heavy chains of protein C and factor X constitute the serine protease parts, as do the carboxy terminal parts of factors VII and IX and prothrombin. Figure 1 illustrates that the

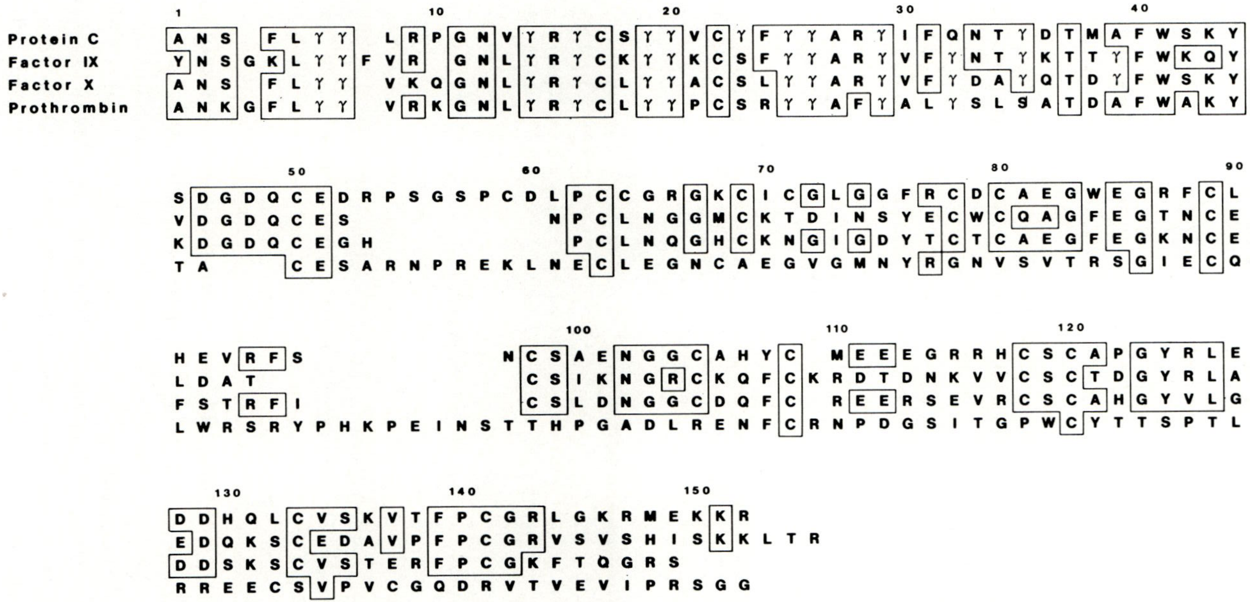

Fig. 1a. Amino acid sequences of the light chains of bovine protein C (18), factor X_1 (19), and the amino terminal parts of bovine factor IX (20) and prothrombin (6). The sequences were aligned for maximal homology. Amino acid residues identical with those that have been found in protein C are enclosed in blocks. The numbering is that of protein C.

Fig. 1b. Alignment of amino acid sequences of heavy chain of bovine protein C and the corresponding parts of factor IX (20), factor X (21), prothrombin (6) and bovine chymotrypsin A (22). The activation peptide of protein C (residues 1 to 14) is included. Otherwise as in Fig. 1a.

serine protease part of protein C has about the same degree of similarity to chymotrypsin A as factor IX, factor X, and prothrombin have. Protein C also seems to have the four intrachain disulfide bridges characteristic of the serine proteases and a half-cystine in position 98 that, in analogy to chymotrypsin A, thrombin, and factor X, links the heavy chain to the light chain. Although the gross architecture of protein C resembles that of factor X, it has a much shorter activation peptide (14 residues) than factor X (51 residues). The amino terminal part of the heavy chain of factor X_a also has one intrachain disulfide bridge not found in protein C.

Up to position 40 all glutamic acid residues in the amino terminal parts of the light chains of protein C and factor X and the amino terminal parts of factors IX and prothrombin are carboxylated to γ-carboxyglutamic acid residues (8,17). The light chain of protein C is highly homologous to that of factor X and to residue 1–44 in factor IX and prothrombin, whereas after residue 44 of prothrombin, the homology with protein C is very weak. In the amino terminal part of factor IX and the light chain of factor X all 15 half-cystine residues can be aligned. Those in position 17 and 22 are linked by intrachain disulfide bridges, whereas the disulfide pairing of the others is unknown. Which half-cystine residue links the light chain to the heavy chain is also unknown. Alignment of the light chain of protein C with the corresponding region in the other molecules reveals that 15 of the 18 half-cystine residues align with the 15 residues in factors IX and X. Protein C, however, has three additional half-cystine residues in positions 59, 64, and 71, which have no counterparts in factors IX and X. The residue in position 59 occurs in an eight-residue-long insertion. One of these three residues cannot be engaged in a disulfide bond. The functional significance of this apparently unpaired half-cystine residue is unknown.

Quite recently, Kisiel (23) isolated protein C from human plasma and showed that it is structurally and functionally similar to bovine protein C. Like its bovine counterpart it is activated to a serine protease by thrombin and after activation it has anticoagulant activity. The anticoagulant activity appeared to be quite species specific in that bovine protein C_a had no anticoagulant activity in human plasma as measured with the activated partial thromboplastin time (13).

In this communication we summarize recent investigations on human and bovine protein C and their substrate factor V.

HUMAN PROTEIN C

Human protein C was purified by a procedure derived in our laboratory, which is a modification of the procedure we used for the purification of bovine protein C (9). The starting material was either outdated plasma from the local blood bank or plasmapheresis plasma from patients with various disorders. The major steps in the purification procedure were barium citrate adsorption followed by two successive chromatographies on DEAE-Sephacel, the first one in the absence of calcium as described previously (19) and the second one in calcium ion containing Tris-buffer. The final purification was on a heparin Sepharose column. Essential to a successful purification procedure is generous use of protease inhibitors like diisopropylfluorophosphate (DFP) and benzamidine in all buffers to prevent proteolytic degradation of protein C. The overall recovery with this procedure varied from 15–25 percent. About 1 mg of purified protein C was obtained per liter plasma.

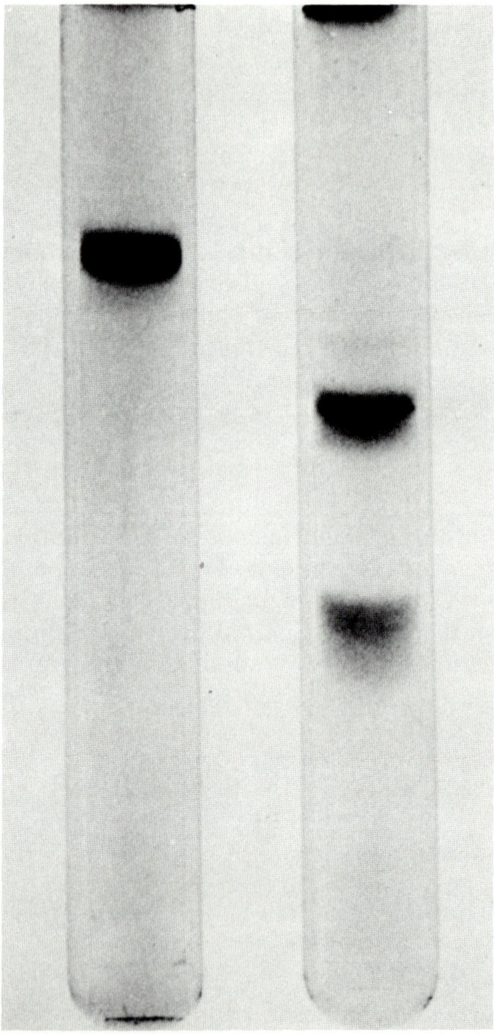

Fig. 2. Electrophoresis of purified human protein C on 10% polyacrylamide gels containing sodium dodecyl sulfate. (Left) Unreduced sample. (Right) Sample reduced with β-mercaptoethanol.

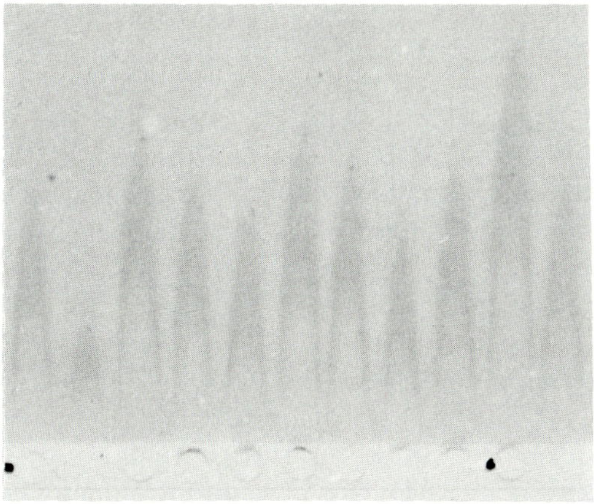

Fig. 3. Quantitation of human protein C in plasma using rabbit anti-human protein C and rocket immunoelectrophoresis. The immunoprecipitates were visualized with peroxidase-labeled swine immunoglobulins against rabbit IgG and using 3-amino-9-ethylcarbazole as a substrate. The two samples to the left are control plasmas.

Human protein C has two polypeptide chains, as does its bovine counterpart (Fig. 2). The amino acid composition agreed well with that reported by Kisiel (23). It also gave an immunoprecipitate with an antiserum raised against human protein C kindly provided by Drs. Kisiel and Davie. Protein C in human plasma was quantitated using a rabbit antiserum against human protein C and a horseradish peroxidase conjugated second antibody to visualize the immunoprecipitates (Fig. 3). The plasma concentration in healthy individuals was 3–5 mg/liter. The enzyme-enhanced rocket immunoelectrophoresis allowed quantitation of protein C down to approximately 0.3 mg/liter. Attempts to develop a radioimmunoassay against protein C were only partly successful. Although the method worked well for the quantitation of protein C in column effluents it was not sufficiently accurate for meaningful measurements in plasma.

FACTOR V

As already mentioned, factor V is a nonenzymatic component of the prothrombinase complex. It was discovered in 1943 by Owren in Norway (24). Despite several attempts, undegraded, biologically active factor V was not purified until 1979 when Esmon (25) and Nesheim *et al.* (26) independently reported its successful purification from bovine plasma. The key to success was rapid handling of the material and generous use of such inhibitors of proteolytic enzymes as DFP and benzamidine. The purified bovine factor V was shown to be a single-chain glycoprotein with a molecular weight of 330,000. Catalytic amounts of thrombin activated the factor V by limited proteolysis with an approximately 80-fold increase in biological activity. It was also shown by Walker *et al.* (27) that bovine protein C_a effectively inactivated bovine factor V_a by limited proteolysis. The rate of inactivation was greatly enhanced by the presence of negatively charged phospholipid and Ca^{2+}.

With bovine factor V purified, the prospect of purifying human factor V, thought to be even more labile than its bovine counterpart, improved considerably. Successful purifications were also reported by Dahlbäck (28), Kane and Majerus (29), and Katzmann *et al.* (30). Starting material for the purification should be freshly drawn platelet-free plasma or freshly frozen platelet-free plasma stored at −70°C prior to purification. As with bovine factor V swift operation and use of inhibitors of proteolytic enzymes are required. The purification procedure of Dahlbäck is outlined in Fig. 4. The original procedure has been modified to increase the concentration of benzamidine and DFP as suggested by Kane and Majerus (29). This has resulted in improved yield of undegraded factor V. Like bovine factor V it is a single-chain glycoprotein with an apparent molecular weight of 330,000.

Activation of human factor V by thrombin was accompanied by an approximately 30-fold increase in factor V activity. The time course of the activation as studied by SDS-polyacrylamide gel electrophoresis is shown in Fig. 5. The final activation products appear to be identical to those seen in bovine factor V (25,31). It is however, apparent that the order of bond cleavages is different. Thus the first peptide bonds cleaved gave rise to intermediate A1, which has no counterpart in bovine factor V activation, and the end product D. Conversion of A1 to B was accompanied by the formation of a weakly stained end product

Plasma
↓
Barium citrate adsorption
↓
Polyethylene glycol precipitation
↓
DEAE - Sepharose CL - 6B chromatography
↓
Ammonium sulfate precipitation
↓
Gelfiltration on Ultrogel ACA 22 or 34

Fig. 4. Outline of the fractionation scheme used for the purification of single-chain factor V from human plasma.

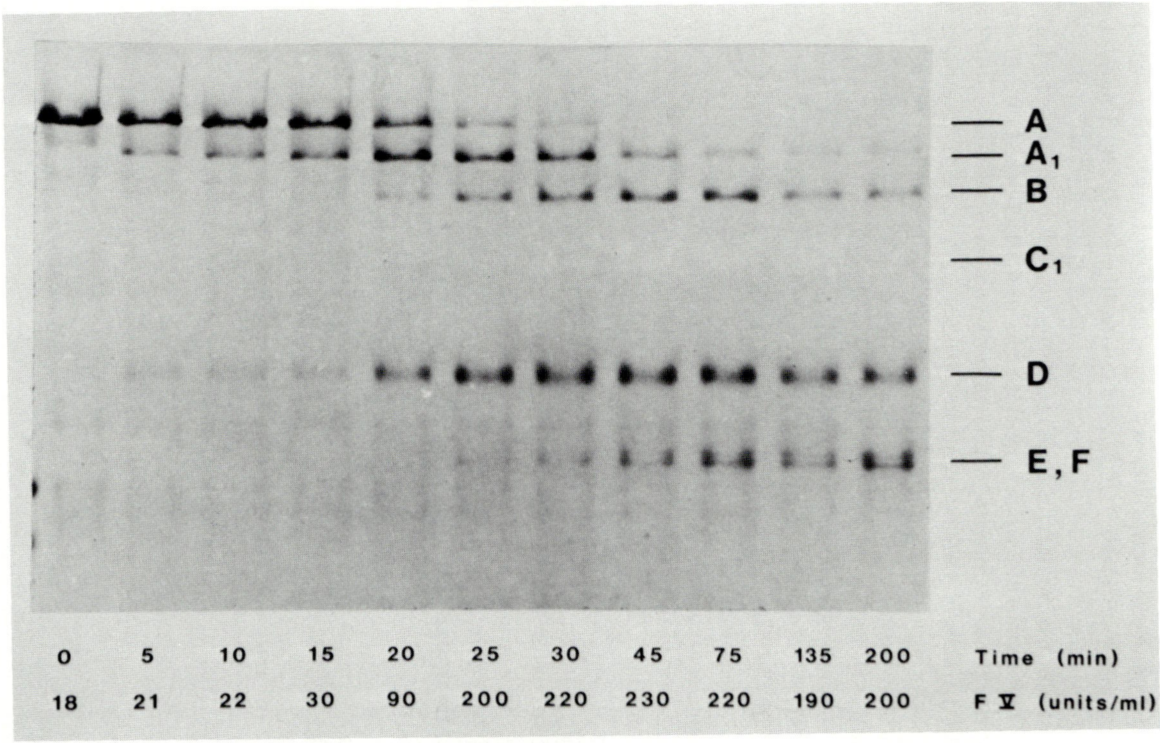

Fig. 5. Activation of single-chain human factor V by thrombin. Human factor V (0.6 mg/ml) was incubated with human thrombin (1 μg/ml) in 50% glycerol, 25 m*M* Tris-HCl, 50 m*M* NH_4Cl, pH 7.5, 5 m*M* $CaCl_2$, 5 m*M* benzamidine, pH 7.5. The reaction mixture was initially kept on an ice bath and after 15 min the temperature was shifted to 37°C. At various intervals aliquots were removed, reduced, and subjected to SDS-gradient (5–15%) polyacrylamide slab gel electrophoresis. The different components are indicated. The position of component C_1 (see text) is given.

E migrating as the doublet F1 F2. Subsequently, B was cleaved to give the two end products C1 and the doublet F1 F2 (subsequently referred to as F). Electrophoresis on agarose gel and ion exchange chromatography indicate that D and F are noncovalently linked and have factor V activity, whereas component C1 is an activation peptide as is component E, which migrates together with F on SDS gels. Component C1 stained very weakly with the Coomassie blue stain but was clearly evident on autoradiography of gels to which a trace amount of ^{125}I-labeled factor V had been added prior to activation. Comparison with the activation schemes and supporting data presented by Esmon (25) and by Nesheim and Mann (31) showed that the main difference between the thrombin-induced activation of human and bovine factor V is that in the human factor V activation, the peptide bond between fragments D and A1 is cleaved prior to the bond between E and B, whereas in the bovine factor V activation the order is reversed. The activation of human factor V by thrombin is summarized in Fig. 6. Human factor V activation with the purified factor V activator from Russel's viper venom gave rise to F concomitant to the increase in factor V activity. The remainder of the factor V molecule termed A2 could be converted to fragments C1, D, and E by thrombin without any further increase in factor V activity. Thus in this instance the increase in factor V activity was concomitant to the cleavage yielding F. This is in marked contrast to the thrombin activation in which the increase in factor V activity appeared to take place when the bond between fragments E and B was cleaved (data not shown).

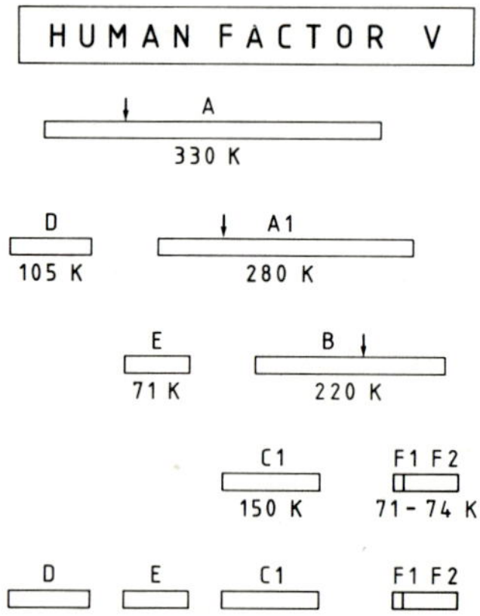

Fig. 6. Proposed schematic model for the thrombin-catalyzed activation of human factor V. The approximate molecular weights are given below the bars. The nomenclature proposed by Nesheim and Mann is adhered to where possible.

PROTEIN C_a AND FACTOR V

Activated protein C inactivates factor V_a by limited proteolysis (13,27). This effect is greatly increased by the presence of Ca^{2+} and negatively charged phospholipid. We found that both bovine and human protein C_a rapidly inactivated human factor V_a. At the concentration of reactants used there was no significant difference between protein C_a from the two species. In contrast to this, human protein C_a prolonged the caolin cephalin clotting time when added to plasma, whereas bovine protein C_a did not have this effect. This was also found by Kisiel (13). Since this assay provides an estimate of the overall efficacy of the steps in the clotting cascade prior to the activation of prothrombin, this observation suggested that there is a species difference with respect to the inactivation of factor VIII by activated protein C.

With both human protein C and human factor V available it is possible to confirm and extend previous experiments made in a bovine system (27) to the human system. We have therefore started studies on the degradation of human factor V_a by protein C_a. The rate of degradation of bovine factor V_a by protein C_a has been shown to be greatly increased by the presence of negatively charged phospholipid and

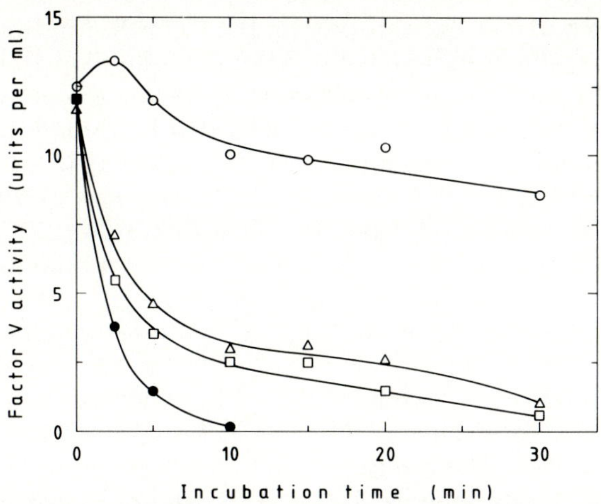

Fig. 7. Inactivation of human factor V_a by human protein C_a. Factor V_a (20 μg/ml) in 50 m*M* Tris-HCl, 0.10 *M* NaCl, 5 m*M* $CaCl_2$, pH 7.5 containing 0.1% bovine serum albumin and 0.1 mg/ml of phospholipid was incubated at 37°C. After 10 min protein C_a was added and the factor V activity measured. The protein C_a concentration was △, 5 ng/ml; □, 20 ng/ml; and ●, 100 ng/ml. ○ indicates 100 ng/ml of protein C_a without phospholipid.

Ca^{2+}. Using human factor V_a and human protein C_a a phospholipid requirement for rapid factor V_a inactivation was also observed (Fig. 7). By comparison unactivated factor V was far less susceptible to the protein C_a (data not shown).

The time course of inactivation of human factor V_a is shown in Fig. 8. The end product D is obviously rapidly degraded. At least nine degradation products can be identified on the gel. The decrease in factor V activity seems to parallel the destruction of fragment D.

PROTEIN C AND PLATELETS

Stimulated platelets have long been known to have procoagulant activity. The platelets are able to provide suitable, i.e., negatively charged, phospholipid for the activation of the vitamin K-dependent clotting factors. When the platelet release reaction is induced by, for example, ADP and/or collagen, the platelets also release clotting factors previously contained within their granules. Among these clotting factors are factor V and fibrinogen.

An important step toward a better understanding of the role of platelets in prothrombin activation was taken when Miletich *et al.* (32) showed that factor X_a bound to isolated platelets after the release reaction had taken place. The receptor for the factor X_a binding was shown to be factor V_a released from the platelet granules (33,34). Platelets can thus provide both the phospholipid and the factor V required for the prothrombinase complex. Factor V activated by thrombin to factor V_a binds to platelets, thereby forming the factor X_a receptor together with the negatively charged phospholipid (33,34). The phospholipid requirement is postulated from the fact that factor X_a lacking the γ-carboxyglutamic acid-containing region does not bind to the platelet factor V_a (35). Prothrombin devoid of γ-carboxyglutamic acid residues (acarboxyprothrombin) does not interact with the platelets and is thus very slowly activated by platelet-bound factor X_a, providing another example of the phospholipid–Ca^{2+}–γ-carboxyglutamic acid requirement for normal interaction of the vitamin K-dependent clotting factors with platelets (36). The binding of factor V_a to platelets takes place on stimulated platelets, whereas it is still a matter of controversy whether factor V_a binds to unstimulated platelets, i.e., whether unstimulated platelets have receptors for factor X_a.

With this new knowledge of the "platelet prothrombinase" at hand, it seemed meaningful to try to answer the question whether protein C had any effect on primary hemostasis, i.e., whether activated protein C

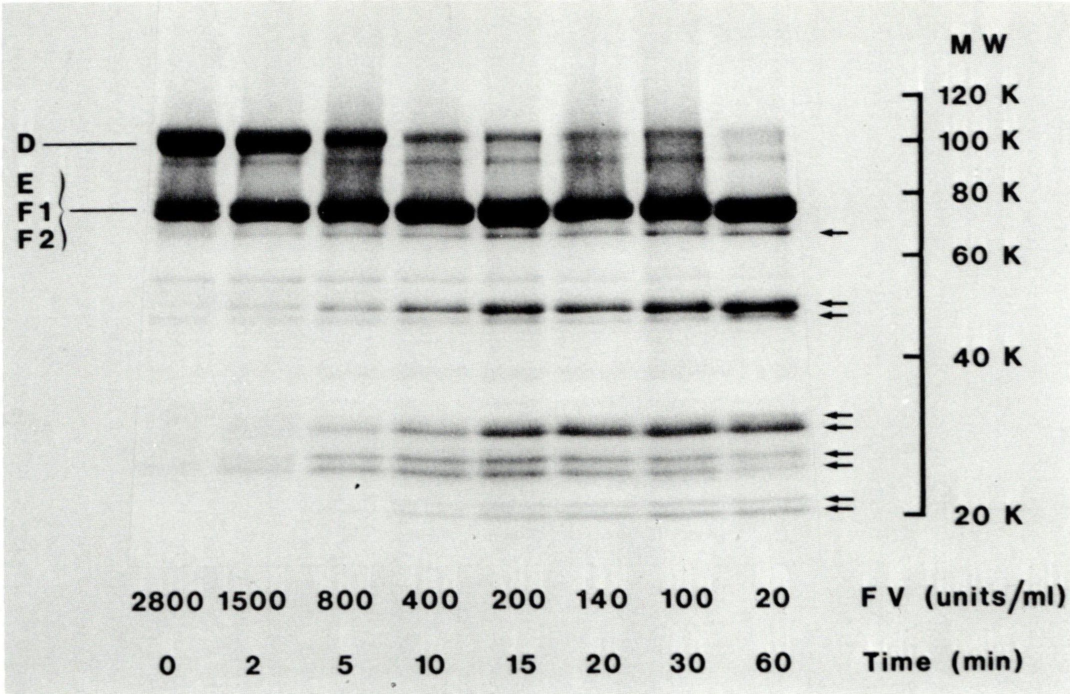

Fig. 8. Human factor V (1.25 mg/ml), in 50 m*M* Tris-HCl, 0.1 *M* NaCl, 5 m*M* $CaCl_2$, pH 7.5, containing 0.35 mg/ml of phospholipid was activated by thrombin (2 U/ml) and then DAPA (dansylarginine-*N*-(3-ethyl-1.5-pentanedinyl)amide) was added to 10 μM to inactivate the thrombin. The solution was incubated at 37°C with human protein C_a (6 μg/ml) and aliquots were drawn for SDS-polyacrylamide slab gel electrophoresis in 5–15% gradient gels. The samples were reduced with β-mercaptoethanol prior to electrophoresis.

would destroy the platelet-bound factor V_a which functions as the factor X_a binding site and thereby acts as a regulator of the local prothrombin activation. Such an inhibitory effect was indeed observed when bovine protein C_a was included in a platelet factor X_a–prothrombin reaction mixture, although relatively high concentrations of protein C_a (1–10 μg/ml) were necessary to obtain significant inhibition (37). The low efficiency of the inactivation of factor V_a by protein C_a alone could be due to a missing cofactor. One candidate for such a cofactor is protein S, as suggested by Walker (38). Protein S is a recently discovered vitamin K-dependent protein (39–41) found in both bovine and human plasma.

We found that the inhibitory effect of protein C_a was due to proteolytic destruction of the factor X_a platelet receptors. Thus, when thrombin-stimulated platelets were preincubated for various periods with protein C_a before the addition of factor X_a and prothrombin, a time-dependent parallel decrease in factor X_a binding and rate of prothrombin activation was found (Fig. 9). This effect was only observed when platelet mixtures were used in which the release reaction had been induced to make factor V_a available before the addition of protein C_a. The zymogen, protein C, or protein C_a inactivated with DFP had no inhibitory effect on the platelet system regardless of whether the release reaction had been induced or not, and we could not demonstrate direct binding to the platelet surface of either protein C_a or protein C. Similar results have been reported independently by Comp and Esmon (42). Available data thus indicate that protein C_a exerts its inhibitory effect by proteolytic degradation of the factor V_a part of the factor X_a receptor on platelets.

Thrombin is probably the activator of protein C *in vivo*. However, *in vitro* this reaction is not particularly effective, and according to Kisiel *et al.* (13) protein C isolated from plasma and serum are indistinguishable and protein C is not activated during coagulation of plasma. The conditions during blood coagulation *in vivo* are different from those in the test tube and the results of their experiments do not preclude a local protein C activation *in vivo*. We observed activation of protein C presumably by thrombin formed during prothrombin activation in a platelet–factor X_a–prothrombin reaction mixture (37). The rate of protein C activation was, however, relatively low, which suggested that a cofactor was missing. The presence of a cofactor for thrombin-mediated activation of protein C has now been elegantly demonstrated by Esmon and Owen (43). They have shown that the cofactor is bound to the surface of endothelial cells and that it causes a dramatic increase

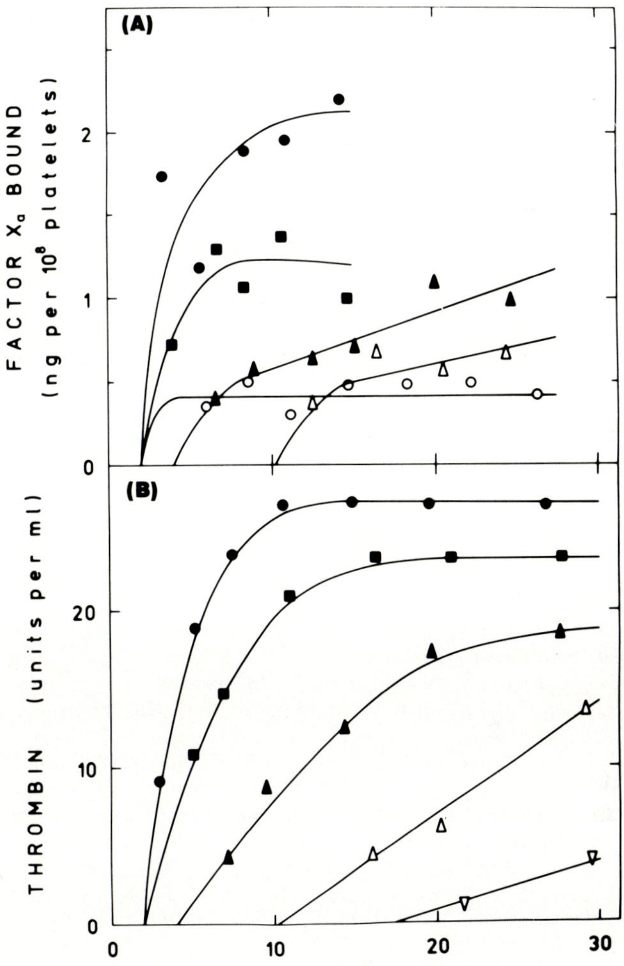

Fig. 9. Effect of protein C_a on the platelet factor X_a receptors and the activation of prothrombin. A platelet suspension (10^8 per milliliter) was treated with 1 unit thrombin per milliliter and then incubated with 1 μg protein C_a/ml at 37°C. At intervals prothrombin (0.1 mg/ml) and labeled factor X_a (20 ng/ml) were added to aliquots of the incubation and factor X_a binding (A) and thrombin formation (B) were measured. (●) Incubation without protein C_a: (■) protein C_a. ^{125}I-labeled factor X_a and prothrombin added simultaneously: (▲) 2-min preincubations: (△) 8-min preincubation: (▽) 15-min preincubation: (○) nonspecific binding of factor X_a.

in the rates of activation of protein C (Esmon and Owen, this volume, and ref. 43).

Evidence accumulated over the last couple of years is all compatible with and has indeed strengthened the assumption that protein C is an important regulator of blood coagulation *in vivo*.

ACKNOWLEDGMENTS

These investigations were supported by grants from the Swedish Medical Research Council (project no. B81-03X-04487-07B), Segerfalks Stiftelse, Greta och Johan Kocks Stiftelse, and Albert Påhlssons Stiftelse. Koji Suzuki has a Visiting Scientist Fellowship from the Swedish Medical Research Council.

REFERENCES

1. Mammen, F. F., Thomas, W. R., and Seegers, W. H. (1960) *Thromb. Diath. Haemorrh.* **5**, 218–250.
2. Seegers, W. H., McLory, L. E., Groben, H. D., Sakuragawa, N., and Agrawal, B. B. L. (1972) *Thromb. Res.* **1**, 443–460.
3. Marciniak, E. (1972) *J. Lab. Clin. Med.* **79**, 921–934.
4. Stenflo, J., Fernlund, P., Eagan, W., and Roepstorff, O. (1974) *Proc. Natl. Acad. Sci. U.S.A.* **71**, 2730–2733.
5. Nelsestuen, G. L., Zytkovic, T., and Howard, J. B. (1974) *J. Biol. Chem.* **249**, 6347–6350.
6. Magnusson, S., Petersen, T. T., Sottrup-Jensen, L., and Claegs, H. (1975) *In* "Proteases and Biological Control" (E. Reich, D. B. Rifkin, and E. Shaw, eds.), pp. 123–149. Cold Spring Harbor Press, Cold Spring Harbor, New York.
7. Stenflo, J., and Suttie, J. W. (1977) *Annu. Rev. Biochem.* **46**, 157–172.
8. Davie, E. W., Fujikawa, K., Kurachi, K., and Kisiel, W. (1979) *Adv. Enzymol.* **48**, 277–318.
9. Stenflo, J. (1976) *J. Biol. Chem.* **251**, 355–363.
10. Esmon, C. T., Stenflo, J., Suttie, J. W., and Jackson, C. M. (1976) *J. Biol. Chem.* **251**, 3052–3056.
11. Seegers, W. H., Novoa, E., Henry, R. L., and Hasouna, H. I. (1976) *Thromb. Res.* **8**, 543–552.
12. Kisiel, W., Ericsson, L. H., and Davie, E. W. (1976) *Biochemistry* **15**, 4893–4900.
13. Kisiel, W., Canfield, W. M., Ericsson, L. H., and Davie, E. W. (1977) *Biochemistry* **16**, 5824–5831.
14. Vehar, G., and Davie, E. W. (1980) *Biochemistry* **19**, 401–410.
15. Marlar, R. A., Kleiss, A. J., and Griffin, J. H. (1980) *Ann. N.Y. Acad. Sci.* 303–310.
16. Marlar, R. A., and Griffin, J. H. (1980) *J. Clin. Invest.* **66**, 1186–1189.
17. Jackson, C. M., and Nemerson, Y. (1980) *Annu. Rev. Biochem.* **49**, 765–811.
18. Fernlund, P., Stenflo, J., and Tufvesson, A. (1978) *Proc. Natl. Acad. Sci. U.S.A.* **75**, 5889–5892.

19. Enfield, D. L., Ericsson, L. H., Fujikawa, K., Walsh, K. A., Neurath, H., and Titani, K. (1980) *Biochemistry* **19,** 659–667.
20. Katayama, K., Ericsson, L. H., Enfield, D. L., Walsh, K. A., Neurath, H., Davie, E. W., and Titani, K. (1979) *Proc. Natl. Acad. Sci. U.S.A.* **76,** 4990–4994.
21. Titani, K., Fujikawa, K., Enfield, D. L., Ericsson, L. H., Walsh, K. A., and Neurath, H. (1975) *Biochemistry* **72,** 3082–3086.
22. Woodbury, R. G., Katunuma, N., Kobayashi, K., Titani, K., and Neurath, H. (1978) *Biochemistry* **17,** 811–819.
23. Kisiel, W. (1979) *J. Clin. Invest.* **64,** 761–769.
24. Owren, P. A. (1947) *Acta Med. Scand., Suppl.* **194.**
25. Esmon, C. T. (1979) *J. Biol. Chem.* **254,** 964–973.
26. Nesheim, M. E., Myrmel, K. H., Hibbard, L., and Mann, K. G. (1979) *J. Biol. Chem.* **254,** 508–517.
27. Walker, F. J., Sexton, P. W., and Esmon, C. T. (1979) *Biochim. Biophys. Acta* **571,** 333–342.
28. Dahlbäck, B. (1980) *J. Clin. Invest.* **66,** 583–591.
29. Kane, W. H., and Majerus, P. W. (1981) *J. Biol. Chem.* **256,** 1002–1007.
30. Katzmann, J. A., Nesheim, M. E., Hibbard, L. S., and Mann, K. G. (1981) *Proc. Natl. Acad. Sci. U.S.A.* **78,** 162–166.
31. Nesheim, M. E., and Mann, K. G. (1979) *J. Biol. Chem.* **254,** 1326–1334.
32. Miletich, J. P., Jackson, C. M., and Majerus, P. W. (1977) *Proc. Natl. Acad. Sci. U.S.A.* **74,** 4033–4036.
33. Miletich, J. P., Jackson, C. M., and Majerus, P. W. (1978) *J. Biol. Chem.* **253,** 6908–6916.
34. Tracy, P. B., Peterson, J. M., Nesheim, M. E., McDuffie, F. C., and Mann, K. G. (1979) *J. Biol. Chem.* **254,** 10354–10361.
35. Morita, T., Kane, W. H., Majerus, P. W., and Jackson, C. M. (1980) *In* "The Regulation of Coagulation" (K. G. Mann and F. B. Tayler, eds.), pp. 187–188. Elsevier/North-Holland, New York.
36. Dahlbäck, B., and Stenflo, J. (1980) *Eur. J. Biochem.* **104,** 549–557.
37. Dahlbäck, B., and Stenflo, J. (1980) *Eur. J. Biochem.* **107,** 331–335.
38. Walker, F. J. (1980) *J. Biol. Chem.* **255,** 5521–5524.
39. DiScipio, R. G., Hermodson, M. A., Yates, S. G., and Davie, E. W. (1977) *Biochemistry* **16,** 698–706.
40. DiScipio, R. G., and Davie, E. W. (1979) *Biochemistry* **18,** 899–904.
41. Stenflo, J., and Jönsson, M. (1979) *FEBS Lett.* **101,** 377–381.
42. Comp, P. C., and Esmon, C. T. (1979) *Blood* **54,** 1272–1281.
43. Esmon, C. T., and Owen, W. G. (1981) *Proc. Natl. Acad. Sci. U.S.A.* **78,** 2249–2252.

Activation of Protein C by a Complex between Thrombin and Endothelial Cell Surface Protein

CHARLES T. ESMON, NAOMI L. ESMON, AND JULIE SAUGSTAD
Department of Pathology
University of Oklahoma Health Sciences Center
Oklahoma City, Oklahoma

WHYTE G. OWEN
The Cardiovascular Center
and the Departments of Pathology and Biochemistry
University of Iowa
Iowa City, Iowa

INTRODUCTION

Control mechanisms involved in localizing clot formation remain obscure. Although several positive and negative regulatory mechanisms have been proposed to exist in the blood (1,2), it is clear that these are insufficient by themselves to limit clot formation, since blood collected *in vitro* will clot to quantitatively convert fibrinogen to fibrin. However, potent anticoagulant activity can be generated from proteins in the blood. As early as 1960, Seegers and his co-workers (3) demonstrated that an anticoagulant could be generated from partially purified prothrombin preparations. The anticoagulant has now been shown to be identical to the activated form of the vitamin K-dependent (4) plasma zymogen (5–8), activated protein C (9).

Activation of protein C by thrombin involves the cleavage of an Arg-Ile bond in the heavy chain of the two-chain zymogen to release a tetradecapeptide (6). The resultant enzyme maintains the two-chain structure with the active site residing in the heavy chain (5) and the Ca^{2+} binding site residing in the light chain (4). Although activated protein C is a potent anticoagulant, little protein C appears to be activated during clot formation (6). Furthermore, removing protein C from

PATHOBIOLOGY OF THE ENDOTHELIAL CELL

ISBN 0-12-521980-6

plasma has little effect on the clotting time (4). Thus, even though activated protein C is a potential anticoagulant, the rate of formation of the activated form in blood or plasma appears inappropriately slow to play a primary role in regulation. Thus the physiological role of protein C has remained questionable.

Recently, many observations have suggested that the endothelial cell plays an active role in the control of clot formation. For instance, endothelial cells can produce prostacyclin in response to a variety of stimuli, including thrombin (10,11). Prostacyclin can function as a potent antagonist of platelet aggregation (11). The endothelium has recently been shown by Lollar and Owen (12) to accelerate thrombin inactivation by antithrombin III. Furthermore, the endothelial cell has a large capacity to bind thrombin (13–15). The concept that the endothelial cell lining might directly function as a major site of the negative regulation of the blood clotting process and the observation that thrombin binding sites are present on the endothelial cell surface prompted us to examine the possibility that complexes of thrombin with a specific cell surface receptor might enhance the ability of thrombin to activate protein C. The existence of such a receptor has now been demonstrated both by perfusion studies (16) and by examination of thrombin-dependent activation of protein C on cultured endothelial cells (17). These studies, which will be reviewed and expanded upon in the results section, demonstrate that the endothelial cell receptor binds thrombin very tightly and enhances the rate of protein C activation several thousandfold. The existence of a receptor for thrombin with the previously stated properties has led us to evaluate both the physiological and pharmacological functions of protein C and activated protein C.

Studies on the mechanism by which activated protein C functions suggest that the anticoagulant activity is dependent on the inactivation of factor V (7,18,19) and factor VIII (19,20). The activated forms of both of these proteins are inactivated much more rapidly than the precursor forms (18,20,21).

Since prothrombin activation can occur on the platelet surface, it was important to investigate the ability of activated protein C to inhibit platelet participation in the process. Factor Va is an essential part of the factor Xa receptor on the platelet (22–25). Recent studies (26,27) have shown that platelet stimulation of factor Xa-dependent prothrombin activation is inhibited in a time-dependent fashion by activated protein C. The ability of the platelet to bind factor Xa decreases in parallel with the loss of prothrombin activation potential. Thus activated protein C is capable of inactivating both plasma factor Va and the platelet factor Xa receptor.

The process of factor Va inactivation is itself regulated. Factor Xa protects factor Va from inactivation both in purified systems (18,28) and on the platelet surface (26,27). The inactivation of factor Va is accelerated by another vitamin K-dependent protein, apparently protein S (29–31), which appears to function as a cofactor in this process (31). The inactivation of factor Va is accelerated further by negatively charged phospholipid surfaces and Ca^{2+} (18,21,28). Regulation of factor Va inactivation is controlled both by the level of factor Xa and by the availability of an appropriate surface. These properties make activated protein C an extremely selective and specific regulator of the clotting process.

The influence of activated protein C is not limited to the coagulation system. Injection of activated protein C into anesthetized dogs enhances fibrinolytic activity (26,32). We have recently demonstrated that this response is dependent on the presence of the active site of the serine protease and that a 20- to 50-fold increase in the rate of whole blood clot lysis is observed following a single injection of activated protein C (33). The response observed is transient and correlates with the appearance of a lysine adsorbable plasminogen activator. The properties of this plasminogen activator are similar to those described for the vascular activator (34), suggesting that activated protein C enhances fibrinolysis by releasing plasminogen activator from the vascular endothelium.

The multiple roles and potency of activated protein C require the presence of some regulatory system to control the levels of the circulating enzyme. Marlar and Griffin (35) have recently demonstrated the presence of an inhibitor to activated protein C that appears distinct from previously described protein protease inhibitors of plasma. In this study they also characterized a group of patients lacking this inhibitor. These patients all exhibited simultaneous factor V and factor VIII deficiency. These studies suggest that protein C activation does occur under physiological conditions and that adequate control of protein C activation and inactivation may be critical to the proper regulation of the clotting process.

This paper will focus on the control and properties of the endothelial cell activator of protein C.

MATERIALS AND METHODS

Proteins used in this study were physically homogeneous and, except where specifically stated, prepared from bovine plasma. Protein C (18), activated protein C (18), and thrombin (36) were purified as described previously.

ASSAYS

Factor Xa one-stage clotting assays were performed as described previously (18). Protein C activation was monitored as either anticoagulant activity or formation of esterase activity. Hydrolysis of N-α-benzoyl arginine ethyl ester was monitored using yeast alcohol dehydrogenase to couple the production of ethanol to reduction of NAD^+, as described previously (17).

CELL CULTURES

Human umbilical cord endothelium was cultured as described previously (14).

PREPARATION OF RAT LUNGS FOR COFACTOR ISOLATION

Rat lungs were excised from freshly sacrificed male rats (100–200 g) and perfused to remove blood with 0.25 *M* sucrose, 0.02 *M* Tris-HCl, 1 m*M* benzamidine-HCl, pH 7.5 at 4°C. The lungs were finely diced and homogenized at 4°C. From 40 rats, 200 ml of homogenate was prepared and sedimented at 30,000 *g* for 40 min. The pellet containing the cofactor was resuspended and homogenized again. The pellet was harvested by centrifugation as described earlier. The cofactor was extracted from the pellet by homogenization in the Tris-sucrose buffer containing Triton X-100 (0.5%). Cofactor activity was separated from the insoluble material by centrifugation, as described earlier. Three extractions were performed on the pellet. The activity was released by Triton X-100 but not by 0.02 *M* Tris-HCl, pH 7.5, containing either 4 *M* NaCl, 3 *M* KCl, or 0.1 *M* NaCl, 0.2 m*M* EDTA.

RESULTS

The possible existence of a cofactor for thrombin-catalyzed activation of protein C was investigated using a perfused heart system (16). The immense capillary bed results in conditions where the vascular-surface-to-plasma volume ratio is high. Since the hypothesized cofactor was presumed to be of endothelial cell origin, these conditions would be anticipated to result in high effective concentrations of thrombin receptors and of the cofactor. Such conditions should favor rapid protein C activation. When the microcirculation of the heart was perfused with protein C in the presence of thrombin, activated protein C was present in the perfusate. Control samples, which were not perfused, failed to generate detectable activated protein C. More activated protein C was formed by 0.1 U/ml thrombin in a single 4-sec pass through the microcirculation than was generated after 20 min in the presence of platelets and 30 U/ml thrombin in a test tube (16,27). Thus the microcirculation increased the rate of protein C activation at least 20,000-fold. When protein C was perfused in the absence of thrombin, no activated protein C was formed. Partial characterization of the activator in the perfused heart was obtained by displacement of the activator with DIP-thrombin. If thrombin bound to the cell surface cofactor were the activator, then DIP-thrombin should increase the rate of thrombin elution from the microcirculation and result in the disappearance of the protein C activator. Protein C activator was formed by perfusing the heart with low levels of thrombin in the presence of protein C. The decay of the protein C activator was studied by continued perfusion of the heart with protein C and thrombin, protein C alone, or protein C and DIP-thrombin (Table I). The half-life of the activator in the microcirculation was estimated graphically. As predicted, the perfusion with DIP-thrombin led to the disappearance of activator activity.

The observed activation of protein C by a thrombin-dependent process in the myocardium suggests that endothelial cells are the source of the cofactor activity. Endothelial cells, smooth muscle cells, and fibroblasts of human umbilical cord origin and foreskin fibroblasts were examined for cofactor activity. Among these structural cell types, cofactor activity was unique to the endothelial cells (17).

Two features of the perfusion system suggested that we could analyze the nature of the cofactor kinetically. First, there was at least a 20,000-fold rate enhancement of protein C activation during perfusion;

TABLE I
Stability of the Vascular Protein C Activator[a]

Perfusion mixture	Rate of decay of the myocardial protein C activator $t_{1/2}$
1. Protein C + thrombin	>4 hr
2. Protein C + thrombin followed by protein C alone	30–50 min
3. Protein C + thrombin followed by protein C and DIP-thrombin	4–8 min

[a] Perfusion was performed as described (16). The rabbit myocardium was perfused at a rate of 1 ml/min with protein C (15 μg/ml), thrombin (0.04 μg/ml), or DIP-thrombin (4 μg/ml).

second, the perfusion system appeared to saturate at very low levels of thrombin (16). Both of these are required in order to detect cofactor participation in the tissue culture cells. This problem of detection arises since the surface-to-volume ratio is low in the culture plates relative to the myocardium, and, as a result, the cofactor concentration is also low. Under these circumstances, greater than 99% of the thrombin is free. Despite these problems, the high affinity of thrombin for the cofactor and the large acceleration of protein C activation resulting from cofactor thrombin interaction made kinetic analysis of this process possible. Several features of the activation of protein C on the endothelial cell surface are noteworthy. First, at constant protein C concentration, a simple hyperbolic plot is obtained when the rate of protein C activation is plotted against the thrombin concentration. Since under these conditions the added thrombin is approximately equivalent to the free thrombin, these data can be transformed into a Lineweaver–Burk plot to give an apparent dissociation constant. The observed K_d for the thrombin–cofactor interaction was 0.48 ± 0.05 n*M*. This value is significantly lower than the binding constant obtained from direct binding studies to endothelium (12,17), suggesting that the cofactor is only one of several thrombin-binding species on the endothelial cells.

Further evaluation of the interaction of thrombin with the cofactor was obtained by examining the ability of DIP-thrombin to inhibit endothelial cell-mediated protein C activation. DIP-thrombin behaved as a competitive inhibitor ($K_i = 0.56$ n*M*). The K_i for DIP-thrombin and K_d for thrombin agree well, suggesting that these values closely approximate the true affinity of thrombin for the cofactor. Protein C behaves as a classical substrate in this system, exhibiting saturation with $K_m = 0.72$ μM. This value suggests that the thrombin cofactor complex is not saturated with respect to protein C under physiological conditions.

These studies indicate that rapid protein C activation only occurs in the presence of the endothelium. They further support the concept that the cofactor is one of the first receptors to be occupied following the initial stages of thrombin formation. Although this information is useful in studying the physiological function of protein C, it provides little information about the nature of the cofactor itself. For example, divalent metals are normally essential in the activation of vitamin K-dependent proteins. However, the role of divalent metals cannot be evaluated in this system, since metals are essential to the maintenance of normal endothelial cell morphology and function.

Further characterization of the cofactor required isolation and physical characterization. Several properties of the cofactor proved useful in the design of an effective isolation procedure. Both the perfusion and tissue culture experiments demonstrated that the cofactor remained bound to the endothelial cell surface during protein C activation, suggesting that the cofactor was tightly bound to the plasma membrane. Direct binding studies indicated that there were too few cofactor molecules per cell for cultured cells to be used as a source for cofactor isolation. However, since the cofactor appears to be unique to the endothelium, it was unnecessary to start the isolation with pure endothelial cells. These considerations suggested that lung would be a good source, since this organ is highly vascular, is reasonably free of protease activity, and can be perfused readily. Preliminary studies with endothelial cells in culture suggested that the cofactor could be extracted in a soluble form with Triton X-100, and therefore Triton was employed to solubilize the lung cofactor.

As discussed previously, DIP-thrombin behaves as a competitive inhibitor of the thrombin–cofactor interaction. Since this interaction does not require the active site on thrombin, DIP-thrombin was chosen for affinity chromatography. Crude rat lung extract was applied to a column of DIP-thrombin agarose, washed extensively, and eluted with 1 *M* NaCl. The cofactor peak was pooled and reapplied to a second DIP-thrombin agarose column (Fig. 1). The details are presented in the legend. The cofactor was eluted as a single peak behind the major protein peak. The middle two-thirds of the activity peak was pooled and concentrated.

Acrylamide gels of the purified protein are shown in Fig. 2. The cofactor activity is resistant to denaturation with sodium dodecyl sulfate (37), which allows assay of cofactor activity directly from slices of the SDS gel. In this system, a major and a minor protein species are observed. Activity is coincident with both and is approximately proportional to the relative mass distribution. These species may represent monomer and SDS-stable dimer forms of the cofactor. All the activity recovered from the sliced native gel comigrated with the protein band. These experiments indicate that the cofactor is a protein with an apparent molecular weight of approximately 115,000 based on SDS gel electrophoresis. Approximately 1 mg of cofactor was obtained from 40 rats with an overall yield of 10–20%. This is undoubtedly an underestimate of the yield, since the membrane phospholipid has been substituted with detergent.

This preparative procedure has proven applicable to canine and

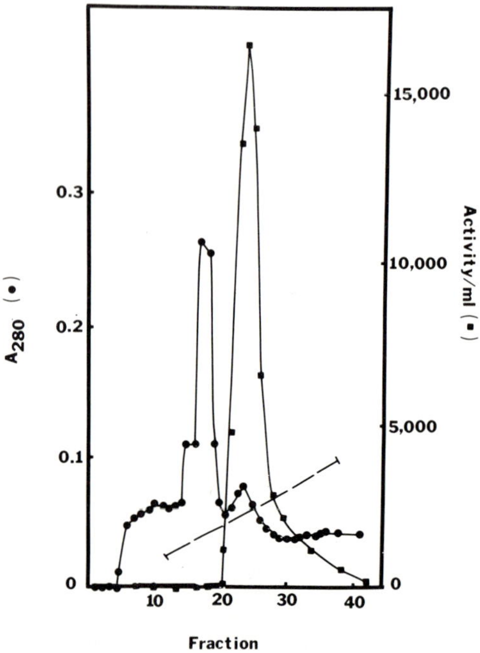

Fig. 1. Chromatography of rat lung protein C activation cofactor on DIP-thrombin agarose. Crude rat lung extract (600 ml) was applied to a DIP-thrombin agarose (1 mg DIP-thrombin/ml) column (2.5 × 8 cm). The column was washed with 0.2 *M* NaCl, 0.02 *M* Tris-HCl, 0.5 m*M* $CaCl_2$, 1 m*M* benzamidine-HCl, 0.5% Lubrol PX, pH 7.5 (2L). The cofactor was eluted with 1 *M* NaCl in the same buffer except that 0.1 m*M* EDTA was substituted for the $CaCl_2$. The activity peak was pooled and dialyzed against 0.1 *M* NaCl, 0.02 *M* Tris-HCl, 0.5 m*M* $CaCl_2$, 1 m*M* benzamidine-HCl, 0.5% Lubrol PX, pH 7.5. This material was applied to a second DIP-thrombin column (0.9 × 30 cm) depicted above equilibrated in the same buffer used for dialysis. After sample application, the buffer was increased to 0.4 *M* NaCl and the column developed with a linear gradient from 0.4 *M* to 1 *M* NaCl (40 ml/reservoir) in 0.02 *M* Tris-HCl, 1 m*M* benzamidine-HCl, 0.1 m*M* EDTA, 0.5% Lubrol PX, pH 7.5.

rabbit lung cofactor (37) as well as rat. The molecular weights of the different species are summarized in Table II.

Isolation of the cofactor allows direct examination of the nature of the cofactor–thrombin interaction. The purified cofactor has an obligatory dependence on Ca^{2+}. The Ca^{2+} optimum (2 m*M*) is similar to that for activation of other vitamin K-dependent zymogens (39). The system is saturable with respect to either cofactor or thrombin. The kinetic properties and requirements are summarized in Table III.

The isolation of the cofactor allows the reexamination of the influence of protein C on standard plasma clotting assays. If the proposed

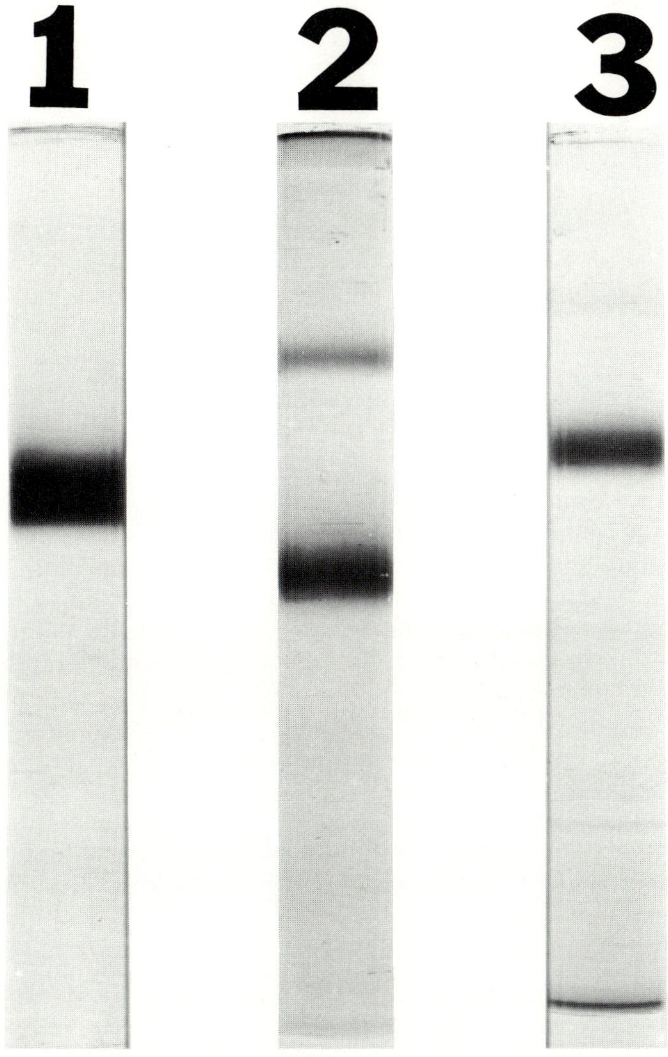

Fig. 2. Acrylamide gel electrophoresis of purified rat protein C activation cofactor. Approximately 20 μg of purified rat cofactor was electrophoresed on the indicated systems and stained for protein. Duplicate gels were sliced and assayed for activity as described previously (37). Lane 1: Alkaline gel system of Davis (37a) containing 0.5% Lubrol PX, 5% acrylamide. Lane 2: SDS gel system of Laemmli (38), 7.5% acrylamide. Lane 3: SDS gel system of Laemmli (38) containing β-mercaptoethanol, 7.5% acrylamide.

TABLE II
Molecular Weights of Purified Lung Protein C Activation Cofactor[a]

Species	$10^3 \times$ molecular weight	
	Unreduced	Reduced
1. Rat	84	115
2. Rabbit	74	105
3. Canine	105	122

[a] Purified protein C activation cofactors were electrophoresed on the SDS gel system of Laemmli (38) in the presence (reduced) and absence (unreduced) of β-mercaptoethanol. Both 7.5% and 10% acylamide were used.

model for protein C activation is correct, then it should be possible to demonstrate protein C-dependent inhibition of plasma clotting in the presence of cofactor. As shown in Fig. 3, as the cofactor concentration is increased, the clotting time of the factor Xa one-stage clotting assay increases. When protein C is inhibited by anti-protein C IgG, the clotting time at 1.5 μg/ml cofactor drops from 50 sec to about 28 sec. Thus most of the inhibition by the cofactor is dependent on the presence of protein C.

DISCUSSION

The recent identification of an endothelial cell cofactor for protein C activation (16) and the identification of patients with bleeding disor-

TABLE III
Comparison of the Purified Protein C Activation Cofactor and the Endothial Cell Surface Cofactor[a]

Property	Purified cofactor	Endothelial cell surface cofactor
1. Stoichiometry[a]	1:1	Unknown
2. K_d for thrombin[a]	<1 nM	0.48 nM
3. K_m for protein C[a]	4 μM	0.72 μM
4. K_d for Ca^{2+}[a]	0.2 mM	Unknown
5. K_i for DIP-thrombin[a]	≤0.5 nM	0.56 nM

[a] Values were determined kinetically.

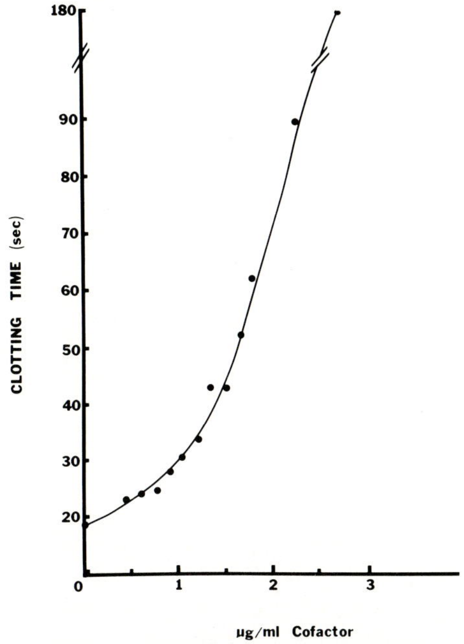

Fig. 3. Effect of protein C activation cofactor on the factor Xa one-stage clotting assay. The factor Xa one-stage was performed as described previously (18). The plasma clotting time was adjusted to 18 sec by adjusting the factor Xa concentration. Purified rabbit lung cofactor was added to give the final concentrations indicated on the figure and the clotting time determined.

ders related to the regulation of activated protein C (35) support the concept that activated protein C may play an important role in the regulation of the clotting cascade. Our studies indicate that protein C activation occurs at the interface between the blood and the endothelium. High-affinity interaction between thrombin and the endothelial cell surface cofactor ensures that the protein C activation complex will form and begin to function at levels of thrombin that are nonthrombogenic. A schematic model illustrating our current view of the basic features of the protein C activation complex is shown in Fig. 4. The basic feature of the model involves the initial formation of a thrombin–cofactor complex on the cell surface. The high affinity between thrombin and the cofactor results in the formation of a stable complex. Protein C with bound Ca^{2+} is the substrate and is rapidly converted to activated protein C. The K_m of protein C in the presence of cofactor (17) is 10 times lower than that observed in the absence of cofactor (40). This implies either that there is a direct cofactor–protein

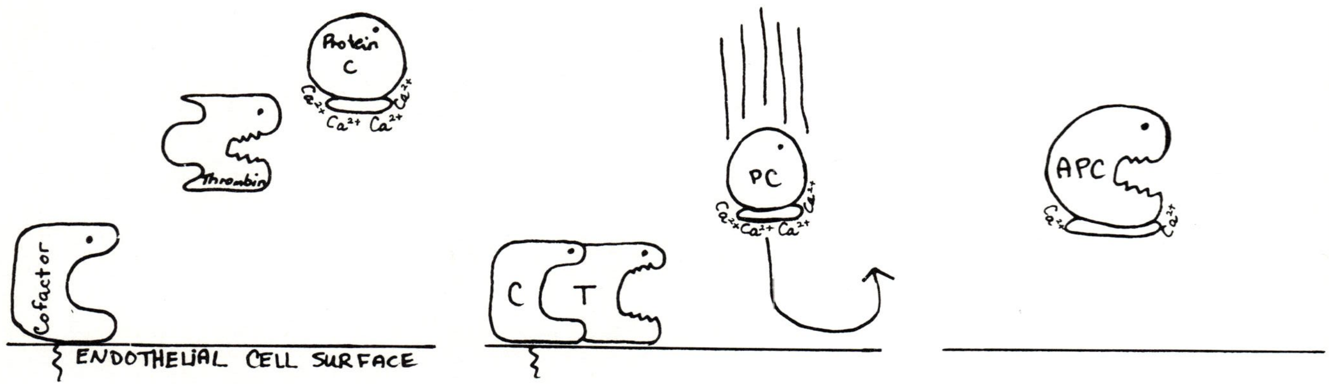

Fig. 4. Model for protein C activation. APC, activated protein C. For discussion, see text.

C interaction or that the interaction of thrombin with the cofactor causes a structural change in thrombin that facilitates protein C binding. Differentiation between these two models will require additional studies.

The cofactor is envisioned as a cell surface protein attached to the endothelial cell plasma membrane by a hydrophobic tail. This form of interaction is supported by the observations that the cofactor activity is not released by treatment with EDTA or high salt but is released in soluble form with low levels of nonionic detergents. The ability of the cofactor to function in the presence of detergents and in the absence of an intact bilayer structure suggests that the active region of the cofactor is not intimately associated with the membrane. Our preliminary studies further support this concept. Rat cofactor preparations have often resulted in a slight shoulder of activity on the trailing edge of the activity peak (Fig. 1). When pooled, this region contains a lower-molecular-weight species that no longer requires nonionic detergents to remain soluble or to retain activity. The simplest explanation of this observation is that this is a proteolytic degradation product of the cofactor from which the hydrophobic region has been cleaved. This model is consistent with similar systems, such as the b-type cytochromes, which can be solubilized in functional form by mild proteolytic cleavage but which require nonionic detergents to solubilize the intact proteins (41,42). An alternative possibility is that the soluble and membrane-bound cofactors are unrelated. More complete chemical and immunochemical characterization of these two forms will be required to resolve this question.

Calcium ions play an interesting role in the regulation of protein C activation. Recently, Ca^{2+} has been shown to inhibit protein C activation by thrombin in the absence of the cofactor (30,40). Amphlett *et al.* have shown that this inhibitory effect is primarily on K_m. The K_m obtained in the presence of either the cells or the purified cofactor is 10–20 times below that observed in the absence of cofactor. Furthermore, the cofactor stimulation of protein C activation has an obligatory requirement for Ca^{2+}. Thus Ca^{2+} functions to stimulate cell-mediated protein C activation while inhibiting the activation in solution. These processes may serve as controls of protein C activation at the site of injury.

One interesting feature of the protein C activation system is that rapid activation occurs on the endothelial cell in the presence of thrombin. Since rapid thrombin formation occurs at the site of injury, excess thrombin could enter the blood flow. Interaction between thrombin and the endothelial cell cofactor probably occurs downstream from this site. Protein C could thus be activated distally from the site of injury. For activated protein C to function in controlling clotting near the site of injury, the activated protein C would have to be cleared slowly from the circulation. This is compatible with current information. Inactivation of activated protein C in plasma is slow ($t_{1/2} \approx 10$ min) (35). The clearance of activated protein C *in vivo* also appears to be slow. When bovine activated protein C was injected into dogs, the half-life was between 15 and 20 min (P. C. Comp and C. T. Esmon, unpublished results). Although questions of a species-specific alteration in clearance rate must be considered, it is likely that *in vivo* clearance of activated protein C is slow. These observations are also compatible with a role for activated protein C as a systemic anticoagulant. In this setting the function may be involved in the inactivation of circulating factor Va and factor VIIIa. By inhibiting these circulating activated clotting factors, activated protein C may serve to help localize clot formation.

The proposed mechanism of protein C activation may provide a link between blood flow and thrombosis. In major vessels the surface-to-volume ratio is low. As a result, the cofactor concentration is probably also low. In stasis, thrombin contact with the endothelium of the major vessels is probably limited by the low availability of endothelial cell surface. Under these circumstances, protein C activation would be slow. Rapid protein C activation probably occurs in the microcirculation. Therefore, adequate blood flow from the major veins is likely to be essential to formation of the protein C activation complex. Failure to activate protein C may contribute to stasis-induced thrombosis.

The activation of protein C by the thrombin–cofactor complex has many features in common with the prothrombin activation complex. Both systems involve a serine protease, a cofactor, and a vitamin K-dependent substrate. The cofactor binds to the enzyme very tightly in both cases (17,39). Both complexes affect the rate of activation of the substrate by lowering the K_m for the substrate and by increasing the V_{max} of the reaction (17,39). Investigations into the mechanism by which the cofactor functions in these processes has proven difficult. In the case of the prothrombin activation complex, the system is complicated by the fact that both the enzyme, factor Xa, and the substrate, prothrombin, bind to phospholipid surfaces in a Ca^{2+}-dependent manner (2). Furthermore, the stability of the cofactor, factor Va, is also dependent on Ca^{2+} (43). These properties have made a simple evaluation of the influence of Ca^{2+} interaction with the substrate difficult. Some of these complications may be more easily studied in the protein C activation complex. Unlike factor Xa, thrombin has no Ca^{2+} binding sites and does not interact directly with the phospholipid surface.

Therefore, the effects of Ca^{2+} are likely to be on the substrate. If this is correct, then the failure of the protein C activation complex to function in the absence of Ca^{2+} is probably due to an improper conformation of the substrate. Ca^{2+} binding to protein C (40) is almost certainly dependent on the formation of γ-carboxyglutamic acid residues, a process that is catalyzed by vitamin K (44) and inhibited by coumarin anticoagulants. Since the vitamin K-dependent proteins formed in the presence of coumarin are unable to bind Ca^{2+} (44), it is likely that protein C in patients on oral anticoagulants will not bind Ca^{2+}. Failure to bind Ca^{2+} will probably result in failure of this form of protein C to be activated by the thrombin–cofactor complex.

These studies suggest many possible biochemical explanations for correlating vascular disease with thrombosis. Understanding the physiological role of protein C in the regulation of clot formation is necessary before clinical evaluation of the relationship between the cofactor and thrombosis can be evaluated. The properties of activated protein C make this protein an ideal candidate for a regulator of clot formation.

ACKNOWLEDGMENTS

We thank Glenna Fry and Dr. John C. Hoak for providing cell cultures and especially acknowledge the expert technical assistance of Susan Hale, Kevin Harris, and Gary Ferrell. We would like to thank the Cornett Packing Company for their generous gift of bovine blood. This work was supported by grants HL-17812-06, HL-22471-03, and HL 14230 (SCOR) from the National Heart Lung and Blood Institute, and by grant No. 79 685 from the American Heart Association with funds contributed in part by the Oklahoma Affiliate. This work was done during the tenure of an Established Investigatorship (CTE) of the American Heart Association, with funds contributed by the Oklahoma Affiliate, and Career Development Award HL-00348-04 (WGO).

REFERENCES

1. Davie, E. W., and Fujikawa, K. (1975) *Annu. Rev. Biochem.* **44,** 799–829.
2. Jackson, C. M., and Nemerson, Y. (1980) *Annu. Rev. Biochem.* **49,** 765–811.
3. Mammen, E. F., Thomas, W. R., and Seegers, W. H. (1960) *Thromb. Diath. Haemmorh.* **5,** 218–249.
4. Stenflo, J. (1976) *J. Biol. Chem.* **251,** 355–363.
5. Esmon, C. T., Stenflo, J., Suttie, J. W., and Jackson, C. M. (1976) *J. Biol. Chem.* **251,** 3052–3056.
6. Kisiel, W., Ericsson, L. H., and Davie, E. W. (1976) *Biochemistry* **15,** 4893–4900.
7. Kisiel, W., Canfield, W. M., Ericsson, L. H., and Davie, E. W. (1977) *Biochemistry* **16,** 5824–5831.
8. Kisiel, W. (1979) *J. Clin. Invest.* **64,** 761–769.
9. Seegers, W. H., Novoa, E., Henry, R. L., and Hassouna, H. I. (1976) *Thromb. Res.* **8,** 543–552.

10. Weksler, B. B., Ley, C. W., and Jaffe, E. A. (1978) *J. Clin. Invest.* **62,** 923–930.
11. Czervionke, R. L., Smith, J. B., Hoak, J. C., Fry, G. L., and Haycraft, D. L. (1979) *Thromb. Res.* **14,** 781–786.
12. Lollar, P., and Owen, W. G. (1980) *J. Clin. Invest.* **66,** 1222–1230.
13. Isaacs, J. D., Gospodarowicz, S. D., Fenton, J. W., II, and Shuman, M. A. (1981) *Biochemistry* **20,** 398–403.
14. Awbrey, B. J., Hoak, J. C., and Owen, W. G. (1979) *J. Biol. Chem.* **254,** 4092–4095.
15. Hoak, J. C., Czervionke, R. L., Fry, G. L., and Smith, J. B. (1980) *Fed. Proc., Fed. Am. Soc. Exp. Biol.* **39,** 2606–2609.
16. Esmon, C. T., and Owen, W. G. (1981) *Proc. Natl. Acad. Sci. U.S.A.* **78,** 2249–2251.
17. Owen, W. G., and Esmon, C. T. (1981) *J. Biol. Chem.* **256,** 5532–5535.
18. Walker, F. J., Sexton, P. W., and Esmon, C. T. (1979) *Biochim. Biophys. Acta* **571,** 333–342.
19. Marlar, R. A., Kleiss, A. J., and Griffin, J. H. (1980) *Protides Biol. Fluids* **28,** 341–344.
20. Vehar, G. A., and Davie, E. W. (1980) *Biochemistry* **19,** 401–410.
21. Canfield, W., Nesheim, M., Kisiel, W., and Mann, K. (1978) *Circulation* **58,** 816 (abstr.).
22. Miletich, J. P., Jackson, C. M., and Majerus, P. W. (1977) *Proc. Natl. Acad. Sci. U.S.A.* **74,** 4033–4036.
23. Miletich, J. P., Jackson, C. M., and Majerus, P. W. (1978) *J. Biol. Chem.* **253,** 6908–6916.
24. Tracy, P. B., Peterson, J. M., Nesheim, M. E., McDuffie, F. C., and Mann, K. G. (1979) *J. Biol. Chem.* **254,** 10354–10361.
25. Dahlback, B., and Stenflo, J. (1978) *Biochemistry* **17,** 4938–4945.
26. Comp, P. C., and Esmon, C. T. (1980) *In* "The Regulation of Coagulation" (K. G. Mann and F. B. Taylor, Jr., eds.), pp. 583–588. Elsevier/North-Holland, New York.
27. Dahlback, B., and Stenflo, J. (1980) *Eur. J. Biochem.* **107,** 331–335.
28. Esmon, C. T., Comp, P. C., and Walker, F. J. (1979) *In* "Vitamin K Metabolism and Vitamin K-dependent Proteins" (J. W. Suttie, ed.), pp. 72–83. University Park Press, Baltimore, Maryland.
29. Walker, F. J. (1980) *J. Biol. Chem.* **255,** 5521–5524.
30. Esmon, C. T., and Owen, W. G. (1981) *In* "Plasma and Cellular Modulatory Proteins" (D. Bing, ed.), pp. 203–214. Center for Blood Research, Inc., Boston, Mass.
31. Walker, F. J. (1981) *Fed. Proc., Fed. Am. Soc. Exp. Biol.* **40,** 1586 (abstr.).
32. Zolton, R. P., and Seegers, W. H. (1973) *Thromb. Res.* **3,** 23–33.
33. Comp, P. C., and Esmon, C. T. (1980) *Circulation* **62,** Suppl. III, Abstr. No. 1288.
34. Radcliffe, R., and Heinze, T. (1978) *Arch. Biochem. Biophys.* **189,** 185–197.
35. Marlar, R. A., and Griffin, J. H. (1980) *J. Clin. Invest.* **66,** 1186–1189.
36. Owen, W. G., Esmon, C. T., and Jackson, C. M. (1974) *J. Biol. Chem.* **249,** 594–605.
37. Esmon, N., Owen, W. G., and Esmon, C. T. (1982) *J. Biol. Chem.,* **257,** 859–864.
37a. Davis, B. J. (1964) *Ann. N.Y. Acad. Sci.* **121,** 404–427.
38. Laemmli, U. K. (1970) *Nature (London)* **227,** 680–685.
39. Nesheim, M. E., Taswell, J. B., and Mann, K. G. (1979) *J. Biol. Chem.* **254,** 10952–10962.
40. Amphlett, G. W., Kisiel, W., and Castellino, F. J. (1981) *Biochemistry* **20,** 2156–2161.
41. von Jagow, G., and Sebald, W. (1980) *Annu. Rev. Biochem.* **49,** 281–314.
42. Spatz, L., and Strittmatter, P. (1971) *Proc. Natl. Acad. Sci. U.S.A.* **68,** 1042–1046.
43. Esmon, C. T. (1979) *J. Biol. Chem.* **254,** 964–973.
44. Stenflo, J., and Suttie, J. W. (1977) *Annu. Rev. Biochem.* **46,** 157–172.

PART III

CELL-PROTEIN INTERACTIONS II

The Role of Factor VIII–von Willebrand Factor in Adhesion of Platelets to the Vessel Wall

JAN J. SIXMA, KJELL S. SAKARIASSEN, AND PIETER A. BOLHUIS
Department of Hematology
University Hospital Utrecht
Utrecht
The Netherlands

INTRODUCTION

Classic von Willebrand disease (VWD) is characterized by the combination of a decrease in factor VIII and prolongation of the bleeding time. The retention of platelets in a glass bead column is also decreased, and the aggregation of platelets with the antibiotic ristocetin is absent or diminished (for a review, see 1,2). Studies in the early 1970s have demonstrated that all these abnormalities can be attributed to the absence of factor VIII–von Willebrand factor (FVIII-VWF).

FVIII-VWF is a multimeric protein, which on reduction appears to have a single subunit with a molecular weight of about 250,000. Under specific, high-ionic-strength conditions dissociation occurs between the coagulant activity (FVIII : C), which is of unknown size and probably present in a small quantity of protein, and the von Willebrand factor (3,4). The latter is at present quantified by the ristocetin cofactor assay (VIII : RiCoF) in which the test sample is compared with normal plasma for its ability to cause aggregation of blood platelets in the presence of the antibiotic ristocetin (5).

Electron microscopic observations in skin wounds showed that the primary *in vivo* defect in VWD consisted of deficient adhesion of platelets to collagen (6). Transfusion studies with various factor VIII

PATHOBIOLOGY OF THE ENDOTHELIAL CELL

ISBN 0-12-521980-6

preparations indicated that these could normalize the bleeding time (7,8), but unequivocal shortening of the bleeding time with purified FVIII-VWF has only been performed in dogs with von Willebrand disease (9). Perfusion studies in which everted pieces of rabbit aorta were mounted on a central rod in an annular perfusion chamber have demonstrated a defect of platelet adhesion to the subendothelium in VWD (10). Similar results were obtained with human renal arteris used in place of rabbit aorta (11). The decrease in FVIII-VWF is responsible for this, as has been confirmed with studies in which rabbit antibodies to FVIII-VWF factor have been used (12).

In this paper we will describe recent results concerning the mechanism by which FVIII-VWF enhances adherence of platelet to the subendothelium.

METHODS

ARTERIES

Human postmortem renal arteries after the first bifurcation were applied as previously described (11). In several experiments we also used umbilical arteries. These arteries were freed from umbilical cords with scissors and a surgical scalpel, pretreated with 0.1 *M* of aspirin in 0.2 *M* Tris-HCl containing penicillin and streptomycin, washed and stored in the same buffer without aspirin at 4°C, and used at least 16 hr after they had been obtained.

PERFUSION SYSTEM

The perfusion system consisted of a nonocclusive roller pump, silastic tubing, a perfusate container, and a perfusion chamber as developed by Baumgartner (13). In many studies we have employed a modified system in which steady flow was obtained by introducing a funnel into the perfusion system (Fig. 1) (14). The advantage of this system is that the pulsatility caused by the roller pump is avoided. This is important because this pulsatility has a higher frequency than exists in the circulation and also increases with increasing shear rates. The perfusion fluids consisted in general of a mixture of washed red blood cells, a plasma substitute, and either normal human plasma, plasma of a patient with severe VWD, or a human albumin solution with or without extra calcium added. Platelets were obtained from citrated blood, labeled with ^{51}Cr, treated with aspirin and washed four times at

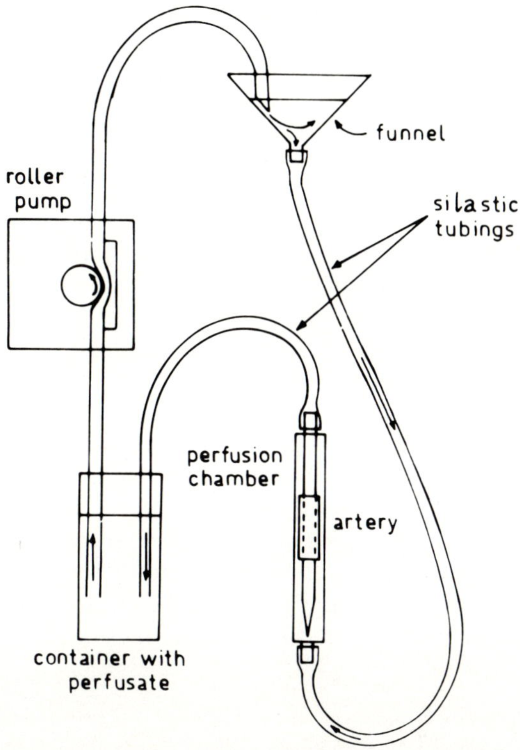

Fig. 1. Perfusion system with steady flow. Blood drains from the funnel by gravity. The velocity of the flow is regulated by the height of the funnel. [From Sakariassen *et al.* (14). Courtesy of Pergamon Press.]

a pH of 6.3. After the final platelet washing, the pellets were resuspended in the plasma or plasma substitute, and kept for 30 min to 2 hr at 20°C before the experiments were performed. FVIII-VWF preparations and washed red cells were added to the platelet suspension just before the prewarming of the perfusates for 5 min at 37°C.

EVALUATION OF THE NUMBER OF ADHERENT PLATELETS

The number of adherent platelets was registered by counting ^{51}Cr and expressed per square centimeter of subendothelium. The morphometric method described by Baumgartner (13) was also used. A good correspondence was obtained between the results of these two techniques.

HUMAN FACTOR VIII–VON WILLEBRAND FACTOR PREPARATIONS

Lyophilized cryoprecipitates were obtained from the Central Laboratory of the Netherlands Red Cross Blood Transfusion Service, and fresh cryoprecipitates from the local central blood bank. Cryoprecipitates were washed three times in Michaelis buffer pH 7.35 containing 0.15% sodium citrate, 10 m*M* benzamidine, and 10 m*M* ϵ-aminocaproic acid, and applied to a Sepharose Cl-4B column at 20°C. The FVIII-VWF containing void volume fractions were pooled and precipitated by dialysis at 4°C against a 35% ammonium sulfate solution adjusted to pH 7.0. The precipitate was obtained at 1000 *g* for 10 min at 4°C and dissolved in about 1 ml of the ammonium sulfate solution. Before addition of the purified FVIII-VWF to perfusates it was centrifuged (10,000 *g* for 2 min at 4°C in an Eppendorf centrifuge). The precipitate was dissolved in 1 ml of human albumin solution and added to the platelet suspension in the perfusate at the proper concentrations. The FVIII R : Ag, VIII : RiCoF, and VIII : C were assayed as previously described (15). The multimeric composition of FVIII-VWF factor was determined by agarose electrophoresis in 0.2% SDS in Tris–Tricine buffer followed by Coomassie blue staining or crossed immunoelectrophoresis according to the technique of Converse & Papermaster (16).

RESULTS AND DISCUSSION

The multimeric composition of FVIII-VWF on SDS-agarose electrophoresis of purified FVIII-VWF gave an extensive multimeric pattern, with at least 10 discernible bands (Fig. 2). The band with the lowest molecular weight was at 3.0×10^6 and more than 50% of the protein had a molecular weight above 6.0×10^6. The interspace between the various bands was 500,000, indicating that the basic physiological subunit is a dimer. The figures for molecular weights should be viewed with caution. We used DNA digests as molecular-weight markers and these may behave different in the SDS-agarose as compared to FVIII-VWF. This also applies however, to cross-linked high-molecular-weight proteins as markers, and our data are consistent with those of Counts *et al.* (17) and Furlan *et al.* (18), who used cross-linking of the 250,000 subunits to obtain information about the size of the basic physiological subunit.

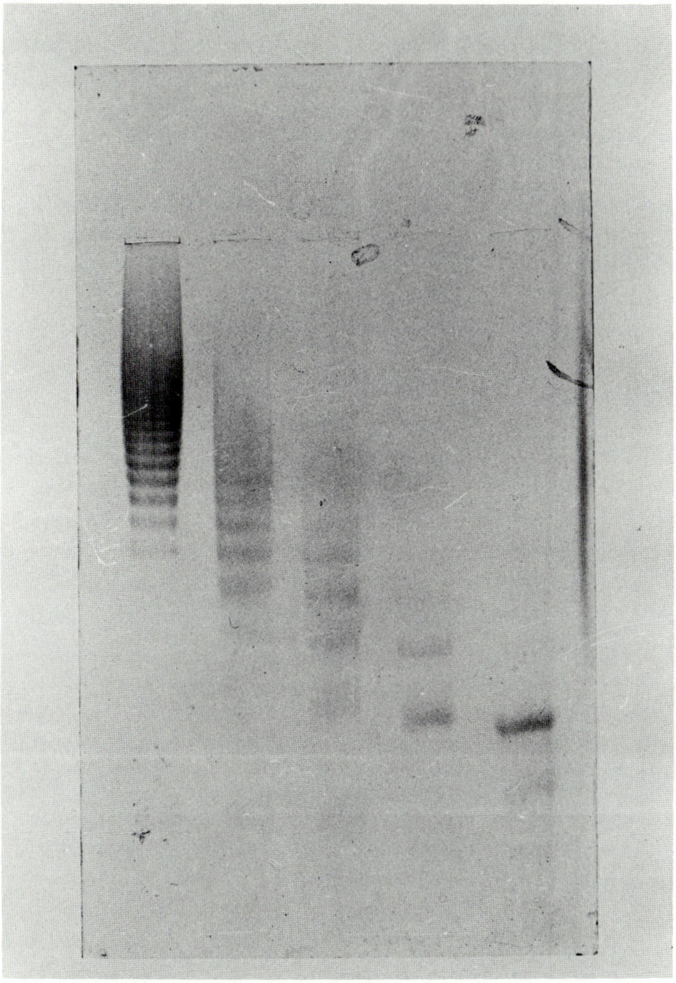

Fig. 2. Multimeric structure of factor VIII-VWF. Purified factor VIII-VWF was subjected to electrophoresis in 1% agarose in Tris-Tricine buffer with 0.1% SDS. The lanes from left to right are reduced FVIII-VWF and are treated with 0.5 mM, 1.0 mM, 2.0 mM, and 10 mM dithioerythritol, respectively. The heavy band in the right lane is the 500K dimer.

DIFFERENCE BETWEEN VON WILLEBRAND PLASMA AND NORMAL PLASMA

The difference between normal plasma and von Willebrand plasma in enhancing platelet adhesion is best seen in a time-course study.

From this it is evident that the initial rate of platelet adhesion is decreased in VWD but that the final coverage of the subendothelium is similar. It should be remembered, however, that for these studies we have used platelets from normal individuals containing FVIII-VWF in a secretable form, as well as pieces of arteries from normal individuals, which contain FVIII-VWF in the subendothelium. Data from pigs with severe von Willebrand disease that lack FVIII-VWF in their platelets and probably also in the subendothelium seem to indicate that these animals have an absolute defect in platelet deposition (19). Most of our studies have been performed at the shear rates that are found in small arteris. Higher shear rates increase the defect of VWD (12).

Addition of purified FVIII-VWF to human albumin, a dextran solution, plasma of patient with severe VWD, and supernatant plasma after cryoprecipitation (which also lacks VIII : RiCoF) showed that in all these cases full restoration of platelet–vessel wall adhesion was obtained (Fig. 3). This indicates that FVIII-VWF is the only plasma factor required for platelet–vessel wall interaction and that there is no inhibitor present in normal plasma under these conditions. This is of importance with regard to the potential role for plasma fibronectin in platelet–vessel wall interaction (20). Our perfusion experiments, as summarized in Fig. 3, do not provide support for such a role. This does not exclude the possibility that fibronectin of the vessel wall is involved, but plasma fibronectin has neither an enhancing nor an inhibiting effect on the adherence of platelets to subendothelium.

INTERACTION OF FACTOR VIII–VON WILLEBRAND FACTOR WITH THE VESSEL WALL

Platelets possess a receptor for FVIII-VWF, which is lacking in the Bernard–Soulier syndrome (21,22). This receptor is a glycoprotein, GP Ib, consisting of an α chain with a molecular weight of 130K and a smaller β chain of 25K, cleaved off by the reduction (23). In normal plasma, FVIII-VWF shows only a very limited binding to platelets, but this is strongly enhanced by the presence of antibiotic ristocetin (24,25). Bovine FVIII-VWF is able to interact directly with GP Ib to cause aggregation of platelets (26). These considerations suggest that FVIII-VWF might change in conformation on interaction with the subendothelium. Platelets would then be able to interact with this changed FVIII-VWF. To test this hypothesis we performed a double perfusion in which we first perfused a piece of renal artery with radiolabeled FVIII : VWF, then rinsed the vessel, and exposed it to

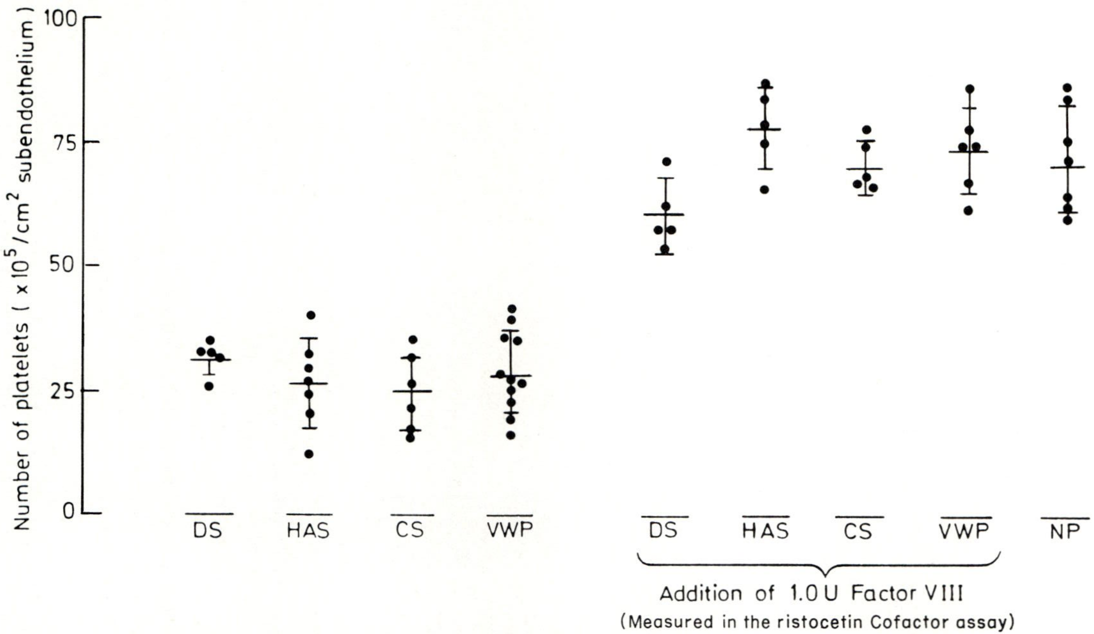

Fig. 3. Platelet adhesion with and without FVIII-VWF (mean ± SC). DS, dextran 40 (3.3%); HAS, human albumin solution (4%); CS, supernatant plasma after cryoprecipitation; VWP, plasma of patient with severe Von Willebrand disease; NP, normal plasma.

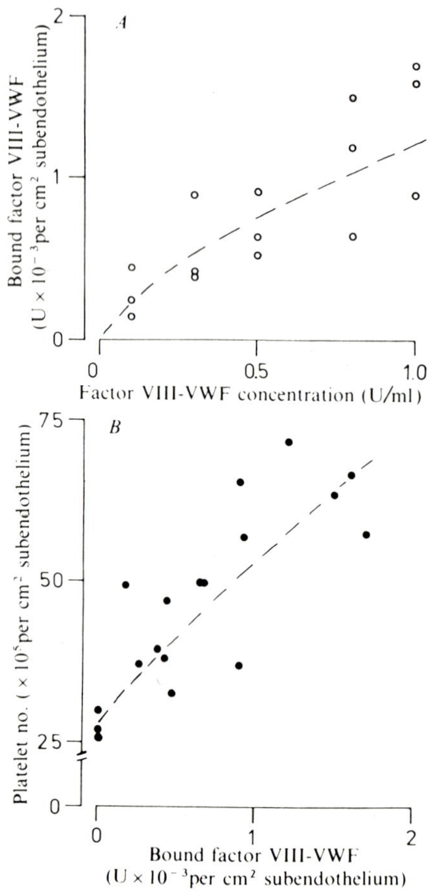

Fig. 4. Factor VIII-VWF binding and subsequent platelet adhesion in a double perfusion experiment. First perfusion was for 5 min with a Krebs-Ringer solution containing 4% human albumin and different concentrations of purified FVIII-VWF, 10% of which was radiolabeled. The system was then rinsed with 135 ml of a 0.2 *M* Tris buffer, pH 7.4. The second perfusion was for 5 min with Krebs-Ringer + 4% human albumin + 2.5×10^5 platelets/liter and a red cell concentration of 40%. [With permission, from Sakariassen *et al.* (11).]

radiolabeled platelets in the absence of FVIII-VWF in the perfusate. (Fig. 4). These studies showed that FVIII-VWF bound to the subendothelium and that subsequent binding of platelets was well correlated with the amount of FVIII-VWF bound in the first perfusion. In all our studies, however, we always found a basic adherence, also in the absence of FVIII-VWF. Rand *et al.* (27) recently reported that normal human and rabbit vessels contained FVIII-VWF in their superficial layers. With immunofluorescent studies with rabbit antibodies against VIII R:Ag, we could confirm this in the postmortem human renal arteries (unpublished data). To demonstrate that binding of FVIII-VWF to subendothelium was important for optimal adhesion, we performed time sequence studies utilizing radiolabeled factor VIII–von Willebrand factor added to normal and von Willebrand plasma (Fig. 5). We could demonstrate that optimal binding of FVIII-VWF preceded the enhanced adhesion rate of platelets found in normal plasma compared to that in plasma of VWF patients (28). Whether the FVIII-VWF bound to the subendothelium influences initial contact of the platelets with the subendothelium or the subsequent spreading of the platelets is still a matter of debate. We favor the idea that spreading of platelets along the subendothelium is influenced by bound FVII-VWF. This interpretation is based on morphometric studies, which indicate that the absolute number of contact platelets in VWD is not decreased or even slightly increased, whereas the percentage of spread platelets is fully responsible for the total decrease in rate of platelet deposition (28) (Fig. 6).

COMPARISON BETWEEN RENAL ARTERIES AND UMBILICAL ARTERIES

The data given in the previous paragraphs were obtained with postmortem human renal arteries. These blood vessels have a considerable amount of subintimal fibrosis, which may influence the adher-

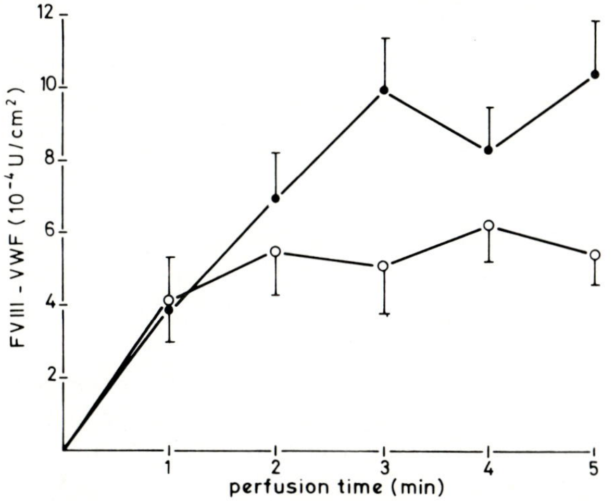

Fig. 5. Factor VIII-VWF binding to subendothelium in plasma: ●——● in the presence of platelets (1.9×10^5/rl plasma); ○——○ in the absence of platelets. The VIII-VWF concentration was 1.0 U/ml; 10% was radiolabeled with ^{125}I.

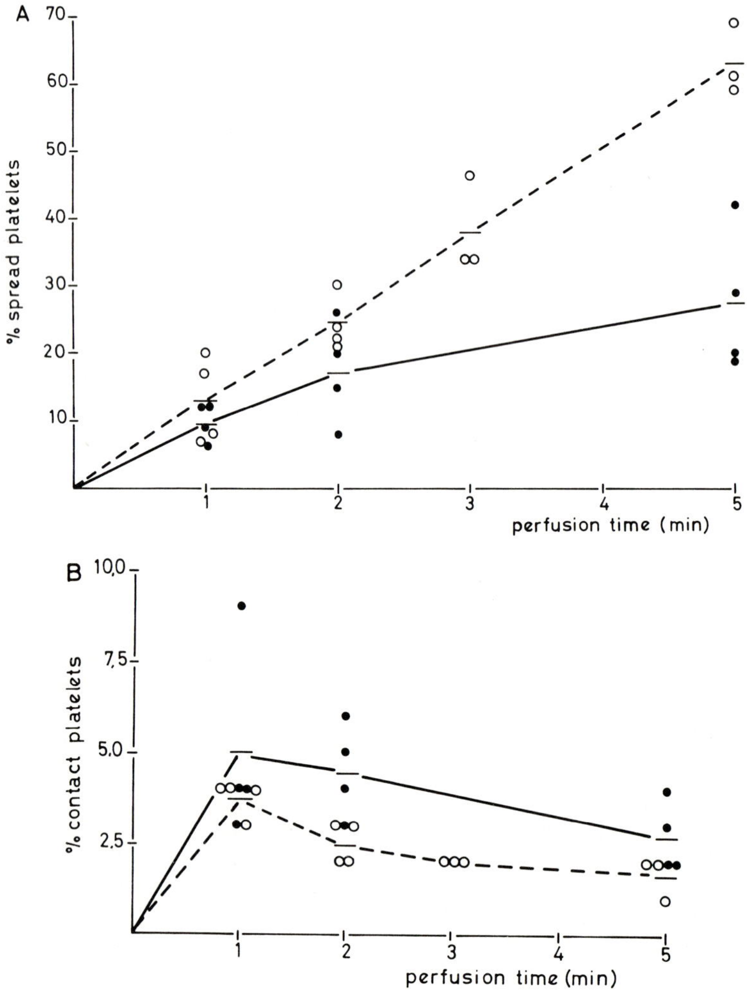

Fig. 6. Adhesion of blood platelets in plasma of a patient with severe VWD (●——●) or in normal plasma (○——○) as evaluated by morphometry. The *y* axis represents % coverage. (A) Percentage of the surface covered by spread platelets. (B) Percentage coverage of platelets in contact with the subendothelium but not spread along it.

ence of blood platelets. We therefore decided to study the adhesion to the subendothelium of umbilical arteries. The dependence on the concentration of FVIII-VWF on renal and umbilical arteries is given in Fig. 7. A plateau level was found at 0.5 U/ml of FVIII : RiCoF, which agrees remarkably well with the lower value of the normal range. No further increase in adhesion was seen even at a concentration of 2.0

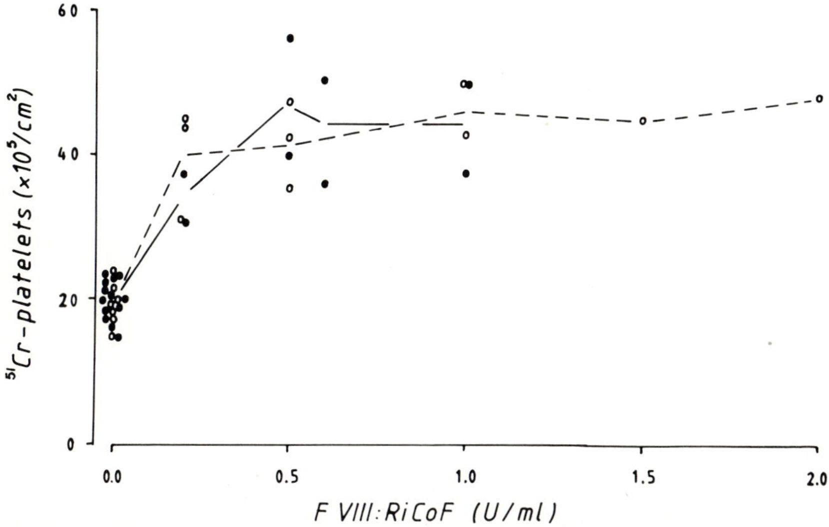

Fig. 7. Comparison of the adherence of platelets to subendothelium of the human umbilical artery (○——○) and postmortem human renal artery (●——●) in response to different concentrations of FVIII-VWF in the perfusion fluids. The FVIII-VWF is measured and expressed by its ristocetin cofactor activity.

U/ml of VIII : RiCoF. These data indicate that the umbilical artery is at least as suitable for the study of platelet vessel wall interaction as postmortem renal arteries. Utilizing the umbilical arteries has the advantage of being readily available and more reproducible in thickness. This is of importance because the thickness of the wall is an important factor in influencing the width of the annulus of the perfusion chamber. A more reproducible thickness of the vessel wall thus results in more reproducible wall shear rates.

THE EFFECT OF VARIOUS FVIII-VWF PREPARATIONS

Most of our studies were performed with FVIII-VWF purified by agarose gel chromatography. Comparison of the effect of cryoprecipitates with that of purified factor VIII showed that they were equally effective. We also performed dissociation of FVIII-VWF in 0.25 *M* $CaCl_2$ into low-molecular-weight (LMW)-FVIII, which has all VIII : C activity and higher molecular weight (HMW-FVIII-VWF). The HMW-FVIII-VWF supported platelet adherence to the vessel wall equally well as the parent FVIII-VWF. Studies with commercial high potency concentrates were also performed. They will be reported in detail elsewhere. The overall conclusion of these studies was that these concentrates which do not correct bleeding time also lack the ability to support the adhesion of blood platelets.

INFLUENCE OF CALCIUM ON FVIII-VWF-INDUCED PLATELET ADHESION

One of the major factors required for optimal platelet–vessel wall adhesion is the presence of divalent cations, such as calcium and magnesium. These cations are required both for spreading (29) as well as for initial contact, but then at a very low concentration ($<10\ \mu M$) (30). One of the possible explanations might be that these cations work by influencing the binding of FVIII-VWF to the vessel wall. We therefore decided to study how the absence of calcium and magnesium ions would influence FVIII-VWF binding. Such an effect was not demonstrated, however. The problem with such studies, however, is that it is not certain whether all bound FVIII-VWF is relevent. Therefore we also studied whether FVIII-VWF that had bound in a calcium-free solution was still able to support platelet adhesion in a calcium-containing medium. This appeared to be the case. This indicates that the binding of relevant FVIII-VWF has no requirement for divalent cations. This is also supported by experiments in which we could show that the effect of FVIII-VWF is still observed at low calcium concentrations at which platelet deposition is considerably decreased.

CONCLUSIONS

These results of the investigations on the role of FVIII-VWF in platelet–vessel wall interaction begin to shape a concept on how platelets do interact with the vessel wall. A great number of interesting questions have to be resolved in the near future. Among these are the problem of determining which components in the vessel wall FVIII-VWF factor does interact, and also which multimers are responsible for von Willebrand factor activity. Another question to be answered is whether the von Willebrand activity of FVIII-VWF is caused by a repetitive structure of the large multimers or whether a specific sequence in the protein is responsible for this activity. These answers to these questions are presently being sought in our laboratory.

ACKNOWLEDGMENTS

The technical help of Mrs. M. Ottenhof-Rovers and N. Beeser-Visser is gratefully acknowledged. Financial support was provided by the Netherlands Heart Foundation (no. 76090) and the Foundation for the Advancement of Pure Scientific Research in the Netherlands (FUNGO).

REFERENCES

1. Hoyer, L. W. (1976) *Prog. Hemostasis Thromb.* **3,** 231–288.
2. Bloom, A. L. (1980) *Semin. Hematol.* **17,** 215–227.
3. Cooper, H. A., Griggs, T. R., and Wagner, R. H. (1973) *Proc. Natl. Acad. Sci U.S.A.* **70,** 2326–2329.
4. Weiss, H. J., and Hoyer, L. W. (1973) *Science* **182,** 1149–1151.
5. Weiss, H. J., Rogers, J., and Brand, H. (1973) *J. Clin. Invest.* **52,** 2697–2707.
6. Hovig, T., and Stormorken, H. (1974) *Acta Pathol. Microbiol. Scand., Suppl.* **248,** 105–122.
7. Cornu, P., Larrieu, M. J., Caen, J., and Bernard, J. (1963) *Br. J. Haematol.* **9,** 189–202.
8. Nilsson, I. M., Blömback, M., Jorpes, E., Blömback, B., and Johansson, S. A. (1957) *Acta Med. Scand.* **159,** 179–188.
9. Bouma, B. N., Dodds, W. J., van Mourik, J. A., Sixma, J. J., and Webster, W. P. (1976) *Scand. J. Haematol.* **17,** 263–275.
10. Tschopp, T. B., Weiss, H. J., and Baumgartner, H. R. (1974) *J. Lab. Clin. Med.* **83,** 296–300.
11. Sakariassen, K. S., Bolhuis, P. A., and Sixma, J. J. (1979). *Nature (London)* **279,** 636–638.
12. Baumgartner, H. R., Tschopp, T. B., and Meyer, D. (1980) *Br. J. Haematol.* **44,** 127–139.
13. Baumgartner, H. R., and Haudenschild, C. (1972) *Ann. N.Y. Acad. Sci.* **201,** 22–36.
14. Sakariassen, K. S., Bolhuis, P. A., and Sixma, J. J. (1980) *Thromb. Res.* **19,** 547–559.
15. Over, J., Sixma, J. J., Doucet-de Bruine, M. H. M., Trieschnigg, A. M. C., Vlooswijk, R. A. A., Beeser-Visser, N. H., and Bouma, B. N. (1978) *J. Clin. Invest.* **62,** 223–234.
16. Converse, C. A., and Papermaster, D. G. (1975) *Science* **189,** 469–472.
17. Counts, R. B., Paskell, S. L., and Elgee, S. K. (1978) *J. Clin. Invest.* **62,** 702–709.
18. Furlan, M., Perret, B. A., and Beck, E. A. (1979) *Biochim. Biophys. Acta* **579,** 325–333.
19. Fuster, V., Bowie, E. J. W., Lewis, J. C., Fass, D. N., Owen, C. A., and Brown, A. L. (1978) *J. Clin. Invest.* **61,** 722–730.
20. Bensusan, H. B., Koh, T. L., Henry, K. G., Murray, B. A., and Culp, L. A. (1978) *Proc. Natl. Acad. Sci. U.S.A.* **75,** 5864–5868.
21. Howard, M. A., Hutton, R. A., and Hardisty, R. M. (1973) *Bri. Med. J.* **2,** 586–588.
22. Nurden, A. T., and Caen, J. P. (1978) *Br. J. Haematol.* **38,** 155–160.
23. Cooper, H. A., Clemetson, K. J., and Lüscher, E. F. (1979) *Proc. Natl. Acad. Sci. U.S.A.* **76,** 1069–1073.
24. Zucker, M. B., Kim, S. J., MacPherson, J., and Grant, R. A. (1977) *Br. J. Haematol.* **35,** 535–549.

25. Schneider-Trip, M. D., Jenkins, C. S. P., Kahlé, L. E., Sturk, A., and Ten Cate, J. W. (1979) *Br. J. Haematol.* **43,** 99–112.
26. Solum, N. O., Hagen, I., and Gjemdal, T. (1977) *Thromb. Haemostasis* **38,** 914–923.
27. Rand, J. H., Sussman, J. J., Gordon, R. E., Chu, S. V., and Solomon, V. (1980) *Blood* **55,** 752–756.
28. Bolhuis, P. A., Sakariassen, K. S., Sander, H. J., Bouma, B. N., and Sixma, J. J. (1981) *J. Lab. Clin. Med.* **97,** 568–576.
29. Baumgartner, H. R., Muggli, R., Tschopp, T. B., and Turitto, V. (1976) *Thromb. Haemostasis* **35,** 124–138.
30. Sakariassen, K. S., Bolhuis, P. A., and Sixma, J. J. (1982). In preparation.

The Factor VIII Complex

LEON W. HOYER
Department of Medicine
University of Connecticut Health Center
Farmington, Connecticut

There is a close relationship of the factor VIII complex and vascular endothelial cells. From the first recognition that these cells contain factor VIII-related antigen (von Willebrand factor) (1), it has been recognized that the endothelial cells is well located to contribute to hemostasis and that its contributions are central. Since our understanding of factor VIII structure and function has evolved during the past decade, the terms used to describe properties of this protein have changed, and the earlier literature may be very confusing. For this reason, a brief review of the properties of the factor VIII complex is necessary to clarify the relationship of this complex to the endothelial cell. A recent review will provide a more complete description of these properties (2).

THE TWO COMPONENTS OF THE FACTOR VIII COMPLEX

The concept that factor VIII has two distinct biological functions—coagulant activity and a role in primary hemostasis—evolved from the dual defect that was recognized in von Willebrand's disease. While these properties could have been due to a deficiency of a bifunctional molecule, recent studies have demonstrated that plasma factor VIII is a complex of two components that have distinct functions, biochemical and immunologic properties and genetic control. The properties of these components are summarized in Table I and the relationship is illustrated in Fig. 1.

One component of the factor VIII complex has antihemophilic factor procoagulant activity and is now usually designated VIII:C. It is

PATHOBIOLOGY OF THE ENDOTHELIAL CELL

ISBN 0-12-521980-6

TABLE I
Components of the Factor VIII Complex

VIII:C.	The factor VIII procoagulant protein; the antihemophilic factor
	Identified as:
	Factor VIII procoagulant activity (VIII:C)
	The procoagulant property of normal plasma that is measured in standard coagulation assays
	Factor VIII procoagulant antigen (VIII:CAg)
	Antigenic determinants closely associated with VIII:C
	Measured by immunoassays carried out with human antibodies
VIIIR.	The factor VIII-related protein: the von Willebrand factor
	A large polymeric protein that is necessary for normal platelet adhesion and bleeding time *in vivo*
	Factor VIII-related antigen (VIIIR:Ag)
	Antigenic determinants on VIIIR that are detected by heterologous antibodies
	Ristocetin cofactor (VIIIR:RC)
	The property of normal plasma VIIIR that supports ristocetin-induced agglutination of washed normal platelets

inactivated by human antibodies and can be measured with these antibodies as VIII:CAg. The other, larger component comprises the majority of the protein mass, interacts with platelets in a way that promotes primary hemostasis, and can be immunoprecipitated by

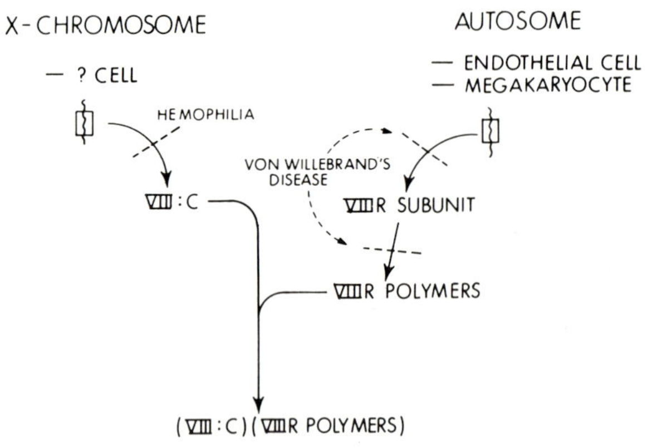

Fig. 1. The factor VIII complex in plasma. This schematic interpretation indicates the interaction of the two components and their genetic control. The effects of hemophilia and von Willebrand's disease are also noted.

heterologous antisera. It is usually designated factor VIII-related protein (VIIIR) or von Willebrand factor.

FACTOR VIII PROCOAGULANT PROTEIN: ANTIHEMOPHILIC FACTOR

BIOCHEMICAL PROPERTIES

There is little published information about the biochemical properties of the factor VIII procoagulant protein per se. The most satisfactory data are for bovine VIII:C that has been purified approximately 300,000-fold from plasma (3). Although the gel filtration properties suggest a molecular weight of 250,000–300,000, sodium dodecyl sulfate/urea-polyacrylamide gel electrophoresis identifies three protein-staining bands that corresponded to estimated molecular weights of 85,000, 88,000, and 93,000. Incubation within thrombin, purified factor X_a, or activated protein C led to proteolytic cleavage and slightly smaller proteins (3). Purified bovine VIII:C has no platelet-aggregating activity.

Although human factor VIII procoagulant protein has not been purified as successfully, its functional and immunologic properties have been determined for material free of VIIIR and most other plasma proteins. Its estimated molecular weight is 285,000 (4), a value very similar to that recently obtained for bovine factor V (5). The correspondence is not surprising, however, for factor V has a cofactor role in prothrombin activation that is very much like that of VIII:C in factor X activation.

IMMUNOLOGIC PROPERTIES

Human anti-VIII:C, obtained from multitransfused hemophilic patients who develop inhibitors and from rare individuals who form autoantibodies, do not form detectable immunoprecipitates with VIII:C or with the factor VIII complex. Nevertheless, these sera can be used to detect VIII:C antigenic determinants by sensitive immunoradiometric assays (6,7).

In general, there is excellent correlation between the plasma factor VIII procoagulant activity and VIII:CAg content. VIII:CAg determinants are more stable than VIII:C activity, however. This is striking in the case of serum in which VIII:CAg values are 60–80% of those in the corresponding plasma, even though there is no residual VIII:C activity (7).

FUNCTION

In the presence of phospholipid and calcium, VIII:C markedly enhances the enzymatic activation of factor X by factor IX_a. In the absence of factor IX_a, it does not have any intrinsic capacity to activate factor X (8). Native VIII:C does not accelerate this reaction, however, and thrombin activation is essential for VIII:C activity (8). Incubation with thrombin cleaves bovine VIII:C (3), and thrombin-activated human VIII:C has gel filtration and ultracentrifugation properties of a 116,000 dalton protein, in contrast to the 285,000 value calculated for the unactivated molecule (4). Factor VIII procoagulant activity is inactivated by higher concentrations of thrombin—or more prolonged exposure to the enzyme—and the (nonfunctional) protein measured by VIII:CAg is approximately 85,000 (4).

FACTOR VIII-RELATED PROTEIN: VON WILLEBRAND FACTOR

BIOCHEMICAL PROPERTIES

As the bulk of the factor VIII complex consists of factor VIII-related protein, data obtained for the intact complex reflect VIIIR properties (9–11). In all of these studies, the most striking property of purified factor VIII is its very large size. It has only recently been appreciated that VIIIR is, in fact, a heterogeneous population of multimers (Fig. 2). The smallest polymers detected in normal human plasma have an apparent molecular weight of 850,000. These molecules would appear to be disulfide-bonded tetramers of a basic 200,000 dalton subunit (9–11), and the larger forms have molecular weights indicating that they are composed of an integral number of these tetramers. Serum obtained from blood clotted in glass tubes at 37°C has the same VIIIR population (12). As many as eight separate bands can be detected in most normal plasmas, and there is, in addition, a population of poorly resolved VIIIR with M_r of approximately $8–12 \times 10^6$.

Purified human VIIIR is a glycoprotein containing 5–6% carbohydrate and the amino acid composition has been determined (9–11).

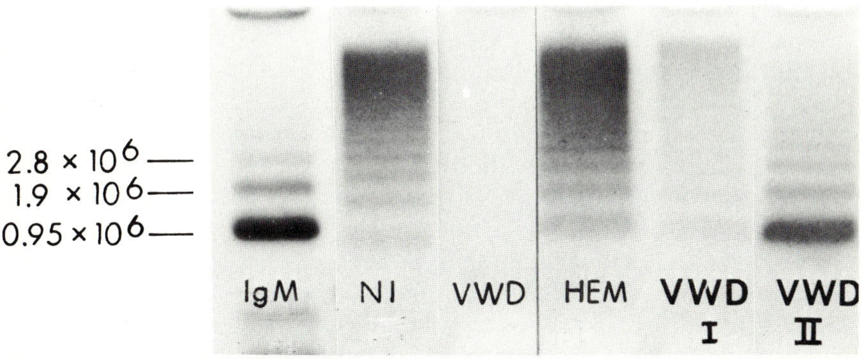

Fig. 2. The polymer pattern of human plasma VIIIR analyzed by SDS-agarose electrophoresis. The migration of IgM and IgM polymers is indicated on the left and their molecular weights are noted. The VIIIR in plasma samples was identified by autoradiography after the gels were incubated with ^{125}I-labeled rabbit anti-VIIIR:Ag (12). From the left, these plasmas are from a normal individual and patients with severe von Willebrand's disease (VIIIR:Ag < 0.01 U/ml), severe hemophilia, mild type I ("classic") von Willebrand's disease, and mild type II ("variant") von Willebrand's disease.

Normal human plasma has between 5 and 10 μg/ml of VIIIR protein, a value that is approximately 100 times greater than the VIII:C content.

IMMUNOLOGIC PROPERTIES

Rabbits immunized with purified human factor VIII form useful immunoprecipitating antibodies (13). Although the sera vary in their ability to inactivate VIII:C, they all form immunoprecipitates with VIIIR and inactivate plasma ristocetin cofactor activity as well as other measures of VIIIR-platelet interaction.

VON WILLEBRAND FACTOR ACTIVITY

Factor VIII-related protein has a central role in normal platelet function, a property that has been designated *von Willebrand factor* activity because it is deficient in patients with von Willebrand's disease.

The prolonged bleeding time in these patients is presumed to be due to the reduced plasma VIIIR content, and it is corrected by transfusion of VIIIR-rich cryoprecipitate. Two *in vitro* platelet assays are usually abnormal in von Willebrand's disease: ristocetin-induced platelet agglutination and retention of platelets in glass bead columns (14). Both are corrected when purified normal factor VIII is added to blood of a patient with severe von Willebrand's disease. While the abnormal platelet function in von Willebrand's disease is due to a plasma deficiency and can be corrected by addition of normal VIIIR, these phenomena also require the presence of a normal platelet surface protein that is deficient in Bernard–Soulier syndrome (15). Only the larger VIIIR forms bind to platelets, however, and it is these large polymers that have VIIIR:RC activity (16).

Ristocetin-induced aggregation is best examined by assays that compare the effect of plasma dilutions on washed normal platelets. These ristocetin cofactor assays can be done with freshly prepared platelets or with formaldehyde-fixed cells. The rate of ristocetin-induced aggregation is related to the amount of plasma added, and the value can be obtained from an aggregometer tracing or by measurement of the time required for platelet agglutination (17).

Although there is a good correlation of ristocetin cofactor activity *in vitro* and the estimated *in vivo* level of von Willebrand factor, as judged by freedom from abnormal bleeding and by bleeding time measurements (18), there are exceptions. In fact, patients with a variant form of von Willebrand's disease have long bleeding times in spite of low-normal ristocetin cofactor values and increased reactivity when ristocetin is added to platelet-rich plasma (19).

THE INTERACTION OF VIII:C AND VIIIR IN THE FACTOR VIII COMPLEX

Although VIII:C and VIIIR have very distinct properties, they interact in plasma through noncovalent bonds to form a stable complex. This is thought to be the basis for the very similar relative concentrations of the two proteins in normal plasmas (Fig. 3) and in most (nonhematologic) disease states (2,20). These data, and the gel filtration studies that detect both VIII:C and VIIIR in void volume fractions, suggest that VIII:C and VIIIR interact in some way. Although we have very little information about the way this complex is formed, the susceptibility to dissociation by high salt buffers implies that electrostatic forces are important. The physiological role of this plasma

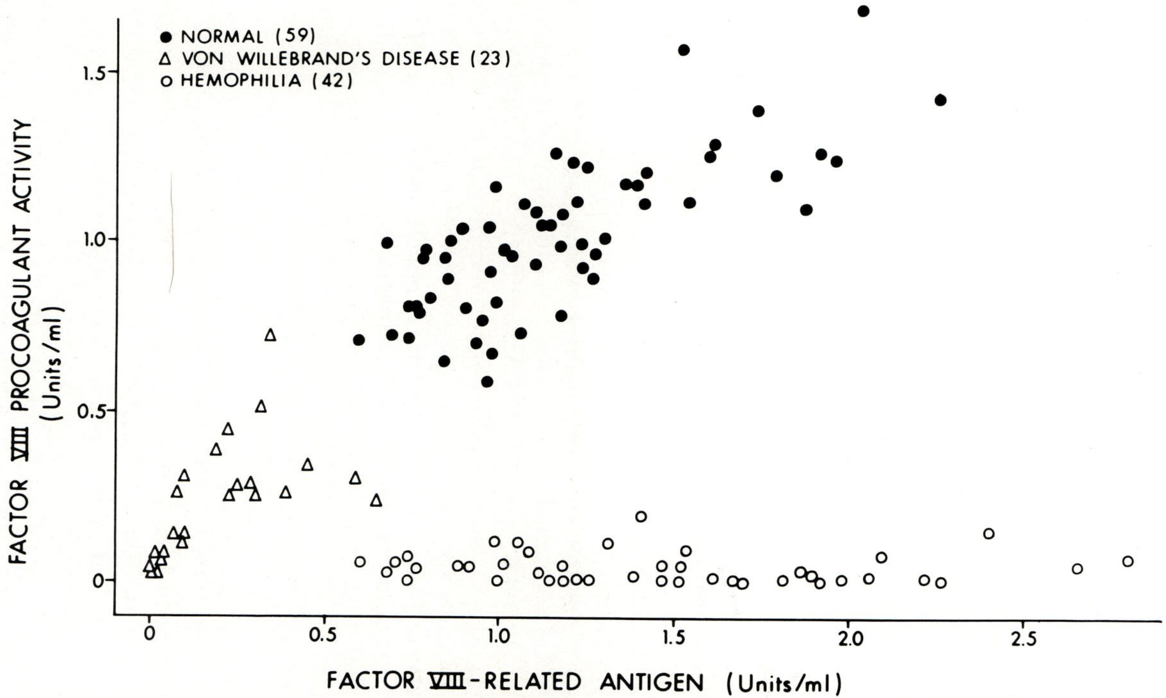

Fig. 3. The relationship of factor VIII procoagulant activity and factor VIII-related antigen in normal individuals and in patients with hemophilia or von Willebrand's disease.

interaction is also uncertain, but some data indicate that VIII:C stability is greater when complexed with VIIIR (21).

THE FACTOR VIII DEFICIENCY DISEASES

HEMOPHILIA A

Although the low factor VIII concentration in plasma has prevented biochemical studies in classic hemophilia (hemophilia A), immunologic techniques have begun to define the molecular defect. These studies have demonstrated both quantitative and qualitative changes in the VIII:C protein, and normal VIIIR concentration and function (Figs. 1 and 3). There is no detectable plasma VIII:CAg in most patients with severe hemophilia (VIII:C less than 1% of normal). There are low VIII:CAg levels (between 1 and 10% of normal) in one-fourth of these plasmas even though there is no detectable VIII:C coagulant activity (2).

Variable VIII:CAg levels are present in plasmas of patients with mild or moderate hemophilia; usually there is slightly more immunoreactive material than VIII:C (6). It would appear that nonfunctional VIII:C-like molecules are synthesized by most hemophilic patients, and normal levels of VIII:CAg are detected in rare instances (2,6): these patients have an X-chromosome mutation that modifies VIII:C structure. The absence of detectable VIII:C protein in the remaining patients may reflect a structural defect that is so severe that antigenic reactivity is lost (as well as coagulant function) or it may indicate that there is no VIII:C-like protein in the plasma.

VON WILLEBRAND'S DISEASE

Immunologic and biochemical assays have also clarified our understanding of von Willebrand's disease. In this case the hemostatic disorder is transmitted by an autosomal locus that affects VIIIR structure and concentration (Fig. 1). Qualitative and quantitative VIIIR defects have been identified in this disease; the reduced VIII:C levels appear to be secondary.

In its most frequent form, von Willebrand's disease is a mild or moderate bleeding disorder in which both components of the factor VIII complex are reduced in quantity and the bleeding time is prolonged (Fig. 3) (14). While the VIIIR multimer pattern has a normal appearance (Fig. 2), the amount of VIIIR protein is reduced.

Severe von Willebrand's disease is a distinct (and unusual) condition in which individuals have very low levels of both factor VIII components, a markedly prolonged bleeding time, and a major bleeding diathesis. Family studies often demonstrate that these patients are homozygous offspring of parents with mild or asymptomatic von Willebrand's disease.

It is not certain why VIII:C is low in von Willebrand's disease, for these patients have the genetic capacity to synthesize this protein. Recent studies suggest that normal VIIIR protects VIII:C from inactivation, and it is possible that VIIIR deficiency permits VIII:C inactivation *in vivo* (21). It is also possible that plasma VIIIR levels may, in some poorly understood way, have an effect on VIII:C synthesis or its release into plasma (22).

Most patients with von Willebrand's disease have similar values for the different factor VIII assays—this has been designated the "classical" pattern. Other patients have been found to have nonfunctional VIIIR, and these individuals have normal or slightly reduced VIII:C and VIIIR:Ag, very low VIIIR:RC, and a prolonged bleeding time. In this case the VIIIR:Ag pattern is abnormal on crossed immunoelectrophoresis because of an increased proportion of small VIIIR and an absence of the largest polymers (Fig. 2). Hemostasis is impaired in these patients since normal platelet adhesion requires large VIIIR. The variant form of von Willebrand's disease has been designated "type II" by some investigators to distinguish it from the classical "type I" pattern in which there is a normal polymer pattern and similar reduction in all factor VIII values.

A distinction has also been made between two subtypes of type II von Willebrand's disease. Platelet-rich plasma from patients with the "type IIB" disorder has increased reactivity when tested for ristocetin-induced platelet aggregation and plasma VIIIR:RC is normal or minimally reduced (16,19). In contrast, patients with the more common "type IIA" disease have markedly reduced VIIIR:RC and their platelets do not aggregate when ristocetin is added.

VIIIR SYNTHESIS BY ENDOTHELIAL CELLS

The site of VIIIR synthesis was established very soon after monospecific rabbit VIIIR:Ag was prepared. Immunofluorescence studies clearly implicated the endothelial cell in factor VIII synthesis (1), and these findings were quickly confirmed (23). A wide variety of human tissues were examined in these studies and only endothelial cells,

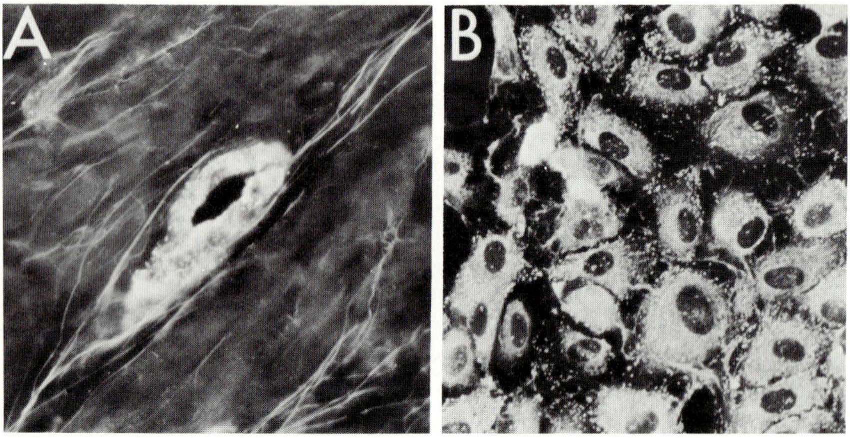

Fig. 4. Immunofluorescence photomicrographs of sections stained with rabbit anti-VIIIR:Ag. (A) Positive staining of endothelial cells of a venule of human cardiac muscle. (B) Positive staining of cultured endothelial cells.

platelets, and megakaryocytes had positive immunofluorescence with rabbit anti-VIIIR:Ag (Fig. 4) (23,24). At approximately the same time that these observations were made, human umbilical cord endothelial cells were first regularly maintained in tissue culture (25). This permitted an examination of isolated endothelial cells, and both VIIIR:Ag and VIIIR:RC activity were detected in the culture media (26,27).

The detection of VIIIR:Ag in the endothelial cell culture medium strongly suggested that these cells synthesize VIIIR, and this interpretation was verified by studies carried out with labeled amino acids. Tritium-labeled amino acids were incorporated into newly synthesized protein, and the culture fluid proteins were subsequently analyzed by immunoprecipitation using rabbit anti-VIIIR:Ag. These experiments demonstrated that the endothelial cells do, in fact, synthesize VIIIR (26). Only $1.5 \pm -0.3\%$ of the protein-bound radioactivity was nonspecifically incorporated in immunoprecipitates when the endothelial cell culture medium was incubated with normal rabbit serum and goat anti-rabbit IgG; the substitution of rabbit anti-VIIIR:Ag led to the precipitation of $7.0 \pm 0.9\%$ of the protein-bound radioactivity (26). Similar evidence for VIIIR synthesis by endothelial cells was obtained by Shearn and co-workers (28). Their studies demonstrated [^{35}S]methionine incorporation in immunoprecipitates obtained when the labeled culture fluid was tested with rabbit anti-VIIIR:Ag.

VIIIR synthesized by cultured human umbilical cord endothelial cells has properties very much like VIIIR purified from human plasma. *In vitro* function appears to be comparable, for ristocetin-mediated platelet agglutination is supported, as is the retention of platelets in glass bead columns (27,28). Endothelial cell VIIIR has the same agarose gel filtration properties as plasma VIIIR (26) and a line of immunologic identity is formed when plasma factor VIII concentrates are tested with rabbit anti-plasma VIIIR:Ag or with rabbit antibodies obtained after immunization with endothelial cell VIIIR (27).

Like human plasma VIIIR, the VIIIR synthesized by cultured endothelial cells is made up of a population of multimers (Fig. 5). A relative increase in both the largest and smallest molecular-weight forms is apparent when the endothelial cell culture fluid is compared to plasma by this method. This analysis clarifies the previous discrepancy between the VIIIR:Ag crossed immunoelectrophoresis pattern

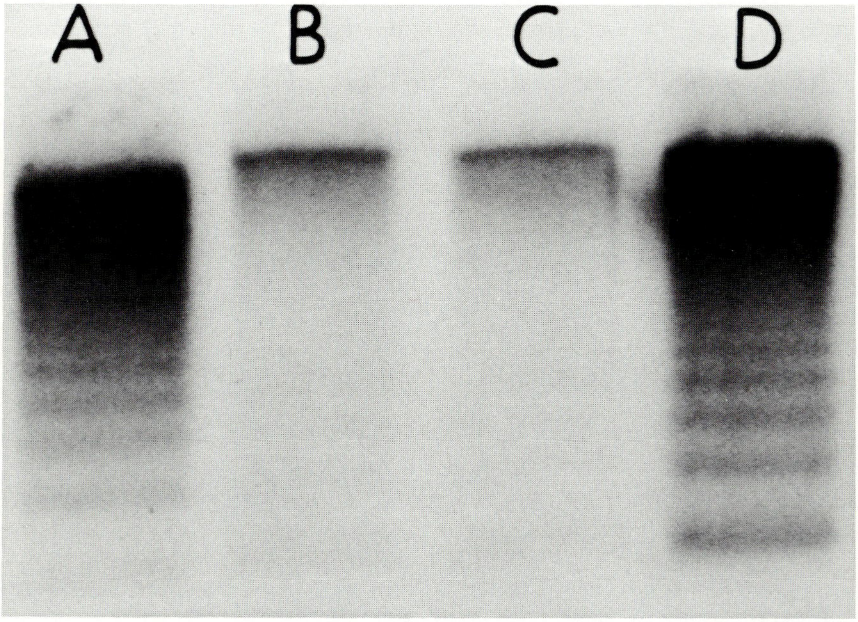

Fig. 5. The VIIIR polymer pattern detected by SDS-agarose electrophoresis. The separation was carried out in the same way as that noted for Fig. 2, and the VIIIR protein was identified by autoradiography after the gels were incubated with ^{125}I-labeled rabbit anti-VIIIR:Ag. The samples include normal human plasma (A), culture fluid from human umbilical cord endothelial cells (B and C), and a 10-fold concentrate of the endothelial cell culture fluid (D).

obtained by Shearn and co-workers (28)—increased anodal mobility when compared to plasma—and that reported by Nachman and co-workers (29)—more prominent slower moving, less anodal VIIIR:Ag. The large and small VIIIR were separately purified in the latter study and were subsequently examined by two-dimensional high voltage electrophoresis and thin layer chromatography of ^{125}I-labeled trypsin digests. As the two fractions had identical properties, it was concluded that the molecular heterogeneity was due to variable association of similar subunits (29). Jaffe and Nachman had previously demonstrated that endothelial cell VIIIR has the same subunit composition as plasma VIIIR (30).

The physiological control of endothelial cell VIIIR synthesis and release has recently been examined in some detail by Tuddenham and co-workers (31). In studies of culture media from multiple replicate cultures at confluence, a wide range of concentrations of several agents failed to alter the rate of VIIIR:Ag release. Among the agents tested were epinephrine, serotonin, 2,3-DPG, cyclic AMP, thyroxin, hydrocortisone, and human growth hormone. None of these compounds had any effect on the rate of VIIIR:Ag accumulation in the culture medium.

VIII:C SYNTHESIS

In contrast to the unequivocal demonstration of VIIIR synthesis by cultured endothelial cells (26,28) and by megakaryocytes (32), the source of VIII:C synthesis is not yet known. Extensive evaluation of endothelial cell culture fluid has not been successful in detecting either VIII:C or VIII:CAg (26,31,33). Although it has been suggested that a protease on the surface of culture endothelial cells degrades VIII:C procoagulant activity (34), control studies have established that the failure to detect VIII:CAg is not due to its instability in the tissue culture medium (31). Thus the failure to detect VIII:C protein is presumed to reflect the lack of VIII:C synthesis by cultured endothelial cells. Immunofluorescence studies using human anti-VIII:C are also negative when this protein is sought in cultured human endothelial cells or in tissue sections.

SUMMARY

The components of the factor VIII complex have a central role in primary hemostasis (VIIIR) and in plasma coagulation (VIII:C). While the former is synthesized by cultured human endothelial cells, and is thus available for immediate release into the circulation, the site of VIII:C synthesis is not yet known. This protein has not been detected in endothelial cells or in products of cultured endothelial cells. Similar studies have been no more successful in the analysis of other cell types, however, and the source of plasma VIIIR:C remains unknown.

ACKNOWLEDGMENTS

These studies were supported in part by research grants HL16626 and HL16872 from the United States Public Health Service.

REFERENCES

1. de los Santos, R. P., and Hoyer, L. W. (1972) *Fed. Proc., Fed. Am. Soc. Exp. Biol.* **31**, 262.
2. Hoyer, L. W. (1981) *Blood* **58**, 1–13.
3. Vehar, G. A., and Davie, E. W. (1980) *Biochemistry* **19**, 401–410.
4. Hoyer, L. W., and Trabold, N. C. (1981) *J. Lab. Clin. Med.* **97**, 50–64.
5. Nesheim, M. E., Myrmel, K. H., Hibbard L., and Mann, K. G. (1979) *J. Biol. Chem.* **254**, 508–517.
6. Lazarchick, J., and Hoyer, L. W. (1978) *J. Clin. Invest.* **62**, 1048–1052.
7. Peake, I. R., Bloom, A. L., Giddings, J. C., and Ludlam, C. A. (1979) *Br. J. Haematol.* **42**, 269–281.
8. Hultin, M. B., and Nemerson, Y. (1978) *Blood* **52**, 928–940.
9. Marchesi, S. L., Shulman, N. R., and Gralnick, H. R. (1972) *J. Clin. Invest.* **51**, 2151–2161.
10. Shapiro, G. A., Andersen, J. C., Pizzo, S. V., and McKee, P. A. (1973) *J. Clin. Invest.* **52**, 2198–2210.
11. Legaz, M. E., Schmer, G., Counts, R. B., and Davie, E. W. (1973) *J. Biol. Chem.* **248**, 3946–3955.
12. Hoyer, L. W., and Shainoff, J. R. (1980) *Blood* **55**, 1056–1059.
13. Zimmerman, T. S., Ratnoff, O. D., and Powell, A. E. (1971) *J. Clin. Invest.* **50**, 244–254.
14. Hoyer, L. W. (1976) *Prog. Hemostasis Thromb.* **3**, 231–287.
15. Howard, M. A., Hutton, R. A., and Hardisty, R. M. (1973) *Br. Med. J.* **2**, 586–588.
16. Ruggeri, Z. M., and Zimmerman, T. S. (1980) *J. Clin. Invest.* **65**, 1318–1325.
17. Sarji, K. E., Stratton, R. D., Wagner, R. H., and Brinkhous, K. M. (1974). *Proc. Natl. Acad. Sci. U.S.A.* **71**, 2937–2941.
18. Weiss, H. J., Hoyer, L. W., Rickles, F. R., Varma, A., and Rogers, J. (1973) *J. Clin. Invest.* **52**, 2708–2716.
19. Ruggeri, Z. M., Pareti, F. I., Mannucci, P. M., Ciavarella, N., and Zimmerman, T. S. (1980) *N. Engl. J. Med.* **302**, 1047–1051.
20. Bloom, A. L. (1977) *In* "Recent Advances in Blood Coagulation" (L. Poller, ed.), pp. 141–181. Churchill-Livingstone, Edinburgh and London.
21. Weiss, H. J., Sussman, I. I., and Hoyer, L. W. (1977). *J. Clin. Invest.* **60**, 390–404.

22. Owen, C. A. Jr., Bowie, E. J. W., and Fass, D. N. (1979) *Br. J. Haematol.* **43**, 307–315.
23. Bloom, A. L., Giddings, J. C., and Wilks, C. J. (1973) *Nature* (*London*) New Biol. **241**, 217–219.
24. Hoyer, L. W., de los Santos, R., and Hoyer, J. R. (1973) *J. Clin. Invest.* **52**, 2737–2744.
25. Jaffe, E. A., Nachman, R. L., Becker, C. G., and Minick, C. R. (1973) *J. Clin. Invest.* **52**, 2745–2756.
26. Jaffe, E. A., Hoyer, L. W., and Nachman, R. L. (1973) *J. Clin. Invest.* **52**, 2757–2764.
27. Jaffe, E. A., Hoyer, L. W., and Nachman, R. L. (1974) *Proc. Natl. Acad. Sci. U.S.A.* **71**, 1906–1909.
28. Shearn, S. A. M., Peake, I. R., Giddings, J. C., Humphrys, J., and Bloom, A. L. (1977) *Thromb. Res.* **11** (1), 43–56.
29. Nachman, R. L., Jaffe, E. A., and Ferris, B. (1981) *Biochim. Biophys. Acta* **667**, 361–369.
30. Jaffe, E. A., and Nachman, R. L. (1975) *J. Clin. Invest.* **56**, 698–702.
31. Tuddenham, E. G. D., Lazarchick, J., and Hoyer, L. W. (1981) *Br. J. Haematol.* **47**, 617–626.
32. Nachman, R., Levine, R., and Jaffe, E. A. (1977) *J. Clin. Invest.* **60**, 914–921.
33. Piovella, F., Giddings, J. C., Peake, I. R., Ricetti, M., Shearn, S. A. M., and Bloom, A. L. (1977) *Lancet* **2**, 889.
34. Stead, N. W., and McKee, P. A. (1979) *Blood* **54**, 560–572.

Fibrinolytic Components of Cultured Endothelial Cells

DAVID J. LOSKUTOFF, EUGENE LEVIN, AND
LUCIANA MUSSONI
Research Institute of Scripps Clinic
La Jolla, California

INTRODUCTION

The fibrinolytic system is the apparent vehicle of plasma for the dissolution of blood clots. It guarantees continued circulation of the blood through an unobstructed vascular bed. Unfortunately, in spite of more than 30 years of research, the identity and origin of the plasminogen activator (PA) in blood responsible for initiating and controlling this process remains controversial (1). The possibility that inhibitors of PA exist has been suggested (1–3) and seems likely. Although these inhibitors probably serve important regulatory roles in controlling the fibrinolytic system, little is known about them, their origin, or their specificity. No inhibitor of PA has been purified. Much of the confusion surrounding the characteristics of the plasma fibrinolytic system results from the presence of multiple, functionally distinct types of PA in blood (1,4), their inherent instability (5), the inability of most assay systems to distinguish among them, and the presence of molecules in blood that inhibit them (3). In an attempt to circumvent these problems, we have begun to characterize the fibrinolytic components of cultured endothelial cells (ECs), since available information suggests that the endothelium may be the site where vascular PA originates (6). The purpose of this review is to summarize (1) studies *in situ* that implicate the endothelium in vascular fibrinolysis, (2) experiments to delineate the fibrinolytic components (both activators and inhibitors) produced by cultured ECs, and (3) observations indicating that the fibrinolytic potential of the endothelium may be modulated by the presence of new information or "signals" brought to the endothelium by the thrombus itself.

PATHOBIOLOGY OF THE ENDOTHELIAL CELL

ISBN 0-12-521980-6

VASCULAR FIBRINOLYSIS

The conversion of fibrinogen to fibrin by thrombin represents the culmination of the coagulation system, the fibrin providing a necessary supporting framework for the hemostatic plug. To ensure hemostasis but not compromise blood vessel patency, the size and duration of the plug must be precisely regulated. Thus as soon as their supporting function is completed, the fibrin deposits are removed by the proteolytic enzyme, plasmin, a process termed fibrinolysis (1,2).

Plasmin is not normally present in blood. Only after PA-mediated cleavage of a single arginine–valine bond in the circulating zymogen, plasminogen, does the plasmin molecule emerge as two disulfide-linked polypeptide chains (Fig. 1). Because plasminogen is present in blood at high concentrations (7), changes in the availability and activity of PA's appear to contribute most significantly to the dynamic regulation of this multicomponent enzyme system. PAs are classified into two distinct groups, the urinary or urokinase-like activators and the tissue activators (TA; 1,8). The presence of both activator types in blood as well as a third, apparently unrelated system [the so-called intrinsic system of fibrinolysis (4,9)] has hampered attempts to characterize the fibrinolytic activity of plasma. Urokinase (UK) has been studied extensively and employed as a preferred agent for thrombolytic fibrinolysis (10). It has been isolated from cultured kidney cells as well as urine (11), generally as two molecular-weight forms (54,000 and 31,000), the smaller being derived from the larger by proteolysis. The TAs are also serine proteinases of variable, but generally of a larger molecular weight than UK. They can be extracted from tissues by chaotropic agents or detergents, are immunologically related to each other but not to UK, and have been purified from a number of different tissues (1,2,5,12,13).

Although each of these activators is an immunologically distinct

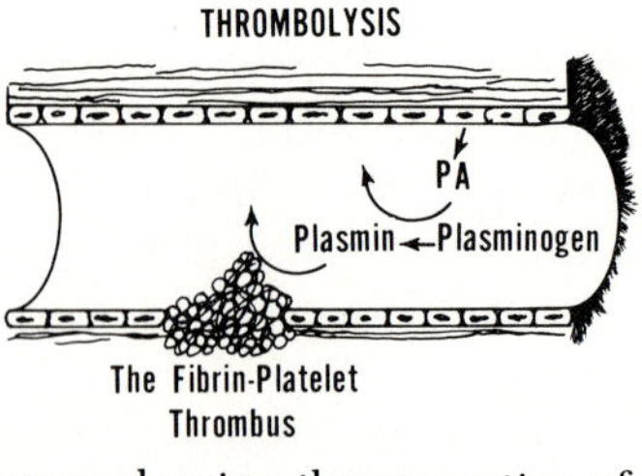

Fig. 1. Schematic diagram showing the generation of thrombolytic activity in a longitudinal section of a typical blood vessel.

1. UK + plasminogen→plasmin→fibrinogenolysis

2. TA + plasminogen→TA + plasminogen

 2 A. TA + fibrin→[TA* – fibrin]

 2 B. [TA* – fibrin] + plasminogen→plasmin→fibrinolysis

Fig. 2. Distinction between UK and TA.

molecule, they both activate plasminogen through cleavage of the same peptide bond. However, the way this activation is achieved differs considerably. This distinction is illustrated in Fig. 2. When UK is mixed with plasminogen in the test tube, active plasmin is formed (reaction 1). However, when TA and plasminogen are mixed, little if any plasmin is generated (reaction 2). TA requires a cofactor before its activity can be expressed, which, interestingly, seems to be fibrin itself (14). Neither fibrinogen nor other proteins (15) can substitute for fibrin in this system. The TA binds to fibrin and is altered in an as yet undefined way to form an active complex (reaction 2A), which in turn can cleave and activate plasminogen (reaction 2B). The effect of fibrin on the cleavage of $[^{125}I]$plasminogen by UK and TA is evident in Fig. 3. Since plasminogen also binds fibrin, these properties guarantee that plasmin will form on the surface of the fibrin molecule itself, not free in the circulation. The ability of TA to initiate fibrinolysis but not fibrinogenolysis suggests that it, not UK, is the physiologically relevant activator for limiting the size and duration of the fibrin–platelet thrombus.

The vascular activator in blood presumably originates from the endothelium. This generally accepted inference comes from a number of indirect observations, including the strategic location of the endothelium at the interface between blood and tissue, and the consistent finding that the fibrinolytic activity of various tissues is associated with their vascular system, particularly the endothelium (6). In addition, a potent PA activity can be recovered when blood vessels are perfused (12), and a good correlation exists between increased blood fibrinolytic activity and decreased PA activity in the endothelium (17). EC's scraped from human inferior vena cava (18) and pure populations of bovine EC's in culture (19) were both fibrinolytically active. However, none of these studies distinguished whether the endothelium synthesizes PA, stores PA, or adsorbs the blood activator. Nevertheless, the idea that the EC activator must be a TA (1,2) may finally have an experimental basis, since antiserum against a TA prepared from the human uterus quenched the fibrinolytic activity of the vascular wall

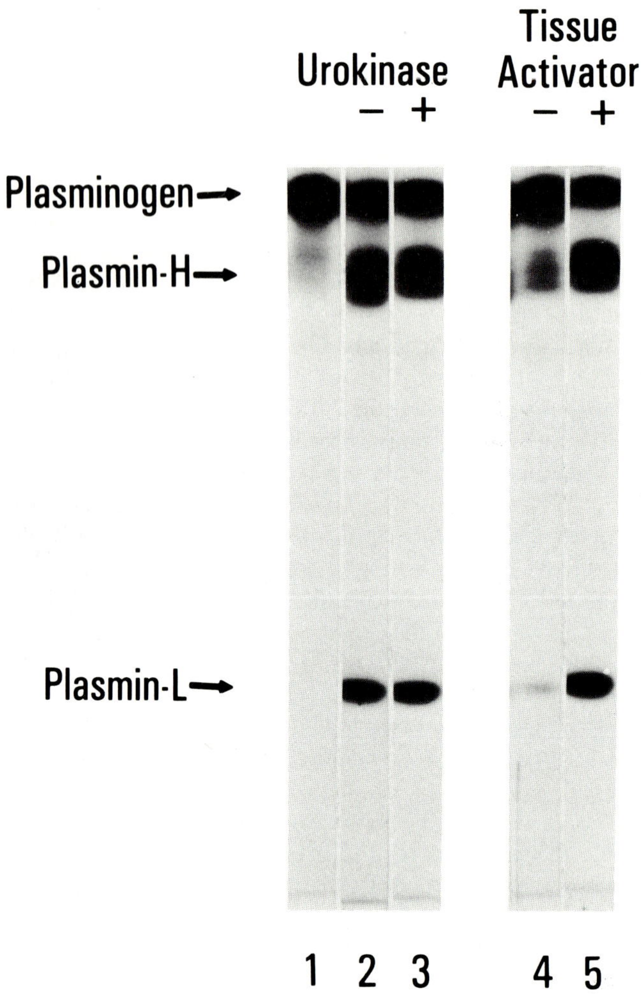

Fig. 3. Autoradiogram showing the effect of fibrin on cleavage of [^{125}I]plasminogen. [^{125}I]plasminogen (2×10^6 cpm/ml; lane 1) was mixed with 10 units/ml UK (lane 2, 3) or TA (lane 4, 5) and was incubated for 60 min at 37°C in the absence (−) and presence (+) of fibrin (1 mg/ml fibrinogen, 1 unit/ml thrombin). The final reaction mixture also contained 0.5% Triton X-100 in 0.1 *M* phosphate buffer, pH 7.4, supplemented with 100 μg/ml albumin and 1000 units/ml trasylol. The samples were boiled in the presence of β-mercaptoethanol and analyzed by electrophoresis on 10% SDS slab gels (SDS-PAGE) according to the method of Laemmli (16). The gel was dried and an autoradiogram prepared.

(20). Antisera against UK had no effect on the activities of either the EC or the plasma activator.

FIBRINOLYTIC COMPONENTS OF CULTURED EC

As the foregoing discussion illustrates, the endothelium appears to be fibrinolytically active, but the molecular basis of this activity remains speculative. To distinguish between synthesis of fibrinolytic enzymes by ECs and adsorption of these molecules to the endothelium from plasma, the fibrinolytic components of ECs cultured for generations *in vitro* were characterized. The prolonged length of time that the cells were in culture minimized the possibility that PA activity resulted from the release of a stored molecule. Figure 4 indicates that when rabbit ECs (REVC) were grown on surfaces coated with [^{125}I]fibrin, the fibrin film was digested in a time-dependent manner. When the cells were cultured in the absence of plasminogen, this fibrinolytic activity was not observed, indicating that it resulted from the presence of a PA. The addition of phorbol myristate acetate (PMA), a tumor-promoting agent, to these cultures stimulated fi-

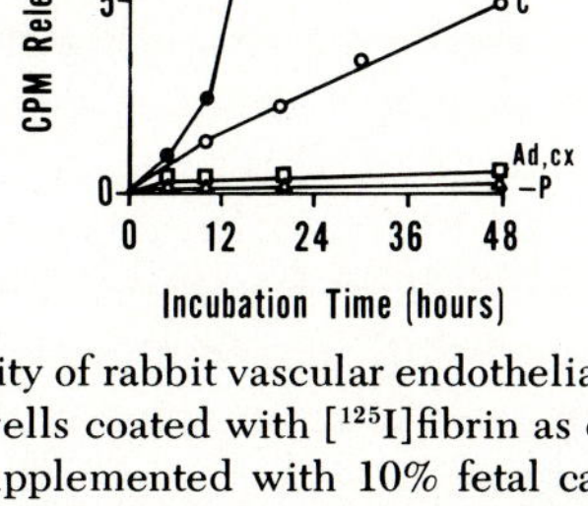

Fig. 4. Fibrinolytic activity of rabbit vascular endothelial cells (REVC). REVC (1 × 10^5) were added to Linbro wells coated with [^{125}I]fibrin as described (21). The growth medium contained MEM supplemented with 10% fetal calf serum depleted of plasminogen. Twenty-four hours later the cultures were washed and the medium was replaced with fresh MEM (△——△) or MEM containing 4 μg/ml plasminogen (○——○) supplemented with either PMA (5 × 10^{-8} *M*; ●——●) or PMA and either (□——□) cycloheximide (cx; 10 μg/ml) or actinomycin D (Ad; 1 μg/ml). Aliquots of media were removed at the indicated times and the radioactivity was determined in a gamma counter. The data are expressed as a percent of the total trypsinizable counts on the surface of each well. C = control.

brinolysis by a factor of 5 to 10. This amplification of activity was not observed in cells pretreated with agents that block RNA or protein synthesis, consistent with the conclusion that ECs have the potential to produce PA. This PA activity was present both in cell extracts, where it was associated with a large, sedimentable, detergent-sensitive structure, and in conditioned media, where it accumulated with time (21). The secreted PA was not sedimentable through sucrose gradients, apparently having lost its association with cellular material.

The PA activity in extracts from a number of cell types was compared to a standard UK dose–response curve to determine the relative activity in each extract in UK units (Table I). Confluent rabbit and human ECs and sparse cultures of bovine EC's were similar in that they contained relatively low levels of PA activity. However, as the bovine cells approached confluency and became quiescent, they began to produce high levels of PA (22–24). Surprisingly, other, nonendothelial cells of the vascular wall also generated considerable activity. These results indicate that the fibrinolytic potential of the vascular wall is not restricted to the endothelium, nor is the endothelium necessarily a rich source of PA. As reported (25), fibroblasts that lacked detectable PA became high producers after viral transformation. The bovine ECs consistently displayed more PA activity than the transformed cells.

PAs are remarkable in that they are not inactivated by sodium dodecyl sulfate (SDS). Granelli-Piperno and Reich exploited this property to develop a fibrin autograph technique for determining the location of PA activity in SDS-polyacrylamide slab gels (26). We have used this approach to characterize further the forms of PA produced by cultured ECs (Fig. 5). The PA profile obtained from bovine cells is complex, with activity at molecular weights of 48,000, 52,000, 74,000, and 100,000. In addition, a broad range of activity was observed in the region of the gel between the two high-molecular-weight forms. These results indicate that the fibrinolytic activity detected in cultured bovine EC's results from at least three molecular-weight forms of PA. Although the significance of this observation is not clear at present, the results described in the next section suggest that these forms include both UK-like and TA-like PAs.

EC's also seem to contain antifibrinolytic activity (Table I) localized to the cell cytosol (21). This inhibitory activity is acid-labile and appears to block plasminogen activation, but not plasmin itself (27). Interestingly, although the inhibitor irreversibly inactivates UK, it shows no activity against purified TA (Fig. 6). In general, ECs contain potent inhibitory potential when compared to other cells (Table I). The cellu-

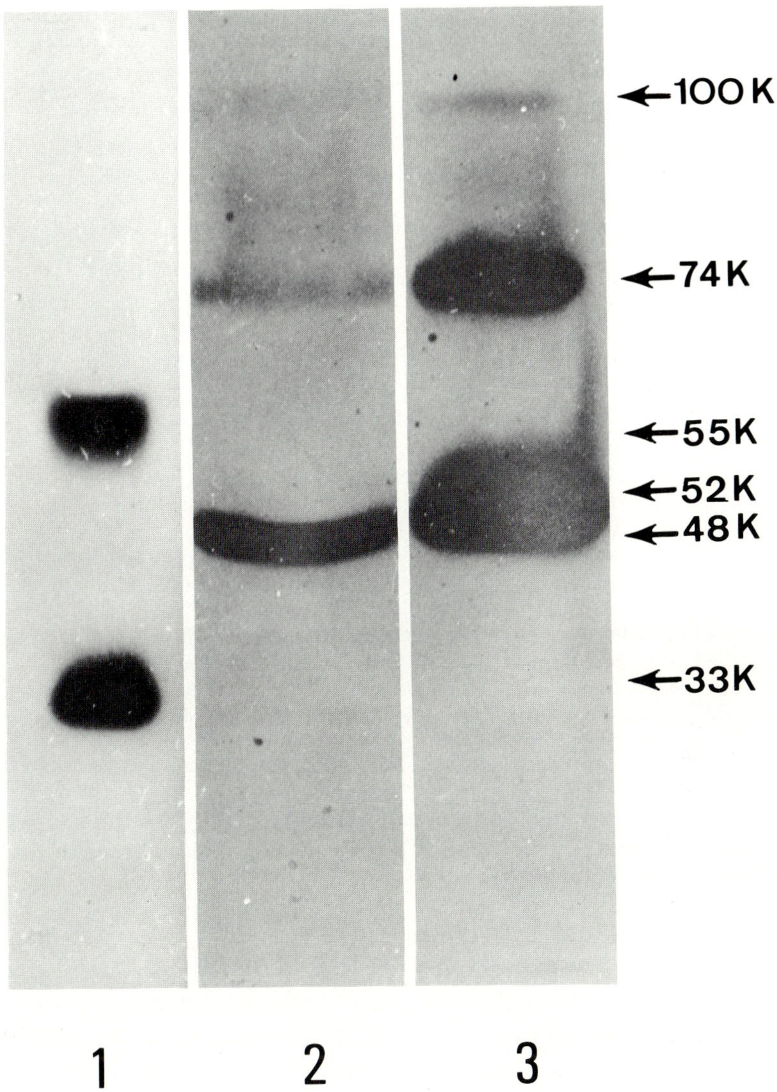

Fig. 5. SDS-PAGE of cell-associated and secreted PA. Urokinase (0.1 unit; lane 1), cell extract (100 μg; lane 2), and conditioned media (100 μl; lane 3) were fractionated by SDS-PAGE, processed to remove the SDS, and then analyzed by the fibrin autograph technique (26). The indicator gel contained 1 mg/ml fibrin and 10 μg/ml human plasminogen in 0.25% agar. The dark zones on the indicator gel correspond to regions of the separating gel containing PA activity.

TABLE I
Fibrinolytic Components Associated with Various Cells

Cell type[a]	PA activity[b]	Inhibitor activity[c]
Endothelial		
Rabbit, vena cava (revc)	0.2	1000
Human, umbilical cord vein	0.2	800
Bovine, aorta (sparse)	0.1	750
Bovine, aorta (confluent)	8.0	N.D.
Smooth muscle		
Rabbit, aorta	3.0	25
Bovine, aorta	5.4	50
Fibroblast		
Chick, embryo	<0.1	100
Chick, embryo: rsv	2.3	50

[a] Cells were inoculated at 2×10^6 cells per 100-mm dish. Seventy-two hours after inoculation the cultures were washed, extracted, and tested for PA activity (22). Extracts of rabbit ECs were acidified prior to analysis (21).

[b] Activity is expressed as units of urokinase activity per mg cell protein as calculated from a urokinase standard curve. The results are the average of duplicate determinations.

[c] Activity is expressed in units of inhibitory activity per mg cell protein. One unit of inhibitory activity is defined as the amount of sample in a 1-ml assay required for a 50% reduction in fibrinolysis induced by 0.025 units of urokinase. The results are the average of duplicate determinations. N.D. = not determined

lar location and unusual degree of specificity distinguish the EC inhibitor from antifibrinolytic agents observed in other cells (29) and in plasma and platelets (3,9).

REGULATION OF EC-MEDIATED FIBRINOLYSIS

It has been suggested (30) that the fibrinolytic system must be in dynamic equilibrium with the coagulation system to achieve and maintain an intact and patent vascular bed. Activation of these thrombotic or thrombolytic pathways must be subject to precise temporal regulation, for any imbalance in these systems results in a tendency toward thrombosis or a hemorrhagic condition (1,2). Although the endothelium is uniquely located to influence this equilibrium, the exact nature and magnitude of endothelial involvement in the control of vascular homeostasis are difficult to establish *in vivo*. It seems likely

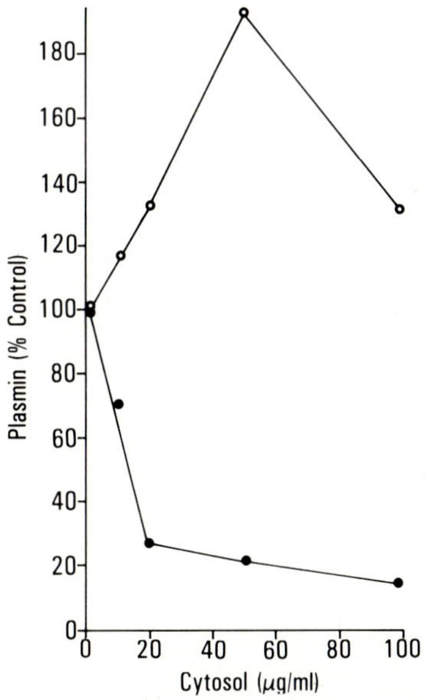

Fig. 6. Effect of cytosol on cleavage of [^{125}I]plasminogen by urokinase (UK) and tissue activator (TA). One unit/ml each of urokinase (●——●) or tissue activator (○——○) were incubated for 5 min at 37°C with increasing amounts of cytosol. [^{125}I]-plasminogen was added to each sample and the incubation continued for another 2 hr at which time the samples were analyzed by SDS-PAGE and autoradiography (28). Quantitation was achieved by excising the radioactive bands in plasminogen, and the plasmin heavy and light chains as localized with the autoradiogram. The figure shows the changes in plasmin (heavy plus light chains) associated radioactivity.

that the thrombus itself may be instrumental in switching the endothelium from a fibrinolytically active structure that maintains vessel patency to a fibrinolytically inert structure that supports thrombosis. To test this possibility, we are examining the effects of molecules found in the thrombus on the PA activity of cultured bovine ECs.

THROMBIN

Thrombin is a highly specific serine protease that not only converts fibrinogen to fibrin but also has dramatic effects on cells (31). Thrombin activates platelets, is mitogenic for some cells, stimulates the production and release of fibronectin from others, and enhances the release of prostacyclin from ECs. The addition of physiological concentration of thrombin to confluent bovine ECs also can cause a rapid and extensive loss of intracellular PA activity and at the same time suppress

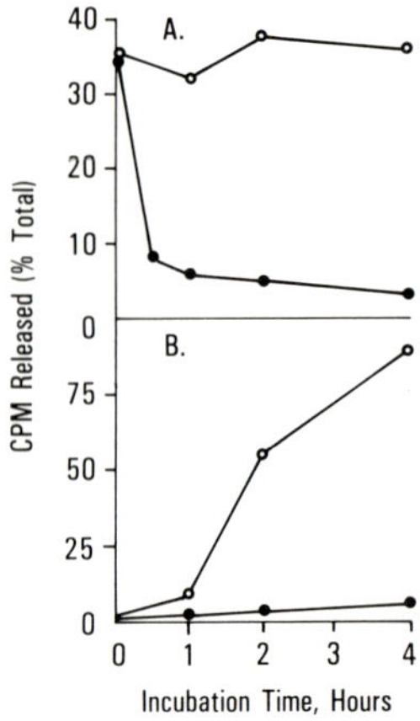

Fig. 7. Effect of thrombin on PA production. Confluent cultures of ECs were washed and then incubated in MEM alone (O——O) or MEM containing 1 unit/ml of thrombin (●——●). At various times (A) the monolayers were washed, and extracted with 0.5% Triton X-100, and then 5 μg of the extract was tested for cell-associated PA activity (31) by the [^{125}I]fibrin plate assay (21,25), or (B), aliquots of the conditioned media were removed from the cells and tested for secreted PA activity.

PA secretion (Fig. 7). The cells were extremely sensitive to thrombin with a half-maximal response observed at 0.1 n*M* thrombin (31). The decrease was essentially complete within 30 min at 37°C, required active thrombin, and did not result from direct inactivation of PA by thrombin itself. The effect of thrombin on PA production was relatively specific, since other proteases (e.g., trypsin, chymotrypsin, and pronase) did not suppress PA production (32). In addition, thrombin was not mitogenic for these cells, did not alter the rate of RNA or protein synthesis, and did not stimulate the release of chromium-51 from cells prelabeled with this molecule. Intracellular PA activity returned to control levels within 6 hr after the addition of the thrombin inhibitor, hirudin. This reversal by hirudin was blocked by cycloheximide but not by actinomycin D, indicating that protein synthesis, but not RNA synthesis, was required. These results indicate that suppression of PA activity by thrombin is dramatic, rapid, reversible, and highly specific. Generation of thrombin at sites of vascular injury may thus protect the growing thrombus against premature, EC-mediated dissolution.

FIBRIN

Fibrinogen is a large, soluble, circulating molecule that is dramatically transformed by thrombin into insoluble fibrin at sites of vascular injury. This transformation not only is the essence of blood coagulation and thrombosis but is also associated with the generation of activities that are not directly related to blood coagulation. For example, fibrinogen, fibrin, and their various physiological derivatives increase microvascular permeability, induce chemotaxis of leukocytes, promote inflammation, and influence platelets, neutrophils, lymphocytes, and other cells (33,34). In addition, the plasmin cleavage products of fibrinogen can suppress the immune response (35). One of the most interesting and specific consequences of converting fibrinogen to fibrin is the appearance of a "cofactor"-like activity that is apparently required to convert previously inactive TA-like PA's into extremely active molecules (14,15). Although the mechanism of this activation is unknown, the fibrin appears to serve as a cofactor to which the TA binds (Fig. 2). In the process, TA is altered to become a much more efficient activator of plasminogen. To investigate whether the multiple forms of PA produced by cultured ECs resemble UK or TA, or in fact include both types, we have begun to study the interaction of these forms with fibrin. One unit each of purified human UK or TA, or the equivalent amount of conditioned medium from confluent cells, was added to a solution of phosphate-buffered saline containing fibrinogen (1 mg/ml). Thrombin (1 U/ml) added to each solution clotted the fibrinogen within 30 sec. After an additional 10 min incubation at 37°C, the clots were removed by centrifugation, dissolved in 0.2% SDS, and tested for PA activity by the standard [^{125}I]fibrin plate assay (21,25). The amount of activity associated with the clot in each instance was compared to that remaining in the supernatant after centrifugation. Self-incubated but noncentrifuged samples served as controls for the calculation of recoveries. As expected, UK did not bind to fibrin under the conditions of this experiment (Table II), whereas TA did. When cellular PA was analyzed in a similar manner, activity was found both in association with the clot and free in the supernatant. These results raise the possibility that cultured EC's produce both UK and TA-like PA's and that these can be recognized by their interaction with fibrin.

The fibrin-autograph technique was then used to evaluate the possibility that distinct molecular forms of PA bind to fibrin and that these differ from forms that do not interact with this molecule (Fig. 8). The clots and supernatant from the experiments shown in Table II were dissolved in SDS, subjected to electrophoresis through 10% SDS gels, and analyzed as described in the legend to Fig. 5. Two classes of PA were, in fact, defined on the basis of the interaction with fibrin. The first class that apparently bound to polymerizing fibrin included only the larger-molecular-weight forms, from 74,000 to 100,000, and, in this

TABLE II
Binding of Various PAs to Fibrin

PA[a]	PA Activity (%)[b]	
	Supernatant	Clot
Urokinase	99.7	0.3
Tissue activator	1.0	99
Conditioned media	45	55

[a] Urokinase (Winkinase) was a kind gift from Sterling–Winthrop Research Institute. Tissue activator was isolated from cultured human melanoma cells and generously provided by D. Collen.

[b] Calculated on the basis of total recovered activity. Fibrin clots were added to the supernatant after fractionation to allow direct comparisons between the two fractions.

respect, resembles TA-like PAs. Whether this association with fibrin also promotes the inherent PA activity of this group of molecules remains to be determined. The second class of molecules produced by these cells did not appear to associate with fibrin and included only the lower molecular weight forms of 48,000 (not shown) and 52,000; thus this second class resembles UK-like PAs.

DISCUSSION

It has been suggested that the coagulation and fibrinolytic systems must maintain a dynamic equilibrium in order to guarantee both an intact vascular bed and vessel patency (30). Yet other studies would seem to indicate that these systems are activated only when needed for specific, local hemostasis (36). Whatever the case, it seems likely that the two systems influence each other. The endothelium is uniquely located to modify this interaction. However, the exact nature and magnitude of endothelial involvement in the control of vascular homeostasis have been difficult to establish *in vivo*. The basic premise of our work is that the endothelium contributes to the fibrinolytic potential of the blood through the synthesis and release of both fibrinolytic activators and fibrinolytic inhibitors, and that the synthesis, release, and activity of these molecules is modified by the presence of "new" information or "signals" brought to the endothelium by the thrombus

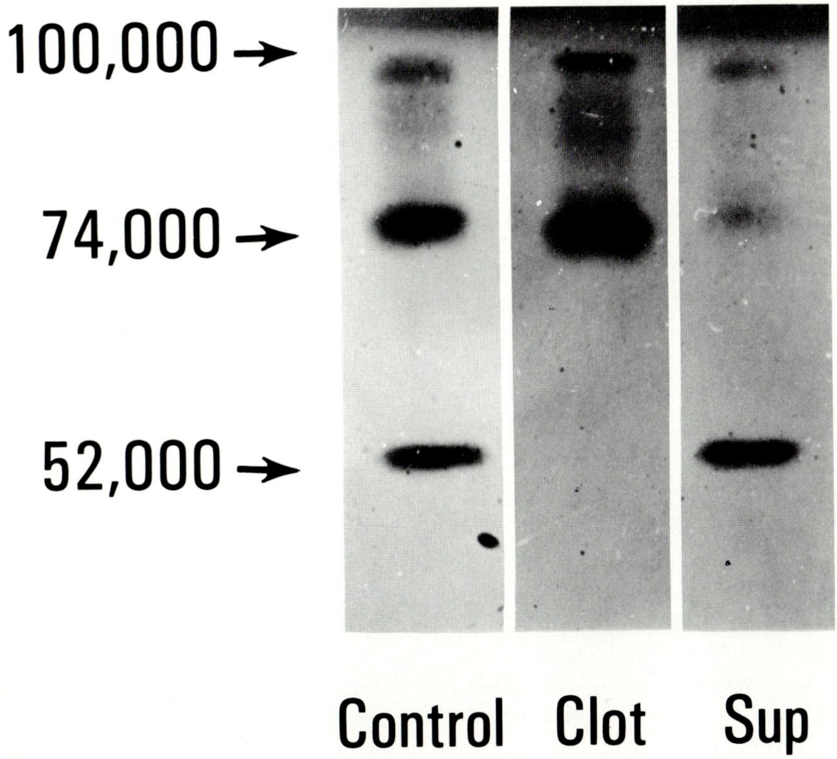

Fig. 8. Binding of PAs to fibrin. Sixteen-hour, serum-free conditioned media from confluent bovine ECs was prepared (22), and 0.5-ml samples were added to tubes containing 1 mg/ml fibrinogen. The fibrinogen was clotted by the addition of thrombin (1 unit/ml) and incubated for 10 min at 37°C. The solution was centrifuged (10,000 *g* for 10 min) and the resulting clot and supernatant (SUP) were extracted with 0.2% SDS and analyzed by SDS-PAGE and fibrin autography as described in the legend to Fig. 5. The control sample was not subjected to centrifugation.

itself. The results summarized in this chapter are consistent with this hypothesis.

First of all, little doubt remains that under some conditions (e.g., cell culture), ECs have the potential to produce PA and respond to exogenous agents such as phorbol myristate acetate and thrombin, by altering the rate of PA production (Figs. 4 and 7). It is also clear that the amount of PA ECs produce varies over a wide range, depending on the species of origin and the growth state (24) of the cells (Table I). These observations correlate generally with *in vivo* findings to support the notion that the endothelium could produce the vascular activator. Definitive proof for this conclusion will require not only the development of techniques to recognize or selectively "mark" PA of the endothelium, but also the ability to then follow its release from the endothelium, its subsequent accumulation in the blood, and finally its association with the thrombus. This sequence of events must be initiated by agents such as epinephrine and nicotine, or events such as stress and vascular occlusion, all known to elevate vascular PA (37). These studies will be technically difficult to conduct because of the exceedingly low concentration of PA in cells, even those relatively high in such activity (38). Moreover, the complexity of blood itself, the large number of tissues that contribute to its composition, and the preliminary findings that many tissues produce similar forms of PA (39,40) belie an early, unambiguous solution to this problem.

We have observed that cells of the vascular wall other than ECs produce substantial amounts of PA in culture (Table I). This behavior may not be apparent *in vivo* (6,41) because antiplasmin activity in the medial, smooth muscle layer of blood vessels (42) masks its activity. Alternatively, the amount of PA these cells produce may increase in culture either because some controlling element disappears when the cells are removed from their normal environment or because of general dedifferentiation in culture.

Cultured ECs from all species examined showed relatively high levels of antifibrinolytic activity (Table I). In the rabbit, where this activity has been most thoroughly detailed (21,27), and perhaps in the human (43), an antiactivator, not an antiplasmin, has been demonstrated as the cause of antifibrinolytic responses. Interestingly, this anti-activator appears to recognize UK, but not TA (Fig. 6). The physiological significance of this latter observation awaits further functional characterization of each activator. The fact that ECs may produce both types of activator adds a new dimension to such considerations.

The suggestion from these *in vitro* studies that the endothelium contains both fibrinolytic activators and inhibitors may account for the wide range of PA activity associated with different regions of the vascular system (6), and for the variable results often found when using the fibrin autograph technique. In addition, these observations suggest that interactions between these molecules contribute to the overall regulation of this system. Therefore, a thorough characterization of the relationship between PA and its inhibitor will be required to understand EC-mediated fibrinolysis. The fact that these molecules are confined to different subcellular compartments raises the possibility that extracellular fibrinolytic activity may be regulated by the selective release of either molecule.

The cell culture studies have also revealed for the first time that EC's produce multiple molecular forms of PA. Vetterlein *et al.* (39) and Wilson *et al.* (40) demonstrated that other cells produce multiple activators as well. That the EC activators are biochemically distinct molecules is suggested by the finding that they fall into two classes: those that bind to fibrin and those that do not. The activity of this latter class of molecules can be blocked by anti-UK but not anti-TA (44), indicating that these are in fact UK-like molecules. This observation differs from that by Rijken *et al.* (20), who detected TA activity alone in the vascular wall. The many possible explanations for this difference include the age of the cells tested, the culture conditions (24), and the difference in species. The binding property of the first class of activator is consistent with its assignment as a TA, but definitive proof awaits demonstration that its activity is promoted by fibrin and blocked with anti-TA but not anti-UK.

ACKNOWLEDGMENTS

The authors wish to thank K. Roegner, D. Lawrence, and M. Sawdey for excellent technical assistance, and L. Simpson for typing the manuscript. We also wish to acknowledge our appreciation to Dr. D. Collen for his generosity in providing us with the purified tissue activator used in these studies. This work was supported by NIH grants HL 22289 and HL 16411 to D. J. L. This publication number is #2493 from the Immunology Departments of the Research Institute of Scripps Clinic.

REFERENCES

1. Collen, D. (1979) *Thromb. Haemostasis* **43,** 77–89.
2. Astrup, T., and Thorsen, S. (1972) *Med. Clin. North Am.* **56,** 153–162.
3. Coccheri, S., de Rosa, V., and Palareti, G. (1977) *In* "Thrombosis and Urokinase" (R. Paoletti and S. Sherry, eds.), pp. 25–36. Academic Press, New York.
4. Kluft, C. (1979) *Thromb. Haemostasis* **41,** 365–383.
5. Radcliffe, R., and Heinze, T. (1978) *Arch. Biochem. Biophys.* **189,** 185–194.
6. Todd, A. S. (1959) *J. Pathol. Bacteriol.* **78,** 281–283.
7. Deutsch, D. G., and Mertz, E. T. (1970) *Science* **170,** 1095–1096.
8. Astrup, T. (1966) *Fed. Proc., Fed. Am. Soc. Exp. Biol.* **25,** 42–51.
9. Kaplan, A., Castellino, F., Collen, D., Wiman, B., and Taylor, F. (1978) *Thromb. Haemostasis* **39,** 263–283.
10. Kakkar, V. V., and Schully, M. F. (1978) *Br. Med. Bull.* **34,** 191–199.
11. White, W. F., Barlow, G. H., and Mosen, M. M. (1966) *Biochemistry* **5,** 2160–2169.

12. Binder, B. R., Spragg, J., and Austen, K. F. (1979) *J. Biol. Chem.* **254,** 1998–2003.
13. Rijken, D. C., Wijngaards, G., Zaal-de Jong, M., and Welbergen, J. (1979) *Biochim. Biophys. Acta* **580,** 140–153.
14. Camiolo, S., Thorsen, S., and Astrup, T. (1971) *Proc. Soc. Exp. Biol. Med.* **138,** 277–280.
15. Wallen, P. (1977) *In* "Thrombosis and Urokinase" (R. Paoletti and S. Sherry, eds.), pp. 91–102. Academic Press, New York.
16. Laemmli, U. K. (1970) *Nature (London)* **227,** 680–681.
17. Nilsson, I. M., and Pandolfi, M. (1970) *Thromb. Diath. Haemorrh.* **30,** 232–242.
18. Kwaan, H. C., and Astrup, T. (1965) *Clin. Res.* **13,** 12.
19. McDonald, R. I., Shepro, D., Rosenthal, M., and Booyse, F. (1973) *Ser. Haematol.* **6,** 469–478.
20. Rijken, D., Wijngaards, G., and Welbergen, J. (1980) *Thromb. Res.* **18,** 815–830.
21. Loskutoff, D. J., and Edgington, T. S. (1977) *Proc. Natl. Acad. Sci. U.S.A.* **74,** 3903–3907.
22. Levin, E. G., and Loskutoff, D. J. (1979) *Thromb. Res.* **15,** 869–878.
23. Loskutoff, D. J., and Levin, E. G. (1980) *In* "The Regulation of Coagulation" (K. Mann and F. B. Taylor, eds.), pp. 589–595. Elsevier/North-Holland, New York.
24. Levin, E. G., and Loskutoff, D. J. (1980) *Cell* **22,** 701–707.
25. Unkeless, J., Tobia, A., Ossowski, L., Quigley, J., Rifkin, D., and Reich, E. (1973) *J. Exp. Med.* **137,** 85–111.
26. Granelli-Piperno, A., and Reich, E. (1978) *J. Exp. Med.* **148,** 223–234.
27. Loskutoff, D. J., and Edgington, T. S. (1981) *J. Biol. Chem.* **256,** 4142–4145.
28. Danø, K., and Reich, E. (1979) *Biochim. Biophys. Acta* **566,** 138–151.
29. Bernik, M. (1978) *Biochim. Biophys. Acta* **525,** 180–185.
30. Astrup, T. (1956) *Blood* **11,** 781–806.
31. Loskutoff, D. J. (1979) *J. Clin. Invest.* **64,** 329–332.
32. Loskutoff, D. J. (1978) *Blood* **52,** 343.
33. Buczko, W., de Gaetano, G., Franco, R., and Donati, M. B. (1976) *Thromb. Haemostasis* **35,** 651–657.
34. Belew, M., Gerdin, B., Lindeberg, G., Poráth, J., Saldeen, T., and Wallin, R. (1980) *Biochim. Biophys. Acta* **621,** 169–178.
35. Germann, G., Pees, H., Schwarze, G., and Scheurlen, P. G. (1976) *Nature (London)* **259,** 399–401.
36. Collen, D., Tutgat, G., Claeys, H., and Piessens, R. (1972) *Br. J. Haematol.* **22,** 681–700.
37. Cash, J. D. (1978) *In* "Progress in Chemical Fibrinolysis and Thrombolysis" (J. F. Davidson, R. M. Rowan, M. M. Samama, and P. C. Desnoyers, eds.), pp. 65–75. Raven, New York.
38. Unkeless, J., Danø, K., Kellerman, G., and Reich, E. (1974) *J. Biol. Chem.* **249,** 4295–4305.
39. Vetterlein, D., Young, P. L., Bell, T. E., and Roblin, R. (1979) *J. Biol. Chem.* **254,** 575–578.
40. Wilson, E. L., Becker, M. L. B., Hoal, E. G., and Dowdle, E. B. (1980) *Cancer Res.* **40,** 933–938.
41. Noordhoek Hegt, V. (1977) *Thromb. Haemostasis* **38,** 407–419.
42. Noordhoek Hegt, V. (1977) *Thromb. Res.* **10,** 121–133.
43. Dosne, A., Dupuy, E., and Bodevin, E. (1978) *Thromb. Res.* **12,** 377–387.
44. Levin, E. G., and Loskutoff, D. J. (1981) *Thromb. Haemostasis* **46,** 240.

Regulation of Fibrinolysis: Plasminogen Activator as a Thrombolytic Agent

D. COLLEN
Department Medical Research
Center for Thrombosis and Vascular Research
University of Leuven, Belgium

INTRODUCTION

The fibrinolytic enzyme system plays a role in the removal of fibrin from the vascular bed and most probably also in several other biological phenomena such as tissue repair (1), malignant transformation (2), macrophage function (2), ovulation (3), and embryo implantation (3). The present review deals only with the mechanism of action of the fibrinolytic system in thrombolysis.

The blood fibrinolytic system consists of three main components: the proenzyme plasminogen, which can be activated by limited proteolysis to plasmin; plasminogen activators, which may be of different origin (blood activator, endothelial activator, tissue activator, or intrinsic plasminogen activator), and inhibitors that rapidly neutralize plasmin or that may interfere with the activation of plasminogen.

The biochemical and biological properties of plasminogen and of α_2-antiplasmin, the main inhibitor of plasmin, have been studied in detail [for references, see ref. (4)]. Much less is known, however, about the "extrinsic" plasminogen activators that are released into the blood, presumably from the vascular wall. During the past few years specific molecular interactions between plasminogen activator and fibrin, between plasminogen and fibrin, and between plasmin and α_2-antiplasmin have been described, on the basis of which a molecular model for the regulation of fibrinolysis *in vivo* was proposed (4,5). In addition, it has recently become possible to purify "extrinsic" plasminogen activator in sufficient amounts to study its biological proper-

PATHOBIOLOGY OF THE ENDOTHELIAL CELL

ISBN 0-12-521980-6

ties in more detail. The present communication will be restricted to a discussion of the mechanisms and interactions regulating fibrinolysis and to recent studies on the biochemical, biological, and thrombolytic properties of an "extrinsic" plasminogen activator purified from a human melanoma cell culture.

REGULATION AND CONTROL OF FIBRINOLYSIS

The regulation and control of fibrinolysis appears to occur at several levels: release of plasminogen activator from the vascular wall, fibrin-associated activation of plasminogen, and inhibition of formed plasmin by α_2-antiplasmin.

RELEASE OF PLASMINOGEN ACTIVATOR

The mechanisms controlling the release of plasminogen activator from the endothelial cells are virtually unknown (6). Cash has speculated that plasminogen activator release may be under neurohumoral control (6). Higher neurogenic centers or specific peripheral afferent organs might stimulate the release of a plasminogen activator releasing hormone (PARH) with a structure similar to vasopressin from the neurohypophyseal region. This PARH would constitute the major pathway for the release of plasminogen activator from the endothelial cells, whereas the catecholamine pathway would only be involved in severely stressful situations (6).

The observation that venous thrombosis is frequently associated with a decreased fibrinolytic activity has stimulated interest in drugs that increase the synthesis and release of plasminogen activator in the vessel wall. The biguanidines and certain anabolic steroids such as ethylestrenol (7) or stanazolol (8) have been reported to produce long-term stimulation of endogenous fibrinolysis by increasing both the synthesis and release of plasminogen activator in the vessel wall (7). A combination therapy with phenformin and ethylestrenol was shown to significantly reduce the occurrence of thrombotic episodes in patients with idiopathic recurrent venous thrombosis (9).

MOLECULAR MECHANISM OF PHYSIOLOGICAL FIBRINOLYSIS

Physiological fibrinolysis appears to be regulated by specific molecular interactions between plasminogen activator and fibrin, between plasminogen and fibrin, and between plasmin and α_2-antiplasmin.

As described in more detail below, extrinsic plasminogen activator has a weak affinity for plasminogen in the absence of fibrin (K_M = 65 μM) but a much higher affinity in the presence of fibrin (K_M between 0.15 and 1.5 μM). This increased affinity appears to be the result of a "surface assembly" of plasminogen activator and plasminogen on the fibrin surface. In this reaction plasminogen binds to fibrin primarily via specific structures called the "lysine-binding site." Thus one way of regulating fibrinolysis is at the level of plasminogen activation localized at the fibrin surface.

Plasmin is extremely rapidly inactivated by α_2-antiplasmin ($k_1 \simeq 10^7$ M^{-1} sec^{-1}) (10); the half-life of free plasmin in the blood is therefore estimated to be approximately 0.1 sec. Plasmin with an occupied lysine-binding site is however inactivated 50 times more slowly by α_2-antiplasmin. Reversible blocking of the active site of plasmin with substrate also markedly reduces the rate of inactivation by α_2-antiplasmin. From these findings one can extrapolate that plasmin

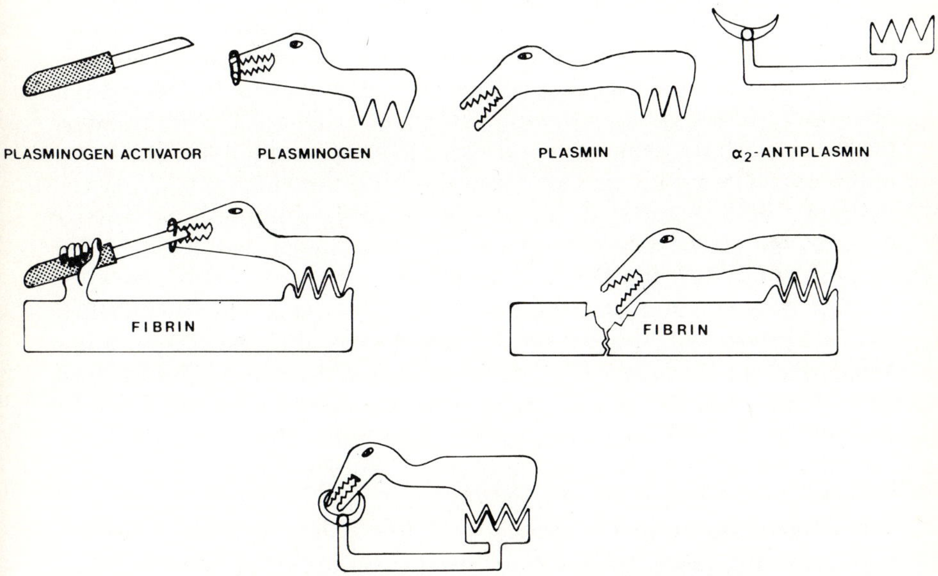

Fig. 1. Schematic visualization of the molecular interactions regulating fibrinolysis. Plasminogen is converted to the proteolytic enzyme plasmin by plasminogen activator, but this conversion only occurs efficiently on the fibrin surface where activator and plasminogen are "assembled." Free plasmin in the blood is very rapidly inactivated by α_2-antiplasmin, but plasmin generated at the fibrin surface is partially protected from inactivation. The lysine-binding sites in plasminogen (represented as the "legs" of the animal) are important for the interaction between plasmin(ogen) and fibrin and between plasmin and α_2-antiplasmin.

molecules generated on the fibrin surface, which are bound to fibrin through their lysine-binding sites and involved in fibrin degradation, are protected from rapid inactivation by α_2-antiplasmin. Plasmin released from the fibrin surface would, however, be rapidly inactivated by α_2-antiplasmin. These interactions are schematically visualized in Fig. 1.

STUDIES ON THE DEVELOPMENT OF HUMAN TISSUE PLASMINOGEN ACTIVATOR AS A THROMBOLYTIC AGENT

The molecular model for the regulation of fibrinolysis described above has important consequences for the development of thrombolytic agents. Indeed, the presently available thrombolytic agents streptokinase and urokinase have no specific affinity for fibrin and therefore activate circulating and fibrin-bound plasminogen relatively indiscriminately. Consequently, plasmin formed in circulating blood will initially be neutralized very rapidly by α_2-antiplasmin and be lost for thrombolysis. Once the inhibitor becomes exhausted, residual plasmin will degrade several plasma proteins (fibrinogen, factor V, factor VIII, etc.) and cause a serious bleeding tendency. This may explain why treatment with streptokinase or urokinase has only a limited efficiency and is associated with serious, sometimes life-threatening side effects.

From this reasoning it appears that specific thrombolysis will be possible only if the activation process of plasminogen can be localized at and confined to the fibrin surface. According to present concepts, this can only be adequately achieved with the use of an activator that, like the physiological activator, adsorbs to the fibrin surface and becomes active *in loco*. Such plasminogen activators have been isolated from tissues (human uterus, pig ovaries, pig heart, etc.). The yield of these procedures is, however, extremely low (to obtain 1 mg of pure activator, 5 kg of human uterus is required), so that it is practically impossible to foresee any therapeutic application of such substances.

We have found that a stable cell line derived from a human melanoma produced a large amount of a plasminogen activator that appeared to be very similar or identical to the physiological plasminogen activator (11). We have developed a relatively simple purification procedure with high yield, which has enabled us for the first time to produce significant quantities of tissue plasminogen activator. Thus on a laboratory bench scale (limited logistics and one full-time laboratory technician) we have produced approximately 50 mg purified plasminogen activator in 1980. This has allowed us to initiate a whole series of studies, not only on the biochemical and biological properties of this activator, but also on its therapeutic possibilities.

The plasminogen activator, isolated from the culture fluid of human melanoma cells, appeared to be a single-chain protein with a molecular weight of approximately 70,000, which is indistinguishable from the plasminogen activator isolated from human uterus on the basis of molecular weight, amino acid composition, kinetic properties toward synthetic substrates, and immunochemical properties (11).

A kinetic analysis of the activation of plasminogen by tissue plasminogen activator obeyed Michaelis–Menten kinetics with $K_M = 65$ μM and $k_{cat} = 0.05$ sec^{-1} in the absence of fibrin and with $K_M = 0.15$ μM and $k_{cat} = 0.05$ sec^{-1} in the presence of fibrin. These findings indicate that fibrin enhances the activation rate of plasminogen by tissue plasminogen activator by increasing the affinity of plasminogen for fibrin-bound tissue plasminogen activator and not by influencing the catalytic efficiency of the enzyme. These data support and provide a molecular explanation for the hypothesis that fibrinolysis is both triggered by and directed toward fibrin (12).

As a preliminary to the study of the thrombolytic properties of tissue plasminogen activator *in vivo,* we have investigated the mechanism of inhibition of this activator in human plasma and its turnover following intravenous injection in rabbits. Our studies on the inhibition of tissue plasminogen activator in human plasma revealed that it is slowly inhibited by α_2-antiplasmin ($t_{1/2} \simeq 60$ min) and by α_2-macroglobulin ($t_{1/2} \simeq 120$ min). We found no evidence for the existence of another significant inhibitor in plasma (13).

Following intravenous injection of mixtures of ^{125}I-labeled and unlabeled tissue plasminogen activator in rabbits, the activator activity disappeared with a $t_{1/2}$ of 2 min. This appeared to be due not to inhibition but to clearing by the liver (14).

In *in vitro* systems composed of mixtures of purified components as well as in total plasma, this activator appeared to have a much higher specific fibrinolytic activity than urokinase. In particular it appeared to be possible to dissolve blood clots hanging in circulating plasma specifically without causing breakdown of the coagulation system in the plasma; this appeared to be impossible with urokinase (15).

The thrombolytic effect of tissue plasminogen activator was investigated in rabbits with an experimental pulmonary embolus. From this study it appeared that on a molar basis tissue plasminogen activator has a more than 30 times higher thrombolytic effect than urokinase and that thrombolysis can be obtained without systemic fibrinolytic activation and breakdown of the hemostasis system (16). Rabbits, however, appear to be very resistant to thrombolysis (10 mg of urokinase infused over 6 hr causes only 10% lysis, whereas 0.4 mg tissue plasminogen activator (35,000 IU) causes 16.5% lysis).

The thrombolytic effect of tissue plasminogen activator was investigated in dogs with an experimental thrombosis of the femoral vein. Urokinase infusion at a rate of 2500 IU per kilogram per hour for 4 hr did not induce significant lysis (17.4 ± 3.7%) as compared to saline infusion (16.3 ± 3.8%). Significant lysis was obtained with 25,000 IU of urokinase per hour for 4 hr (40.6 ± 4.8%) but this was associated with defibrinogenation. Infusion of 2500 urokinase equivalent units of plasminogen activator per hour for 4 hr caused significant lysis (33.5 ± 7.8% with single-chain activator and 60.1 ± 10.8% with two-chain activator) without causing any fibrinogen breakdown.

In summary it appears that recent studies have led to a better insight into the regulation and control of the fibrinolytic system *in vivo* and into the purification and characterization of a plasminogen activator from human cell cultures, which is most probably identical with the physiological plasminogen activator in the blood. This plasminogen activator has the biochemical and biological properties of a specific thrombolytic agent and appears to be effective in experimental thrombosis in animals. Therefore it is possible and even probable that, eventually, extrinsic (tissue-type) plasminogen activator may constitute a more selective thrombolytic agent than the presently employed urokinase and streptokinase.

ACKNOWLEDGMENTS

The recent studies on plasminogen activator from our laboratory reported in this review were carried out in collaboration with Drs. D. C. Rijken (Leuven, Belgium), O. Matsuo (Myazaki, Japan), M. Hoylaerts (Leuven, Belgium), and C. Korninger (Vienna, Austria) and with the technical assistance of A. Van Nuffelen, J. M. Stassen, M. Verstreken, E. Demarsin, and G. Lemmens.

REFERENCES

1. Astrup, T. (1978) *Prog. Chem. Fibrinolysis Thrombolysis* **3**, 1–57.
2. Reich, E. (1975) *Cold Spring Harbor Conf. Cell Proliferation* **2**, 333–341.
3. Strickland, S. (1978) *In* "Regulatory Proteolytic Enzymes and Their Inhibitors" (S. Magnusson, M. Ottesen, B. Foltman, K. Danø, and H. Neurath, eds.), pp. 181–185. Pergamon, Oxford.
4. Collen, D. (1980) *Thromb. Haemostatis* **43**, 77–89.
5. Wiman, B., and Collen, D. (1978) *Nature (London)* **272**, 549–550.

6. Cash, J. D. (1978) *Prog. Chem. Fibrinolysis Thrombolysis* **3,** 65–75.
7. Nilsson, I. M. (1978) *Prog. Chem. Fibrinolysis Thrombolysis* **3,** 77–89.
8. Davidson, J. F., Lochhead, M., McDonald, G. A., and McNicol, G. P. (1972) *Br. J. Haematol.* **22,** 543–559.
9. Nilsson, I. M. (1975) *Prog. Chem. Fibrinolysis Thrombolysis* **1,** 1–12.
10. Wiman, B., and Collen, D. (1978) *Eur. J. Biochem.* **84,** 573–578.
11. Rijken, D. C., and Collen, D. (1981) *J. Biol. Chem.* **256,** 7035–7041.
12. Hoylaerts, M., Rijken, D. C., Lijnen, H. R., and Collen, D. (1982). *J. Biol. Chem.* (in press).
13. Korninger, C., and Collen, D. (1981) *Thromb. Haemostasis* **46,** 662–665.
14. Korninger, C., and Collen, D. (1981) *Thromb. Haemostasis* **46,** 658–661.
15. Matsuo, O., Rijken, D. C., and Collen, D. (1981) *Thromb. Haemostasis* **45,** 225–229.
16. Matsuo, O., Rijken, D. C., and Collen, D. (1981) *Nature* (*London*) **291,** 590–591.
17. Korninger, C., Matsuo, O., Suy, R., Stassen, J. M., and Collen, D. (1982) *J. Clin. Invest.* (in press).

Proteases and Angiogenesis: Production of Plasminogen Activator and Collagenase by Endothelial Cells

DANIEL B. RIFKIN, JANET L. GROSS, AND
DAVID MOSCATELLI
Department of Cell Biology
New York University Medical School
New York, New York

ERIC JAFFE
Department of Medicine
Division of Hematology–Oncology
Cornell University Medical School
New York, New York

INTRODUCTION

The growth of new blood vessels or angiogenesis must involve the elaboration of specific proteases since the capillary endothelial cells are capable of penetrating into and through surrounding connective tissue. The most likely source of these proteases is the endothelial cells themselves, although it is conceivable that other cell types such as mast cells (1), which are sometimes found associated with newly forming blood vessels, may be a source of the lytic enzymes. We have proposed that during angiogenesis, endothelial cells secrete high levels of two proteases, plasminogen activator (PA) and latent collagenase, in response to angiogenic factors (2,3). The latent collagenase would be activated to collagenase via plasmin formed by the action of PA on plasminogen. This would generate two proteases, plasmin and collagenase, one with a broad substrate specificity and the other with a limited specificity, both of which would be required

PATHOBIOLOGY OF THE ENDOTHELIAL CELL

ISBN 0-12-521980-6

for the invasive behavior of endothelial cells. This model leads to at least two predictions. First, capillary endothelial cells should respond to angiogenic stimuli by secreting enhanced amounts of PA and latent collagenase. Second, since neovascularization does not occur from large blood vessels but only from the microvasculature, only capillary endothelial cells and not aortic endothelial cells should respond to these stimuli. In this paper we describe several experiments designed to test these predictions.

METHODS

CELLS

The isolation and culture of bovine capillary endothelial (BCE) cells were performed as described previously (2,4). Cells were only used between the fifth and tenth passage. Bovine aortic endothelial (BAE) cells were isolated from the aortae of calves and propagated as described previously (5). The cells were positive when tested for factor VIII antigen production by indirect immunofluorescence (2).

ENZYME ASSAYS

The PA content of detergent-lysed cells was assayed as described by Gross *et al.* (2). Collagenase was assayed according to the method of Moscatelli *et al.* (6) after activation by trypsin. The trypsin was inactivated before assay by the introduction of a 10-fold excess by weight of soybean trypsin inhibitor.

REAGENTS

Bovine retinal extract (7) and mouse adipocyte conditioned medium (8) were generously given to us by those individuals who first described their angiogenic potential. Human hepatoma lysate was prepared from cultured cells after disruption by sonication (9). 12-*O*-Tetradecanoyl phorbol-13-acetate was obtained from Consolidated Midlands (Brewster, N.Y.).

RESULTS

In our initial experiments, we used the phorbol ester tumor promoter 12-*O*-tetradecanoyl phorbol-13-acetate (TPA) as a potential stimulator of endothelial cell proteases because TPA had been reported previously to be an effective stimulator of neutral protease production in a variety of cells and because this compound had been shown to be similar in its action to certain growth factors (2,6,10,11). When BCE cells were exposed to TPA, there was a rapid morphological change like that previously observed with other types of endothelial cells (6,10) (Fig. 1). There was also an increase in the level of cell-associated PA and secreted collagenase concomitant with the morphological change (Table I). As had been observed with other endothelial cells (6), all of the collagenase was in a latent form and could only be assayed after activation with either trypsin or mercurials. The response of the cells followed a typical dose response curve with maximal stimulation occurring at 2×10^{-8} *M* TPA. The stimulation of PA production occurred rapidly, as an increase over the basal level could be observed as early as 4 hr after exposure of the cells to TPA (Fig. 2). The induction of collagenase production occurred more slowly than that of PA, since an increase over the basal level of collagenase production could not be detected until 12–18 hr after TPA addition to the cultures (Fig. 2).

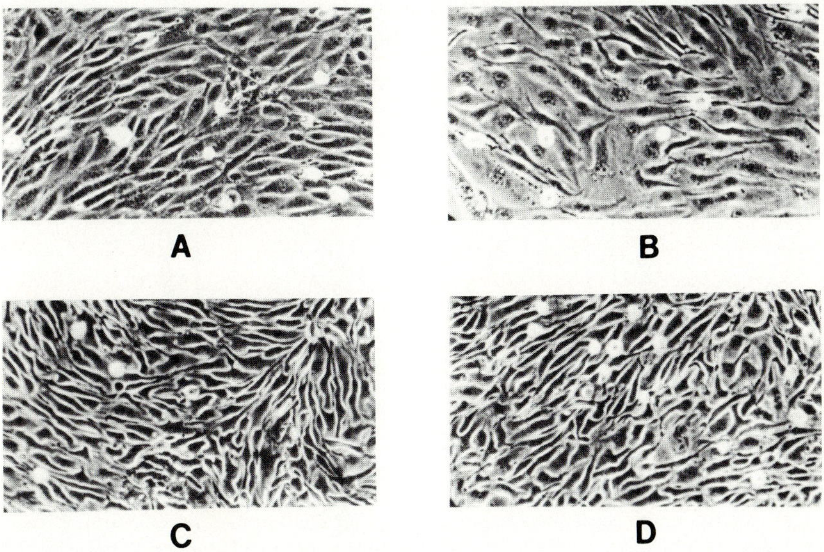

Fig. 1. The morphology of unstimulated and TPA-stimulated bovine endothelial cells. BCE cells were grown to confluence and incubated with or without TPA. Photographs were taken using a phase-contrast microscope (magnification, 125×) 2.5 hr after the addition of TPA. (A) BCE cell control; (B) TPA, 10^{-9} *M;* (C) TPA, 10^{-8} *M;* (D) TPA, 10^{-7} *M*.

TABLE I
Effect of TPA on PA and Collagenase Production by BCE Cells[a]

Addition	PA activity (m units/μg)	Collagenase (m units/culture)
None	5.8	13.0
TPA	18.1	62.0

[a] All cells were grown to confluence in complete tumor conditioned medium with 10% donor calf serum. Duplicate cultures were treated for 23 hr with 2×10^{-7} M TPA in complete medium. The culture fluids were collected, activated with trypsin, and assayed for collagenase. Cell layers were assayed for PA. The data represent the average of duplicate measurements.

The two enzymes produced after TPA stimulation were partially characterized. The PA had an apparent molecular weight of 42,000, was a serine protease, was stable to SDS, and was the urokinase type of PA. The collagenase was a typical vertebrate interstitial collagenase (2); it gave the appropriate $\frac{3}{4}$- and $\frac{1}{4}$-length cleavage products of native

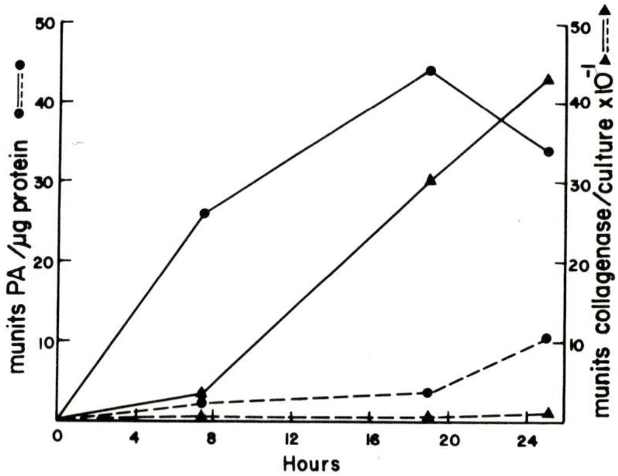

Fig. 2. TPA-induced PA and collagenase production by BCE cells. BCE cells (passage 8) were grown to confluence in complete tumor-conditioned medium and prepared for experimental use, as described (2). The cultures were incubated for 23 hr in DMEM with 5% calf serum (depleted of plasminogen) and various concentrations of TPA. The culture fluids from duplicate cultures were collected, activated with trypsin, and assayed for collagenase. The cells from the same cultures were extracted with detergent and the lysates were assayed for PA. ---, controls; ——, 2×10^{-7} M TPA.

skin collagen and was sensitive to EDTA. It was capable of degrading types I, II, and III collagens but was inactive against collagen types IV and V.

Since TPA is not known to be angiogenic, we tested several preparations that had been assayed by others and shown to be rich in angiogenic activity for their ability to stimulate BCE PA and latent collagenase production. These included an extract of human hepatoma cells cultured *in vitro*, an extract from bovine retinae, and culture medium conditioned by mouse adipocytes. The ability of these preparations to enhance PA and latent collagenase production by BCE cells is shown in Table II. All three preparations were capable of stimulating the production of both enzymes. The most active preparation was the hepatoma lysate, which was an even more effective stimulus than TPA. However, the degree of stimulation varied among the three preparations. It is not clear whether this indicates different mechanisms of action, quantitative differences in the amounts of inducing factors, or other variables.

A second prediction of our model is that capillary cells but not aortic cells should respond to these inducing factors. To test this, the re-

TABLE II
Effect of Angiogenic Stimulators on PA and Collagenase Production by BCE Cells[a]

Addition	PA	Collagenase
	(Fold stimulation over control)	
Hepatoma lysate (0.86 mg)	9.9	55.3
Retinal extract (200 μg)	3.9	6.3
Adipocyte medium (150 μl/ml)	3.3	6.0
TPA (2×10^{-7} *M*)	4.8	17.5

[a] BCE cells were grown to confluence in complete tumor-conditioned medium with 10% donor calf serum and treated as previously described (2). Cultures were maintained for 24 hr in DMEM containing 5% calf serum (depleted of plasminogen) with or without the additions described above. The culture fluids (1.5 ml/plate) and cell layers were assayed for collagenase and PA, respectively, as in the legend to Table I.

TABLE III

Addition	PA		Collagenase	
	(Fold stimulation over control)			
	BCE	BAE	BCE	BAE
Hepatoma lysate (0.86 mg)	9.9	4.3	55.3	2.4
Retinal extract (200 μg)	3.9	1.0	6.3	0
Adipocyte medium (150 μl/ml)	3.3	0.8	6.0	0
TPA (2×10^{-7} *M*)	4.8	2.3	17.5	0.9

[a] BCE cells were grown to confluence in complete tumor-conditioned medium with 10% donor calf serum. BAE cells were maintained until confluence in DMEM with 10% donor calf serum. The cells were then treated as described in the note to Table II.

sponses of BCE and bovine aortic endothelial (BAE) cells to TPA, retinal extract, hepatoma extract, and adipocyte conditioned medium were examined. In each case, only the BCE cells responded with both enhanced latent collagenase and PA production (Table III). In some instances the extracts stimulated PA production in BAE cells; however, latent collagenase production was never induced in these cells.

DISCUSSION

In this paper we have presented data from experiments designed to test some of the predictions of our model of angiogenesis. This model states that during neovascularization the invasion by endothelial cells requires the production of two specific proteases, PA and latent collagenase. We show here that, indeed, a number of inducers of angiogenesis do stimulate the production of these two proteases. Moreover, the production of these enzymes is specific, as only capillary cells respond in this manner. Endothelial cells cultured from bovine aortae respond differently and do not produce increased amounts of latent collagenase.

It should also be noted that our model for endothelial cell invasion can be translated to describe tumor cell invasion with only minor modifications. It may be of some significance that, first, many tmor cells produce high levels of PA (12) and, second, cartilage, a tissue resistant to neovascularization, is also relatively resistant to tumor invasion (13).

The results presented here do not prove our proposed model. Hopefully, they do provide a basis for future experimentation. In addition the use of induction of PA and latent collagenase as a quantitative biochemical assay should allow the isolation of defined molecules whose activities and roles can be rigorously assessed.

ACKNOWLEDGMENTS

This work was supported by grants CA23753 and CD-77 from the NIH and American Cancer Society. D. Moscatelli was supported by training grants CA09161 and 06515 from the NIH. J. Gross was supported by a fellowship (SPI-7914824) from the NSF.

REFERENCES

1. Azizkhan, R. G., Azizkhan, J. C., Zetter, B. R., and Folkman, J. (1980) *J. Exp. Med.* **152**, 931–944.
2. Gross, J. L., Moscatelli, D., Jaffe, E., and Rifkin, D. B. (1982) *J. Cell Biol.* (submitted for publication).
3. Moscatelli, D., Jaffe, E., Isseroff, R., and Rifkin, D. B. (1980) *In* "Proteinases and Tumor Invasion" (P. Strauli, ed.), pp. 143–152. Raven, New York.
4. Folkman, J., Haudenschild, C. C., and Zetter, B. R. (1979) *Proc. Natl. Acad. Sci. U.S.A.* **76**, 5217–5221.
5. Schwartz, S. M. (1978) *In Vitro* **14**, 966–980.
6. Moscatelli, D., Jaffe, E., and Rifkin, D. B. (1980) *Cell* **20**, 343–351.
7. Glaser, B. M., D'Amore, P. A., Michels, R. G., Patz, A., and Fenselau, A. (1980) *J. Cell Biol.* **84**, 298–304.
8. Castellot, J. J., Karnovsky, M. J., and Spiegelman, B. M. (1980) *Proc. Natl. Acad. Sci. U.S.A.* **77**, 6007–6011.
9. Folkman, J., personal communication.
10. Levin, E. G., and Loskutoff, D. J. (1980) *Cell* **22**, 701–707.
11. Lee, L. S., and Weinstein, I. B. (1979) *Proc. Natl. Acad. Sci. U.S.A.* **76**, 5168–5172.
12. Reich, E., Rifkin, D. B., and Shaw, E., eds. (1975) "Proteases and Biological Control." Cold Spring Harbor Press, Cold Spring Harbor, New York.
13. Kuettner, K. E. (1972) *In* "The Comparative Molecular Biology of Extracellular Matrices" (H. C. Slavkin, ed.), p. 358. Academic Press, New York.

PART IV

ENDOTHELIAL PROTEIN SYNTHESIS

Synthesis and Secretion of Thrombospondin and Fibronectin by Cultured Human Endothelial Cells

DEANE F. MOSHER AND MARY JEAN DOYLE
Department of Medicine
University of Wisconsin–Madison
Madison, Wisconsin

ERIC A. JAFFE
Department of Medicine
Cornell University Medical College
New York, New York

INTRODUCTION

Soluble fibronectin is a large [dimer or multimer of ~200 kilodalton (kd) subunits] glycoprotein that is found in blood and other body fluids. Insoluble fibronectin is found in connective tissues and associated with basement membranes. Fibronectin has been known by many names, including cold-insoluble globulin, α_2-surface binding opsonic protein, antigelatin factor, large external transformation-sensitive (LETS) protein, and cell surface protein. In cell culture, fibronectin is secreted into the media of substrate-attached cells, including endothelial cells (1–3), and deposited in the extracellular connective tissue matrix. Collagen, a second component of the matrix, binds to fibronectin. Sulfated proteoglycans, heparin, hyaluronic acid, fibrin, gangliosides, and components of eukaryotic cell surfaces and bacterial cell walls also interact with fibronectin. Fibronectin has been detected in very early embryos of species as distant as the chick, the mouse, and the sea urchin. Thus fibronectin may have an organizing role in the formation of the extracellular matrix, may act as an adhesive

PATHOBIOLOGY OF THE ENDOTHELIAL CELL

ISBN 0-12-521980-6

protein for the orderly growth and positioning of cells, and may be an opsonic protein for tissue debris, soluble fibrin, and gram-positive bacteria. A number of reviews have appeared recently that describe fibronectin in detail (4–8).

Thrombospondin, also known as thrombin-sensitive protein (9,10) or glycoprotein G (11), is a major platelet α-granule glycoprotein that is secreted and then partially bound to platelet membranes when human platelets aggregate in response to thrombin (12–16). Studies by Lawler and co-workers indicate that thrombospondin is a 450 kd filamentous protein of dimensions 7 × 65 nm, has a pI of 4.7, and is composed of three large disulfide-linked subunits (17,18). During the process of aggregation, thrombin-stimulated platelets develop a membrane-bound lectin activity (19–21) that originates from α-granules and appears to play an important role in mediating platelet aggregation by binding to a specific receptor on other platelets (22,23). We recently found that purified human platelet thrombospondin has lectin activity (i.e., agglutinates fixed trypsinized sheep erythrocytes) and blocks the agglutination of thrombin-treated platelets. Therefore we suggested that thrombospondin is the endogenous lectin of human platelets (24). This may explain why platelets from patients with the gray platelet syndrome, which lack α-granules and α-granule constituents, including thrombospondin, aggregate poorly in response to thrombin (25).

In addition to fibronectin, cultured endothelial cells synthesize and secrete large amounts of a second glycoprotein, GP-160, which is a disulfide-bonded multimer of 160 kd subunits and resists digestion with bacterial collagenase (1,26). A protein with the same properties is produced by cultured HT-1080 human sarcoma cells (27). Because of the similarities between thrombospondin and GP-160, we undertook experiments to explore whether GP-160 is identical or nearly identical to thrombospondin. We report the results of the experiments in this chapter as well as evidence that fibronectin synthesized by endothelial cells is slightly larger than plasma fibronectin. In the discussion, we compare fibronectin and thrombospondin to vWF, a third platelet and plasma glycoprotein that is synthesized by endothelial cells.

METHODS

CULTURE OF ENDOTHELIAL CELLS

Human endothelial cells were derived from umbilical cord veins and cultured in medium 199 and 20% human serum using methods and materials described previously (1,28). Radioactively labeled proteins synthesized and secreted by confluent endothelial cells were prepared by incubating washed 75 cm^2 cell monolayers for 24 hr at 37°C in leucine-free minimal essential medium, 10 ml per flask containing 200 μCi of L-[3,4,5-^{3}H]leucine. In experiments in which tritiated fibronectin and GP-160 were isolated by gelatin-agarose chromatography, the cells were labeled in medium containing 20% pooled human serum. In experiments in which tritiated GP-160 was immunoisolated using rabbit antithrombospondin, the cells were labeled in medium containing either 20% rabbit serum or 0.5% tryptose phosphate broth. In experiments in which the accumulation of thrombospondin antigen was quantitated by radioimmunoassay, endothelial cells were cultured in medium 199 containing 20% human serum in 2 cm^2 wells of multiwell plates. When the cells were confluent, the cell layer was washed, and medium containing 20% rabbit serum was placed over the cells. After an overnight incubation, the cell layer was washed, and the cells were placed in 1 ml fresh medium 199 containing 20% rabbit serum. At various times after the second medium change, postculture medium was removed and clarified by centrifugation at 8000 *g* for 2 min in a microfuge. Samples were frozen for further analysis. Cells were dispersed with 0.02% (w/v) collagenase and 0.01% (w/v) EDTA in 10 m*M* Hepes, 140 m*M* sodium chloride, pH 7.4, and enumerated with an electronic cell counter.

PURIFICATION AND IODINATION OF THROMBOSPONDIN

Platelet concentrates were obtained from Badger Red Cross, Madison, WI. Platelets were washed and reacted with thrombin as described by Lawler *et al.* (17,18). The releasate was frozen in a dry-ice–ethanol bath and thawed in a 37°C bath. A small fibrin clot formed in the thawing solution and was gently removed with a stir rod. The supernatant was applied to a column of Bio-Gel P-300, equilibrated, and eluted with TBS-EDTA. The first protein peak was applied to a column of heparin-agarose equilibrated in TBS-EDTA. The heparin-agarose column was washed with TBS-EDTA, and thrombospondin was eluted with 0.55 *M* sodium chloride in 10 m*M* Tris, 1 m*M* EDTA, pH 7.4. Peak fractions from the heparin-agarose column were pooled, and portions were frozen and stored at −70°C. The amount of thrombospondin in the initial platelet releasate, determined by radioimmunoassay (see below), was typically 1 mg per platelet unit, and the yield of purified thrombospondin was typically 0.2 mg per unit (20% yield). The purity of thrombospondin, estimated by densitometry after SDS-PAGE, was >95%. The purified protein was labeled by the chloramine T-catalyzed iodination, mixed with PMSF-treated albumin, 1 mg/ml, snap frozen, and stored in portions at −70°C. Specific activity was approximately 0.5 mCi/mg.

PRODUCTION AND CHARACTERIZATION OF RABBIT ANTITHROMBOSPONDIN

Thrombospondin was further purified by preparative electrophoresis after reduction on SDS-PAGE slab gels (29). The final product, 10–20 μg in 0.3 ml of 0.1% SDS, was emulsified with an equal volume of complete Freund's adjuvant and injected subcutaneously into a rabbit. The injections in complete adjuvant were repeated twice at monthly intervals. Two weeks later, 1 mg of protein from the heparin-affinity column was given without adjuvant. Antiserum was collected over the next 2 months. Immunoglobulin from the antiserum, when loaded onto protein A-bearing staphylococci (30), bound iodinated thrombospondin and not iodinated fibronectin. In experiments in which immune complexes were collected with beads coated with goat antirabbit immunoglobulin or with protein A-bearing staphylocci, maximum binding of 75 ng of [^{125}I]thrombospondin was obtained with 1 ml of a 1/1000 dilution of antiserum.

RADIOIMMUNOASSAY OF THROMBOSPONDIN

Protein A-bearing staphylocci were "armed" with antithrombospondin as described by O'Keefe and Bennett (30). Approximately 90% of [^{125}I]thrombospondin bound when mixed with an equal volume of a 10% (w/v) suspension of "armed" staphylococci. Standard or unknown in physiological saline solution, 100 μl, was mixed with [^{125}I]thrombospondin, approximately 2×10^4 cpm in 100 μl, and then with 100 μl of a suspension that consisted of one part 10% "armed" staphylococci and nine parts 10% "unarmed" carrier staphylococci. After 3 hr at 4°C, the mixture was centrifuged for 1 min at 10,000 *g* in a microfuge, the supernatant was discarded, and the pellet was analyzed in a gamma counter. The assay was sensitive to 0.7 μg/ml thrombospondin. Fibronectin, 200 μg/ml, did not inhibit in the assay.

IMMUNOISOLATION AND ANALYSIS OF [^{3}H]LEUCINE-LABELED ENDOTHELIAL CELL PROTEINS

Tritiated leucine-labeled endothelial cell medium was made 3% in SDS and 6 *M* in urea, heated in a boiling water bath for 5 min, dialyzed against two changes of 3% SDS in PBS, diluted 10-fold with 0.12 *M* sodium chloride, 33 m*M* Tris acetate, pH 8.5, and made 1.7% in Triton X-100 (31). The mixed micellar solution, 30 ml, was incubated with 5 μl of rabbit antithrombospondin, preimmunization serum from the antithrombospondin rabbit, rabbit antifibronectin (1), or rabbit antiovalbumin (1) previously coupled to 10 μl of protein A–Sepharose. After overnight incubation with end-over-end rotation at 20°C, the beads were centrifuged; washed eight times with 2 *M* urea, 0.1 *M* glycine, 1% Triton X-100 (31); and eluted by boiling in 2% SDS, 6 *M* urea. Immunoisolated labeled proteins were analyzed by electrophoresis in SDS in continuous 3% acrylamide–0.5% agarose gels with and without reduction (1). Unlabeled purified fibronectin and thrombospondin were analyzed in simultaneously run gels. Unlabeled proteins were visualized by staining with Coomassie brilliant blue. The gels containing radioactive proteins were sectioned into 2 mm slices, and each slice was processed for liquid scintillation counting as previously described (1).

PARTIAL PURIFICATION OF [^{3}H]LEUCINE-LABELED GP-160 AND FIBRONECTIN ON GELATIN-AGAROSE

Purified plasma fibronectin, 12 mg, was added to 25 ml of [^{3}H]leucine-labeled endothelial cell postculture medium, and the mixture was applied to a 1.5 × 12 cm column of gelatin-agarose. The column support consisted of gelatin coupled to cyanogen bromide-activated BioGel A-15M by the method of March *et al.* (32). Unlabeled fibronectin, tritiated fibronectin, and tritiated GP-160 bound to the column. The column was washed with PBS-EDTA, and bound proteins were eluted with 1 *M* sodium bromide, 20 m*M* sodium acetate, pH 5.0. The eluted proteins were dialyzed against TBS, snap frozen, and kept at −70°C until further studies were performed.

POLYACRYLAMIDE SLAB GEL ELECTROPHORESIS

One-dimensional SDS-PAGE was performed using the discontinuous slab gel system of Ames (33). Size markers included reduced fibronectin, 210 kd; β′ subunit of RNA polymerase, 150 kd; β-galactosidase, 116 kd; phosphorylase, 93 kd; albumin, 68 kd; ovalbumin, 43 kd; and chymotrypsinogen, 24.5 kd. Two-dimensional gel electrophoresis, in which reduced proteins were separated by isoelectric focusing in 8 *M* urea in the first dimension and SDS-PAGE in the second dimension, was performed as described by Anderson and Anderson (34). Nonreduced–reduced two-dimensional SDS-PAGE was performed by the method of Phillips and Poh Agin (35) with slight modifications. Nonreduced plasma fibronectin was assumed to have a size of 420 kd. Slab gels were stained with Coomassie brilliant blue, prepared for fluorography by incubation in Enhance, and dried onto filter paper. The corners of the paper were marked with ink that contained [^{14}C]glucose before the gel was placed against film. After the fluorogram was developed, alignment of the radioactive ink lines allowed exact comparisons of radioactive spots and protein-staining spots.

RESULTS

Cultured human endothelial cells were incubated with [^{3}H]leucine, and the metabolically labeled proteins that were secreted into the medium were analyzed by SDS-PAGE with and without reduction (Fig. 1A). Without reduction 31% of the [^{3}H]leucine migrated in a peak that had an apparent size of 450 kd. With reduction the peak at 450 kd disappeared and was replaced by peaks with apparent sizes of 220 and 160 kd. The 220-kd peak, which accounted for 17% of the labeled proteins, could be immunoisolated with antifibronectin (Fig. 1D). The 160 kd peak accounted for 14% of the labeled proteins.

Purified human platelet thrombospondin, used as a marker, migrated in SDS-PAGE with the 160 kd peak when analyzed after reduction and with the 450 kd peak when analyzed without prior reduction (Fig. 1). When antithrombospondin serum was reacted with [^{3}H]leucine-labeled postculture medium, the antithrombospondin specifically immunoisolated ^{3}H-labeled protein with apparent sizes of 450 kd when analyzed without reduction (Fig. 1B) and 160 kd when analyzed with reduction (Fig. 1C). No labeled proteins were specifically isolated in control experiments with antiovalbumin (Figs. 1B and 1C) or with preimmune serum from the antithrombospondin rabbit. Immunoisolation of the 160 kd protein was not blocked by preadsorption of the antithrombospondin antiserum (5 μl) with purified fibronectin (100 μg).

In the presence of purified fibronectin (0.5 mg/ml in PBS), [^{3}H]leucine-labeled endothelial cell fibronectin and GP-160 bound to gelatin-agarose and could be eluted with acetate-buffered 1 *M* NaBr, pH 5. SDS-PAGE of unbound and bound fractions and calculations of yields of labeled proteins indicated that affinity chromatography on gelatin-agarose in the presence of unlabeled fibronectin resulted in

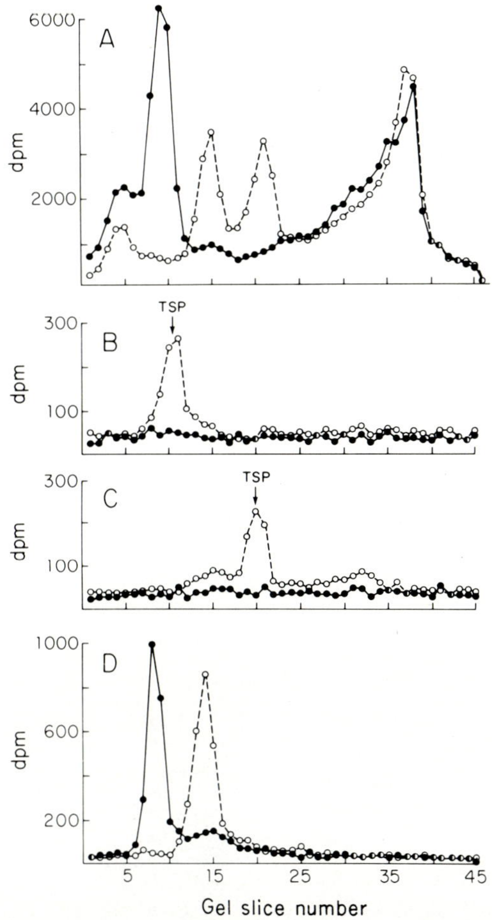

Fig. 1. Analysis of [^{3}H]leucine-labeled postculture medium and immunoisolated [^{3}H]leucine-labeled GP-160. The following samples were analyzed by SDS-PAGE: (A) unfractionated postculture medium without (solid line) or with (dashed line) prior reduction; (B) labeled protein that was immunoisolated with antithrombospondin (dashed line) or antiovalbumin (solid line) and analyzed without reduction; (C) labeled protein that was isolated with antithrombospondin (dashed line) or antiovalbumin (solid line) and analyzed with reduction; and (D) labeled protein that was isolated with antifibronectin and analyzed without (solid line) or with (dashed line) reduction. *Arrows* point to the positions of migration of nonreduced (450 kd) and reduced (160 kd) marker thrombospondin (TSP).

nearly quantitative recovery of [^{3}H]leucine-labeled fibronectin and [^{3}H]leucine-labeled GP-160. To make further comparisons between thrombospondin and [^{3}H]leucine-labeled GP-160, thrombospondin was added to the mixture of unlabeled fibronectin, [^{3}H]leucine-labeled endothelial cell fibronectin, and [^{3}H]leucine-labeled endothelial cell GP-160. The four proteins were then analyzed by two different two-dimensional electrophoretic systems: isoelectric focusing of reduced proteins in 8 *M* urea followed by polyacrylamide gel electrophoresis in SDS (Fig. 2A) and polyacrylamide gel electrophoresis in SDS without reduction followed by polyacrylamide gel elec-

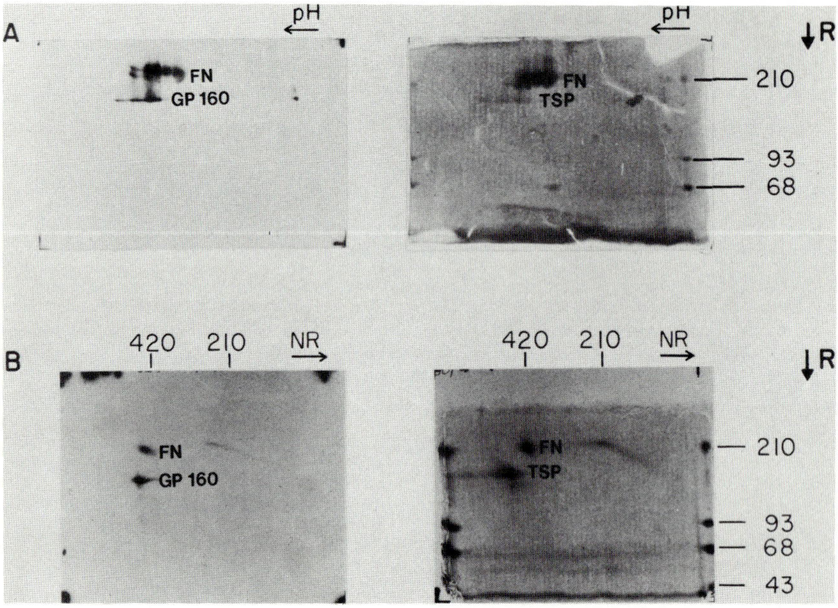

Fig. 2. Two-dimensional analyses of fibronectin (FN), GP-160, and thrombospondin (TSP). The following mixture of proteins was analyzed: plasma fibronectin, platelet thrombospondin, and [^{3}H]leucine-labeled endothelial cell fibronectin and GP-160. Two systems were used: (A) isoelectric focusing (pH, acid side to the left) in 8 *M* urea followed by electrophoresis through 6% polyacrylamide in SDS after reduction (R); and (B) electrophoresis through 4% polyacrylamide in SDS without reduction (NR) followed by electrophoresis through 6% polyacrylamide in SDS after reduction (R). The slabs were stained for protein and dried onto filter paper (right) and analyzed by fluorography (left). Apparent size in kilodaltons is indicated on the right. The figure has been made to allow one-for-one comparisons of fluorographic and protein-staining patterns. We could make a more critical evaluation by overlaying the dried stained gel with the developed fluorography film.

TABLE I
Accumulation of Thrombospondin Antigen[a] in Culture Medium of Human Umbilical Vein Endothelial Cells over 24 Hr as Ascertained by Radioimmunoassay

μg/ml	μg/10^6 cells
4.4 ± 0.2	49 ± 4

[a] $\bar{x}$ ± SD, n = 3.

trophoresis in SDS after reduction (Fig. 2B). GP-160 was acidic (Fig. 2A, left), had an apparent size of 450 kd in its unreduced form (Fig. 2B, left), and was appaently a disulfide-bonded trimer of 160 kd subunits (Fig. 2B, left). GP-160 migrated at exactly the same positions as thrombospondin in both two-dimensional systems. GP-160 and thrombospondin exhibited the same limited isoelectric heterogeneity (Fig. 2A). Labeled endothelial cell fibronectin was slightly larger than plasma fibronectin in its reduced (220 kd vs. 210 kd) and unreduced forms and thus migrated slightly higher and more to the left in the nonreduced-reduced system (Fig. 2B). There were also slight differences in isoelectric heterogeneity between endothelial cell and plasma fibronectins (Fig. 2A).

Thrombospondin antigen was detectable by radioimmunoassay after conditioning of medium for 8 hr by confluent endothelial cells, and approximately 50 μg per 10^6 cells was present after 24 hr (Table I). The curves for inhibition of [^{125}I]thrombospondin binding to rabbit antibodies were similar for thrombospondin and conditioned medium. The value of 4.4 μg thrombospondin per ml of 24 hr conditioned medium is comparable to the value of 5.2 μg fibronectin per ml of 72 hr conditioned medium that we determined earlier (1).

DISCUSSION

Several different comparisons of GP-160 synthesized and secreted by cultured endothelial cells and thrombospondin secreted by thrombin-stimulated platelets indicate that these two proteins are the same. GP-160 and thrombospondin migrated identically in two different two-dimensional gel systems. Therefore it seems likely that the differences between GP-160 and thrombospondin, if present, are minor. In contrast, fibronectin secreted by cultured human umbilical vein endothelial cells appears to be slightly larger and more acidic than plasma fibronectin.

Although thrombospondin and GP-160 are disulfide-bonded trimers of 160 kd subunits, there is no evidence that thrombospondin and GP-160 are collagenous. We have been unable to degrade human endothelial cell GP-160 using purified bacterial collagenase (unpublished experiments carried out in both of our laboratories). Similarly, Sage *et al.* (26) were unable to degrade bovine endothelial cell GP-160 by purified bacterial collagenase under conditions in which type III procollagen was readily degraded. By amino acid analysis, thrombospondin is not rich in proline or glycine (10,17).

We are unsure of the determinants of binding of [^{3}H]leucine-labeled GP-160 to gelatin-agarose in the presence of unlabeled fibronectin. In preliminary studies, we have found that >60% of [^{125}I]thrombospondin binds to gelatin-agarose, underived agarose, or BioGel P-300. Because of this apparently nonspecific binding, a better experimental system than affinity chromatography is needed to investigate specific binding.

We can only speculate on the function(s) of endothelial cell thrombospondin. As described earlier, thrombospondin is secreted and bound by thrombin-activated platelets (9–16) and apparently mediates

TABLE II
Comparison of vWF (von Willebrand Factor), Fibronectin (FN), and Thrombospondin (TSP)[a]

Property	Protein		
	vWF	FN	TSP
Size (kd)	800–20,000	420	450
Subunits (kd)	n × 200	2 × 210	3 × 160
Plasma concentration, μg/ml	10	400	~1
Platelet concentration, μg/ml	1	2	15
Percent in platelets	10	0.5	~90
Binds to heparin	Yes	Yes	Yes
Binds to collagen	?Yes	Yes	?
Binds to fibrin	No	Yes	?
Binds to platelets	Yes	Yes	Yes
Adhesive for platelets	Yes	Yes	?
Synthesized by endothelial cells	Yes	Yes	Yes
In subendothelium	Yes	Yes	?

[a] For references, see text and Ref. 36.

thrombin-induced aggregation (19–24). Two other proteins, von Willebrand factor and fibronectin (Table II), also found in platelet α-granules (37–40), bind to the surface of thrombin-stimulated platelets (41–43) and are synthesized by endothelial cells (1–3,44,45). Von Willebrand factor and fibronectin are found in plasma and in the subendothelium (46,47). Von Willebrand factor mediates platelet adhesion to subendothelium (48,49), whereas the role of fibronectin in platelet function is unknown. While plasma thrombospondin levels as measured by radioimmunoassay are quite low (~1 μg/ml, our unpublished data), endothelial cells may, when appropriately stimulated, release thrombospondin into the local microenvironment and thus support platelet interactions with endothelial cells and the subendothelium. Thrombospondin, because of its lectin activity, may also play important roles in the interaction of endothelial cells with each other and with the underlying extracellular matrix. Experiments to test these hypotheses are currently under way.

ACKNOWLEDGMENTS

This work was supported from grants by the National Institutes of Health, HL-18828, HL-21644, and HL-24885. It was done during the tenure of an Established Investigatorship from the American Heart Association and its Wisconsin Affiliate to Deane F. Mosher and a National Institutes of Health Career Development Award and a Career Scientist Award from the Irma T. Hirschl Trust to Eric A. Jaffe. We thank Douglas Armellino, Linda Griese, and Peter Schad for their excellent technical assistance.

* Abbreviations: kd, kilodalton; PBS, 10 m*M* sodium phosphate, 140 m*M* sodium chloride, pH 7.4; PBS-EDTA, PBS containing 1 m*M* EDTA; PMSF, phenylmethylsulfonyl fluoride; SDS, sodium dodecyl sulfate; SDS-PAGE, polyacrylamide gel electrophoresis in 0.1% SDS; TBS, 10 m*M* Tris, 140 m*M* sodium chloride, pH 7.4; TBS-EDTA, TBS containing 1 m*M* EDTA; and vWF, von Willebrand Factor or Factor VIII-related protein.

REFERENCES

1. Jaffe, E. A., and Mosher, D. F. (1978). *J. Exp. Med.* **147,** 1779–1791.
2. Birdwell, C. R., Gospodarowicz, D., and Nicolson, G. L. (1978) *Proc. Natl. Acad. Sci. U.S.A.* **75,** 3273–3277.
3. Macarak, E. J., Kirby, E., Kirk, T., and Kefalides, N. A. (1978) *Proc. Natl. Acad. Sci. U.S.A.* **75,** 2621–2652.
4. Mosesson, M. W., and Amrani, D. L. (1980) *Blood* **56,** 145–158.
5. Mosher, D. F. (1980) *Prog. Hemostasis Thromb.* **5,** 111–151.

6. Pearlstein, E., Gold, L. I., and Garcia-Pardo, A. (1980) *Mol. Cell. Biochem.* **29,** 103–128.
7. Wartiovaara, J., Leivo, I., and Vaheri, A. (1980) *In* "The Cell Surface: Mediator of Developmental Processes" (S. Subtelny and N. K. Wessells, eds.), pp. 305–324. Academic Press, New York.
8. Yamada, K. M., and Olden, K. (1978) *Nature (London)* **275,** 179–184.
9. Baenziger, N. L., Brodie, G. N., and Majerus, P. W. (1971) *Proc. Natl. Acad. Sci. U.S.A* **68,** 240–243.
10. Baenziger, N. L., Brodie, G. N., and Majerus, P. W. (1972) *J. Biol. Chem.* **247,** 2723–2731.
11. George, J. N. (1978) *J. Lab. Clin. Med.* **92,** 430–446.
12. Hagen, I. (1975) *Biochim. Biophys. Acta* **392,** 242–254.
13. Hagen, I., Olsen, T., and Solum, N. O. (1976) *Biochim. Biophys. Acta* **455,** 214–225.
14. Lawler, J. W., Chao, F. C., and Fang, P.-H. (1977) *Thromb. Haemostasis* **37,** 355–357.
15. George, J. N., Lyons, R. M., and Morgan, R. K. (1980) *J. Clin. Invest.* **66,** 1–9.
16. Phillips, D. R., Jennings, L. K., and Prasanna, H. R. (1980) *J. Biol. Chem.* **255,** 11629–11632.
17. Lawler, J. W., Slayter, H. S., and Coligan, J. E. (1978) *J. Biol. Chem.* **253,** 8609–8616.
18. Margossian, S. S., Lawler, J. W. and Slayter, H. S. (1981) *J. Biol. Chem.* **256,** 7495–7500.
19. Gartner, T. K., Williams, D. C., and Phillips, D. R. (1977) *Biochem. Biophys. Res. Commun.* **79,** 592–599.
20. Gartner, T. K., Williams, D. C., Minion, F. C., and Phillips, D. R. (1978) *Science* **200,** 1281–1283.
21. Gartner, T. K., Phillips, D. R., and Williams, D. C. (1980) *FEBS Lett.* **113,** 196–200.
22. Gartner, T. K., Gerrard, J. M., White, J. G., and Williams, D. C. (1981) *Nature (London)* **289,** 688–700.
23. Gartner, T. K., Gerrard, J. M., White, J. G., and Williams, D. C. (1981) *Blood* **58,** 153–157.
24. Jaffe, E. A., Leung, L. L. K., Nachman, R. L., Levin, R. I., and Mosher, D. F. (1982) *Nature (London)* **295,** 246–248.
25. Gerrard, J. M., Phillips, D. R., Rao, G. H. R., Plow, E. F., Walz, D. A., Ross, R., Harker, L. A., and White, J. G. (1980) *J. Clin. Invest.* **66,** 102–109.
26. Sage, H., Crouch, E., and Bronstein, P. (1979) *Biochemistry* **18,** 5433–5442.
27. Alitalo, K., Vaheri, A., Krieg, T., and Timpl, R. (1980) *Eur. J. Biochem.* **109,** 247–255.
28. Jaffe, E. A., Nachman, R. L., Becker, C. G., and Minick, C. R. (1973) *J. Clin. Invest.* **52,** 2745–2756.
29. Lazarides, E. (1976) *J. Supramol. Struct.* **5,** 531–563.
30. O'Keefe, E., and Bennett, V. (1980) *J. Biol. Chem.* **255,** 561–568.
31. Erickson, A. H., and Blobel, G. (1979) *J. Biol. Chem.* **254,** 11771–11774.
32. March, S. C., Parikh, I., and Cuatrecasas, P. (1979) *Anal. Biochem.* **60,** 149–152.
33. Ames, G. F.-L. (1974) *J. Biol. Chem.* **249,** 634–644.
34. Anderson, L., and Anderson, N. G. (1977) *Proc. Natl. Acad. Sci. U.S.A.* **74,** 5421–5425.
35. Phillips, D. R., and Poh Agin, P. (1977) *J. Biol. Chem.* **252,** 2121–2126.
36. Mosher, D. F. (1981) *In* "Interaction of Blood with Natural and Artificial Surfaces" (E. W. Salzman, ed.) pp. 85–101. Dekker, New York.
37. Nachman, R. L., and Jaffe, E. A. (1975) *J. Exp. Med.* **141,** 1101–1112.

38. Koutts, J., Walsh, P. N., Plow, E. F., Fenton, J. W., II, Bouma, B. N., and Zimmerman, T. S. (1978) *J. Clin. Invest.* **62,** 1255–1263.
39. Plow, E. F., Birdwell, C., and Ginsberg, M. H. (1979) *J. Clin. Invest.* **63,** 540–543.
40. Zucker, M. B., Mosesson, M. W., Broekman, M. J., and Kaplan, K. L. (1979) *Blood* **54,** 8–12.
41. Ginsberg, M. H., Painter, R. G., Forsyth, J., Birdwell, C., and Plow, C. F. (1980) *Proc. Natl. Acad. Sci. U.S.A.* **77,** 1049–1053.
42. Plow, E. F., and Ginsberg, M. H. (1981) *J. Biol. Chem.* **256,** 9477–9482.
43. Hawiger, J., Fujimoto, T., and Ohara, S. (1981) *Clin. Res.* **29,** 571.
44. Jaffe, E. A., Hoyer, L. W., and Nachman, R. L. (1973) *J. Clin. Invest.* **52,** 2757–2764.
45. Jaffe, E. A., and Nachman, R. L. (1975) *J. Clin. Invest.* **56,** 698–702.
46. Stenman, S., and Vaheri, A. (1978) *J. Exp. Med.* **147,** 1054–1064.
47. Rand, J. H., Sussman, I. I., Gordon, R. E., Chu, S. V., and Solomon, V. (1980) *Blood* **55,** 752–756.
48. Tschopp, T. B., Weiss, H. J., and Baumgartner, H. R. (1974) *J. Lab. Clin. Med.* **83,** 296–300.
49. Weiss, H. F., Baumgartner, H. R., Tschopp, T. B., Turitto, V. T., Cohen, D. (1978) *Blood* **51,** 267–279.

Synthesis and Secretion of Structural Macromolecules by Endothelial Cells in Culture

PAUL BORNSTEIN, JOHN McPHERSON, AND HELENE SAGE

Departments of Biochemistry and Medicine
University of Washington
Seattle, Washington

INTRODUCTION

Endothelial cells form a continuous inner lining throughout much of the vasculature and, as such, present a nonthrombogenic surface and a permeability barrier to the blood (1). The capillary endothelium of some organs is discontinuous or fenestrated, however, and in these regions the subendothelial matrix may contribute to the control of vascular permeability and thrombosis. The extent to which endothelial cells are capable of performing their functions reflects, in part, the ability of these cells to adhere to the subendothelial stroma and to direct secretion of matrix macromolecules in a polarized fashion. Thus endothelial cell injury and/or loss is thought to represent a crucial step in the chain of events leading to the formation of a mural thrombosis in larger blood vessels and, under some circumstances, to the development of an atherosclerotic plaque. Cellular injury could also disturb the structural and functional polarization of endothelial cells and alter the characteristics of the luminal cell surface that normally tend to discourage platelet release and aggregation and the ensuing steps in the coagulation cascade.

There is good evidence that endothelial cells produce and secrete a number of macromolecules that function in the regulation of hemostasis and in adhesion to a substratum. Much of this information has come from studies of cultured endothelial cells *in vitro;* the extent to

PATHOBIOLOGY OF THE ENDOTHELIAL CELL

ISBN 0-12-521980-6

which this information can be applied to the endothelium *in vivo* is not known, although there are increasing indications that even complex phenomena such as angiogenesis can be simulated by cells in culture under appropriate conditions (2).

We will focus in this brief review on a critical appraisal of studies describing the synthesis and secretion of macromolecules that contribute to the formation of a basal lamina and that may be important in endothelial cell attachment. The complex role of endothelial cells in hemostasis and associated vascular phenomena, as exemplified by production and/or binding and uptake of factor VIII antigen (von Willebrand's factor) (3), plasminogen activator (4), angiotensin converting enzyme (5), or the prostaglandins (6) will not be considered here.

COLLAGEN SYNTHESIS

The bovine aortic endothelial (BAE) cell may be considered to be a prototype of endothelial cells in culture. These cells grow well *in vitro* (7) and their secretory products have been extensively characterized (8–14); however, as we will note, significant differences exist in the synthetic products of cells from different vascular beds (15). BAE cells secrete fibronectin, several different collagen types, a glycoprotein that is identical or very similar to platelet thrombospondin (16), and proteoglycans, of which proteoheparan sulfate is the major component (17).

The collagens synthesized by aortic endothelial cells in culture have been characterized by a variety of techniques, including sodium dodecyl sulfate slab gel polyacrylamide gel electrophoresis (SDS-PAGE) and ion exchange chromatography (18). When cells are labeled with [^{3}H]proline for 24 hr and the culture medium is subjected to SDS-PAGE after reduction with dithiothreitol, three major bands are observed (Fig. 1, lane 2). One of these is selectively lost after digestion with bacterial collagenase, indicating its collagenous nature (Fig. 1, lane 3). We have identified this band as the proα1(III) chain, derived from type III procollagen, by several criteria, the most definitive of which is one-dimensional mapping of CNBr-produced fragments (10). Although type III procollagen is also found in the matrix associated with the cell layer, small amounts of types IV and V collagens are found exclusively in this compartment *in vitro*, a finding that is compatible with the proposed role of these collagens as basement membrane- and pericellular-associated collagens (12). Madri *et al.* (9) have also described types IV and V collagens in the cell layers of BAE cells

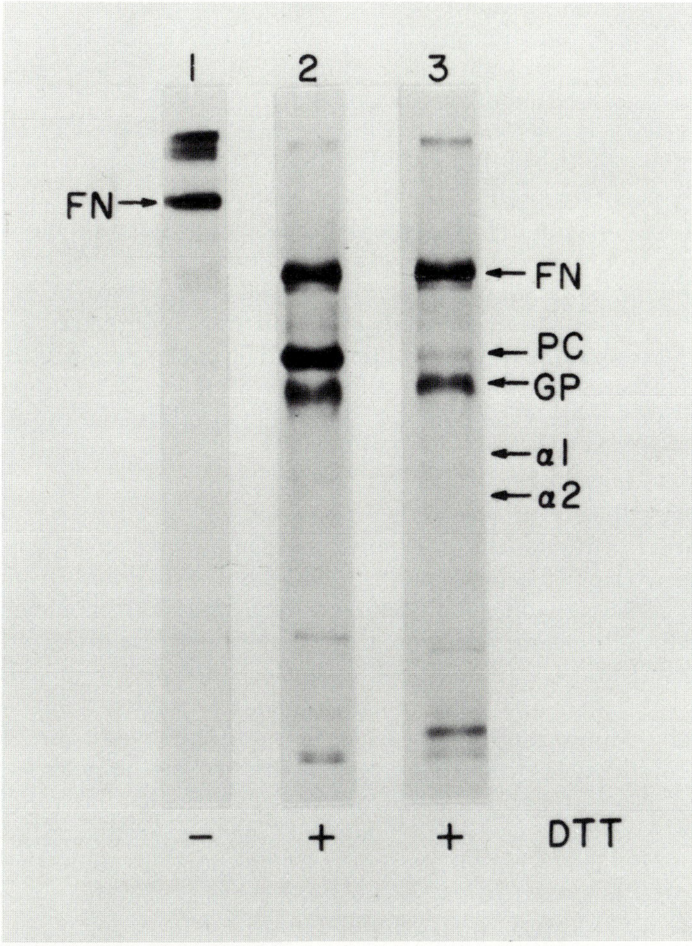

Fig. 1. SDS-PAGE of culture medium proteins secreted by BAE cells. Nearly confluent cultures were pulse labeled with L-2,3-[^{3}H]proline (50 μCi/ml) for 24 hr in serum-free DMEM supplemented with sodium ascorbate (50 μg/ml) and β-aminopropionitrile (64 μg/ml). Proteins were precipitated using 20% ammonium sulfate and subjected to electrophoresis on composite 6%/10% slab gels in the presence and absence of 50 m*M* DTT. Radiolabeled proteins were visualized by fluorescence autoradiography. Lane 1: unreduced proteins; lane 2: reduced proteins; lane 3: proteins after incubation with bacterial collagenase (5 μg/ml) for 2 hr at 37°C. Fibronectin (FN), procollagen (PC), and the glycoprotein, thrombospondin (GP) are indicated as well as the positions of migration of the bovine α1(I) and α2(I) collagen chains.

in culture, a finding that correlates well with their immunofluorescence and immunoelectron microscopic studies of the aortic subendothelium. They attribute the nonthrombogenicity of the endothelial lining to the presence of type V collagen on endothelial luminal cell surfaces *in vivo*. These authors were, however, unable to identify any procollagens in the medium of cultured cells. The reason for this discrepancy is not apparent at the present time, since we have identified type III procollagen in the medium of BAE cells under a wide variety of different culture conditions (10,18). Howard *et al.* (8) originally identified the major secreted product of BAE cells as type IV collagen based on the apparent molecular weight and hydroxyproline and glycosylated hydroxylysine content of the pepsin-treated protein. However, these criteria are not sufficiently specific and could well be fulfilled by a type III collagen with an increased degree of peptidyl hydroxylation. Peptide maps of collagen chains are required for unequivocal identification.

Although the major secreted components can be identified by the technique illustrated in Fig. 1, minor macromolecules may be missed. When the [^{3}H]proline-labeled culture medium of BAE cells is precipitated in 20% $(NH_4)_2SO_4$ and the resulting material chromatographed on DEAE-cellulose in 6 *M* urea, three major peaks are observed (Fig. 2). Peaks II and III represent fibronectin and type III procollagen, respectively. The first peak, which elutes unretarded from the ion exchange column, represents a hitherto unidentified collagen that we have termed EC (11). As shown by the inset to Fig. 2, peak I can be resolved into three bands by SDS-PAGE. The molecular weights of EC1, EC2, and EC3 are approximately 177,000, 125,000, and 100,000, respectively. These chains are not disulfide-bonded, since the same pattern was observed in the presence or absence of reducing agents. EC2 and EC3 appear to be derived from EC1, as suggested by a number of observations, including peptide mapping, but the manner in which this process occurs, or its significance, is unclear (11). EC collagen is unusually susceptible to cleavage by pepsin and several neutral serine proteases under conditions in which most collagens are relatively protease resistant. This raises the possibility that the triple helix in this protein may be partially unfolded even under physiological conditions. Both protease and CNBr peptide maps indicate that EC is distinct from other known collagen types. It is not clear whether this collagen is the product of other cell types in culture, such as fibroblasts or smooth muscle cells, although endothelial cells from other sources can be shown to synthesize EC (see the following discussion). The distribution and function of this collagen therefore remain unknown.

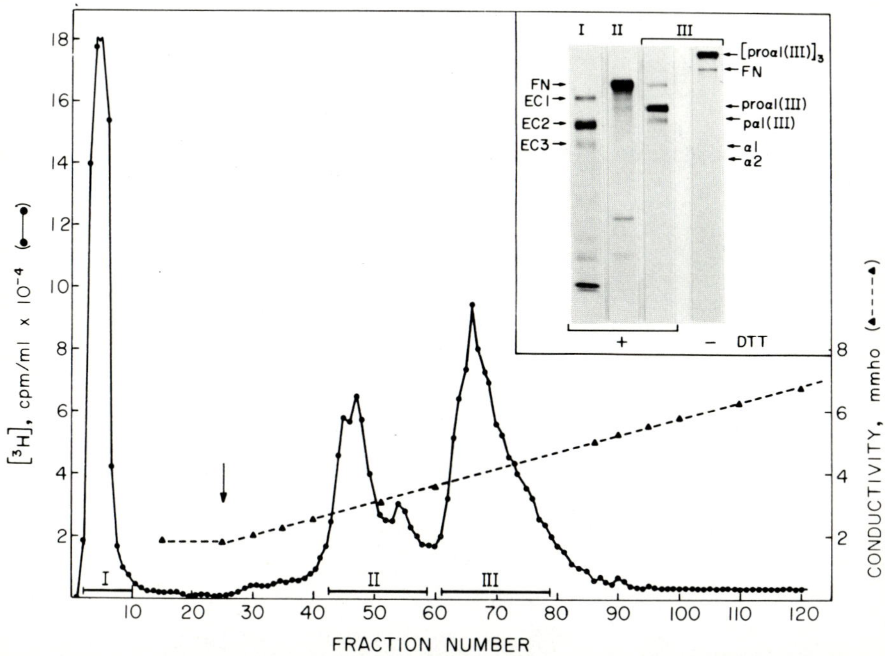

Fig. 2. DEAE-cellulose chromatography of BAE cell culture medium proteins. $[^3H]$proline-labeled proteins were precipitated from culture medium using 20% ammonium sulfate and chromatographed on DE-52 cellulose in 6 *M* urea and 50 m*M* Tris-HCl, pH 8.0, containing protease inhibitors at 4°C. Gradient elution was started as indicated by the arrow and performed from 0 to 200 m*M* NaCl over a volume of 400 ml. Inset: SDS-PAGE of peaks separated by DEAE-cellulose chromatography. Roman numerals correspond to the pooled fractions indicated on the chromatogram. Fibronectin (FN), EC, and pro**α**1(III) chains are identified, and the positions of migration of **α**1(I) and **α**2 collagen chains are indicated.

A COMPARISON OF THE SECRETORY PHENOTYPES OF ENDOTHELIAL CELLS

One approach to an understanding of the nature of the subendothelial basal lamina produced *in vivo* has been to examine the pericellular and extracellular stromal components synthesized by different endothelial cells *in vitro,* although such parallels must be drawn with caution. In a recent study we analyzed the proteins secreted by bovine endothelial cells cultured from large vessels, capillaries, and cornea (15). As shown in Fig. 3A, the aortic, venous, and corneal cells resembled one another in containing the three major secretory components, fibronectin, type III procollagen, and thrombospondin, although the relative proportions of these proteins differed. However, the capillary

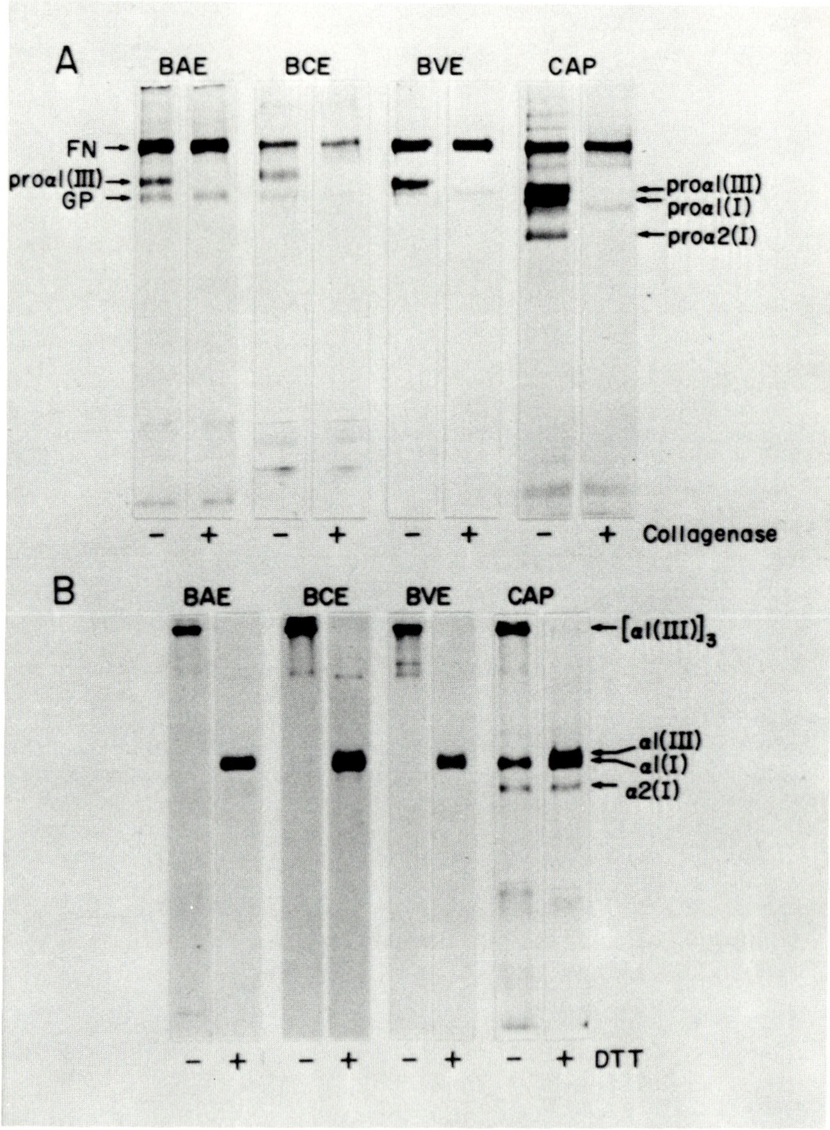

Fig. 3. Comparison of proteins secreted into the culture medium by bovine endothelial cells. Aortic cells were prepared as previously described (7,10). Cells from the superior vena cava were isolated according to Gimbrone (19). Corneal cells were cultured from Descemet's membrane (15). Capillary cells were cloned from adrenal glands as described by Folkman *et al.* (20). All cultures stained positively for factor VIII antigen. Aortic, venous, and corneal cells were grown in Dulbecco modified Eagle's medium (DMEM). Capillary cells were grown on collagen in tumor-conditioned medium

cells were of considerable interest in that they synthesized, in addition, a substantial amount of type I procollagen indicated by the presence of proα1(I) and proα2(I) chains after reduction. When the culture media, illustrated in Fig. 3A, were treated with pepsin under conditions in which noncollagenous proteins are digested to low-molecular-weight fragments and procollagens are converted to collagens, type III collagen was found to be the major collagen in the media of aortic, venous, and corneal cells, whereas the media of capillary cells contained both type I and III collagens (Fig. 3B). Since EC collagen is pepsin-sensitive, we fractionated the culture media of corneal and venous cells by DEAE-cellulose chromatography, as was done for the aortic cells. EC collagen was shown to be present in these cell types by the criterion of CNBr peptide mapping. We have not had sufficient capillary cells to determine whether or not these cells make EC.

MODULATION OF PROTEIN SECRETION BY ENDOTHELIAL CELLS

The significance of the ability of capillary cells to produce type I collagen in culture is not known, but the finding is clearly of interest in relation to the angiogenic properties of these cells *in vitro,* as described by Folkman and Haudenschild (2). We have for some time been interested in what may be a related phenomenon, that is, the ability of BAE cells occasionally to undergo a secondary growth pattern sometimes referred to as sprouting, in analogy with the branching of narrow cords of capillary cells from an existing vessel wall (21). As described by a number of laboratories (7,22,23), some endothelial cell strains at confluence spontaneously develop an elongated mycelial-like pattern of cells that undergrow the endothelial monolayer (Fig. 4). Various lines of evidence indicate that these morphologically atypical cells represent a variant of endothelium rather than the outgrowth of contaminating fibroblasts or smooth muscle cells (7,24).

---

(20). Nearly confluent cultures were incubated with [^{3}H]proline for 24 hr in serum-free medium supplemented with β-aminopropionitrile and sodium ascorbate. Radiolabeled samples were resolved by SDS-PAGE on composite 6%/10% slab gels in the presence or absence of 50 m*M* DTT and were visualized by fluorescence autoradiography. (A) Culture medium protein from each cell type is shown before and after collagenase treatment. Samples were reduced with DTT. (B) Culture medium protein from each cell type was digested with pepsin and pepsin-resistant proteins purified by precipitation with NaCl and molecular sieve chromatography on 6% agarose. BAE: bovine aortic endothelial; BCE: bovine corneal endothelial; BVE: bovine venous endothelial; CAP: bovine capillary endothelial. Fibronectin (FN), thrombospondin (GP), and type I and III procollagen and collagen chains are identified.

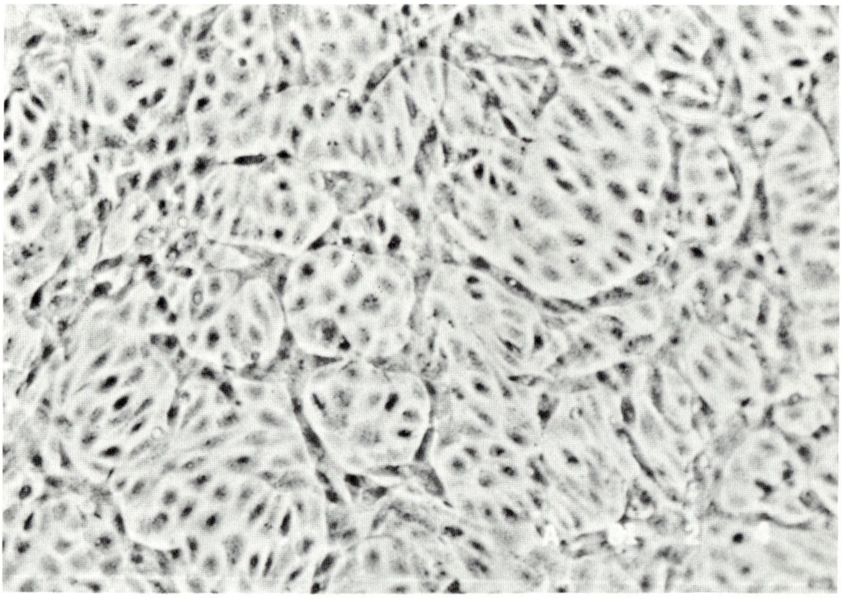

Fig. 4. Phase contrast micrograph of a confluent culture of BAE cells in which a secondary growth pattern (sprouting) appeared. BAE cells cultured as described by Schwartz (7) and Sage *et al.* (10). The morphologically altered cells are spindle shaped, are located beneath the monolayer, and are often aligned end to end to form a mycelial structure. (Micrograph courtesy of Drs. C. Gajdusek and S. Schwartz, University of Washington.)

When a culture containing a large proportion of phenotypically altered cells was labeled with [^{3}H]proline and culture medium proteins examined by slab gel SDS-PAGE fluorescence autoradiography, the pattern was substantially different from that obtained with typical endothelial cells (Fig. 5). Sprouting cultures synthesized less fibronectin and thrombospondin than BAE cells, and, in addition to type III procollagen, these cultures synthesized and secreted type I procollagen. The presence of type I procollagen in these cultures was quite apparent after the culture medium proteins were digested with pepsin and the $\alpha 1$ and $\alpha 2$ chains of type I collagen identified by SDS-PAGE (24). The secretory phenotype of sprouting cultures therefore resembles that of smooth muscle cells (Fig. 5) and capillary endothelial cells (see Fig. 3) more than it does that of BAE cells. The resemblance to capillary endothelial cells is of particular interest and raises the possibility that the tendency of aortic endothelial cells to develop an altered

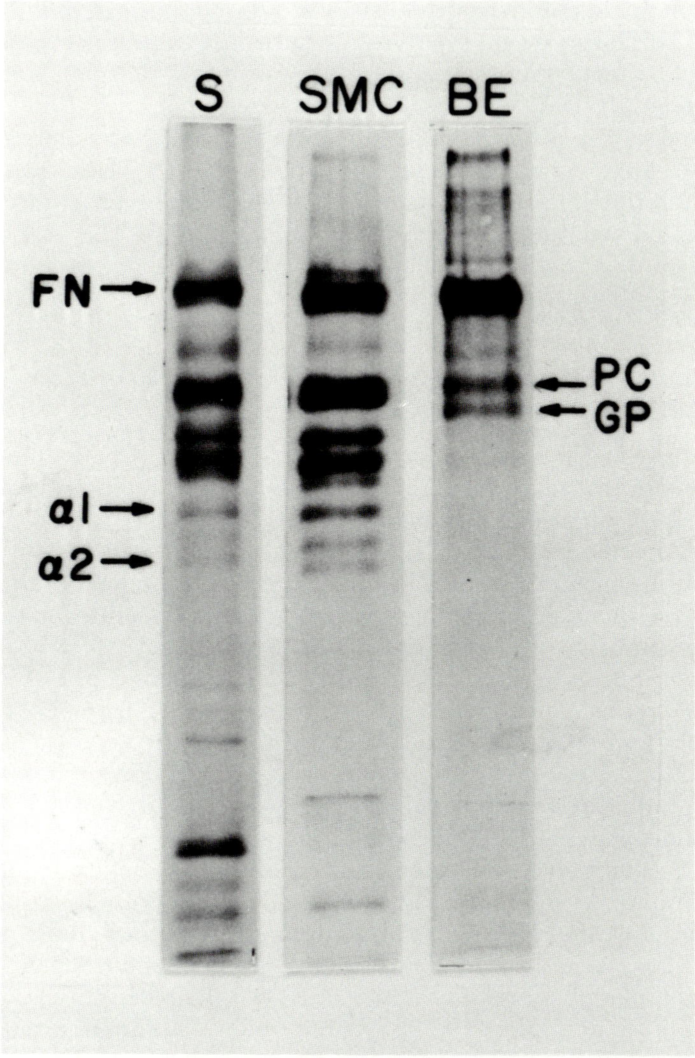

Fig. 5. SDS-PAGE of culture medium proteins secreted by a sprouting culture of endothelial cells (S), smooth muscle cells (SMC), and bovine aortic endothelial cells (BE). Cultures were labeled with [^{3}H]proline in the absence of serum as described in the legend to Fig. 1. Proteins were precipitated with trichloroacetic acid, run on a composite 6%/10% slab gel in the presence of 50 m*M* DTT, and subjected to fluorescence autoradiography. The positions of migration of fibronectin (FN), type III procollagen (PC), thrombospondin (GP), and the $\alpha 1$ and $\alpha 2$ chains of type I collagen are indicated.

morphology in culture represents an *in vitro* counterpart to the sprouting of capillaries from larger vessels in tissues. We do not understand the factors that control the onset of the sprouting phenomenon. However, recently Delvos *et al.* (25) have shown that typical endothelial cell monolayers can be induced to adopt a mycelial-like pattern when the cells are overlaid with a layer of type I collagen. These morphologically altered cells are not damaged, as shown by their ability to migrate into a wound and to grow well after replating. The biosynthetic profile of these cells and other aspects of this phenotypic modulation are currently under investigation.

THE RELATION OF AN ENDOTHELIAL CELL GLYCOPROTEIN TO PLATELET THROMBOSPONDIN

We have been interested in factors that might mediate endothelial cell attachment to a substratum, migration, and thromboresistance. In this regard we turned our attention to a glycoprotein secreted by endothelial cells (Fig. 1; 10). The glycoprotein was purified to homogeneity from culture medium by fractionation with ammonium sulfate and by sequential chromatography on gelatin-Sepharose (to which it did not bind), Sephacryl S-300, and hydroxyapatite (16). The protein was found to have an apparent subunit molecular weight of 190,000 after reduction of disulfide bonds and to exist as an oligomer of three or possibly four chains in its native state.

Stimulated by a report by Doyle *et al.* (26) indicating a close resemblance between a 145,000-dalton endothelial cell glycoprotein and an α-granule platelet protein, thrombospondin (27–29), we have found that the endothelial cell protein and thrombospondin are identical or very similar. As shown in Fig. 6, antibodies raised to the endothelial cell protein cross-react with thrombospondin both in a whole platelet lysate (lane 2) and in a mixture of proteins released from thrombin-activated platelets in the presence of EDTA (lane 3). EDTA is necessary in order to prevent Ca^{2+}-mediated binding of thrombospondin to platelet membranes (32). The antibodies also detect a 160,000 K protein that, on the basis of structural studies, appears to be a derivative of thrombospondin.

A close structural similarity between the endothelial cell glycoprotein and thrombospondin was further shown by peptide mapping (Fig. 7). We used a technique whereby protein bands, identified by Coomassie blue staining, are radioiodinated with the Bolton Hunter reagent within the gel and are digested with proteinase K. The resulting peptides are resolved by two-dimensional electrophoresis

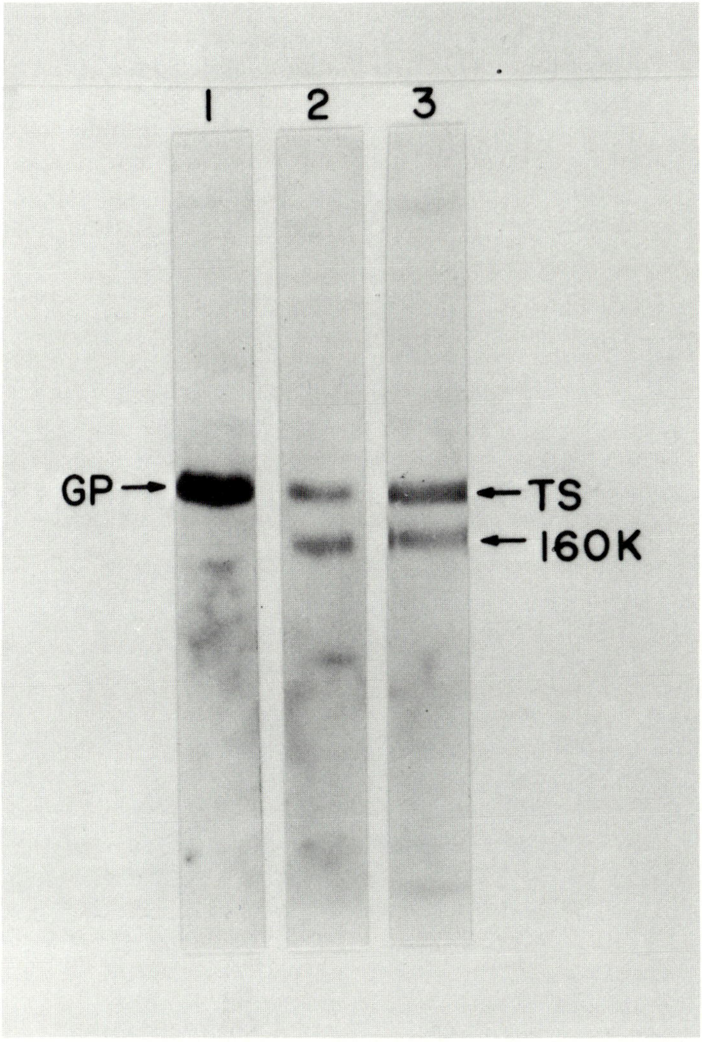

Fig. 6. Fluorescence autoradiogram showing reactivity of anti-endothelial cell glycoprotein toward bovine thrombospondin (TS). Purified endothelial cell glycoprotein (lane 1), washed bovine platelets (lane 2), and thrombin-released protein from bovine platelets (lane 3) were resolved on a 5% SDS polyacrylamide slab gel in the presence of 50 m*M* DTT. Following electrophoresis, protein was transferred from the slab gel to nitrocellulose paper and subsequently exposed to anti-glycoprotein antibodies and ^{125}I-labeled protein A (30,31). Following this treatment the nitrocellulose paper was dried and exposed to X-ray film.

and chromatography on thin-layer cellulose plates (33). It is apparent that the map of the bovine endothelial cell protein shown in Fig. 7A and the maps of the human and bovine platelet proteins in Figs. 7B and 7C are very similar. Furthermore, the map of the 160 K component of bovine thrombospondin (Fig. 7D) resembles that of the higher molecular form (Fig. 7C) and thereby supports the immunologic data.

Several important questions remain regarding thrombospondin. We do not know whether platelets acquire the protein from plasma or

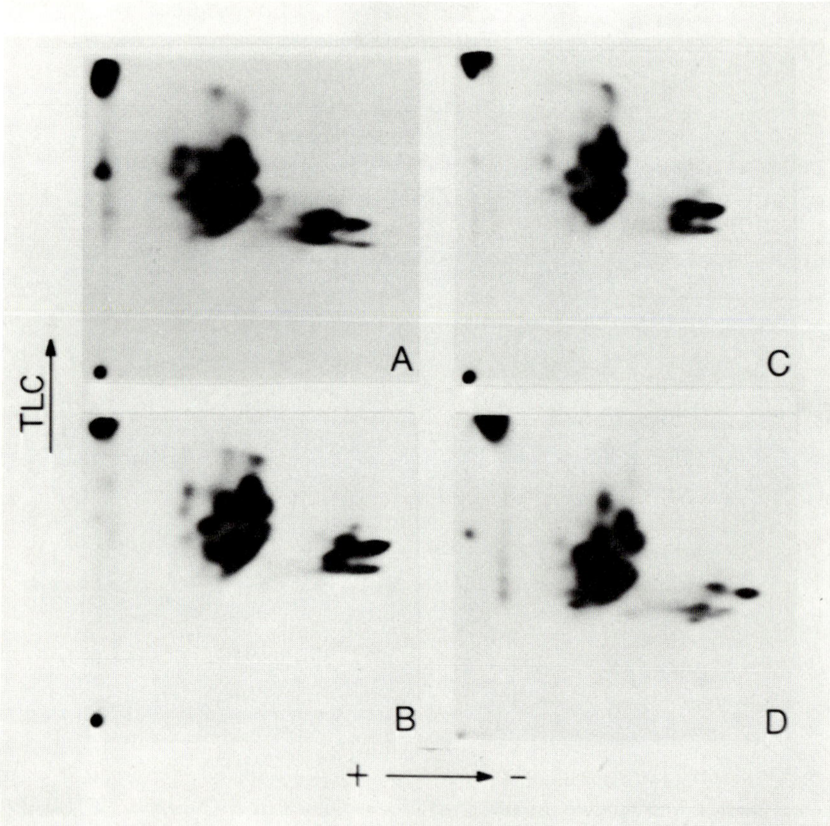

Fig. 7. Fluorescence autoradiogram of two-dimensional peptide maps of the endothelial cell glycoprotein and platelet thrombospondin. Proteins were initially resolved by SDS-PAGE and visualized by staining with Coomassie blue. The bands of interest were excised and prepared for two-dimensional peptide mapping on thin-layer cellulose plates, as described in the text. Samples are (A) bovine endothelial cell glycoprotein; (B) human platelet thrombospondin; (C) bovine platelet thrombospondin; (D) a 160 K protein from thrombin-activated bovine platelets that cross-reacts with antibodies to the endothelial cell protein.

whether megakaryocytes also synthesize it. Its function is also essentially unknown, although several possibilities, including a role in cell attachment or in hemostasis, are under investigation.

CONCLUDING REMARKS

We have for some time hypothesized that the structural macromolecules synthesized and secreted by mesenchymal cells in a connective tissue matrix not only reflect the phenotype of the cell but in some manner also influence cellular functions. In effect, then, there exists, in our view, a dynamic reciprocity in which secreted macromolecules continually modulate cellular behavior. The manner in which such an effect might be mediated is not known but a good deal of attention has recently been given to the existence of an external protein meshwork composed of collagens, putative attachment proteins such as fibronectin and laminin, and proteoglycans (34). This meshwork could interact with the internal cytoskeleton to regulate cellular shape and movement and affect metabolic function by, for example, modulating ionic fluxes or protein phosphorylation/dephosphorylation.

If this view has any merit, the analysis of proteins secreted by connective tissue cells should give us some concept of the manner in which the extracellular matrix can influence cellular behavior. Endothelial cells, because of their polarity and their tendency to secrete an extracellular matrix in a basal direction both *in vitro* and *in vivo*, serve as a useful model system to test some of these concepts. These studies derive additional importance because of the interest in understanding the factors that play a role in the attachment of endothelial cells to the subendothelium in blood vessels.

ACKNOWLEDGMENTS

Original work from this laboratory was supported by NIH grants AM 11248, DE 02600 and HL 18645. J. McP. is a recipient of National Research Service Award F 32 A 706553; H.S. is a recipient of a fellowship from R. J. Reynolds Industries, Inc.

REFERENCES

1. Mason, R. G., Mohammad, S. F., Saba, H. I., Chuang, H. Y. K., Lee, E. L., and Balis, J. V. (1979) *Pathol. Annu.* 9, 1–48.
2. Folkman, J., and Haudenschild, C. (1980) *Nature (London)* **288**, 551–556.

3. Jaffe, E. A., Hoyer, L. W., and Nachman, R. L. (1973) *J. Clin. Invest.* **52**, 2757–2764.
4. Loskutoff, D. J., and Edgington, T. S. (1977) *Proc. Natl. Acad. Sci. U.S.A.* **74**, 3903–3907.
5. Hial, V., Gimbrone, M. A., Peyton, M. P., Wilcox, G. M., and Pisano, J. J. (1979) *Microvasc. Res.* **17**, 314–329.
6. Baenziger, N. L., Becherer, P. R., and Majerus, P. W. (1979) *Cell* **16**, 967–974.
7. Schwartz, S. (1978) *In Vitro* **14**, 966–980.
8. Howard, B. V., Macarak, E. J., Gunson, D., and Kefalides, N. A. (1976) *Proc. Natl. Acad. Sci. U.S.A.* **73**, 2361–2364.
9. Madri, J. A., Dreyer, B., Pitlick, F. A., and Furthmayr, H. (1980) *Lab. Invest.* **43**, 303–315.
10. Sage, H., Crouch, E. C., and Bornstein, P. (1979) *Biochemistry* **18**, 5433–5442.
11. Sage, H., Pritzl, P., and Bornstein, P. (1980) *Biochemistry* **19**, 5747–5755.
12. Sage, H., Pritzl, P., and Bornstein, P. (1981a) *Biochemistry* **20**, 436–442.
13. Gospodarowicz, D., and Tauber, J.-P. (1980) *Endocr. Rev.* **1**, 201–227.
14. Greenburg, G., Vlodavsky, I., Foidart, J.-M., and Gospodarowicz, D. (1980) *J. Cell. Physiol.* **103**, 333–347.
15. Sage, H., Pritzl, P., and Bornstein, P. (1981b) *Arteriosclerosis* **1**, 427–442.
16. McPherson, J., Sage, H., and Bornstein, P. (1981) *J. Biol. Chem.* **256**, 11330–11336.
17. Oohira, A., Wight, T. N., McPherson, J., and Bornstein, P. (1982) *J. Cell Biol.* (in press).
18. Sage, H. (1981) *In* "The Biology of Endothelial Cells in Culture" (E. A. Jaffe, ed.), Nijhoff, The Netherlands (in press).
19. Gimbrone, M. A. (1976) *Prog. Hemostasis Thromb.* **3**, 1–28.
20. Folkman, J., Haudenschild, C., and Zetter, B. R. (1979) *Proc. Natl. Acad. Sci. U.S.A.* **76**, 5217–5221.
21. Ausprunk, D. H., and Folkman, J. (1977) *Microvasc. Res.* **14**, 53–65.
22. Gospodarowicz, D., and Mecher, A. L. (1978) *Natl. Cancer Inst. Monogr.* **48**, 109–130.
23. McAuslan, B. R., and Reilly, W. (1979) *J. Cell. Physiol.* **101**, 419–430.
24. Cotta-Pereira, G., Sage, H., Bornstein, P., Ross, R., and Schwartz, S. (1980) *J. Cell. Physiol.* **102**, 183–191.
25. Delvos, V., Gajdusek, C., Sage, H., Harker, L., and Schwartz, S. M. (1982) *Lab. Invest.* (in press).
26. Doyle, M. J., Mosher, D. F., and Jaffe, E. A. (1980) *J. Cell Biol.* **87**, 306a.
27. Baenziger, N. S., Brodie, G. N., and Majerus, P. W. (1971) *Proc. Natl. Acad. Sci. U.S.A.* **68**, 240–243.
28. Hagen, I., Olsen, T., and Solum, N. O. (1976) *Biochim. Biophys. Acta* **455**, 214–225.
29. Lawler, J. W., Slayter, H. S., and Coligan, J. E. (1978) *J. Biol. Chem.* **253**, 8609–8616.
30. Towbin, H., Theophil, S., and Gordon, J. (1979) *Proc. Natl. Acad. Sci. U.S.A.* **76**, 4350–4354.
31. Burnette, W. N. (1981) *Anal. Biochem.* **112**, 195–203.
32. Phillips, D. R., Jennings, L. K., and Prasanna, H. R. (1980) *J. Biol. Chem.* **255**, 11629–11632.
33. Sage, H., Pritzl, P., and Bornstein, P. (1981c) *Collagen Rel. Res.* **1**, 3–15.
34. Bornstein, P., Duksin, D., Balian, G., Davidson, J. M., and Crouch, E. (1978) *Ann. N.Y. Acad. Sci.* **312**, 93–105.

Glycoprotein Synthesis and Endothelial Cell Function: Influence of Thrombin and Melittin on Release of Glycoconjugates, Activity of Converting Enzyme, and Cell Morphology

ALICE R. JOHNSON, WERNER W. SCHULZ,
AND AARON HEIFETZ
Departments of Pharmacology, Pathology, and Biochemistry
The University of Texas Health Science Center at Dallas
Dallas, Texas

INTRODUCTION AND REVIEW

The vascular endothelium is a metabolically active tissue intimately concerned with the transfer of solutes between blood and tissues, regulation of vasoactive materials in circulation, and maintenance of a continuous nonthrombogenic surface for the vascular tree. These functional properties are determined by the luminal localization of the endothelium, the nature of intercellular junctions, the juxtaposition of cells within the vascular wall, and the composition of the extracellular matrix.

Numerous glycoproteins, including blood group antigens, alloantigens (1), factor VIII antigen (2,3) angiotensin I converting enzyme (4,5), fibronectin (6,7), and collagens (8–12) are associated with cultured endothelial cells. The distribution of these components *in vivo* clearly helps to determine their activities. For example, angiotensin I converting enzyme (kininase II) is concentrated at the luminal surface of endothelium where it has access to circulating peptides (13–15).

PATHOBIOLOGY OF THE ENDOTHELIAL CELL

ISBN 0-12-521980-6

Structural proteins, such as fibronectin and collagens, are localized primarily at the abluminal surface where they anchor and orient the monolayer of endothelial cells (8,16–18).

The glycoproteins and proteoglycans of the endothelial extracellular matrix are important structural components of the vessel wall. The composition of the extracellular matrix, and particularly the distribution of glycosaminoglycans, can affect such functions as transport, permeability, and response to circulating mediators. The proteoglycans are important for pathogenic mechanisms as well (19). A lipoprotein-complexing proteoglycan from human arterial media that interacts with low-density lipoproteins has been implicated in the atherogenic process (20), although vascular glycosaminoglycans are believed to inhibit thrombus formation (21).

Glycosaminoglycans, the carbohydrate moieties of proteoglycans, are major components of the glycocalyx of mammalian cells (22). These polyanionic compounds, heparan sulfate, chondroitin sulfate, dermatan sulfate, and hyaluronic acid, are synthesized and secreted by cells of the vessel wall where they influence cell shape, adhesion, and cell–cell interactions. Glycosaminoglycans have been implicated in embryonic development (23–25) and cell recognition and association phenomena (26,27), and they influence the permeability of connective tissues (28). Glycosaminoglycans are altered in virally transformed cells (29), and these molecules may regulate growth of both normal and transformed cells in culture (30,31).

Studies with cultured cells have facilitated understanding of how glycosaminoglycans are synthesized and released and have aided in definition of their influence on cell functions. Not all glycosaminoglycans are secreted in equal amounts. In 10 established mammalian cell lines of lymphoid, epithelial, or fibroblastic origin, the rates of total mucopolysaccharide production, as well as the proportion of the individual products, differed according to the cell type (32). The rate of synthesis, however, appears to depend, in part, on cell density (33).

Glycosaminoglycans can be released from cell surfaces by proteases (34) or glucuronidases (35), but they are not easily released by changes in the environment, such as altered pH or ionic strength (36). Some newly synthesized compounds are released spontaneously, whereas others remain bound to the cell surface (37). The major glycosaminoglycan released by human glial cells (38) and fibroblasts (39) is dermatan sulfate. The primary glycosaminoglycan associated with endothelial cells is heparan sulfate (39–42).

We found that human pulmonary arterial endothelial cells in culture can synthesize and release several classes of sulfated glycoconjugates, including heparan sulfate, chondroitin sulfate, and dermatan sulfate, as well as sulfated oligosaccharides that are resistant to enzymatic degradation by hyaluronidase and chondroitinase. These glycopeptides are less charged than the endothelial cell glycosaminoglycans, and they are not precipitated by cetylpyridinium chloride (43). The ratio of glycosaminoglycans synthesized and secreted by the human endothelial cells was similar to that reported by Gamse and associates for bovine aortic endothelial cells (44). These authors found that endothelial cells had a high proportion of heparan sulfate and chondroitin sulfate, but very little dermatan sulfate (44).

One of the basic characteristics of endothelium is its nonreactivity to circulating cells and coagulation factors. This property is related to several factors, including the anionic charge of the cells (45,46), production of antithrombogenic prostanoids (47–49), and uptake and degradation of thrombogenic materials from circulation (15,50,51). The anionic nature of endothelial glycosaminoglycans suggests that they contribute to the antithrombotic potential of the luminal surface.

The heparan sulfate associated with endothelial cells (39,40) is similar to the anticoagulant material, heparin, with respect to the distribution of *N*-sulfated residues and susceptibility to heparatinase (44). Gamse and associates noted that although the metabolism of sulfated glycosaminoglycans was similar in endothelial and smooth muscle cells from bovine aortas, endothelial cells had the most rapid turnover of intracellular and pericellular components (44).

The interrelationships of either heparan sulfate or heparin with endothelium, however, may have complex effects on the antithrombotic potential of the vessels. Heparan sulfate, like heparin, reduces platelet adherence to collagen-coated surfaces *in vitro*, and this glycosaminoglycan probably prevents platelet adherence to normal endothelium (42). Circulating heparin is taken up by endothelium *in vivo* (52) and could exert an anticoagulant effect or bind other molecules. Antithrombin III, which inhibits thrombin and several other activated procoagulant proteins (53), was identified by immunofluorescence in endothelial cells from human umbilical veins. This material is released into cell culture medium, suggesting that it is synthesized by the endothelial cells (54).

Heparin sulfate also catalyzes the reaction between antithrombin III and thrombin, and endothelial cells have specific binding sites for thrombin (55,56). The heparin sulfate of the endothelial pericellular matrix could thus function as a catalyst and binding site. Exogenous heparin is taken up by endothelial cells (57) and can bind antithrombin III and platelet factor 4 (58). It was suggested that platelet factor 4 (whose function is not known) could interfere with the association of thrombin and antithrombin III to facilitate hemostasis (58). An interaction between heparin and alpha-2-macroglobulin has also been described (59).

The uptake of heparin by endothelium would enhance the negative charge at the vascular lumen, but it may have other important effects on endothelial functions. Heparin is reported to neutralize the inhibitory effects of prostacyclin (PGI_2) on platelet aggregation (60,61), and it can block the generation of prostaglandin by thrombin (62). Heparin from mast cells can stimulate migration of capillary endothelial cells *in vitro*, suggesting that it may affect angiogenesis. This chemokinetic effect is specific, since other polyamines with similar charge density, including heparan sulfate, do not stimulate migration (63).

The composition of the endothelial surface coat has a significant influence on the adherence of platelets (64). Platelets do not adhere to normal endothelium, even when formation of prostacyclin is inhibited (65), but they adhere readily to virally transformed cells (18,65). Wechezak and associates (66) found that contraction of endothelial cells within a monolayer culture resulted in increased adherence of platelets to a subcellular network of filaments exposed by contraction.

Thrombin treatment of endothelial cells enhances platelet adherence (67–69), as does injury to endothelium by mechanical or chemical means (70–72). Thrombin binds to specific receptors on human venous endothelial cells (73) and bovine corneal endothelial cells (55,56). This protease can cause contraction of cells in human umbilical veins and changes in the morphology of the cell surface membranes (64). Thrombin–receptor complexes on bovine endothelial cells are internalized and subsequently degraded in lysosomes, which may be one means of regulating the procoagulant activity of thrombin (55).

The collagenous components of the subendothelium and surface coat are also important determinants of thromboresistance (12). Many different laboratories have studied the composition of the subendothelial matrix, but there is still disagreement on the types of collagen produced by endothelial cells. Human venous endothelial cells (6) and bovine aortic cells in culture (9) synthesize type IV, or basement membrane collagen. Other investigators found that types I and III, interstitial collagens, are also produced by endothelial cells (10,74–76). Sage and associates reported that endothelial cells from bovine aortas secrete primarily type III procollagen (10). They later found a unique, pepsin-sensitive material that they designated endothelial collagen (77). Type V collagen was also found in bovine endothelial cells

(12,76) and in cells from human umbilical cord veins (12). This type of collagen, which is preferentially associated with cell surfaces and basement membranes (78), appears to be part of the pericellular matrix of endothelial cells (76).

Basement membrane collagens IV and V do not aggregate platelets *in vitro* (79,80), but types I and III do. Madri and co-workers identified types I and III by immunofluorescence in smooth muscle cells and fibroblasts but did not find these proteins in endothelial cells (12).

Other constituents of the extracellular matrix can influence interactions between the endothelium and its environment. Human endothelial cells synthesize and secrete fibronectin (6,7,17). This polymorphic, high-molecular-weight glycoprotein forms part of the subendothelial matrix structure, and it has potentially important interactions with other proteins. Soluble fibronectin (cold insoluble globulin) binds to collagen (81), probably to anchor cells to each other and to the subcellular matrix. It can promote uptake of fibrin–collagen complexes (82) and bacteria (83) by phagocytic cells. Fibronectin is released from platelets during aggregation, and it can potentiate the action of plasminogen activator (84). It has also been implicated in the formation of atherosclerotic lesions (84).

Endothelial matrix proteins can influence cell growth and nutritional requirements in culture (85), and they may also regulate organizational and migratory activities required for tissue repair and neovascularization (86). Altered matrix protein synthesis might explain some of the differences between normal and transformed endothelial cells (65,87), and the aberrant morphology of variant, or sprouting, endothelial cells (18,88–90) may be due to changes in the distribution of matrix proteins. The distribution of fibonectin over the upper (luminal) surface was noted in a line of varient endothelial cells from bovine aorta, and these cells also leave deposits of fibronectin when they migrate (91). There may be other causes of altered morphology, however, since the sprouting bovine aortic cells studied by Cotta-Pereira and associates (90) synthesized less fibronectin than normal cells but more type I collagen.

Exposure of an endothelial monolayer to fibrin either in the form of a clot or in a gel with collagen results in disorganization of the monolayer into individual migratory cells (92). This observation led Kadish *et al.* (92) to suggest that endothelial migration was a means of delivering plasminogen activator to areas of thrombosis. The affected cells showed rapid and reversible changes in morphology as a result of interaction with fibrin, which further supports the concept that the endothelial cell is capable of a variety of responses to its environment.

STUDIES WITH MATRIX COMPONENTS AND ENDOTHELIAL CELLS

Clearly the acellular matrix on which endothelium rests influences a number of its functions. We and others (93–97) found that thrombin is a strong stimulus for formation and release of prostacyclin from human umbilical cord cells. This protease also stimulates striking changes in the surface morphology of the endothelial cells (64,97) In intact vessels thrombin causes contraction of the endothelium and focal exposure to subendothelial components to which platelets can adhere (64,98). We already have obtained some evidence that the substratum for cultured endothelial cells can affect cell functions. Melittin, which also stimulates prostaglandin formation and changes in cell morphology (96,97), has less effect on cells on a gelatin substratum than those on plastic (96).

Human pulmonary arterial cells grown on collagen or on an extracellular matrix from smooth muscle cells synthesized different proportions of glycosaminoglycans and released less of their total glycosaminoglycans into the culture medium than cells grown on untreated plastic dishes (43). Converting enzyme activity of cells grown for 2–4 days on a smooth muscle matrix was consistently lower than that of cells grown on plastic alone (43).

Because of the increasing evidence that endothelial cell matrix components can affect cell functions, we wanted to determine how the synthesis and release of glycoconjugates interfaced with cell morphology and enzyme activities. We were also concerned with the question of how membrane-active agents, such as thrombin and melittin, affected these parameters.

Endothelial cells from human umbilical vessels were used to determine whether there was a difference in the synthesis and release of glycosaminoglycans and glycopeptides by arterial and venous cells. The release of glycoconjugates, converting enzyme activity, and cell surface morphology were observed in control cells and in cells treated with either melittin or thrombin.

METHODS

CELL CULTURES

Endothelial cells were cultured from veins and arteries of human umbilical cords as previously described (4,5,99). Cells were grown in 25-cm^2 Corning flasks in medium 199 supplemented with 10% human serum (type 0^+) and 15% fetal calf serum. Cells from vessels of several cords were pooled for each experiment. The pooled cells were plated in replicate flasks and used without passage after 3–5 days in culture.

TREATMENT OF CELLS

Monolayer cultures were treated with either Hepes buffer A (100) alone or with thrombin (5 u/ml) or melittin (2.5 μg/ml) in buffer, for 30 min at 37°C. The flasks were gently rocked on a platform in the culture incubator during this time. After incubation, the medium was removed and saved for analysis of glycoconjugates or measurement of converting enzyme activity. The cells were washed once, and the wash fluid was combined with the original medium. The cell monolayers were retained for measurement of enzyme activity or glycoconjugates.

ELECTRON MICROSCOPY

Cells to be examined by scanning electron microscopy were plated from the original cell pool on small glass coverslips. They were treated as described above, washed once with medium, and fixed with 2% glutaraldehyde in 0.1 *M* phosphate buffer. They were dehydrated in an acetone series of 25, 40, 65, 90, and 100%, dried by carbon dioxide to critical point, and sputter-coated with gold palladium. The specimens were examined and photographed with a JEOL JSM 35 microscope.

ENZYME ASSAYS

Replicate flasks of endothelial cells were treated as described with Hepes buffer, thrombin, or melittin. After removal of the medium and wash fluids, the cells were gently scraped into 1 ml of Hepes buffer and transferred to a separate tube. The flasks were rinsed with an additional 1 ml of buffer which was added to the cell samples. The material was dispersed by sonication for 15 sec with a Branson probe sonifier, and aliquots from each flask were assayed in triplicate for angiotensin I converting enzyme using a Ventrex kit assay. This method utilizes the hydrolysis of [^{3}H]hippuryl diglycine to determine enzyme activity. The specificity of the reaction for converting enzyme was established by inclusion of SQ 14225 (captopril) in triplicate samples from each flask (5). The amount of DNA in each cell sample was used to determine the cell number (101), and enzyme activity was expressed as nmole/hr/10^6 cells.

Converting activity was also measured in a soluble preparation of enzyme from human lung following treatment with thrombin, melittin, or varying concentrations of heparin sulfate (Nutritional Biochemicals) or chondroitin sulfate (Sigma). Incubation with these agents was for 30 min at 37°C prior to determination of the enzyme activity.

RADIOLABELING OF CELLS

Cultures of endothelial cells were double-labeled with 5 μCi/ml of D-[6-^{3}H]glucosamine and 50 μCi/ml of [^{35}S]H_2SO_4 (New England Nuclear) in Ham's F-12 medium for 18–24 hr. After labeling, the radioactive medium was removed, the monolayers were washed three times, and Hepes buffer was added for the treatments indicated above. After incubation with buffer, thrombin, or melittin, the medium was collected, the cells were washed, and the pooled medium and wash fluids were treated with 10% cold trichloracetic acid to isolate acid-insoluble glycoconjugates. The cells were removed by scraping in 2 ml of methanol. Trichloracetic acid-insoluble material from cells or incubation media were extracted with chloroform–methanol–water as we previously described (43). Pronase digests of the material were separated by chromatography on Sephadex G-50 in 0.1 *M* ammonium acetate (25). The labeled glycosaminoglycans eluted with the excluded volume (V_0) and the oligosaccharide side chains of the glycoproteins eluted in the retarded fractions. These two major components were examined in pooled fractions from the column. Pool A (fractions 25–49) contained glycosaminoglycans, and pool B (fractions 50–80) contained the glycoprotein side chains (glycopeptides). These pools were lyophilized and reconstituted in 1 ml of distilled water before further analysis.

The labeled glycosaminoglycans were further analyzed by treatment with chondroitinase ABC (Miles Laboratories) as described by Hart and Lennarz (102). The components resistant to enzymatic cleavage were separated from smaller degradation products by precipitation by cetylpyridinium chloride (CPC). This procedure precipitates charged glycosaminoglycans. The CPC-precipitable material resistant to chondroitinase ABC was identified as heparan sulfate.

GEL ELECTROPHORESIS

In experiments to determine whether thrombin or melittin released glycoproteins from the cells, endothelial cells were labeled for 24 hr with [^{3}H]glucosamine and then treated as described above. The media from control and treated cells was applied to polyacrylamide gels. Polyacrylamide slab gel electrophoresis was as described by Laemmli (103) with 4% stacking gel and a separating gel containing a linear gradient of polyacrylamide from 4–12% and of glycerol from 0–15%. The gels were fixed, impregnated with Enhance (New England Nuclear), and dried, and high-molecular-weight glycoproteins were identified by autoradiography.

RESULTS

CELL MORPHOLOGY

Scanning electron microscopy revealed striking changes in the appearance of cells that were treated with either thrombin or melittin. Thrombin contracted individual cells slightly, but the most pronounced effect was the formation of numerous small blisters at the cell surface (Fig. 1). There was no obvious detachment of thrombin-treated cells or decrease in cell viability.

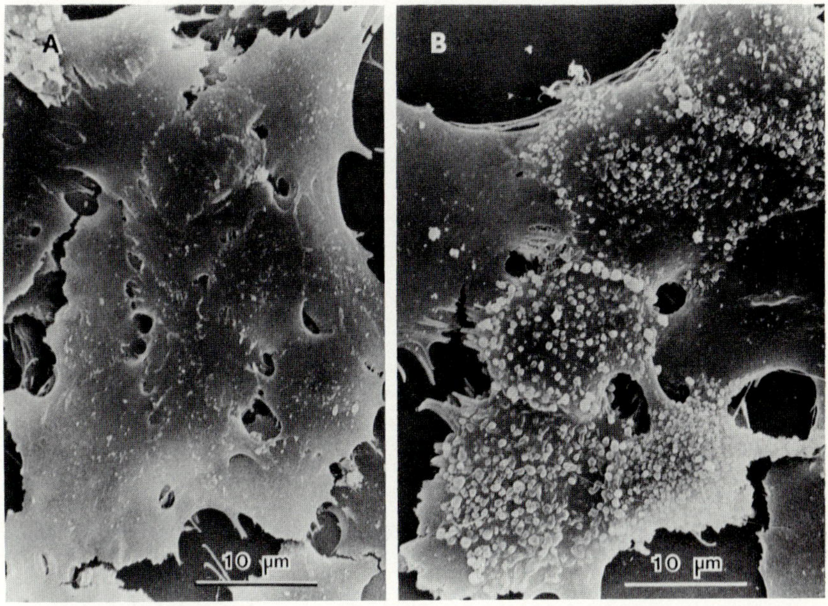

Fig. 1. Scanning electron micrograph of human endothelial cells. (A) A monolayer of control (buffer-treated) cells. (B) Cells that were treated for 30 min with 5 U/ml of thrombin.

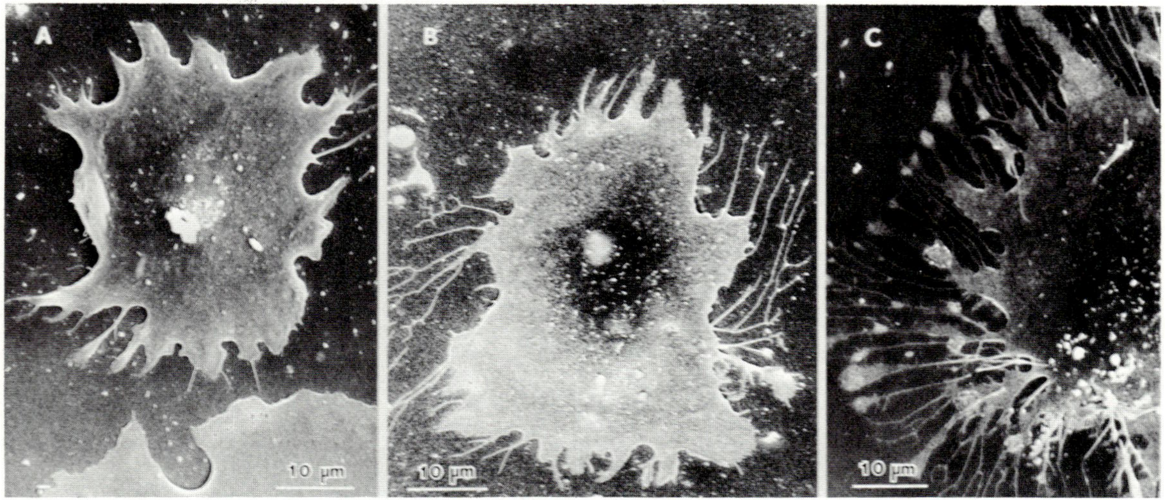

Fig. 2. Scanning electron micrograph of endothelial cells. (A) Cells treated with buffer alone. (B) Cells treated with 1.0 μg/ml of melittin. (C) Cells treated with 2.5 μg/ml of melittin.

Melittin caused changes in both the cell shape and adherence to the substratum. There was an increased number of long membranous processes extending from melittin-treated cells compared to controls. These filipodia sometimes extended beyond the original periphery of the cell, and they were usually attached to the coverslip surface (Fig. 2). In some specimens it could be seen that these extensions had broken off, leaving "footprints" where they were originally attached. The extensions from the cell surface stimulated by melittin formed an intricate network of fibrillar material that often reached far beyond the body of the cell (Fig. 3). It is not yet clear whether these projections result from contraction of the cells or whether they are extended in response to stimulation by the polypeptide.

CONVERTING ENZYME ACTIVITY

Endothelial cells from human umbilical cord veins that were incubated with either thrombin or melittin for 30 min had more converting enzyme activity than cells that were incubated with buffer alone (Table I). In six separate experiments untreated cells had an activity of 44 nmol/hr 10^6 cells. In four experiments with thrombin the activity in-

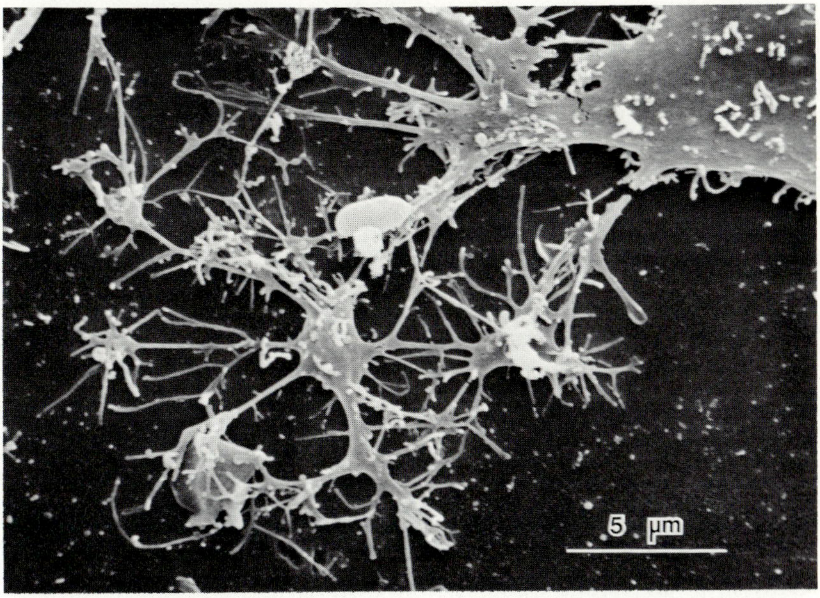

Fig. 3. Projections from endothelial cells stimulated by melittin (2.5 μg/ml).

TABLE I
Converting Enzyme Activity in Human Venous Endothelial Cells: Influence of Thrombin and Melittin[a]

Treatment	No. experiments	Enzyme activity (nmol/hr 10^6 cells)
None (Hepes buffer)	6	43.8 ± 5.1
Thrombin (5 u/ml)	4	62.3 ± 4.8*
Melittin (2.5 μg/ml)	5	62.7 ± 5.3*

[a] Endothelial monolayers were incubated with buffer alone, thrombin, or melittin for 30 min at 37°C, washed, and the amount of enzyme associated with the cells was measured by hydrolysis of [^{3}H]hippuryl diglycine. Data are means ± S.E.M. of four to six individual experiments. Asterisks indicate significant difference from control ($p \leq .01$).

creased by approximately 50% to 63 nmol/hr 10^6 cells. In five experiments with melittin, there was a comparable increase in enzyme activity over control cells. Assay of the enzyme in medium from the cell monolayers did not show any difference between control or treated cells, and the release was never more than 10% of the total activity (Table I).

Soluble converting enzyme from human lung that was preincubated with thrombin or melittin had the same activity as enzyme incubated with buffer alone. However, when the enzyme was incubated with either heparin sulfate or chondroitin sulfate (5–500 μg/ml), there was a decrease in enzyme activity related to the concentration of sulfated mucopolysaccharide. Five μg/ml of heparin sulfate reduced the enzyme activity to 68% of control, 50 μg/ml reduced it to 59%, and 500 μg/ml reduced it to 43%. Chondroitin sulfate (50 and 500 μg/ml) reduced activity to 73% and 53% of the control activity.

RELEASE OF GLYCOCONJUGATES

Endothelial cells incubated for 18–24 hr with [^{3}H]glucosamine and [^{35}S]H_2SO_4 incorporated both labels into glycosaminoglycans and glycoprotein side chains. As we previously reported for pulmonary arterial cells (43), the sulfated glycoconjugates that were released and retained by the cells were similar.

Figure 4 shows labeled glycoconjugates released from untreated venous cells (panel A) and from cells treated with thrombin (B) and melittin (C). The corresponding panels D, E, and F show the materials released from arterial cells. There was no obvious difference in either the amount or type of labeled material released by buffer or thrombin from venous and arterial cells. Melittin stimulated the release of

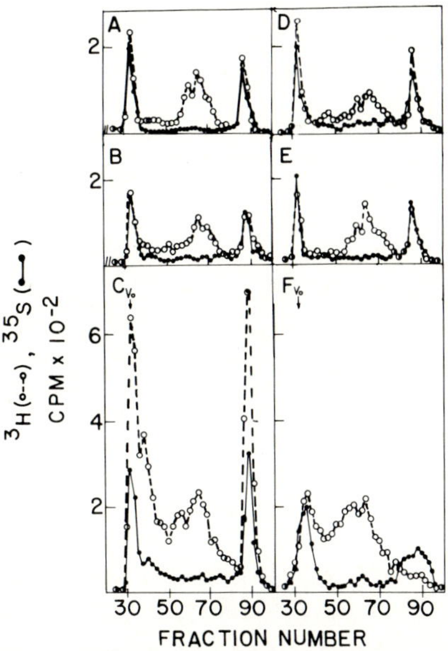

Fig. 4. Release of glycosaminoglycans and glycopeptides from endothelial cells. (A), (B), and (C) are counts per minute of ^3H or ^{35}S released from umbilical venous cells into the medium. (A) Media from control cells. (B) Media from thrombin-treated cells. (C) Media from melittin-treated cells. (D), (E), and (F) are counts per minute released by arterial cells. (D) is media from control cells, (E) is from thrombin-treated cells, and (F) is from melittin-treated cells. The fraction number of material separated by Sephadex G-50 is indicated on the abcissa. V_0 is the void volume.

glycosaminoglycans from venous cells, and increased release of smaller fragments as well. The release of glycosaminoglycans was most pronounced in venous cells, and in arterial cells most of the labeled material was in the glycopeptide fraction (Fig. 4). While only 1–2% of the total labeled material was released from either type of cell by buffer or thrombin, melittin stimulated release of approximately 10% of the labeled glycoconjugates.

Analysis of radiolabeled materials in the incubation media showed that melittin stimulated a fourfold increase in the total glycosaminoglycan release from venous endothelial cells and doubled the amount released by arterial cells. The total amount of glycopeptides released by melittin-treated cells also increased 2- to 5-fold (Table II).

Enzymatic digestion and precipitation with cetylpyridinium chloride was used to identify heparan sulfate in the media from endothelial cells. The material from pooled fraction A from venous endothelial cells contained a large amount of heparan sulfate. Most of the

TABLE II
Release of Glycosaminoglycans and Glycopeptides from Human Endothelial Cells[a]

		(A) Glycosaminoglycans		(B) Glycopeptides	
Cell source	Treatment	^3H (cpm)	^{35}S (cpm)	^3H (cpm)	^{35}S (cpm)
Umbilical	None	3,700	2,500	3,800	560
vein	Melittin	12,200	3,200	8,000	1,400
Umbilical	None	2,300	2,000	2,000	550
artery	Melittin	4,000	1,400	10,000	530

[a] Cells were double-labeled for 18 hr, the media was removed, and the cells were treated with plain medium or medium containing melittin (2.5 μg/ml) for 30 min. The released glycosaminoglycans and glycopeptides were separated from the media samples by chromatography on Sephadex G-50. Data are the total counts per minutes in pooled fractions (see text) A and B.

glycosaminoglycan (CPC-precipitable material) released was in the form of heparan sulfate. Arterial cells appeared to have less heparan sulfate than venous cells. Melittin stimulated release of heparan sulfate from venous cells but not from the arterial cells. However, this polypeptide caused a 5- to 10-fold increase in the release of glycopeptides from both types of endothelial cells (Table III). The cetylpyridinium chloride-soluble material released by melittin from either arterial or venous endothelial cells resembled polyglycosyl chains con-

TABLE III
Glycosaminoglycans Released from Endothelial Cells

	CPC-precipitable material[b]					
	Total		Heparan-SO_4		CPC-soluble material[c]	
Media[a]	^3H (cmp)	^{35}S (cpm)	^3H (cpm)	^{35}S (cpm)	^3H (cpm)	^{35}S (cpm)
V (C)	2700	2400	2300	1300	1000	100
(M)	7000	3100	5600	1800	5200	140
A (C)	2000	1900	1000	800	320	120
(M)	500	1400	400	600	3500	64

[a] V = media from umbilical venous cells; A = media from umbilical arterial cells.

[b] Total counts were determined in CPC-precipitable material from control (C) and melittin-treated (M) cell media. Heparan SO_4 was determined in material resistant to digestion with chondroitinase ABC.

[c] CPC-soluble material has characteristics of polyglycosyl chains containing *N*-actyl glucosamine galactose disaccharide repeating units.

taining repeating units of *N*-acetyl glucosamine → galactose disaccharide (25).

Application of [^{3}H]glucosamine-labeled media from melittin-treated endothelial cells to polyacrylamide gels showed that there were several high-molecular-weight glycoproteins, which appear as diffuse bands on the autoradiograph shown in Fig. 5. These products were not apparent in material from cells treated with either buffer or thrombin (Fig. 5). Comparison with molecular-weight markers indicated that the glycoproteins released by melittin were 100,000 daltons or more.

DISCUSSION

We wanted to determine how morphological and biochemical properties of human endothelial cells were affected by cell matrix components, particularly the glycosaminoglycans that are synthesized and secreted by the cells. As a first step, we measured the release of glycoconjugates from control cells and those that were treated with membrane-active agents.

Thrombin, a protease involved in the final stage of blood coagulation, and melittin, a polypeptide from bee venom (104), are two agents that stimulate prostaglandin formation in human endothelial cells (96,97). These two apparently unrelated substances also affect other properties of endothelial cells. Both thrombin and melittin caused an increase in the activity of cell-associated angiotensin I converting enzyme, and both caused changes in the morphology of the cells. The structural changes were different, however, for the two agents. Thrombin did not affect the pattern of glycoconjugates released from either arterial or venous cells, but melittin stimulated the release of heparan sulfate from venous endothelial cells and the release of polysaccharide side chains from both arterial and venous cells.

The mechanism through which thrombin enhances converting enzyme activity of endothelial cells is probably quite different from that of melittin. Thrombin combines with surface receptors on umbilical cord cells (73) or corneal endothelial cells (55,56), and it stimulates changes in the surface of endothelium *in situ* (64,98). Other actions of thrombin include the stimulation of prostacyclin formation (93–97), suppression of fibrinolytic activity (105), and potentiation of polypeptide growth factors (106).

Some, if not all, of these actions appear to involve receptors on the cell surface. For example, thrombin potentiates the effects of growth

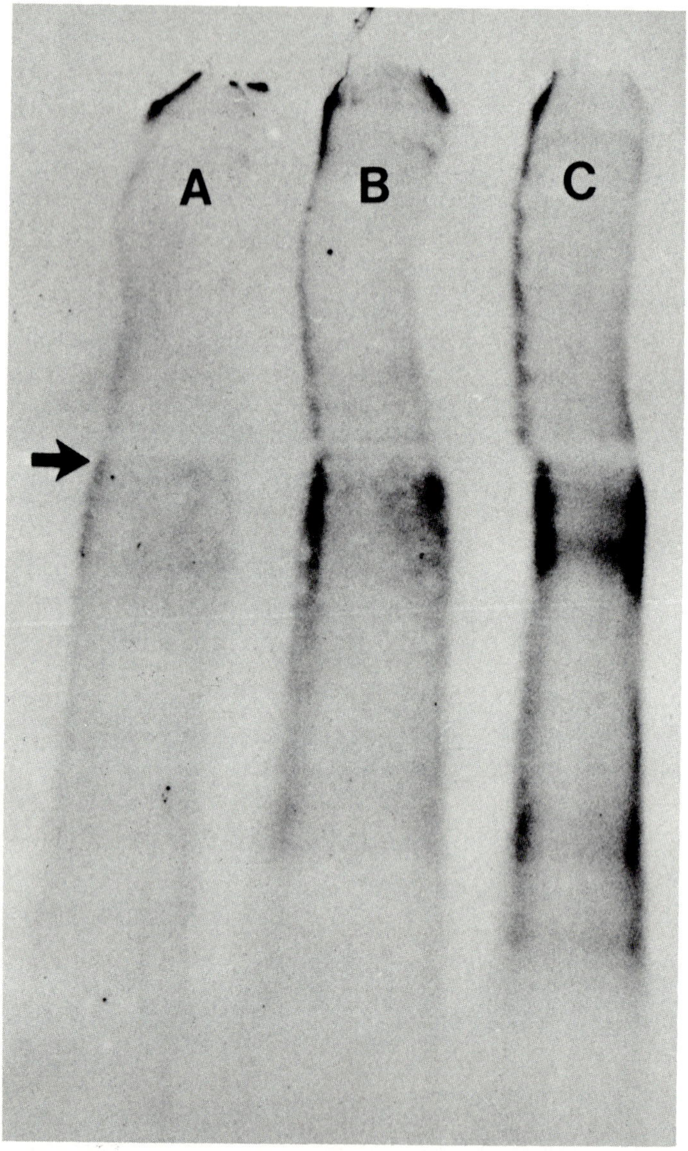

Fig. 5. Autoradiograph of ^{3}H-labeled material separated by polyacrylamide slab gel electrophoresis. (A) Gel containing media from control cells. (B) Gel with media from thrombin-treated cells. (C) From melittin-treated cells. The arrow indicates the interface between the stacking gel and the running gel.

factors on human umbilical venous endothelial cells but not those from bovine aorta (106). While thrombin is a potent stimulator of prostacyclin formation in umbilical cord cells (93–97), it has only marginal effects on bovine aortic cells (94), and it does not stimulate prostaglandin formation in human pulmonary arterial or venous cells at all (96). Possibly, differences in the endothelial matrix can account for the lack or availability of thrombin receptors.

In our experiments, thrombin stimulated both an increase in converting enzyme activity and a striking change in the surface morphology of the cells. The increased surface area produced by the blister formation at the cell surface could possibly account for the observed increase in converting enzyme activity. This cell surface enzyme is generally more active in cells from arteries than in cells from veins (5), and we have suggested previously that this difference could be due to the increased number of surface projections found on arterial cells (99). Although this hypothesis remains unproved, the present study provides evidence that supports it. Since we did not find any difference in release of glycoconjugates from control and thrombin-treated cells, it is unlikely that the loss of cell matrix materials could account for the increase in enzyme activity.

The enhanced converting enzyme activity in melittin-treated cells, however, may be related to a loss of components from the cell surface. The morphological changes caused by melittin also include a potential increase in membrane surface through the elongated extensions from the cell. It is still not clear whether these processes are extended by the cell, or whether they are remnants of cellular and extracellular materials exposed by cell movement.

Since soluble converting enzyme was partially inhibited by heparin sulfate and chondroitin sulfate, and since melittin releases glycosaminoglycans from endothelial cells, control of the cell-bound enzyme by glycosaminoglycans seems a possibility. Heparan sulfate has many of the properties of heparin (42,55,56), and the heparan sulfate of endothelial cells is reported to have a higher charge density and *N*-sulfate content than that in other types of cells (44). Thus the increased activity of converting enzyme in endothelial cells following treatment with melittin might be due to loss of heparan sulfate. The fact that arterial cells have less heparan sulfate might explain the slightly higher activity of converting enzyme observed in those cells (5).

Melittin can activate membrane-bound kallikrein and membrane-bound renin in the rat kidney, possibly by different mechanisms (107,108). Melittin is a polypeptide of approximately 2800 MW with basic and hydrophobic amino acids (104). It is a potent activator of membrane phospholipase (109,110), and it releases histamine from mast cells. Although it can release converting enzyme from rat lung (111) or pulmonary arterial endothelial cells (96), the concentrations used in this study did not release the enzyme from umbilical venous cells.

Although we cannot exclude a direct effect of melittin on the cell-bound converting enzyme, melittin did not stimulate the activity of purified converting enzyme from lung. The release of glycosaminoglycans from endothelial cells coupled with enhanced enzymatic activity strongly suggests that the glycosaminoglycans inhibit the cell-bound converting enzyme in untreated or uninjured cells.

The release of glycopeptides from arterial and venous endothelial cells by melittin is also interesting. The oligosaccharide side chains are similar to the novel glycoprotein saccharides associated with glycoproteins and glycolipids of human erythrocytes (112,113). These products have been identified in early organogenesis-stage mouse embryos (25), and they may be components of many different tissues.

Several conclusions can be made from these experiments. First, arterial and venous endothelial cells synthesize and secrete similar glycoconjugates. The amount released is small in proportion to that remaining in the cells. Second, the enhanced converting enzyme activity measured in melittin-treated cells may be related to the loss of glycosaminoglycans from the cell surface. Third, thrombin-induced changes in the cell surface membrane might account for an increase in converting enzyme activity, but this increase is not due to a loss of glycosaminoglycans. Finally, release of glycopeptides with repeating units of *N*-acetyl glucosamine galactose by melittin may indicate a selective loss of glycoproteins from the cell membrane.

ACKNOWLEDGMENTS

We thank Kay Wolens, Donna Allen, and Anita White for their excellent technical assistance. We are grateful to Dr. Ervin Erdös of the Department of Pharmacology for supplying the human lung converting enzyme.

This work was supported by HL 18826 and HL 25937 from the NHLBI of NIH.

REFERENCES

1. Moraes, J. R., and Stastny, P. (1977) *J. Clin. Invest.* **60,** 449–454.
2. Jaffe, E. A., Hoyer, L. W., and Nachman, R. L. (1973) *J. Clin. Invest.* **52,** 2757–2764.

3. Jaffe, E. A., Hoyer, L. W., and Nachman, R. L. (1974) *Proc. Natl. Acad. Sci. U.S.A.* **71,** 1906–1909.
4. Johnson, A. R., and Erdös, E. G. (1977) *J. Clin. Invest.* **59,** 684–695.
5. Johnson, A. R. (1980) *J. Clin. Invest.* **65,** 841–850.
6. Jaffe, E. A., and Mosher, D. F. (1978) *J. Exp. Med.* **147,** 1779–1791.
7. Macarak, E. J., Kirby, E., Kirk, T., and Kefalides, N. A. (1978) *Proc. Natl. Acad. Sci. U.S.A.* **75,** 2621–2625.
8. Jaffe, E. A., Minick, C. R., Adelman, B., Becker, C. G., and Nachman, R. L. (1976) *J. Exp. Med.* **144,** 209–225.
9. Howard, B. V., Macarak, E. J., Gunson, D., and Kefalides, N. A. (1976) *Proc. Natl. Acad. Sci. U.S.A.* **73,** 2361–2364.
10. Sage, H., Crouch, E., and Bornstein, P. (1979) *Biochemistry* **18,** 5433–5442.
11. Raj, C. V. S., Freeman, I. L., Church, R. L., and Brown, S. I. (1979) *Invest. Ophthalmol. Visual Sci.* **18,** 75–84.
12. Madri, J. A., Dreyer, B., Pitlick, F. A., and Furthmayr, H. (1980) *Lab. Invest.* **43,** 303–315.
13. Caldwell, P. R. B., Seegal, B. C., and Hsu, K. C. (1976) *Science* **191,** 1050–1051.
14. Ryan, J. W., Ryan, U. S., Schultz, D. R., Whitaker, C., and Chung, A. (1975) *Biochem. J.* **146,** 497–499.
15. Ryan, J. W., and Ryan, U.S. (1977) *Fed. Proc., Fed. Am. Soc. Exp. Biol.* **36,** 2683–2691.
16. Zetter, B. R., Martin, G. R., and Birdwell, C. R. (1978) *Ann. N.Y. Acad. Sci.* **312,** 299–316.
17. Birdwell, C. R., Gospodarowics, D., and Nicolson, G. L. (1978) *Proc. Natl. Acad. Sci. U.S.A.* **75,** 3273–3277.
18. Zetter, B. R., Johnson, L. K., Shuman, M. A., and Gospodarowicz, D. (1978) *Cell* **14,** 501–509.
19. Hollander, W. (1976) *Exp. Mol. Pathol.* **25,** 106–120.
20. Camejo, G., Cortez, M. M., Lopez, F., Starosta, R., Mosquera, B., and Cocorro, L. (1980) *Acta Med. Scand., Suppl.* **642,** 159–164.
21. Izuka, K., and Murata, K. (1972). *Atherosclerosis* **16,** 217–224.
22. Luft, J. H. (1976) *Int. Rev. Cytol.* **45,** 291–382.
23. Hay, E. D. (1977) *In* "International Cell Biology" (B. R. Brinkley and K. R. Porter, eds.), pp. 50–57. Rockefeller Univ. Press, New York.
24. Meier, S., and Hay, E. D. (1974) *Proc. Natl. Acad. Sci.* **71,** 2310–2313.
25. Heifetz, A., Lennarz, W. J., Libbus, B., and Hsu, Y.-C. (1980) *Dev. Biol.* **80,** 398–408.
26. Pessac, B., and Defendi, V. (1972) *Science* **175,** 898–900.
27. Underhill, C., and Dorfman, A. (1978) *Exp. Cell Res.* **117,** 155–164.
28. Maroudas, A. (1976) *J. Anat.* **122,** 335–347.
29. Roblin, R., Albert, S. O., Gelb, N. A., and Black, P. H. (1975) *Biochemistry* **14,** 347–357.
30. Terry, A. H., and Culp, L. A. (1974) *Biochemistry* **13,** 414–425.
31. Cohn, R. H., Cassiman, J. J., and Bernfield, M. R. (1976) *J. Cell Biol.* **71,** 280–294.
32. Saito, H., and Uzman, B. G. (1971) *Biochem. Biophys. Res. Commun.* **43,** 723–728.
33. Hronowski, L., and Anastassiades, T. P. (1980) *J. Biol. Chem.* **255,** 10091–10099.
34. Kraemer, P. M. (1971) *Biochemistry* **10,** 1437–1445.
35. Vogel, K. G., and Kelley, R. O. (1977). *J. Cell. Physiol.* **92,** 469–480.
36. Vogel, K. G., and Dolde, J. (1979) *Biochim. Biophys. Acta* **552,** 194–200.
37. Vogel, K. G., and Kendall, V. F. (1980) *J. Cell. Physiol.* **103,** 475–487.

38. Glimelius, B., Norling, B., Westermark, B., and Wasteson, A. (1978) *Biochem. J.* **172,** 443–456.
39. Buonassisi, V. (1973) *Exp. Cell Res.* **76,** 363–368.
40. Buonassisi, V., and Root, M. (1975) *Biochim. Biophys. Acta* **385,** 1–10.
41. Wasteson, A., Glimelius, B., Busch, C., Westermark, B., Heldin, C. H., and Norling, B. (1977) *Thromb. Res.* **11,** 309–321.
42. Busch, C., Ljungman, C., Heldin, C. M., Waskson, E., and Obrink, B. (1979) *Haemostasis* **8,** 142–148.
43. Heifetz, A. H., and Johnson, A. R. (1981) *J. Supramol. Struct. and Cell. Biochem.* **15,** 359–367.
44. Gamse, G., Fromme, H. G., and Kresse, H. (1978) *Biochim. Biophys. Acta* **544,** 514–528.
45. Skutelsky, E., Rudich, A. and Danon, D. (1975) *Thromb. Res.* **7,** 623–634.
46. Pelikan, P., Gimbrone, M. A., and Cotran, R. S. (1979) *Atherosclerosis (Shannon, Irel.)* **32,** 69–80.
47. Needleman, P., Wyche, A., and Raz, A. (1979) *J. Clin. Invest.* **63,** 345–349.
48. Moncada, S., and Vane, J. R. (1979) *Pharm. Rev.* **30,** 293–321.
49. Gorman, R. R. (1979) *Fed. Proc., Fed. Am. Soc. Exp. Biol.* **38,** 83–88.
50. Dosne, A. M., Legrand, C., Bavouis, B., Bodevin, E., and Caen, J. P. (1978) *Biochem. Biophys. Res. Commun.* **85,** 183–189.
51. Glasgow, J. G., Schade, R., and Pitlick, F. A. (1978) *Thromb. Res.* **13,** 255–266.
52. Hiebert, L. M., and Jaques, L. B. (1976) *Artery* **2,** 26–37.
53. Teien, A. U. N., Abildgaard, U., and Höök, M. (1976) *Thromb. Res.* **8,** 859–867.
54. Chan, V, and Chan, T. K. (1979) *Thromb. Res.* **15,** 209–213.
55. Savion, N., Isaacs, J. D., Gospodarowicz, D., and Shuman, M. A. (1981). *J. Biol. Chem.* **256,** 4514–4519.
56. Isaacs, J. D., Savion, N., Gospodarowicz, D., Fenton, J. W., and Shuman, M. A. (1981) *Biochemistry* **20,** 398–403.
57. Glimelius, B., Busch, C., and Höök, M. (1978) *Thromb. Res.* **12,** 773–782.
58. Busch, C., Dawes, J., Pepper, D. S., and Wasteson, A. (1980) *Thromb. Res.* **19,** 129–137.
59. Schanberge, J. N., and Sridhar, C. N. (1980) *Circulation* **62,** Suppl. III, 52.
60. Eldor, A., and Weksler, B. B. (1979) *Thromb. Res.* **16,** 617–628.
61. Saba, H. I., Saba, S. R., Blackburn, C. A., Hartmann, R. C., and Mason, R. G. (1979) *Science* **205,** 499–501.
62. Buchanan, M. R., Dejana, E., Cazenave, J. P., Mustard, J. F., and Hirsch, J. (1979) *Thromb. Res.* **16,** 551–555.
63. Azizkhan, R. G., Asiskhan, J. C., Zetter, B. R., and Folkman, J. (1980) *J. Exp. Med.* **152,** 931–944.
64. Barnhart, M. I., and Chen, S. T. (1978) *Semin. Thromb. Hemostasis* **52,** 112–155.
65. Curwen, K. D., Gimbrone, M. A., and Handin, R. I. (1980) *Biochim. Biophys. Acta* **544,** 514–528.
66. Wechezak, A. R., Holbrook, K. A., Way, S. A., and Mansfield, P. B. (1979) *Blood Vessels* **16,** 35–42.
67. Czervionke, R. L., Hoak, J. C., and Fry, G. L. (1978) *J. Clin. Invest.* **62,** 847–856.
68. Hoak, J. C., Czervionke, R. L., Fry, G. L., and Smith, J. B. (1980) *Fed. Proc., Fed. Am. Soc. Exp. Biol.* **39,** 2606–2609.
69. Cazenave, J. P., Dejana, E., Kinlough-Rathbone, R., Packham, M. A., and Mustard, J. F. (1979) *Haemostasis* **8,** 183–192.
70. Ross, R., Glomset, J., and Harker, L. (1977) *Am. J. Pathol.* **86,** 675–684.

71. Lewis, J. C., and Kottke, B. A. (1977) *Science* **196,** 1007–1009.
72. Schwartz, S. M. (1977) *Cardiovasc. Med.* **2,** 991–1002.
73. Awbry, B. J., Hoak, J. C., and Owen, W. G. (1979) *J. Biol. Chem.* **254,** 4092–4095.
74. Barnes, M. J., Morton, L. R., and Levene, C. I. (1978) *Biochem. Biophys. Res. Commun.* **84,** 646–653.
75. Mayne, R., Vail, M. S., and Miller, E. J. (1978) *Biochemistry* **17,** 446–452.
76. Sage, H., Pritzl, P., and Bornstein, P. (1981) *Biochemistry* **20,** 436–442.
77. Sage, H., Pritzl, P., and Bornstein, P. (1980) *Biochemistry* **19,** 5747–5755.
78. Madri, J. A., and Furthmayr, H. (1979) *Am. J. Pathol.* **94,** 323–330.
79. Huang, T. W., and Benditt, E. P. (1978) *Am. J. Pathol.* **92,** 99–108.
80. Trelstad, R. L., and Carvalho, A. C. (1979) *J. Lab. Clin. Med.* **93,** 499–505.
81. Engvall, E., and Ruoslahti, E. (1977) *Int. J. Cancer* **20,** 1–5.
82. Blumenstock, F. A., Saba, T. M., Weber, P., and Laffin, R. (1978) *J. Biol. Chem.* **253,** 4287–4290.
83. Mosher, D. F., and Proctor, R. A. (1980) *Science* **209,** 927–929.
84. Stenman, S., von Smitten, K., and Vaheri, A. (1980) *Acta Med. Scand., Suppl.* **642,** 165–170.
85. Vlodavsky, I., Lui, G. M., and Gospodarowicz, D. (1980) *Cell* **19,** 607–616.
86. Folkman, J., and Haudenschild, C. (1980) *Nature (London)* **288,** 551–556.
87. Gimbrone, M. A., and Fareed, G. C. (1976) *Cell* **9,** 685–693.
88. Schwartz, S. M. (1978) *In Vitro* **14,** 966–980.
89. McAuslan, B. R., and Reilly, W. (1979) *J. Cell. Physiol.* **101,** 419–430.
90. Cotta-Pereira, G., Sage, H., Bornstein, P., Ross, R., and Schwartz, S. (1980) *J. Cell. Physiol.* **102,** 183–191.
91. McAuslan, B. R., Hannan, G. N., Reilly, W., and Stewart, F. H. C. (1980) *J. Cell. Physiol.* **104,** 177–186.
92. Kadish, J. L., Butterfield, C. E., and Folkman, J. (1979) *Tissue Cell* **11,** 99–108.
93. Weksler, B. B., Ley, C. W., and Jaffe, E. A. (1978) *J. Clin. Invest.* **62,** 923–930.
94. Hong, S. L. (1980) *Thromb. Res.* **18,** 787–795.
95. Czervionke, R. L., Smith, J. B., Hoak, J. C., Fry, G. L., and Haycraft, D. L. (1979) *Thromb. Res.* **14,** 781–786.
96. Johnson, A. R., Callahan, K. S., Tsai, S. J., and Campbell, W. B. (1981) *Bull. Eur. Physiopathol. Respr.* 17, 531–551.
97. Johnson, A. R., Tsai, S. C., Reynolds, R. C., Shulz, W. W., and Campbell, W. B. (1981) *Fed. Proc., Fed. Am. Soc. Exp. Biol.* **40,** 705.
98. Mohammad, S. F., and Mason, R. G. (1981) *Arch. Pathol. Lab. Med.* **105,** 62–66.
99. Johnson, A. R., Schulz, W. W., Noguiera, L. A., and Erdös, E. G. (1980) *Clin. Exp. Hypertens.* **2,** 659–674.
100. Weksler, B. B., Marcus, A. J., and Jaffe, E. A. (1977) *Proc. Natl. Acad. Sci., U.S.A.* **74,** 3922–3926.
101. Burton, K. (1968) *in* "Methods in Enzymology" (L. Grossman and K. Moldave, eds.), Vol. 12, Part B, pp. 163–166. Academic Press, New York.
102. Hart, G. W., and Lennarz, W. J. (1978) *J. Biol. Chem.* **253,** 5795–5801.
103. Laemmli, J. K. (1970) *Nature (London)* **227,** 680–685.
104. Habermann, E. (1972) *Science* **177,** 314–322.
105. Loskutoff, D. J. (1979) *J. Clin. Invest.* **64,** 329–332.
106. Gospodarowicz, D., Brown, K. D., Birdwell, C. R., and Zetter, B. R. (1978) *J. Cell Biol.* **77,** 774–788.
107. Nishimura, K., Ward, P., and Erdös, E. G. (1980) *Hypertension* **2,** 538–545.
108. Nishimura, K., Alhenc-Gelas, F., White, A., and Erdös, E. G. (1980) *Proc. Natl. Acad. Sci., U.S.A.* **77,** 4975–4978.

109. Hassid, A., and Levine, L. (1977). *Res. Commun. Chem. Pathol. Pharmacol.* **18,** 507–517.
110. Shier, W. T. (1979) *Proc. Natl. Acad. Sci. U.S.A.* **76,** 195–199.
111. Igic, R., Nakajima, T., Yeh, H. S. J., Sorrells, K., and Erdös, E. G. (1973) *In* "Kinin Peptides" (G. H. Acheson, ed.), pp. 307–319. Karger, Basel.
112. Krusius, T., Rinne, J., and Pauvala, H. (1978) *Eur. J. Biochem.* **92,** 289–300.
113. Fukuda, M. N., Fukuda, M., and Hakomori, S. (1979) *J. Biol. Chem.* **254,** 5458–5465.

PART V

PROSTAGLANDINS

The Role of Prostaglandins in Platelet-Vessel Wall Interactions

S. MONCADA AND J. R. VANE
Wellcome Research Laboratories
Beckenham, Kent, England

INTRODUCTION

The term *prostaglandin* was introduced by von Euler (1935) to describe the vasodepressor and smooth-muscle stimulating activity present in human semen and accessory genital glands (1,2). The activity was due to several acidic lipids, and it was not until 1960 that Bergström and Sjövall (3) isolated in pure form prostaglandin E_1 (PGE_1) and prostaglandin $F_{1\alpha}$ ($PGF_{1\alpha}$) and later several other stable prostaglandins (4,5).

In 1964 Bergström *et al.* (6) and Van Dorp *et al.* (7) independently demonstrated that prostaglandins are biosynthesized from the polyunsaturated fatty acids dihomo-γ-linolenic acid (C20 : 3ω6), arachidonic acid (C20 : 4ω6), and eicosapentaenoic acid (C20 : 5ω3). These give rise to the mono-, bis-, or trienoic prostaglandins, respectively. There is little evidence that monoenoic prostaglandins are made by mammalian tissues, apart from their presence in seminal fluid.

Arachidonic acid, the precursor of all bisenoic prostaglandins, is the most common fatty acid present in cellular phospholipids and can be obtained directly from the diet or by desaturation and chain elongation from dietary linoleic acid (C18 : 2ω6). Arachidonic acid is liberated from cellular phospholipids by the action of phospholipases, which can be activated by changes in their chemical environment (8,9). Simple mechanical stimulation can result in generation of prostaglandins, as shown in lung (10), spleen (11,12), and platelets (13–15). The enzymes that synthesize prostaglandins are present in almost all organs so far studied, but some tissues, such as seminal vesicles, kidneys, and lungs, have a greater capacity for prostaglandin synthesis than others (16).

PATHOBIOLOGY OF THE ENDOTHELIAL CELL

ISBN 0-12-521980-6

Much of the earlier work concentrated on prostaglandins E_2 and $F_{2\alpha}$ as these possessed potent and diverse biological activities and were available in pure form. Since 1973, important discoveries have been made on the nature of unstable metabolites of arachidonic acid (Fig. 1). These include the prostaglandin endoperoxides (PGG_2 and PGH_2), thromboxane A_2 (TXA_2), and prostacyclin (PGI_2), which, despite their instability, have potent biological activities. Investigations into the sites of production of these compounds and their differing and, in some cases, opposing, biological activities have led to the development of new concepts of vascular homeostasis. This review will concentrate predominantly on prostacyclin and thromboxane A_2 with some discussion of the other intermediates. Prostacyclin has now been assigned the approved name epoprostenol, but the trivial name will be used throughout this review.

PATHWAYS OF ARACHIDONIC ACID METABOLISM

Once released from the membrane phospholipids, arachidonic acid is metabolized by two enzymes. The term *eicosanoids* is applied to all the 20-carbon derivatives, whereas *prostanoids* refers only to those with a prostanoic acid skeleton. The cyclooxygenase enzyme forms the prostaglandin endoperoxide PGG_2 (17). This is converted to PGH_2, which then isomerizes enzymically or nonenzymically to the stable PGE_2, $PGF_{2\alpha}$, and PGD_2 (see Fig. 1). A 17-carbon hydroxy acid called 12-hydroxy-5,8,10-heptadecatrienoic acid (HHT) is also formed, together with malondialdehyde (MDA). The prostaglandin endoperoxides are also transformed enzymically into two other unstable products, prostacyclin and TXA_2. Unlike PGE_2, D_2, or $F_{2\alpha}$, these products are not formed by chemical breakdown of PGH_2.

LIPOXYGENASES

The enzyme cyclooxygenase (sometimes called prostaglandin synthetase) seems to be present in all cell types (except erythrocytes), whereas 5- or 12-lipoxygenases have so far been identified in platelets, lungs, white cells, blood vessels, and epicardium (17–21).

There were several indications in the last few years that slow-reacting substance in anaphylaxis (SRS-A) could be a product of arachidonic acid (22). This has now been confirmed by the structural elucidation of slow-reacting substances (SRS's) derived from arachidonic acid via the 5-lipoxygenase pathway. The formation of a

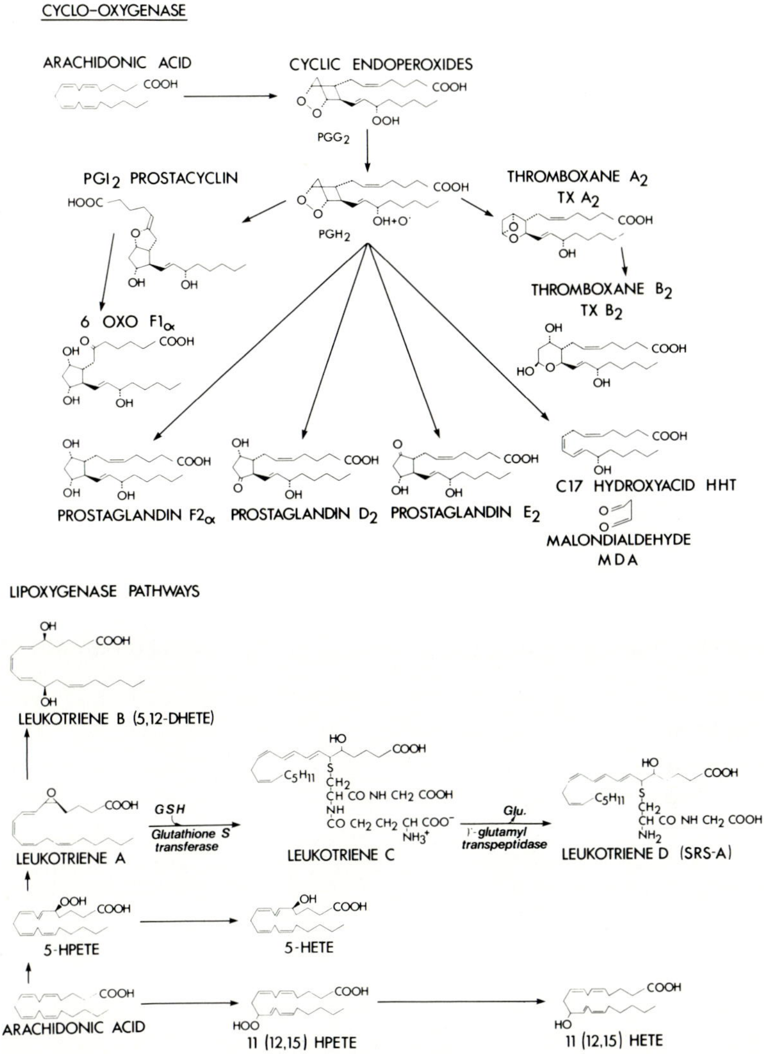

Fig. 1. Pathways of arachidonic acid metabolism. (Upper panel) Conversion of arachidonic acid via the cyclooxygenase pathway to prostaglandins of the 2-series. (Lower panel) Conversion of arachidonic acid via the lipoxygenase pathway to leukotrienes.

5,6-epoxide of arachidonic acid (23) leads to the biosynthesis of a group of SRS's that have been renamed leukotrienes (24) (see Fig. 1). Leukotrienes (LT's) are characterized by three conjugated double bonds. The epoxide (LTA) can be converted to a 5,12-dihydroxy acid (LTB) or, alternatively, with the addition of glutathione in the 5 position, can be converted to an unusual lipid–peptide structure (LTC) (25). Further metabolism of LTC with the removal of glutamic acid by γ-glutamyl transpeptidase yields LTD, the cysteinylglycine derivative (25–27). The further enzymic removal of glycine gives the cysteinyl derivative (LTE) (25). Slow-reacting substance in anaphylaxis (SRS-A) is probably a mixture of LTC and LTD.

Leukotriene B is a potent chemotactic agent for leukocytes *in vitro* (28,29) and *in vivo* (30,31). Leukotriene B also causes plasma exudation in rabbit skin, and this activity is increased by combination with vasodilator prostaglandins, bradykinin, or histamine (32). Leukotrienes are potent spasmogens of guinea pig and human airway smooth muscle (33–35). These findings are beginning to highlight the importance of the products of 5-lipoxygenase in cell migration, inflammation, and other conditions such as increased mediator release and airway reactivity during anaphylaxis. The biological importance of the products of the 12-lipoxygenase pathway is poorly understood.

THROMBOXANE SYNTHETASE

The prostaglandin endoperoxides, PGG_2 and PGH_2 were shown by Hamberg *et al.* (36) to induce platelet aggregation. This was accompanied by the formation of an unstable vasoconstrictor substance identified as thromboxane A_2 (37). Thromboxane A_2 has a chemical half-life at body pH and temperature of 30 sec. The activity of the "rabbit aorta-contracting substance," or RCS, described first by Piper and Vane (38) (see later) can be accounted for by TXA_2. The enzyme that synthesizes TXA_2 from PG endoperoxides was first localized in the high-speed particulate fraction of human and horse blood platelets (39,40). The enzyme has been solubilized and separated from the cyclooxygenase (41–43) and detailed studies of human and bovine platelet thromboxane synthetase have been published (44–46).

Other tissues capable of synthesizing TXA_2 include rabbit and human polymorphonuclear leukocytes (47,48); mouse, rat, and guinea pig macrophages (49,50); human lung fibroblasts (51); rabbit and cat spleen (52); rabbit iris and conjunctiva (53); guinea pig lungs (18); human umbilical artery (54); rabbit pulmonary artery (55); and rabbit and rat kidney (56,57). Interestingly, Needleman and his associates

(56) have shown that kidneys made hydronephrotic begin to synthesize TXA_2, suggesting the possibility that tissue damage unmasks in some way the ability to form this substance. However, it may also be in some instances that the production of TXA_2 by tissues is due to the presence of platelets or migratory cells.

Although MDA and HHT can be formed nonenzymically by spontaneous degradation of PGH_2 (42), the biosynthetic pathways of TXA_2 and HHT are linked, for a purified enzyme that synthesizes TXA_2 also catalyzes HHT formation (41,43).

PROSTACYCLIN SYNTHETASE

Prostacyclin is generated by blood vessel microsomes or fresh vascular tissue from the prostaglandin endoperoxide, PGH_2 (58,59). It is chemically unstable, with a half-life of 2–3 min, breaking down to 6-oxo-$PGF_{1\alpha}$. Prostacyclin synthetase displays a broad pH optimum and catalyzes a rapid conversion of saturating concentrations of PGH_2 at 37°C (60). Wlodawer and Hammarström (61) demonstrated that prostacyclin synthetase in porcine aorta microsomes was stimulated by the high-speed supernatant and that the stimulant factor was soluble, nondialysable, and resistant to boiling. Unlike the formation of TXB_2 and HHT, formation of prostacyclin and HHT by porcine aorta microsomes is not coupled (61). Various tissues have been shown to generate prostacyclin or its stable breakdown product 6-oxo-$PGF_{1\alpha}$ (see Table I).

PROSTACYCLIN

GENERATION BY AND EFFECTS ON THE VESSEL WALL

Prostacyclin is the main product of arachidonic acid in all arteries and veins so far tested. Not much is known about the microcirculation, but Goehlert *et al.* (95) have demonstrated that microvessels, mainly capillaries, isolated from rat cerebrum generate predominantly prostacyclin.

The ability of the large vessel wall to synthesize prostacyclin is greatest at the intimal surface and progressively decreases toward the adventitia (96). Production of prostacyclin by cultured cells from vessel walls also shows that endothelial cells are the most active producers of prostacyclin (68,69); moreover, this production persists after numerous subcultures *in vitro* (67).

TABLE I
Tissues That Have Been Shown to Generate Prostacyclin or Its Stable Breakdown Product 6-Oxo-PF$G_{1\alpha}$

	Reference
Pig aortic microsomes	58,62
Rabbit microsomes	58,62
Pig mesenteric artery microsomes	58,62
Rabbit coeliac and mesenteric arteries	63
Human mesenteric arteries and veins	64
Bovine coronary artery	65
Hamster aorta	66
Cultured endothelial cells; human umbilical and bovine aortic	67,68
Pig aortic endothelial cells	69
Pig postcaval vein endothelium	69
Pig smooth muscle	69
Pig adventitial fibroblasts	69
Human arterial smooth muscle	70
Human skin fibroblasts	70
Mouse macrophages	71
Rabbit eye, iris, and conjunctiva	53
Bovine iris and ciliary body	72
Guinea pig perfused lung	73
Cat lung	74
Rabbit lung	75
Rabbit pleura, pericardium, and peritoneum	21
Guinea pig heart	76
Rabbit heart	77,78
Ductus arteriosus (fetal lamb)	79
Ductus arteriosus (fetal calf)	80
Fetal calf aorta	80
Rat kidney	57
Rabbit kidney	78
Human renal cortex	81
Pig kidney	82
Rabbit renal collecting tubule cells	83
Cat spleen	84
Rat stomach fundus microsomes	62
Rat stomach	85
Rat gastric mucosa	86
Rat small intestine	87
Uterus decidual tissue of pregnant rat	88
Myometrium of pregnant rat	89
Homogenates of pseudopregnant rat uterus	90
Human placenta	91
Human chorion, amnion, and decidua	92
Ram seminal vesicles	93
Rat inflammatory exudate	94
Rat cerebral microvessels and choroid plexus	95

Initially it was demonstrated that vessel microsomes in the absence of cofactors could utilize prostaglandin endoperoxides, but not arachidonic acid, to synthesize prostacyclin (58). Later it was shown that fresh vascular tissue could utilize both precursors, although the endoperoxides are much better substrates (59). Moreover, vessel microsomes, fresh vascular rings, or endothelial cells treated with indomethacin could, when incubated with platelets, generate a prostacyclin-like antiaggregating activity (59,62,97). The release of this substance was inhibited by 15-hydroperoxy arachidonic acid (15-HPAA) and other fatty acid hydroperoxides known to be selective inhibitors of prostacyclin formation (58,60,62). From all these data we concluded that the vessel wall can synthesize prostacyclin not only from its own endogenous precursors, but also from prostaglandin endoperoxides released by the platelets, thus suggesting a biochemical cooperation between platelet and vessel wall (98). Other observations support this conclusion. For instance, incubation of platelet-rich plasma (PRP) with fresh, indomethacin-treated arterial tissue leads to an increase in platelet cyclic AMP (cAMP) (99) which parallels the inhibition of the aggregation and can be abolished by previous treatment of the vascular tissue with tranylcypromine, a less active inhibitor of prostacyclin formation. Undisturbed endothelial cell monolayers can also readily transform PGH_2 to prostacyclin (15).

However, the hypothesis was challenged by Needleman and associates (78), who demonstrated that while arachidonic acid was rapidly converted to prostacyclin by perfused rabbit hearts and kidneys, PGH_2 was not readily transformed. They concluded that some degree of vascular damage is necessary for the endoperoxide to be utilized by prostacyclin synthetase. Needleman *et al.* (100) and Hornstra *et al.* (101), using vessel microsomes or fresh vascular tissue, also concluded that endoperoxides from platelets cannot be utilized by other cells under their experimental conditions. However, Marcus *et al.* (102,103) showed that feeding of endoperoxides to endothelial cells suspended in PRP takes place *in vitro*, but only when the platelet number corresponds to normal blood levels. Too high a platelet concentration induces a platelet-to-platelet interaction that limits the platelet–endothelial cell reaction. It should be stressed, however, that the possibility of endoperoxides released from platelets being utilized by endothelial cells has not yet been tested *in vivo*. Adherence of the platelet to the vessel wall, known to be one of the first responses to injury, could well provide the close proximity that would be needed for such "cooperation."

It is also possible that formed elements of blood such as white cells, which produce endoperoxides and TXA_2 (47,48,104), could interact with the vessel wall to promote formation of prostacyclin. Moreover, leukocytes themselves generate prostacyclin in whole blood, especially in the presence of thromboxane synthetase inhibitors (105). Thus prostacyclin might regulate white cell behavior and help control white cell activity during the inflammatory response (106,107).

As in man, prostacyclin is hypotensive in dog, rabbit, or rat (108) and is more potent than PGE_2. Prostacyclin is at least 100 times more active than its degradation product, 6-oxo-$PGF_{1\alpha}$ (108).

In isolated Langendorff-perfused hearts of the guinea pig, rabbit, and rat, prostacyclin not only is a potent vasodilator but also is the predominant metabolite of arachidonic acid (76–78, 109, 110). It is also a coronary dilator in the intact heart of open-chest dogs (108,111). Local injection of prostacyclin (50–500 ng) into the coronary circulation increased coronary blood flow without systemic effects and it was more potent than PGE_2. Furthermore, prolonged coronary vasodilatation was elicited by prostacyclin absorbed through the myocardium after dripping a solution of 20–100 μg onto the surface of the left ventricle (111).

Interesting results are beginning to appear on the effects of prostacyclin on infarct size after coronary occlusion. In the anesthetized dog an infusion of prostacyclin (320 ng/kg/min) for 6 hr starting 17 min after coronary occlusion did not alter infarct size (112), but infusion at 420 ng/kg/min for 6 hr in anesthetized dogs starting 3 min after occlusion reduced the infarct size (113). Prostacyclin in the rat increased arrhythmia and ventricular fibrillation following coronary occlusion but decreased infarct size in the survivors (114). Ohlendorf *et al.* (115) found that platelet counts in cats decreased by 30–35% in the first 20 min after ligation of the left anterior descending coronary artery. This decrease was reversed by intravenous prostacyclin. At the same time, prostacyclin also prevented the loss of cathepsin-D and creatine phosphokinase activities in the infarcted area and inhibited ST segment elevation. Lefer *et al.* (116) and Ogletree *et al.* (117) have also demonstrated beneficial effects of prostacyclin in acute myocardial ischemia, and more recently Ribeiro *et al.* (118), using the dog, have shown that prostacyclin given as an intravenous infusion after coronary ligation decreases myocardial oxygen demands, maintains ischemic blood flow while reducing nonischemic blood flow and greatly reduces early mortality.

There is increasing evidence that prostacyclin can mediate the release of renin. Arachidonic acid, prostaglandin endoperoxides, and prostacyclin all stimulate renin release from slices of rabbit renal cortex, but PGE_2 has no such effect (119,120). Furthermore, indomethacin reduces renin release in animals and man (121–123). Prostacyclin-like activity and 6-oxo-$PGF_{1\alpha}$ have been identified in incubates of PGG_2 or PGH_2 with renal cortical microsomes (57,81,124). Thus prostacyclin may be the obligatory endogenous mediator of renin secretion by the kidney.

Prostacyclin is also a strong vasodilator in the mesenteric, hind limb, and gastric mucosal circulations of the dog (where TXA_2 is a vasoconstrictor) (125–127) and on the precapillary side of the microcirculation of the hamster cheek pouch (128), where it also reverses epinephrine-induced vasoconstriction. In this preparation 6-oxo-$PGF_{1\alpha}$ had one-twentieth the vasodilator activity of prostacyclin and was more potent than PGE_2. In the pulmonary circulation of the dog, prostacyclin is the only product of arachidonic acid that produces strong vasodilatation (129,130). It also dilates the pulmonary vascular bed of the fetal lamb, where its potency is greater than PGE_1 but less than PGE_2 (131). In the cat, prostacyclin is also a potent pulmonary vasodilator. Furthermore, it inhibits the pulmonary hypertensive and platelet-aggregating effects of infused adenosine diphosphate (132). An important study has been made by Utsunomiya *et al.* (133) of the effects of prostacyclin on an experimental pulmonary embolism in dogs. One hour after embolization with an autologous blood clot labeled with [^{125}I]fibrin, half the dogs were infused with prostacylin. Almost twice the amount of fibrin degradation products appeared in the blood of the treated group than in the controls and ^{125}I activity rapidly disappeared from the lungs of the prostacyclin group. Many of the cardiopulmonary abnormalities were also reversed, including the hypoxia. Thus prostacyclin reversed in 15–30 min the physiological dead space and shunting associated with pulmonary embolism, maintained cardiac output, and enhanced fibrinolysis.

Another study of the effects of prostacyclin on lung injury has been made in sheep by Demling *et al.* (134). In seven of the eight animals treated, they found that prostacyclin protected the lungs against injury induced by endotoxin.

EFFECTS ON PLATELETS

Prostacyclin is the most potent endogenous inhibitor of platelet aggregation yet discovered. It is 30 to 40 times more potent than PGE_1 (58) and more than 1000 times more active than adenosine (135). *In vivo*, prostacyclin applied locally in low concentrations inhibits thrombus formation due to ADP in the microcirculation of the hamster cheek pouch (136), and given systemically to the rabbit it prevents electrically induced thrombus formation in the carotid artery and increases bleeding time (137). The duration of these effects *in vivo* is short; they disappear within 30 min of administration. Prostacyclin disaggregates platelets *in vitro* (58,137), in extracorporeal circuits where platelet clumps have formed on collagen strips (74,138), and in the circulation of man (139). Moreover, it inhibits thrombus formation in a coronary artery model in the dog when given locally or systemically (140) and protects against sudden death (thought to be due to platelet clumping) induced by intravenous arachidonic acid in rabbits (141).

As mentioned earlier, prostacyclin is unstable and its activity disappears within 15 sec on boiling or within 10 min at 22°C at neutral pH. In blood at 37°C, the activity of prostacyclin (as measured by bioassay on vascular smooth muscle) has a half-life of 3 min (142,143). It has recently been reported that prostacyclin has an extended stability in plasma and in blood (144) and that this may be associated with binding to albumin or with metabolism to 6-oxo-PGE_1 (145). The relevance of these observations to the actual biological activity remains unclear. Alkaline pH increases the stability of prostacyclin (146,147) so that at pH 10.5 at 25°C, it has a half-life of 100 hr. It is stabilized as a pharmaceutical preparation by freeze drying and can be reconstituted for use in man in an alkaline glycine buffer.

MECHANISM OF ACTION

Prostacyclin inhibits platelet aggregation by stimulating adenylate cyclase, leading to an increase in cAMP levels in the platelets (148,149). In this respect prostacyclin is much more potent than either PGE_1 or PGD_2. 6-oxo-$PGF_{1\alpha}$ has weak antiaggregating activity and is almost devoid of activity on platelet cAMP (149).

Not only is prostacyclin more potent than PGE_1 and PGD_2 in elevating cAMP, but the elevation persists longer. Prostacyclin, PGE_1, and PGD_2 increase adenylate cyclase activity by acting on two distinct receptors on the platelet membrane (150,151). PGE_1 and prostacyclin act on one, whereas PGD_2 acts on another. These and other results suggest that the previously recognized PGE_1 receptor in platelets (152) is in fact the prostacyclin receptor.

There have not been many detailed studies on the mechanism of action of prostacyclin. In contrast to TXA_2 it enhances Ca^{2+} sequestration (153). Moreover, an inhibitory effect on platelet phospholipase (152,154) and platelet cyclooxygenase have been described (155). All these effects are related to its ability to increase cAMP in platelets. Moreover, prostacyclin inhibits endoperoxide-induced aggregation, suggesting additional sites of action, still undefined, but dependent on the cAMP effect (152). These observations have extended and given important biological significance to the original observation (156) that substances such as PGE_1 that increase cAMP in platelets inhibit the release of TXA_2 (measured as RCS) in platelets. Prostacyclin, by inhibiting several steps in the activation of the arachidonic acid metabolic cascade, exerts an overall control of platelet aggregability *in vivo*.

Prostacyclin increases cAMP levels in cells other than platelets including cultured human fibroblasts (157), human fat cell ghosts (158), and guinea pig lung homogenates (159). Thus there is the possibility that in these cells an interaction with the thromboxane system could lead to a similar control of cell behavior to that observed in platelets, suggesting that the PGI_2–TXA_2 system has wider biological significance in cell regulation and the definition of cell receptors for prostaglandins. Recently, Hopkins and Gorman (160) have shown that prostacyclin increases cAMP in the endothelial cell itself and have suggested that this may act as a negative feedback control for prostacyclin production by the endothelium.

THROMBOXANE A_2

BIOLOGICAL PROPERTIES

The half-lives of RCS (<2 min) (38) and PG endoperoxides (<5 min) (17) were known to be different. Also, the amount of PG endoperoxides released from platelets during aggregation or from guinea pig lungs by arachidonic acid was not sufficient to account for the observed rabbit aorta contracting activity. These observations led Samuelsson's group to look for the presence of an additional substance with RCS activity. They identified, and named as thromboxane A_2, a highly unstable intermediate in the conversion of PGG_2 to "PHD" (a very polar compound, the hemiacetal derivative of 8-(1-hydroxy-3-oxopropyl)-9,12L-dihydroxy-5,10-heptadecaenoic acid) now known as thromboxane B_2 (TXB_2) (18,19) (Fig. 1). This unstable metabolite strongly contracted the aortic strip and induced platelet aggregation.

RCS from guinea pig lungs as described by Piper and Vane (38) was, then, a mixture of prostaglandin endoperoxides and TXA_2. In this organ the activity is mostly due to TXA_2 (63,161). However, the composition of RCS from other organs may vary; for example, the RCS activity obtained by incubating arachidonic acid with ram seminal vesicles is due entirely to prostaglandin endoperoxides (162). In platelets, as in lung, there is high conversion of endoperoxides to TXA_2, but in polymorphonuclear cells, we and others have observed TXA_2 production alongside that of other prostaglandins (47,48,104).

As originally described by Vane's group (163), RCS released from the guinea pig lungs contracted all vascular tissues tested. In many of these preparations, both the endoperoxides and TXA_2 contract the tissues, but TXA_2 is much more potent. Such tissues include rabbit aorta (39,40), human umbilical artery (54), guinea pig trachea (164), bovine and pig coronary artery (65,165,166), bovine cerebral arteries (167), and lamb ductus arteriosus (168). *In vivo* TXA_2 is a constrictor of vascular beds in the dog (125,127) and cat (169); moreover, it increases the tracheal inflation pressure in the anaesthetized guinea pig (164).

PROSTACYCLIN AND THROMBOXANE A_2 INTERACTIONS IN HEMOSTASIS AND THROMBOSIS

Prostaglandin endoperoxides are at the crossroads of arachidonic acid metabolism, for they are precursors of substances with opposing biological properties (see Fig. 1). On the one hand, TXA_2 produced by the platelets is a strong contractor of large blood vessels and induces platelet aggregation. On the other hand, prostacyclin produced by the vessel wall is a strong vasodilator and the most potent inhibitor of platelet aggregation known. Each substance has opposing effects on cAMP concentrations, thereby giving a balanced control mechanism, which will, therefore, affect thrombus and hemostatic plug formation.

A number of diseases have now been related to an imbalance in the prostacyclin–TXA_2 system. Platelets from patients with arterial thrombosis, deep venous thrombosis, or recurrent venous thrombosis produce more PG endoperoxides and TXA_2 than normal and have a shortened survival time (170). Platelets from rabbits made atherosclerotic by dietary manipulation (171) and from patients who have survived myocardial infarction (172) are abnormally sensitive to aggregating agents and produce more TXA_2 than controls. Elevated TXB_2 levels have been demonstrated in the blood of patients with Prinzmetal's angina (173). Hirsh *et al.* (174) also studied TXB_2 levels in coronary sinus blood of patients with unstable angina. They concluded that local thromboxane A_2 release is associated with recent episodes of angina but were unable to distinguish whether the release was cause or effect.

Platelets from rats made diabetic display an increased release of TXA_2, whereas their blood vessels show a reduced production of prostacyclin (175,176); these effects are reversed by chronic insulin treatment (175). Prostacyclin production by blood vessels from patients with diabetes is depressed (177) and circulating levels of 6-oxo-$PGF_{1\alpha}$ are reduced in diabetic patients with proliferative retinopathy (178). Davis *et al.* (179) tended to confirm that vessels taken from diabetics produced less prostacyclin than normals. However, their results did not support an association between reduced prostacyclin production and diabetic retinopathy.

Thrombocytopenic purpura (TTP), like diabetes, is associated with formation of microvascular thromboemboli, and a deficiency in prostacyclin production may be responsible for the increased platelet consumption that occurs in TTP (180–182). This deficiency is postulated to be secondary to a lack of a "plasma factor" that normally stimulates prostacyclin production (69). A patient with TTP had an undetectable level of 6-oxo-$PGF_{1\alpha}$ (<60 pg/ml), whereas the mean value in control subjects was 154 ± 48 pg/ml (183).

An increased prostacyclin production, resulting from an accumulation of the "plasma factor" which stimulates prostacyclin synthesis, has been suggested to explain the hemostatic defect in uremic patients (184). Patients with Bartter's syndrome excrete in the urine about four times as much 6-oxo-$PGF_{1\alpha}$ as controls (185). This has led to the suggestion that overproduction of prostacyclin mediates both the hyperreninemia and the hyporesponsiveness to pressor agents observed in these patients (185). Enhanced prostacyclin production by blood vessels of spontaneously hypertensive rats has also been demonstrated (186). However, Grose *et al.* (187) have described a diminished excretion of 6-oxo-$PGF_{1\alpha}$ in the urine of patients with essential hypertension. This could reflect diminished prostacyclin production by the kidney itself or, less likely, by the body as a whole.

In general, it seems that in diseases where there is a tendency for thrombosis to develop, TXA_2 production is elevated and/or prostacyclin production is reduced, whereas the opposite is found in some diseases associated with an increased bleeding tendency.

FACTORS AFFECTING PROSTACYCLIN FORMATION

CYCLOOXYGENASE INHIBITION

A further consequence of the discovery of prostacyclin has been the need to reexamine the use of aspirin as an antithrombotic compound.

Because aspirin inhibits cyclooxygenase it can prevent production of thromboxane A_2 and of prostacyclin. Aspirin irreversibly acetylates the active site of cyclooxygenase and therefore inhibits the enzyme in platelets for their entire lifespan because platelets are unable to synthesize new protein (188). Aspirin also has an effect on megakariocytes in the marrow (189). Inhibition of the vascular cyclooxygenase, however, may persist for a much shorter period because of the generation of new enzyme. Also, the cyclooxygenase of human skin fibroblasts and of arterial smooth muscle cells appeared less sensitive to inhibition by aspirin than that of human platelets (70). However, Preston *et al.* (190), using vein samples from five subjects who had ingested low dose aspirin, concluded that there is little difference in the initial (8-hr) substantial inhibition of cyclooxygenase seen in platelets or vessel wall.

In rabbits aspirin has a biphasic effect on cutaneous bleeding time and on the formation of platelet clumps in an extracorporeal system. Low doses increase the bleeding time and are antithrombotic, but with higher doses neither effect occurs (191,192). This dose-dependent effect has been attributed to the low dose affecting only platelets and allowing prostacyclin production to continue, while both systems are inhibited by high dose aspirin. O'Grady and Moncada (193) showed that in healthy volunteers ingestion of a single dose of aspirin of 0.3 g significantly prolonged bleeding time but a dose of 3.9 g had no significant effect on bleeding time. These results have been confirmed by some (194) but not by others (195). Godal *et al.* (195) measured bleeding time in healthy volunteers before and 2 hr after 0.425 or 3.875 g aspirin orally. Bleeding time increased to a similar extent after both high and low dose aspirin. However, the volunteers in the study by O'Grady and Moncada (193) were aged between 20 and 34 whereas those in the study by Godal *et al.* (195) were older, aged between 21 and 56. Subsequently, Jorgensen *et al.* (196) showed that bleeding time in man decreased with age and confirmed that higher doses of aspirin produced a significantly shorter bleeding time than low doses in the age groups 18 to 22 and 28 to 32 years but not in the age group 66 to 70 years. These findings are consistent with a decrease in prostacyclin production with age.

Amezcua *et al.* (197) showed in man that following a single high oral dose of aspirin (3.9 g) the bleeding time is unchanged 2 hr after ingestion, although marked inhibition of platelet function and thromboxane A_2 release is demonstrable. Twenty-four and 72 hr after aspirin, the bleeding time increased substantially, returning to normal levels one week after treatment. This suggests that shortly after high dose aspirin the production of prostacyclin and thromboxane A_2 was blocked. As the recovery of the endothelial cyclooxygenase proceeded faster than the release of new platelets, the ratio of prostacyclin to thromboxane A_2 was changed in favor of prostacyclin at 24 and 72 hr after treatment, thus explaining the increased bleeding time and its slow recovery over one week. Aspirin 0.3 g daily and 0.6 g twice daily administered for seven days to man both prolonged the bleeding time measured 2 hr after the first and last doses (198). However, at the higher dose the bleeding time was significantly shorter after the last as compared to the first aspirin dose, and this could reflect inhibition of prostacyclin generation by chronic aspirin dosing. How these short-term effects relate to effects over much longer periods is unknown. However, Harter *et al.* (199) have shown that low dose aspirin (160 mg/day) significantly reduced thrombus formation in arteriovenous shunts in patients on chronic hemodialysis.

THROMBOXANE SYNTHETASE INHIBITORS

Theoretically, a selective inhibitor of thromboxane synthetase should prove to be a superior antithrombotic agent to aspirin by allowing prostacyclin formation by vessel walls or other cells either from their own endoperoxides or from those released from platelets. Needleman *et al.* (100) observed that when platelets were treated with a thromboxane synthetase inhibitor *in vitro*, endoperoxides were available for utilization by the vessel wall. Interestingly, in the presence of a thromboxane synthetase inhibitor, arachidonic acid or collagen added to blood *in vitro* led to the formation of 6-oxo-$PGF_{1\alpha}$ rather than TXB_2. Platelets cannot synthesize prostacyclin, so some other blood cell must have done so (105,200). We have shown that injection of heterologous blood into anesthetized cats causes hypotension, respiratory distress, and frequently death and that this is accompanied by a sharp rise in blood levels of TXB_2 (Fig. 2). Pretreatment with either a thromboxane synthetase inhibitor (an imidazole derivative) or a cyclooxygenase inhibitor such as aspirin prevented death. However, after inhibition of thromboxane synthetase the blood levels of 6-oxo-$PGF_{1\alpha}$ rose about five times more than after the shock in control animals, suggesting diversion of the platelet prostaglandin endoperoxides away from TXA_2 production toward prostacyclin production (Fig. 2).

Other work on selective thromboxane synthetase inhibitors is beginning to appear (201), including the first publications on administration of one of these compounds to man (202,203).

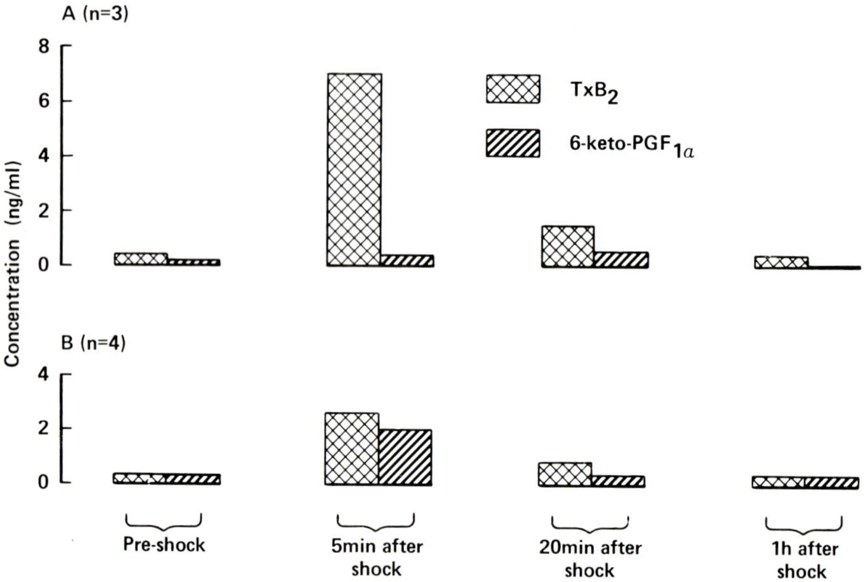

Fig. 2. Thromboxane B_2 and 6-keto-$PGF_{1\alpha}$ plasma levels after heterologous blood shock in (A) control cats; (B) cats given thromboxane synthetase inhibitor (6 mg/kg) 30 min before shock. Inhibition of thromboxane synthetase was demonstrated by a reduction in TXB_2 levels by more than 50%. Levels of 6-oxo $PGF_{1\alpha}$ in the treated group, however, were increased approximately fivefold, suggesting a diversion of platelet endoperoxides toward conversion by the vessel wall to prostacyclin.

LIPID PEROXIDES

High concentrations of lipid peroxides have been demonstrated in advanced atherosclerotic lesions (204). Lipid peroxidation induced by free radical formation is known to occur in vitamin E deficiency, the aging process, and perhaps also hyperlipidaemia accompanying atherosclerosis (205). It has been shown (59,62,206) that 15-HPAA, a lipid peroxide, is a potent (IC_{50} 0.48 μg/ml) and selective inhibitor of prostacyclin generation by vessel wall microsomes or by fresh vascular tissue. Other fatty acid peroxides and their methyl esters behave similarly (60). Accumulation of lipid peroxides in, for example, atheromatous plaques could predispose to thrombus formation by inhibiting generation of prostacyclin by the vessel wall without affecting thromboxane A_2 production by platelets. Moreover, platelet aggregation is induced by 15-HPAA and this aggregation is not inhibited by adenosine or PGE_1 (207). D'Angelo *et al.* (208) reported that human atheromatous plaques from three patients were incapable of prostacyclin production. Prostacyclin generation by atherosclerotic arterial tissue has been shown to be significantly lower than from normal arterial tissue but no difference was found between early and advanced atherosclerotic lesions (209). This suggests that the early "fatty streak" may be a biochemically critical stage of the atherosclerotic process. Bourgain *et al.* (210), using a model of thrombosis *in vivo* in the rat, demonstrated that application of 15-HPAA to the outside of mesenteric vessels increased the rate of thrombus formation in response to superfusion with ADP. All these results, therefore, suggest that it would be worth exploring whether attempts to reduce lipid peroxide formation by inhibiting peroxidation influence the development of atherosclerosis and arterial thrombosis. Vitamin E acts as an antioxidant and perhaps the enthusiasm for its use in arterial disease in the past (211–213) had in fact a biochemical rationale.

Raised concentrations of low density lipoprotein (LDL) are regarded as one of the risk factors associated with ischemic heart disease (214–216), whereas high density lipoprotein (HDL) is thought to protect against the disease (215,216). Nordoy *et al.* (217) were the first to show that LDL reduced the release of a prostacyclin-like substance by human endothelial cells. Beitz and Förster (218) extended these observations by showing that LDL inhibited, whereas HDL stimulated, prostacyclin synthesis. A mixture of low LDL and high HDL also stimulated prostacyclin synthesis. Gryglewski and Szczeklik (219) have confirmed that LDL inhibits prostacyclin synthesis. More important, they also discovered from an analysis of lipoproteins taken from a group of hyperlipidemics that the LDL fraction (but not the HDL) contained lipid peroxides at a concentration several times higher than those in the total serum. Thus the interesting possibility arises that it is the lipid peroxide associated with LDL that inhibits prostacyclin synthesis, and this inhibition explains the correlation between LDL and ischemic heart disease.

Cell proliferation *in vitro* is inhibited by substances that stimulate cyclic AMP formation (220). Cell growth in tissue culture (221,222), including vascular smooth muscle cell culture (223), is inhibited by PGE_1. Possibly prostacyclin has a role in the regulation of cell growth in the vascular wall. Smooth muscle proliferation in atherosclerotic plaques (224) might be a consequence of inhibition of prostacyclin generation by lipid peroxides.

MacIntyre *et al.* (69) observed that prostacyclin production by cultures of vascular endothelial cells was enhanced by platelet poor plasma. Later, Defreyn *et al.* (225) confirmed the presence of a factor in plasma that stimulates prostacyclin production and showed that the platelet-poor plasma from six out of seven patients with chronic uremia caused more stimulation of prostacyclin production by cultured endothelial cells than the platelet-poor plasma from control volunteers. They suggested that increased levels of circulating prostacyclin could well contribute to the bleeding tendency frequently encountered in uremic patients. Conversely, deficiency of a plasma factor regulating prostacyclin has been reported in hemolytic–uremic syndrome, which is associated with widespread thrombotic occlusions (180).

Thrombin, calcium ionophore A23187, and trypsin all stimulate prostacyclin production by cultured endothelial cells (226). It is interesting that all these substances cause platelet aggregation and probably act by stimulating phospholipases that liberate free arachidonic acid from cellular phospholipids. The action of thrombin in stimulating postacyclin production is indeed inhibited by mepacrine, a phospholipase inhibitor (227). Generation of thrombin and subsequent stimulation of prostacyclin formation at a site of vascular damage and hemostatic plug formation may be a controlling factor limiting thrombus size (226). Trypsin, but not chymotrypsin, stimulates prostacyclin production by cultured cells, an effect inhibited by soybean trypsin inhibitor (226). Release of trypsinlike enzymes by both platelets and white cells has been proposed as a mechanism by which prostacyclin production by the vessel wall might be regulated (228). Investigating the prostacyclin stimulating activity of serum, Coughlin *et al.* (229) concluded that at least 80% of the activity was derived from platelets. The platelet-dependent activity was nondialysable, stable at 56°C, and at least partially stable at 100°C. Because these properties were similar to those of platelet-derived growth factor (PDGF) and PDGF had been proposed to be the substance present in serum capable of activating phospholipase A_2 in Swiss mouse 3T3 fibroblasts (230), they tested PDGF for its ability to stimulate prostacyclin production in bovine cell cultures. PDGF reversibly stimulated prostacyclin production and at a concentration of 8.0 ng/ml stimulated synthesis 74-fold in aortic endothelial cells, 84-fold in smooth muscle, and 15-fold in capillary endothelial cell cultures. The mitogenic activity of PDGF was only observed with smooth muscle cells indicating dissociation between the two activities (229). Control of prostacyclin production is possible, therefore, by a variety of agents acting locally at the level of the cell–vessel wall interaction, but the precise mechanisms acting at any time have yet to be clarified.

Stimulation of prostacyclin release from the vessel wall by an antithrombotic compound Bay g 6575 has been reported (231) and confirmed (232). In addition, Bay g 6575 increased prostacyclin production in rats already having lowered prostacyclin synthesizing capacity as a result of streptozotocin-induced diabetes (232). Bendrofluazide used to treat patients with essential hypertension caused significant increases in circulating levels of 6-oxo-PGF$_{1\alpha}$, suggesting a possible contribution of prostacyclin to the reduction in peripheral resistance seen with thiazide treatment (233). Similarly, the vasodilator nitroglycerine in clinically attainable concentrations has been shown to enhance prostacyclin production by cultured human endothelial cells (234).

PROSTACYCLIN IN MAN

EFFECTS ON PLATELETS

Prostacyclin has a potent effect on platelets in man. During intravenous infusion of prostacyclin in healthy volunteers for 60 min at doses ranging from 2 to 16 ng/kg/min there was a dose-related inhibition of platelet aggregation in whole blood and in platelet-rich plasma (139,235). Inhibition of platelet aggregation at 15 or 45 min after the start of the infusion was similar. At a dose of 8 ng/mg/min partial inhibition of aggregation was demonstrable for up to 105 min after the end of infusion, and this persistence of effect on platelets has been confirmed (236). The longevity of the effects of prostacyclin on platelets despite its short biological half-life might be accounted for by persistent elevation of platelet cyclic AMP.

In a double blind controlled study of the cardiovascular effects of prostacyclin (0–4 ng/kg/min) Warrington and O'Grady (237) concluded that prostacyclin caused arteriolar vasodilation, which would be expected to lower diastolic and mean blood pressure and thus reflexly increase heart rate and contractility.

Headache has been reported by several subjects when doses greater than 8 ng/kg/min are administered (139,235,238). Colicky central abdominal discomfort has been less frequently experienced but was reproducible in one subject (235). Ill-defined sensations of unease and restlessness have been experienced by subjects receiving higher doses (8–20 ng/kg/min) of prostacyclin (139,235,236).

THERAPEUTIC POTENTIAL OF PROSTACYCLIN

EXTRACORPOREAL CIRCULATIONS

The circulation of blood through extracorporeal systems brings blood into contact with artificial surfaces that cannot generate prostacyclin. In the course of such procedures thrombocytopenia and loss of platelet hemostatic function occur and make an important contribution to the bleeding problems following charcoal hemoperfusion and prolonged cardiopulmonary bypass in man (239–241). Formation of microemboli during cardiopulmonary bypass may also contribute to cerebral complications that sometimes follow this procedure (242). In animals subjected to experimental renal dialysis (243), charcoal hemoperfusion (244) and cardiopulmonary bypass (245), infusion of prostacyclin during the procedure prevented this platelet damage and thrombocytopenia, thus increasing the biocompatibility of the procedure. Platelet aggregates in the blood returning to the animals, as measured by Swank screen filtration pressure, were also reduced by infusion of prostacyclin (Fig. 3). These findings have been confirmed in patients with fulminant hepatic failure undergoing charcoal hemoperfusion (246). Prostacyclin infusion prevented the fall in platelet count and elevation of β-thromboglobulin seen in the control

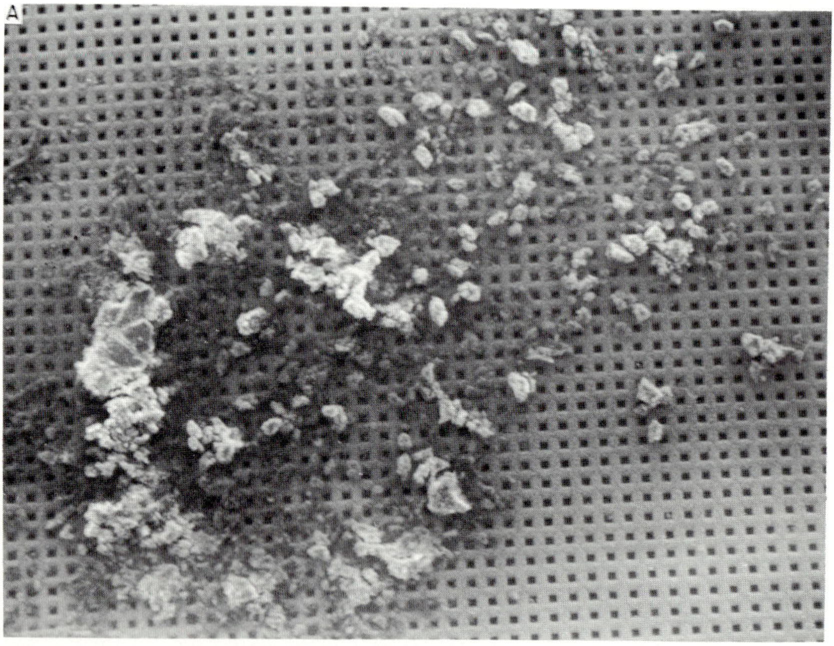

Fig. 3. Scanning electron micrograph photographs (×130) of stainless steel screens used in Swank screen filtration pressure apparatus. Blood samples (5 ml) were taken from the dialyser outline lines of dogs having undergone 90 min of dialysis. (a) Heparin only.

patients. In addition, two of the control patients developed marked hypotension during the procedure, in one associated with a marked rise in Swank screen filtration pressure, whereas this did not occur in the prostacyclin-treated patients. A study of the influence of serial hemoperfusion with prostacyclin on the survival rate of patients with fulminant hepatic failure is now in progress (247).

Longmore *et al.* (248), Bunting *et al.* (249), Walker *et al.* (250), Bennett *et al.* (251), and Chelly *et al.* (252) have published double blind clinical trials of prostacyclin in cardiopulmonary bypass. The treatment groups showed a preservation of platelet number and function, with a reduction in the blood loss in the first 18 hr after operation. In the trial by Longmore *et al.* (248) the blood loss was halved and the reduction was statistically significant. The heparin-sparing effect of prostacyclin was confirmed and the vasodilator effects were not troublesome.

The observation that prostacyclin potentiates the effects of heparin

(b) heparin plus prostacyclin. Prostacyclin clearly reduced the amount of particulate material (mainly platelet aggregates) present in the blood. (Scanning electron micrographs by Dr. N. Read, Department of Pathology, Wellcome Research Laboratories.) From unpublished experiments of Woods *et al.* (243).

(244) led to further studies on this interaction. These demonstrated that prostacyclin has a small indirect anticoagulant effect. Indeed, platelets stimulated by low doses of aggregating agents accelerate clotting by providing a surface on which coagulation factors can combine and react more efficiently (15). Prostacyclin, by preventing platelet activation, inhibits the shortening of clotting time produced when either kaolin or collagen is incubated with platelet-rich plasma (253,254). Platelets release antiheparin activity, which reduces the anticoagulant effect of heparin *in vitro*. Prostacyclin, by inhibiting this release and by preventing the development of procoagulant activity, can enhance the action of heparin by as much as 100% (253,254).

Heparin therapy in some patients is complicated by thrombocytopenia and thromboembolic episodes (255,256) and *in vitro* can cause platelet aggregation and potentiate aggregation to other aggregating agents (257). Perhaps because the stimulus to coagulation is milder, we showed that during hemodialysis in dogs, prostacyclin could be used alone to prevent platelet loss and coagulation. Heparin was not needed. This surprising result has now been confirmed in 10 patients by Zusman *et al.* (258). Prostacyclin was infused intravenously for 10 min before dialysis and into the arterial line during dialysis. The rate of infusion was adjusted to prevent prostacyclin-induced hypotension. They concluded that prostacyclin can safely replace heparin as the sole antithrombotic agent during hemodialysis and that it may be more advantageous if anticoagulation is contraindicated. This work has been confirmed by Turney *et al.* (259) in five patients.

OTHER APPLICATIONS

Clinical assessment of prostacyclin is still in its infancy, with many trials in progress. Open studies and individual case reports have been described where both the platelet inhibitory activity and vasodilator properties of prostacyclin have been utilized. The results in many cases are therefore preliminary; nevertheless they point the way to conditions in which prostacyclin therapy may be useful.

Szczeklik *et al.* (260) reported striking and prolonged benefits following intraarterial infusion of prostacyclin in five patients with advanced atherosclerotic lower limb peripheral vascular disease. Rest pain disappeared, previously refractory ulcers healed, and the muscle blood flow, as measured by xenon^{133} clearance, was significantly increased for at least six weeks after prostacyclin infusion. This group has now reported on 55 patients with advanced peripheral arterial disease of the lower extremities (261). In summary, 42% of patients treated showed a persistent, long-lasting improvement. In 40% of patients the improvement lasted no longer than 2 months, and in the remaining 18% of patients the results were virtually negative. The authors believe that these figures do not represent the true efficacy of prostacyclin therapy in advanced peripheral arterial disease and that successful treatment largely depends on choice of patients, localization of the vascular lesions, and the advancement of the disease (261). Carefully conceived and executed trials are therefore required to resolve this problem. Other reports also suggest that prostacyclin may have beneficial effects in peripheral artery disease (262,263). Zygulska-Mach *et al.* (264) infused prostacyclin into three patients with sudden blockage of central retinal veins. Improvement was observed in those two patients who were treated within the first 48 hr.

Prostacyclin has been successfully used in cases of pulmonary hypertension. One, an 8-year-old-girl with severe idiopathic pulmonary artery hypertension, received a prostacyclin infusion of 8–44 ng/kg/min. Pulmonary vascular resistance was reduced during the period of infusion and no adverse effects were reported (265). In a group of patients with pulmonary hypertension secondary to mitral valve stenosis, prostacyclin caused a dose-dependent pulmonary vasodilatation with no observed side effects (266). In both of the preceding studies prostacyclin was shown to be more effective than PGE_1. Single case studies have suggested that prostacyclin may be useful in the treatment of patent ductus arteriosus (267) and preeclamptic toxemia (268).

Bergman *et al.* (269) gave an intravenous infusion of prostacyclin to patients with coronary artery disease and showed that doses between 2 and 8 ng/kg/min for 10 min had no deleterious effects. Heart rate and cardiac index were increased and mean blood pressure and systemic and pulmonary resistance all fell. Mean atrial pacing time to angina rose from 142 to 241 sec. They concluded that acute administration of prostacyclin was beneficial in angina, having effects similar to short-acting nitrates. Hall and Dewar (270) concluded from their study of five patients with coronary artery disease that prostacyclin can safely be infused directly into diseased coronary arteries, and Szczeklik and Gryglewski (261) found a beneficial effect of intravenous prostacyclin infusions in patients with unstable angina.

A prostacyclin deficiency has been reported in thrombotic thrombocytopenic purpura (183), but infusion of prostacyclin into two patients with TTP did not produce an increase in circulating platelet count (183,271). However, Fitzgerald *et al.* (272) have reported an increase in platelet count and an improvement in the neurological status of one such patient during 18 days of prostacyclin infusion. They were sufficiently encouraged to conclude that the controlled evaluation of prostacyclin in TTP was warranted.

Clearly, there are many clinical conditions that may respond to prostacyclin treatment, and its place in therapeutics (or that of stable analogs) will be defined in the next few years. Some of these conditions are preeclamptic toxemia (268), hemolytic uremic syndrome (273), peptic ulceration (127), and the thrombotic complications associated with transplant rejection (274).

ACKNOWLEDGMENTS

We gratefully acknowledge the help given in the preparation of this manuscript by Ann Higgs and Gill Henderson.

REFERENCES

1. Goldblatt, M. W. (1933) *J. Soc. Chem. Ind., London* **52,** 1056–1057.
2. von Euler, U.S. (1934) *Arch. Exp. Pathol. Pharmakol.* **175,** 78–84.
3. Bergström, S., and Sjövall, J. (1960) *Acta Chem. Scand.* **14,** 1693–1701.
4. Bergström, S., Ryhage, R., Samuelsson, B., and Sjövall, J. (1963) *J. Biol. Chem.* **238,** 3555–3564.
5. Samuelsson, B. (1963) *J. Am. Chem. Soc.* **85,** 1878–1879.
6. Bergström, S., Danielson, H., and Samuelsson, B. (1964) *Biochim. Biophys. Acta* **90,** 207–210.
7. Van Dorp, D. A., Beerthuis, R. K., Nugteren, D. H., and Vonkeman, H. (1964) *Biochim. Biophys. Acta* **90,** 204–207.
8. Vonkeman, H., and Van Dorp, D. A. (1968) *Biochim. Biophys Acta* **164,** 430–432.
9. Flower, R. J., and Blackwell, G. J. (1976). *Biochem. Pharmacol.* 25, 285–291.
10. Piper, P. J., and Vane, J. R. (1971) *Ann. N.Y. Acad. Sci.* **180,** 363–385.
11. Ferreira, S. H., and Vane, J. R. (1967) *Nature (London)* **216,** 868–873.
12. Gryglewski, R. J., and Vane, J. R. (1972) *Br. J. Pharmacol.* **46,** 449–457.
13. Salzman, E. W., Lindon, N. J., and Rodvien, R. (1976) *J. Cyclic Nucleotide Res.* **2,** 25–37.
14. Flower, R. J. (1978) *Adv. Prostaglandin Thromboxane Res.* **3,** 105–112.
15. Marcus, A. J. (1978) *J. Lipid Res.* **19,** 793–826.
16. Christ, E. J., and Van Dorp, D. A. (1972) *Adv. Biosci.* **9,** 35–38.
17. Hamberg, M., Svensson, J., and Samuelsson, B. (1974) *Proc. Natl. Acad. Sci. U.S.A.* **71,** 3824.
18. Hamberg, M., and Samuelsson, B. (1974) *Biochem. Biophys. Res. Commun.* **61,** 942.
19. Hamberg, M., and Samuelsson, B. (1974) *Proc. Natl. Acad. Sci. U.S.A.* **71,** 3400–3404.
20. Nugteren, D. H. (1975) *Biochim. Biophys. Acta* **380,** 299–307.

21. Herman, A. G., Claeys, M., Moncada, S., and Vane, J. R. (1978) *Arch. Int. Pharmacodyn. Ther.* **236,** 303–304.
22. Bach, M. K., Brashler, J. R., and Gorman, R. R. (1977) *Prostaglandins* **14,** 21–38.
23. Murphy, R., Hammarström, S., and Samuelsson, B. (1979) *Proc. Natl. Acad. Sci. U.S.A.* **76,** 4275–4279.
24. Samuelsson, B., Borgeat, P., Hammarström, S., and Murphy, R. (1979) *Prostaglandins* **17,** 785–787.
25. Samuelsson, B., and Hammarström, S. (1980) *Prostaglandins* **19,** 645–648.
26. Morris, H., and Taylor, G. (1980) *Prostaglandins* **19,** 185–201.
27. Morris, H., Taylor, G., Piper, P., and Tippins, J. (1980). *Nature* (*London*) **285,** 104–106.
28. Ford-Hutchinson, A., Bray, M., Doig, M., Shipley, M., and Smith, M. (1980) *Nature* (*London*) **285,** 264–265.
29. Palmer, R., Stepney, R., Higgs, G., and Eakins, K. (1980) *Prostaglandins* **20,** 411–418.
30. Bhattacherjee, P., Eakins, K. E., and Hammond, B. (1981) *Br. J. Pharmacol.* **73,** 254P–255P.
31. Carr, S., Higgs, G., Salmon, J., and Spayne, J. (1981) *Br. J. Pharmacol.* **73,** 253P–254P.
32. Eakins, K., Higgs, G., Moncada, S., Salmon, J., and Spayne, J. (1980) *J. Physiol.* (*London*) **307,** 71P.
33. Dahlen, S.-E., Hedqvist, P., Hammarström, S., and Samuelsson, B. (1980) *Nature* (*London*) **288,** 484–486.
34. Drazen, J., Austen, K., Lewis, R., Clark, D., Goto, G., Marfat, A., and Corey, E. (1980) *Proc. Natl. Acad. Sci. U.S.A.* **77,** 4354–4358.
35. Palmer, M., Piper, P., Samhoun, M., and Tippins, J. (1981) *Br. J. Pharmacol.* **73,** 213P.
36. Hamberg, M., Svensson, J., Wakabayashi, T., and Samuelsson, B. (1974) *Proc. Natl. Acad. Sci. U.S.A.* **71,** 345–349.
37. Hamberg, M., Svensson, J., and Samuelsson, B. (1975) *Proc. Natl. Acad. Sci. U.S.A.* **72,** 2994–2998.
38. Piper, P. J., and Vane, J. R. (1969) *Nature* (*London*) **223,** 29–35.
39. Moncada, S., Needleman, P., Bunting, S., and Vane, J. R. (1976) *Prostaglandins* **12,** 323–325.
40. Needleman, P., Moncada, S., Bunting, S., Vane, J. R., Hamberg, M., and Samuelsson, B. (1976) *Nature* (*London*) **261,** 558–560.
41. Diczfaluzy, U., Falardeau, P., and Hammarström, S. (1977) *FEBS Lett.* **84,** 271–274.
42. Hammarström, S., and Falardeau, P. (1977) *Proc. Natl. Acad. Sci. U.S.A.* **74,** 3691–3695.
43. Yoshimoto, T., Yamamoto, S., Okuma, M., and Hayaishi, O. (1977) *J. Biol. Chem.* **252,** 5871–5874.
44. Ho, P. P. K., Walters, C. P., and Sullivan, H. R. (1976) *Prostaglandins* **12,** 951–970.
45. White, H. L., and Glassman, A. T. (1976) *Prostaglandins* **12,** 811–828.
46. Sun, F. F. (1977) *Biochem. Biophys. Res. Commun.* **74,** 1432–1440.
47. Higgs, G. A., Bunting, S., Moncada, S., and Vane, J. R. (1976) *Prostaglandins* **12,** 749–757.
48. Goldstein, I. M., Malmsten, C. L., Kaplan, H. B., Kindahl, H., Samuelsson, B., and Weissman, G. (1977) *Clin. Res.* **25,** 518A.
49. Brune, K., Glatt, M., Kalin, H., and Peskar, B. A. (1978) *Nature* (*London*) **274,** 261–263.

50. Murota, S.-I., Kawamura, M., and Morita, I. (1978) *Biochim. Biophys. Acta* **528,** 507–511.
51. Hopkins, N. K., Sun, F. F., and Gorman, R. R. (1978) *Biochem. Biophys. Res. Commun.* **85,** 827–836.
52. Nijkamp, F. P., Moncada, S., White, H. L., and Vane, J. R. (1977) *Eur. J. Pharmacol.* **44,** 179–187.
53. Bhattacherjee, P., Kulkarni, P. S., and Eakins, K. E. (1979) *Invest. Ophthalmol.* **18,** 172–178.
54. Tuvemo, T., Strandberg, K., Hamberg, M., and Samuelsson, B. (1976) *Adv. Prostaglandin Thromboxane Res.* **1,** 425–428.
55. Salzman, P., Salmon, P., and Moncada, S. (1980) *J. Pharmacol. Exp. Ther.* **215,** 240–247.
56. Morrison, A. R., Nishikawa, K., and Needleman, P. (1978) *J. Pharmacol. Exp. Ther.* **205,** 1–8.
57. Zenser, T. V., Herman, C. A., Gorman, R. R., and Davis, B. B. (1977) *Biochem. Biophys. Res. Commun.* **79,** 357–363.
58. Moncada, S., Gryglewski, R. J., Bunting, S., and Vane, J. R. (1976) *Nature (London)* **263,** 663–665.
59. Bunting, S., Gryglewski, R., Moncada, S., and Vane, J. R. (1976) *Prostaglandins* **12,** 897–913.
60. Salmon, J. A., Smith, D. R., Flower, R. J., Moncada, S., and Vane, J. R. (1978) *Biochim. Biophys. Acta* **523,** 250–262.
61. Wlodawer, P., and Hammarström, S. (1979) *FEBS Lett.* **97**(1), 32–36.
62. Gryglewski, R. J., Bunting, S., Moncada, S., Flower, R. J., and Vane, J. R. (1976) *Prostaglandins* **12,** 685–714.
63. Bunting, S., Moncada, S., and Vane, J. R. (1976) *Br. J. Pharmacol.* **57,** 462P.
64. Moncada, S., Higgs, E. A., and Vane, J. R. (1977) *Lancet* **1,** 18–21.
65. Dusting, G. J., Moncada, S., and Vane, J. R. (1977) *Prostaglandins* **13,** 3–15.
66. Higgs, E. A., Higgs, G. A., Moncada, S., and Vane, J. R. (1978) *Br. J. Pharmacol.* **63,** 535–539.
67. Christofinis, G. J., Moncada, S., Bunting, S., and Vane, J. R. (1979) *In* "Prostacyclin" (J. R. Vane and S. Bergström, eds.), pp. 77–84. Raven, New York.
68. Weksler, B. B., Marcus, A. J., and Jaffe, E. A. (1977) *Proc. Natl. Acad. Sci. U.S.A.* **74,** 3922–3926.
69. MacIntyre, D. E., Pearson, J. D., and Gordon, J. L. (1978) *Nature (London)* **271,** 549–551.
70. Baenziger, N. L., Dillender, M. J., and Majerus, P. W. (1977) *Biochem. Biophys. Res. Commun.* **78,** 294–301.
71. Humes, J. L., Bonney, R. J., Pelus, L., Dahlgren, M. E., Sadowski, S. J., Kuehl, F. A., and Davis, P. (1977) *Nature (London)* **269,** 149–151.
72. Kulkarni, P. S., Eakins, H. M. T., Saber, W. L., and Eakins, K. E. (1977) *Prostaglandins* **14,** 689–700.
73. Dawson, W., Boot, J. R., Cockerill, A. F., Mallen, D. N. B., and Osborne, D. J. (1976) *Nature (London)* **262,** 699–702.
74. Gryglewski, R. J., Korbut, R., and Ocetkiewicz, A. C. (1978) *Nature (London)* **273,** 765–767.
75. Moncada, S., Korbut, R., Bunting, S., and Vane, J. R. (1978) *Nature (London)* **273,** 767–768.
76. Schror, K., Moncada, S., Ubatuba, F. B., and Vane, J. R. (1978) *Eur. J. Pharmacol.* **47,** 103–114.

77. De Dekere, E. A. M., Nugteren, D. H., and Ten Hoor, F. (1977) *Nature* (*London*) **268,** 160–163.
78. Needleman, P., Bronson, S. D., Wyche, A., Sivakoff, M., and Nicolaou, K. C. (1978) *J. Clin. Invest.* **61,** 839–849.
79. Pace-Asciak, C., and Rangaraj, G. (1977) *Biochim. Biophys. Acta* **486,** 583–585.
80. Powell, W. S., and Solomon, S. (1977) *Biochem. Biophys. Res. Commun.* **75,** 815–822.
81. Remuzzi, G., Cavenaghi, A. E., Mecca, G., Donati, M. B., and de Gaetano, G. (1978) *Thromb. Res.* **12,** 363–366.
82. Terragno, N. A., Terragno, D. A., Early, J. A., Roberts, M. A., and McGiff, J. C. (1978) *Clin. Sci. Mol. Med.* **55,** 199–202s.
83. Grenier, F. C., and Smith, W. L. (1978) *Prostaglandins* **16,** 759–772.
84. Amezcua J-L., Higgs, E. A., Moncada, S., Salmon, J. A., and Vane, J. R. (1978) *Abstr. 7th Int. Congr. Pharmacol., 1978* p. 341.
85. Pace-Asciak, C. (1976) *J. Am. Chem. Soc.* **98,** 2348–2349.
86. Moncada, S., Salmon, J. A., Vane, J. R., and Whittle, B. J. R. (1977) *J. Physiol.* (*London*) **275,** 4P–5P.
87. Whittle, B. J. R. (1978) *Br. J. Pharmacol.* **64,** 438P.
88. Williams, K. I., and Downing, I. I. (1977) *Prostaglandins* **14,** 813–817.
89. Williams, K. I., Dembinska-Kiec, A., Zmuda, A., and Gryglewski, R. J. (1978) *Prostaglandins* **15,** 343–350.
90. Fenwick, L., Jones, R. L., Naylor, B., Poyser, N. L., and Wilson, N. H. (1977) *Br. J. Pharmacol.* **59,** 191–196.
91. Myatt, L., and Elder, M. G. (1977) *Nature* (*London*) **268,** 159–160.
92. Mitchell, M. D., Bibby, J. G., Hicks, B. R., and Turnbull, A. C. (1978) *Prostaglandins* **16,** 931–937.
93. Cottee, F., Flower, R. J., Moncada, S., Salmon, J. A., and Vane, J. R. (1977) *Prostaglandins* **14,** 413–423.
94. Chang, W-C., Murota, S.-I., and Tsurufuji, S. (1977) *Prostaglandins* **13,** 17–24.
95. Goehlert, U. G., Ng Ying Kin, N. M. K., and Wolfe, L. S. (1981) *J. Neurochem.* **36,** 1192–1201.
96. Moncada, S., Herman, A. G., Higgs, E. A., and Vane, J. R. (1977) *Thromb. Res.* **11,** 323–344.
97. Bunting, S., Moncada, S., and Vane, J. R. (1977) *Lancet* **2,** 1075–1076.
98. Moncada, S., and Vane, J. R. (1978) *Br. Med. Bull.* **34,** 129–135.
99. Best, L. C., Martin, T. J., Russell, R. G. G., and Preston, F. E. (1977) *Nature* (*London*) **267,** 850–851.
100. Needleman, P., Wyche, A., and Raz, A. (1979) *J. Clin. Invest.* **63,** 345–349.
101. Hornstra, G., Haddeman, E., and Don, J. A. (1979) *Nature* (*London*) **279,** 66–68.
102. Marcus, A. J., Weksler, B. B., Jaffe, E. A., and Broekman, M. J. (1979) *Blood* **54,** Suppl. 1, 290a, Abstr. 803.
103. Marcus, A. J., Broekman, M. J., Weksler, B. B., Jaffe, E. A., Safier, L. B., Ullman, H. L., and Tack-Goldman, K. (1981) *Philos. Trans. R. Soc. London, Ser. B* 294, 343–353.
104. Davidson, E. M., Ford-Hutchinson, A. W., Smith, M. J. H., and Walker, J. R. (1978) *Br. J. Pharmacol.* **63,** 407P.
105. Flower, R. J., and Cardinal, D. C. (1979) *In* "Prostacyclin" (J. R. Vane and S. Bergström, eds.), pp. 211–220. Raven, New York.
106. Weksler, B. B., Knapp, J. M., and Jaffe, E. A. (1977) *Blood* **50**(5), Suppl. 1, 287.
107. Higgs, G. A., Moncada, S., and Vane, J. R. (1978) *J. Physiol.* (*London*) **280,** 55P–56P.

108. Armstrong, J. M., Chapple, D. J., Dusting, G. J., Hughes, R., Moncada, S., and Vane, J. R. (1977) *Br. J. Pharmacol.* **61,** 136P.
109. Schror, K., and Moncada, S. (1979) *Prostaglandins* **17,** 367–373.
110. Wennmalm, A. (1979) *Acta Physiol. Scand.* **106,** 47–52.
111. Dusting, G. J., Chapple, D. J., Hughes, R., Moncada, S., and Vane, J. R. (1978) *Cardiovasc. Res.* **12,** 720–730.
112. Ribeiro, L. G. T., Reduto, L. A., Brandon, T. A., Taylor, A. A., Hopkins, D. G., and Miller, R. R. (1979) *Clin. Res.* **27,** 199A.
113. Jugdutt, B. F., Hutchins, G. M., Bulkley, B. H., and Becker, L. C. (1979) *Clin. Res.* **27,** 177A.
114. Au, T. L. S., Collins, G. A., Harvie, C. J., and Walker, M. J. A. (1980) *Adv. Prostaglandin Thromboxane Res.* **7,** 647–649.
115. Ohlendorf, R., Perzborn, E., and Schror, K. (1980) *Thromb. Res.* **I9,** 447–453.
116. Lefer, A. M., Ogletree, M. L., Smith, J. B., Silver, M. J., Nicolaou, K. C., Barnette, W. E., and Gasic, G. P. (1978) *Science* **200,** 52–54.
117. Ogletree, M. L., Lefer, A. M., Smith, J. B., and Nicolaou, K. C. (1979) *Eur. J. Pharmacol.* **56,** 95–103.
118. Ribeiro, L. G. T., Brandon, T. A., Hopkins, D. G., Reduto, L. A., Taylor, A. A., and Miller, R. R. (1981) *Am. J. Cardiol.* **47,** 835–840.
119. Weber, P. C., Larsson, C., Anggard, E., Hamberg, M., Corey, E. J., Nicolaou, K. C., and Samuelsson, B. (1976) *Circ. Res.* **39,** 868–874.
120. Whorton, A. R., Misono, K., Hollifield, J., Frolich, J. C., Inagami, T., and Oates, J. A. (1977) *Prostaglandins* **14,** 1095–1104.
121. Larsson, C., Weber, P., and Anggard, E. (1974) *Eur. J. Pharmacol.* **28,** 391–394.
122. Data, J. L., Crump, W. J., Hollifield, J. W., Frolich, J. C., and Nies, A. S. (1976) *Clin. Res.* **24,** 397A.
123. Frolich, J. C., Hollifield, J. W., Dormois, J. C., Frolich, B. L., Seyberth, H., Michelakis, A. M., and Oates, J. A. (1976) *Circ. Res.* **39,** 447–452.
124. Whorton, A. R., Smigel, M., Oates, J. A., and Frolich, J. C. (1977) *Prostaglandins* **13,** 1021.
125. Dusting, G. J., Moncada, S., and Vane, J. R. (1978b) *Eur. J. Pharmacol.* **49,** 65–72.
126. Kauffman, G. L., Jr., and Whittle, B. J. R. (1981) *Gastroenterology* **80,** 1189A.
127. Whittle, B. J. R., Kauffman, G. L., and Moncada, S., (1981) *Nature (London)* **292,** 472–474.
128. Higgs, G. A., Cardinal, D. C., Moncada, S., and Vane, J. R. (1979) *Microvasc. Res.* **18,** 245—254.
129. Kadowitz, P. J., Chapnick, B. M., Feigen, L. P., Hyman, A. L., Nelson, P. K., and Spannhake, E. W. (1978) *J. Appl. Physiol.* **45,** 408–413.
130. Mullane, K. M., Dusting, G. J., Salmon, J. A., Moncada, S., and Vane, J. R. (1979) *Eur. J. Pharmacol.* **54,** 217–228.
131. Leffler, C. W., and Hessler, J. R. (1979) *Eur. J. Pharmacol.* **54,** 37–42.
132. Hyman, A. L., and Kadowitz, P. J. (1979) *Circ. Res.* **45,** 404–409.
133. Utsunomiya, T., Krausz, M. M., Valeri, C. R., Shepro, D., and Hechtman, H. B. (1980) *Surgery* **88,** 25–30.
134. Demling, R. H., Smith, M., Gunther, R., Gee, M., and Flynn, J. (1981) *Surgery* **89,** 257–263.
135. Born, G. V. R. (1962) *Nature (London)* **194,** 927–929.
136. Higgs, G. A., Moncada, S., and Vane, J. R. (1977) *Br. J. Pharmacol.* **61,** 137P.
137. Ubatuba, F. B., Moncada, S., and Vane, J. R. (1979) *Thromb. Haemostasis* **41,** 425–434.

138. Gryglewski, R. J., Korbut, R., Ocetkiewicz, A. C., and Stachura, T. (1978). *Naunyn Schmiedeberg's Arch. Pharmacol.* **302,** 25–30.
139. Szczeklik, A., Gryglewski, R. J., Nizankowski, R., Musial, J., Pieton, R., and Mruk, J. (1978) *Pharmacol. Res. Commun.* **10,** 545.
140. Aiken, J. W., Gorman, R. R., and Shebuski, R. J. (1979) *Prostaglandins* **17,** 483–494.
141. Bayer, B. L., Blass, K. E., and Förster, W. (1979) *Br. J. Pharmacol.* **66,** 10–12.
142. Dusting, G. J., Moncada, S., and Vane, J. R. (1977) *Br. J. Pharmacol.* **62,** 414–415P.
143. Dusting, G. J., Moncada, S., and Vane, J. R. (1978) *Br. J. Pharmacol.* **64,** 315–320.
144. Gimeno, M. J., Sterin-Borda, L., Borad, E. S., Lazzari, M. A., and Gimeno, A. L. (1980) *Prostaglandins* **19,** 907–916.
145. Blasko, G., Nemesanszky, E., Szabo, G., Stadier, I., and Palos, L. A. (1980) *Thromb. Res.* **17,** 673–681.
146. Johnson, R. A., Morton, D. R., Kinner, J. H., Gorman, R. R., McGuire, J. C., Sun, F. F., Whittaker, N., Bunting, S., Salmon, J., Moncada, S., and Vane, J. R. (1976) *Prostaglandins* **12,** 915–928.
147. Cho, M. J., and Allen, M. A. (1978) *Prostaglandins* **15,** 943–954.
148. Gorman, R. R., Bunting, S. S., and Miller, O. V. (1977) *Prostaglandins* **13,** 377–388.
149. Tateson, J. E., Moncada, S., and Vane, J. R. (1977) *Prostaglandins* **13,** 389–399.
150. Whittle, B. J. R., Moncada, S., and Vane, J. R. (1978) *Prostaglandins* **16,** 373–388.
151. Miller, O. V., and Gorman, R. R. (1979) *J. Pharmacol. Exp. Ther.* **210,** 134–140.
152. Minkes, M., Stanford, M., Chi, M., Roth, G., and Raz, A. (1977) *J. Clin. Invest.* **59,** 449–454.
153. Kazer-Glanzman, R., Jakabova, M., George, J., and Luscher, E. (1977) *Biochim. Biophys. Acta* **466,** 429–440.
154. Lapetina, E. G., Schmitges, C. J., Chandrabose, K., and Cuatrecasas, P. (1977) *Biochem. Biophys. Res. Commun.* **76,** 828–835.
155. Malmsten, C., Granstrom, E., and Samuelsson, B. (1976) *Biochem. Biophys. Res. Commun.* **68,** 569–576.
156. Vargaftig, B. B., and Chignard, M. (1975) *Agents Actions* **5,** 137–144.
157. Gorman, R. R., Hamilton, R. D., and Hopkins, N. K. (1979) *J. Biol. Chem.* **254,** 1671–1676.
158. Kather, H., and Simon, B. (1979) *Res. Exp. Med.* **176,** 25–29.
159. Macdermot, J., and Barnes, P. J. (1980) *Eur. J. Pharmacol.* **67,** 419–425.
160. Hopkins, N. K., and Gorman, R. R. (1981) *J. Clin. Invest.* **67,** 540–546.
161. Svensson, J., Hamberg, M., and Samuelsson, B. (1975) *Acta Physiol. Scand.* **94,** 222–228.
162. Rose, J. C., Kot, P. A., Ramwell, P. W., Doykos, M., and O'Neill, W. P. (1976) *Proc. Soc. Exp. Biol. Med.* **153,** 209–212.
163. Palmer, M. A., Piper, P. J., and Vane, J. R. (1973) *Br. J. Pharmacol.* **49,** 226–242.
164. Hamberg, M., Hedqvist, P., Strandberg, K., Svensson, J., and Samuelsson, B. (1975) *Life Sci.* **16,** 451–462.
165. Svensson, J., and Hamberg, M. (1976) *Prostaglandins* **12,** 943–950.
166. Dusting, G. J., Moncada, S., and Vane, J. R. (1977b) *Eur. J. Pharmacol.* **45,** 301–304.
167. Ellis, E. F., Nies, A. S., and Oates, J. A. (1977) *Stroke* **8,** 480.
168. Coceani, F., Bishai, I., White, E., Bodach, E., and Olley, P. M. (1978) *Am. J. Physiol.* **234,** H117–H122.
169. Svensson, J., and Fredholm, B. (1977) *Acta Physiol. Scand.* **101,** 366–368.
170. Lagarde, M., and Dechavanne, M. (1977) *Lancet* **1,** 88.
171. Shimamoto, T., Kobayashi, M., Takahashi, T., Takashima, Y., Sakamoto, M., and Morooka, S. (1978) *Jpn. Heart J.* **19,** 748–753.

172. Szczeklik, A., Gryglewski, R. J., Musial, J., Grodzinska, L., Serwonska, M., and Marcinkiewicz, E. (1978) *Thromb. Diath. Haemorrh.* **40,** 66–74.
173. Lewy, R. I., Smith, J. B., Silver, M. J., Wiener, L., and Walinsky, P. (1979). *Prostaglandins Med.* **2,** 243–248.
174. Hirsh, P. D., Hillis, L. D., Campbell, W. D., Firth, B. G., and Willerson, J. T. (1981) *N. Engl. J. Med.* **304,** 685–691.
175. Harrison, H. E., Reece, A. H., and Johnson, M. (1978) *Life Sci.* **23,** 351–356.
176. Johnson, M., Reece, A. H., and Harrison, H. E. (1978). *Proc. 7th, Int. Cong. Pharmacol., 1978* Abstracts, p. 342.
177. Johnson, M., Harrison, H. E., Raftery, A. T., and Elder, J. B. (1979) *Lancet* **1,** 325–326.
178. Dollery, C. T., Friedman, L. A., Hensby, C. N., Kohner, E., Lewis, P. J., Porta, M., and Webster, J. (1979) *Lancet* **2,** 1365.
179. Davis, T. M. E., Brown, E., Finch, D. R., Mitchell, M. D., and Turner, R. C. (1981) *Br. Med. J.* **282,** 1259–1262.
180. Remuzzi, G., Misiani, R., Marchesi, D., Livio, M., Mecca, G., de Gaetano, G., and Donati, M. B. (1978) *Lancet* **2,** 871–872.
181. Remuzzi, G., Rossi, E. C., Misiani, R., Marchesi, D., Mecca, G., de Gaetano, G., and Donati, M. B. (1980) *Semin. Thromb. Hemostasis* **6,** 391–394.
182. Cocchetto, D. M., Cook, L., Cato, A. E., and Niedel, E. (1981) *Semin. Thromb. Hemostasis* **7,** 43–51.
183. Hensby, C. N., Lewis, P. J., Hilgard, P., Mufti, G. J., Hows, J., and Webster, J. (1979b) *Lancet* **2,** 748.
184. Remuzzi, G., Cavenaghi, A. E., Mecca, G., Donati, M. B., and de Gaetano, G. (1977) *Thromb. Res.* **11,** 919–920.
185. Gullner, H-G., Cerletti, C., Bartter, F. C., Smith, J. B., and Gill, J. R., Jr. (1979) *Lancet* **2,** 767–768.
186. Pace-Asciak, C. R., Carrara, M. C., Rangaraj, G., and Nicolaou, K. G. (1978) *Prostaglandins* **15,** 1005–1012.
187. Grose, J. H., Lebel, M., and Gbeassor, F. M. (1980) *Clin. Sci.* **59,** 121–123s.
188. Steiner, M. (1970) *Experientia* **26,** 786–789.
189. Bye, A., Lewis, Y., and O'Grady, J. (1979) *Br. J. Clin. Pharmacol.* **7,** 283–286.
190. Preston, F. E., Whipps, S., Jackson, C. A., French, A. J., Wyld, P. J., and Stoddard, C. J. (1981) *N. Engl. J. Med.* **304,** 76–79.
191. Amezcua, J.-L., Parsons, M., and Moncada, S. (1978) *Thromb. Res.* **13,** 477–488.
192. Korbut, R., and Moncada, S. (1978) *Thromb. Res.* **13,** 489–500.
193. O'Grady, J., and Moncada, S. (1978) *Lancet* **2,** 780.
194. Rajah, S. M., Penny, S., and Kester, R. (1978) *Lancet* **2,** 1104.
195. Godal, H. C., Eika, C., Dybdahl, J. H., Daae, L., and Larsen, S. (1979) *Lancet* **1,** 1236.
196. Jorgensen, K. A., Dyerberg, J., Olesen, A. S., and Stoffersen, E. (1980) *Thromb. Res.* **19,** 799–805.
197. Amezcua, J.-L., O'Grady, J., Salmon, J. A., and Moncada, S. (1979) *Thromb. Res.* **16,** 69–74.
198. Treacher, D., Warlow, C., and McPherson, K. (1978) *Lancet* **2,** 1378.
199. Harter, H. R., Burch, J. W., Majerus, P. W., Stanford, N., Delmez, J. A., Anderson, C. B., and Weerts, C. A. (1979) *N. Engl. J. Med.* **301,** 577–579.
200. Blackwell, G. J., Flower, R. J., Russell-Smith, N., Salmon, J. A., Thorogood, P. B., and Vane, J. R. (1978) *Br. J. Pharmacol.* **64,** 436P.
201. Feuerstein, N., and Ramwell, P. W. (1981) *Eur. J. Pharmacol.* **69,** 533–534.
202. Tyler, H. M., Saxton, C. A. P. D., and Parry, M. J. (1981) *Lancet* **1,** 629–632.

203. Vermylen, J., Carreras, L. O., Schaeren, J. V., Defreyn, G., Machin, S. J., and Verstraete, M. (1981) *Lancet* **1,** 1073–1075.
204. Glavind, J., Hartmann, S., Clemmesen, J., Jessen, K. E., and Dam, H. (1952) *Acta Pathol. Microbiol. Scand.* **30,** 1.
205. Slater, T. F. (1972) "Free Radical Mechanisms in Tissue Injury." Pion Ltd., London.
206. Moncada, S., Gryglewski, R. J., Bunting, S., and Vane, J. R. (1976) *Prostaglandins* **12,** 715–733.
207. Mickel, H. S., and Horbar, J. (1974) *Lipids* **9,** 68–71.
208. D'Angelo, V., Villa, S., Mysliwiec, M., Donati, M. B., and de Gaetano, G. (1978) *Thromb. Haemostasis* **39,** 535–536.
209. Sinzinger, H., Feigl, W., and Silberbauer, K. (1979) *Lancet* **2,** 469.
210. Bourgain, R. H., Six, F., and Andries, R. (1980) *Artery* **8,** 96–100.
211. Marks, J. (1962) *Vitam. Horm. (N.Y.)* **20,** 573–598.
212. Boyd, A. M., and Marks, J. (1963) *Angiology* **14,** 198–208.
213. Haeger, K. (1968) *Vasc. Dis.* **5,** 199–213.
214. Medalie, J. H., Kahn, H. A., Neufeld, H. N., Riss, E., and Gouldbourt, U. (1973) *J. Chronic Dis.* **26,** 325–349.
215. Streja, D., Steiner, G., and Kwiterovich, P. O. (1978) *Ann. Intern. Med.* **89,** 871–880.
216. Kannel, W. B., Castelli, W. P., and Gordon, T. (1979) *Ann. Intern. Med.* **90,** 85–91.
217. Nordoy, A., Svensson, B., Wiebe, D., and Hoak, J. (1978) *Cir. Res.* **43,** 527–534.
218. Beitz, J., and Förster, W. (1980) *Biochim. Biophys. Acta* **620,** 352–355.
219. Gryglewski, R. J., and Szczeklik, A. (1981) *In* "Clinical Pharmacology of Prostacyclin" (P. J. Lewis and J. O'Grady, eds.), pp. 89–95. Raven, New York.
220. Pastan, I. H., Johnson, G. S., and Anderson, W. B. (1975) *Annu. Rev. Biochem.* **44,** 491–522.
221. Johnson, G. S., Pastan, I., Peery, C. V., Otten, J., and Willingham, M. C. (1972) *In* "Prostaglandins in Cellular Biology and the Inflammatory Process" (P. W. Ramwell and B. B. Pharriss, eds.), pp. 195–205. Plenum, New York.
222. Santoro, M. G., Philpott, G. W., and Jaffe, B. M. (1976) *Nature (London)* **263,** 770–777.
223. Huttner, J. J., Gwebu, E. G., Panganamala, R. V., Milo, G. E., Cornwell, D. G., Sharma, H. M., and Geer, J. C. (1977) *Science* **197,** 289–291.
224. Bendit, E. P. (1977) *Sci. Am.* **236,** 74–85.
225. Defreyn, G., Vergara Dayden, M., Machin, S. J., and Vermylen, J. (1980) *Thromb. Res.* **19,** 695–699.
226. Weksler, B. B., Ley, C. W., and Jaffe, E. A. (1978) *J. Clin. Invest.* **62,** 923–930.
227. Weksler, B. B., Ley, C. W., and Jaffe, E. A. (1978) *Clin. Res.* **26,** 619A.
228. Weksler, B. B., Reinus, J., and Eldor, A. (1981) *In* "Prostaglandins and Cardiovascular Disease" (R. Johnsson Hegyeli, ed.), pp. 125–138. Raven, New York.
229. Coughlin, S. R., Moskowitz, M. A., Zeiter, B. R., Antoniades, H. N., and Levine, L. (1980) *Nature (London)* **288,** 600–602.
230. Shier, W. T. (1980) *Proc. Natl. Acad. Sci. U.S.A.* **77,** 137–141.
231. Vermylen, J., Chamone, D. A. F., and Verstraete, M. (1979) *Lancet* **1,** 518–520.
232. Carreras, L. O., Chamone, D. A. F., Klerckx, P., and Vermylen, J. (1980) *Thromb. Res.* **19,** 663–670.
233. Webster, J., Dollery, C. T., and Hensby, C. N. (1980) *Clin. Sci.* **59,** 125–128s.
234. Levin, R. I., Jaffe, E. A., Weksler, B. B., and Tack-Goldman, K. (1981) *J. Clin. Invest.* **67,** 762–769.
235. O'Grady, J., Warrington, S., Moti, M. J., Bunting, S., Flower, R. J., Fowle, A. S. E.,

Higgs, E. A., and Moncada, S. (1979) *In* "Prostacyclin" (J. R. Vane and S. Bergström, eds.), pp. 409–417. Raven, New York.

236. Chierchia, S., Ciabattoni, G., Cinotti, G., Maseri, A., Patrono, C., Publiese, F., Distante, A., Simonetti, I., and Bernini, W. (1979) *Circulation* **59–60,** Suppl. II, Abstr. 318.
237. Warrington, S. J., and O'Grady, J. (1980) *Adv. Prostaglandin Thromboxane Res.* **7,** 619–624.
238. Fitzgerald, G. A., Friedman, L. A., Miyamore, I., O'Grady, J., and Lewis, P. J. (1979) *Life Sci.* **25,** 665.
239. Moriau, M., Masure, R., Hurlet, A., Debeys, C., Chalant, C., Ponlot, R., Jaumain, P., Servaye-kestens, Y., Ravaux, A., Louis, A., and Goenen, M. (1977) *Vox Sang.* **32,** 41–51.
240. Weston, M. J., Rubin, M., Hanid, A., Langley, P. G., Westaby, S., and Williams, R. (1977) *In* "Thromboembolism—A New Approach to Therapy" (J. R. A. Mitchell and J. G. Domenet, eds.), pp. 76–82. Academic Press, New York.
241. Friedenberg, W. R., Myers, W. O., Plotka, E. D., Beathard, J. N., Kummer, D. J., Gatlin, P. F., Stoiber, D. L., Ray, J. F., and Sautter, R. D. (1978) *Ann. Thorac. Surg.* **25,** 298–305.
242. Patterson, R. H., and Kessler, J. (1969) *Circulation* **40,** Suppl. 3, 160.
243. Woods, H. F., Ash, G., Weston, M. J., Bunting, S., Moncada, S., and Vane, J. R. (1978) *Lancet* **2,** 1075–1077.
244. Bunting, S., Moncada, S., Vane, J. R., Woods, H. F., and Weston, M. J. (1979) *In* "Prostacyclin" (J. R. Vane and S. Bergström, eds.), pp. 361–369. Raven, New York.
245. Longmore, D. B., Bennett, G., Gueirrara, D., Smith, M., Bunting, S., Reed, P., Moncada, S., Read, N. G., and Vane, J. R. (1979) *Lancet* **1,** 1002–1005.
246. Gimson, A. E. S., Hughes, R. D., Mellon, P. J., Woods, H. F., Langley, P. G., Canalese, J., Williams, R., and Weston, M. J. (1980) *Lancet* **1,** 173–175.
247. Gimson, A. E. S., Canalese, J., Hughes, R. D., Langley, P. G., Mellon, P. J., and Williams, R. (1981) *In* "Clinical Pharmacology of Prostacyclin" (P. J. Lewis and J. O'Grady, eds.), pp. 211–218. Raven, New York.
248. Longmore, D. B., Bennett, J. G., Hoyle, P. M., Smith, M. A., Gregory, A., Osivand, T., and Jones, W. A. (1981) *Lancet* **1,** 800–804.
249. Bunting, S., O'Grady, J., Fabiani, J.-N., Terrier, E., Moncada, S., Vane, J. R., and Dubost, C. (1981) *In* "Clinical Pharmacology of Prostacyclin" (P. J. Lewis and J. O'Grady, eds.), pp. 181–194. Raven, New York.
250. Walker, I. D., Davidson, J. F., Faichney, A., Wheatley, D., and Davison, K. (1981) *In* "Clinical Pharmacology of Prostacyclin" (P. J. Lewis and J. O'Grady, eds.), pp. 195–199. Raven, New York.
251. Bennett, J. G., Longmore, D. B., and O'Grady, J. (1981) *In* "Clinical Pharmacology of Prostacyclin" (P. J. Lewis and J. O'Grady, eds.), pp. 201–208. Raven, New York.
252. Chelly, J., Tricot, A., Garcia, A., Boucherie, J.-C., Fabiani, J.-N., Passalecq, J., and Dubost, C. (1981) *In* "Clinical Pharmacology of Prostacyclin" (P. J. Lewis and J. O'Grady, eds.), p. 209. Raven, New York.
253. Bunting, S., and Moncada, S. (1980) *Br. J. Pharmacol.* **69,** 268P–269P.
254. Bunting, S., Simmons, P. M., and Moncada, S. (1981) *Thromb. Res.* **21,** 89–102.
255. Cimo, P. L., Moake, J. L., Weinger, R. S., Ben-Menachem, Y., and Khalil, K. G. (1979) *Am. J. Hematol.* **6,** 125–133.
256. Hussey, C. V., Bernhard, V. M., McLean, M. R., and Fobian, J. E. (1979) *Ann. Clin. Lab. Sci.* **9,** 487–493.
257. Salzman, E. W., Rosenberg, R. D., Smith, M. H., Lindon, J. N., and Favreau, L. (1980) *J. Clin. Invest.* **65,** 64–73.

258. Zusman, R. M., Rubin, R. H., Cato, A. E., Cocchetto, D. M., Crow, J. W., and Tolkoff-Rubin, N. (1981) *N. Engl. J. Med.* **304,** 934–939.
259. Turney, J. H., Dodd, N. J., and Weston, M. J. (1981) *Lancet* **1,** 1101.
260. Szczeklik, A., Nizankowski, R., Skawinska, S., Szczeklik, J., Gluszko, P., and Gryglewski, R. J. (1979) *Lancet* **1,** 1111–1114.
261. Szczeklik, A., and Gryglewski, R. J. (1981) *In* "Clinical Pharmacology of Prostacyclin" (P. J. Lewis and J. O'Grady, eds.), pp. 159–167. Raven, New York.
262. Olsson, A. G. (1980) *Lancet* **32,** 1076.
263. Pardy, B. J., Lewis, J. D., and Eastcott, H. H. G. (1980) *Surgery* **88,** 826–832.
264. Zygulska-Mach, H., Kostka-Trabka, E., Niton, A., and Gryglewski, R. J. (1980) *Lancet* **2,** 1075.
265. Watkins, W. D., Peterson, M. B., Crone, R. K., Shannon, D. C., and Levine, L. (1980) *Lancet* **1,** 1083.
266. Szczeklik, J., Szczeklik, A., and Nizankowski, R. (1980). *Lancet* **2,** 1076.
267. Lock, J. E., Olley, P. M., Coceani, F., Swyer, P. R., and Rowe, R. D. (1979) *Lancet* **1,** 1343.
268. Fidler, J., Ellis, C., Bennett, M. J., de Swiet, M., and Lewis, P. J. (1981) *In* "Clinical Pharmacology of Prostacyclin" (P. J. Lewis and J. O'Grady, eds.), pp. 141–144. Raven, New York.
269. Bergman, G., Daly, R., Atkinson, L., Rothman, M., Richardson, P. J., Jackson, G., and Jewitt, D. E. (1981) *Lancet* **1,** 569–572.
270. Hall, R. J. C., and Dewar, H. A. (1981) *Lancet* **1,** 949.
271. Budd, G. T., Bukowski, R. M., Lucas, F. V., Cato, A. E., and Cocchetto, D. M. (1980) *Lancet* **2,** 915.
272. Fitzgerald, C. A., Roberts, L. J., II, Maas, D., Brash, A. R., and Oates, J. A. (1981) *In* "Clinical Pharmacology of Prostacyclin" (P. J. Lewis and J. O'Grady, eds.), p. 81. Raven, New York.
273. Webster, J., Borysiewicz, L. K., Rees, A. J., and Lewis, P. J. (1981) *In* "Clinical Pharmacology of Prostacyclin" (P. J. Lewis and J. O'Grady, eds.), pp. 77–80. Raven, New York.
274. Mundy, A. R., Bewick, M., Moncada, S., and Vane, J. R. (1980) *Prostaglandins* **19,** 595–603.

Prostaglandins, Thromboxanes, and Leukotrienes

E. GRANSTRÖM AND P. HEDQVIST
Department of Physiological Chemistry and Department of Physiology
Karolinska Institutet
Stockholm, Sweden

The prostaglandins, thromboxanes, and leukotrienes belong to a large family of compounds which are all oxygenated metabolites of certain polyunsaturated fatty acids (1,2).

The first members of this family to be identified were the prostaglandins, so named because they were believed to be biosynthesized mainly in the prostate gland (3). They were, however, later found to be of nearly ubiquitous occurrence: Almost all cell types in the body possess the capacity for prostaglandin biosynthesis. Originally, the prostaglandins were regarded as hormones mainly of importance in reproductive physiology. With their later identification in other tissues and organs, this concept was abandoned, and the prostaglandins are now regarded as modulators of many different cell functions, both in physiological and pathological processes.

Chemically, the prostaglandins are all derivatives of a C_{20} carboxylic acid, prostanoic acid, which contains a central cyclopentane ring (C-8 to C-12) and two side chains with 7 and 8 carbons, respectively (see Fig. 1). During biosynthesis of the prostaglandins, various oxygen-containing substituents are introduced into the fatty acid molecule. These substituents largely determine the biological potency of the compounds: the most important positions seem to be at C-9, C-11, and C-15. Most naturally occurring prostaglandins have a Δ^{13} trans double bond and an L-hydroxyl group at C-15; if these structures are metabolically or chemically altered, the compound may lose its biological activity (1).

The nature of the substituents in the ring determines the type of prostaglandin and is of fundamental importance qualitatively for its biological effects. A keto group at C-9 and an α-hydroxyl at C-11 gives

PATHOBIOLOGY OF THE ENDOTHELIAL CELL

ISBN 0-12-521980-6

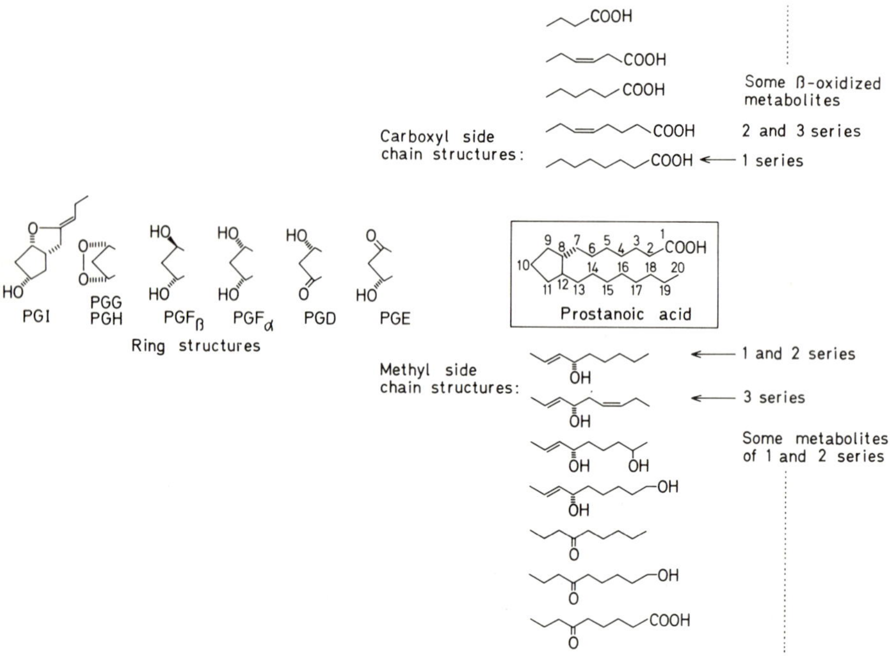

Fig. 1. Structure of prostanoic acid and the prostaglandins.

a prostaglandin of the E type; the reverse situation is characteristic for the D type of prostaglandins; and having hydroxyl groups at both positions is typical for PGF compounds. An index (α or β) indicates the stereochemistry of the hydroxyl group at C-9 (Fig. 1).

The different classes of prostaglandins are further subgrouped according to the degree of unsaturation, and an index indicates the number of double bonds. Prostaglandins of the "1" series contain only the previously mentioned Δ^{13}trans double bond; PG_2 compounds contain also a Δ^5cis double bond, and PG_3 and additional Δ^{17}cis double bond. Thus $PGF_{2\alpha}$, for example, is 9α,11α,15-trihydroxyprosta-5(*cis*), 13(*trans*)-dienoic acid.

Prostaglandins are biosynthesized from certain polyunsaturated fatty acids. Arachidonic acid (5,8,11,14-eicosatetraenoic acid) is the precursor of prostaglandins of the 2 series: the Δ^5cis double bond of the precursor is found unaltered in the products. The corresponding parent compounds for the 1 and 3 series are 8,11,14-eicosatrienoic acid (dihomo-γ-linolenic acid) and 5,8,11,14,17-eicosapentaenoic acid (timnodonic acid), respectively.

The polyunsaturated fatty acids generally do not occur free in the cell. Normally they occur esterified in phospholipids, most commonly in position 2′. Prostaglandin biosynthesis must start by release of the precursor fatty acids, most often by the action of a phospholipase A_2 (for a review, see, e.g., Ref. 4).

Studies on the conversion of 8,11,14-eicosatrienoic acid into PGE_1 showed that the two oxygens in the five-membered ring of the prostaglandin were derived from the same oxygen molecule (5). This finding, together with the occurrence of 12L-hydroxy-8,10-heptadecadienoic acid and malondialdehyde as by-products of the biosynthesis, strongly indicated the existence of an endoperoxide intermediate oxygenated at C-15 (Fig. 1) (5). This was also supported by the simultaneous formation of PGE, PGFα, and PGD compounds in certain systems.

The prostaglandin endoperoxide had already been postulated as an intermediate in prostaglandin biosynthesis in 1965 (5); however, not until 8 years later was the compound actually identified. An unstable compound, reducible to the corresponding PGF_{α}, could be detected in short-time incubations of precursor acids with a prostaglandin synthetase preparation. This labile intermediate was isolated from incubations with arachidonic acid and found to be 15-hydroxyprostaglandin endoperoxide (PGH_2, 15-hydroxy-9α,11α-peroxidoprosta-5,13-dienoic acid). Later, an additional endoperoxide was found: 15-hydroperoxy-9α,11α-peroxidoprosta-5,13-dienoic acid (PGG_2) (6–8).

The structures of these unstable compounds could not be established directly, but were assigned mainly by a number of chemical transformations into previously known prostaglandins. Thus, treatment of either endoperoxide with stannous chloride or triphenylphosphine resulted in the formation of $PGF_{2\alpha}$. Treatment of PGG_2 with lead tetraacetate, followed by reduction with stannous chloride, yielded 15-keto-$PGF_{2\alpha}$ (8).

The two endoperoxides also possess pronounced biological activity, both on several smooth muscle preparations and as agents capable of inducing platelet aggregation (8). Both endoperoxides are unstable in aqueous medium ($t_{1/2}$ about 4–5 min). The main product formed from PGG_2 is 15-hydroperoxy-PGE_2, whereas PGH_2 is mainly converted into PGE_2. *In vivo*, the endoperoxides are the precursors of several different products. The earliest compounds detected were the prostaglandins of the E, F, and D type, as well as a C_{17} monohydroxy fatty acid (12-hydroxy-5,8,10-heptadecatrienoic acid, HHT, from arachidonic acid) and malondialdehyde (MDA).

Later it was discovered that in certain tissues and cells, such as platelets and lung tissue, arachidonic acid is converted only to a minor extent into the classic prostaglandins. The major products formed are

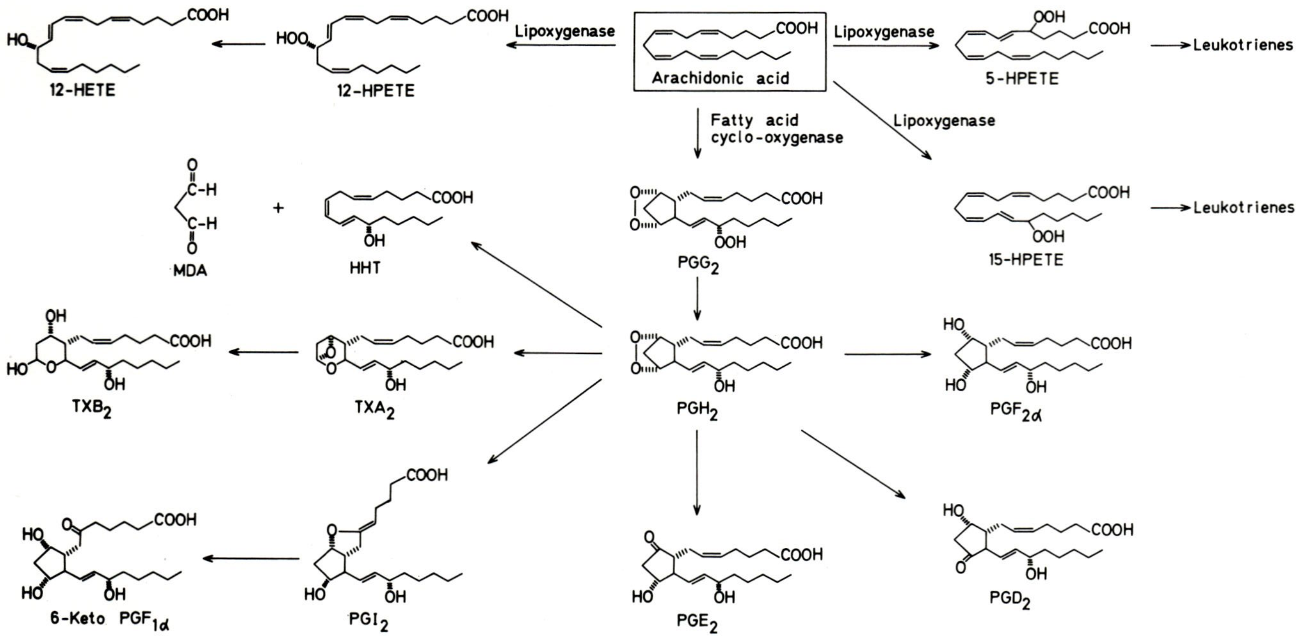

Fig. 2. Metabolism of arachidonic acid into prostaglandins, thromboxanes, prostacyclin, and leukotrienes.

instead 12-hydroxy-5,8,10,14-eicosatetraenoic acid (HETE), HHT, and a nonprostanoic compound, the hemiacetal derivative of 8-(1-hydroxy-3-oxopropyl)-9,12-dihydroxy-5,10-heptadecadienoic acid, now called thromboxane B_2 (TXB_2) (9,10) (Fig. 2).

In human platelets indomethacin and aspirin block the formation of HHT and TXB_2 but increase the formation of HETE (11). It was also demonstrated that PGG_2 added to a suspension of human platelets was rapidly converted into HHT and TXB_2. These findings indicated that the latter two compounds are formed from PGG_2, whereas HETE is formed by a separate pathway. Isotope experiments, in which arachidonic acid was incubated with platelets under $^{18}O_2$, suggested that TXB_2 was formed from PGG_2 by rearrangement and subsequent incorporation of one molecule of water (12). The rearranged intermediate was possible to trap by the addition of nucleophilic reagents after short-time incubations. When excess amounts of methanol, ethanol, or sodium azide were added after 30 sec incubations, mono-*O*-methyl TXB_2, mono-*O*-ethyl TXB_2, and an azido derivative of TXB_2 were formed, respectively. These products were not seen when the reaction was interrupted in this way after 5 min instead.

These trapping experiments demonstrated the presence of a very unstable intermediate in the conversion of PGG_2 into TXB_2. The half-life of this intermediate in aqueous medium was determined by two different methods and was found to be 30–40 sec.

The unstable intermediate has been assigned an oxane–oxetane ring structure, shown in Fig. 2, and is called thromboxane A_2 (TXA_2) (12). Two other alternate structures could be ruled out either from experiments with deuterium-labeled precursor or from the results of the half-life determination.

According to the recently introduced thromboxane nomenclature in which the basic thromboxane skeleton is named thrombanoic acid (13), the structure of TXA_2 is $9\alpha,11\alpha$-epoxy-15-hydroxythromba-5,13-dienoic acid and that of TXB_2 9α,11,15-trihydroxythromba-5,13-dienoic acid.

Thromboxane A_2 was found to possess very strong biological activity. It is even more potent than the endoperoxides in many biological systems, such as induction of platelet aggregation and contraction of smooth muscle from several organ systems (14,15). Thromboxane biosynthesis was first demonstrated in platelets and in lung tissue, and in both it represents a major pathway in arachidonic acid conversion (9,10). The earlier discovered rabbit aorta contracting substance (RCS) (16) has been shown to consist mainly of TXA_2 (17). Thromboxane biosynthesis has later been demonstrated in spleen, brain, leukocytes, uterus, umbilical artery, and many other cells and tissues (2).

Shortly after the discovery of the thromboxane pathway, a novel unstable intermediate in arachidonic acid metabolism was discovered, which was generated from aortic microsomes (18–21). Like TXA_2 it was formed from the endoperoxide PGH_2, but its biological activities were the opposite: relaxation of vascular smooth muscle and inhibition or reversal of platelet aggregation (21). The compound had a half-life of about 2 min in aqueous solution at physiological pH and at 37°C but was relatively stable at alkaline pH. Its structure was established as 9-deoxy-6,9α-epoxy-Δ^5-$PGF_{1\alpha}$ [PGI_2, or prostacyclin (22)]. The Δ^5 double bond was later found to have the Z configuration (23).

The spontaneously formed hydrolysis product is 6-keto-$PGF_{1\alpha}$ (22,23), which can occur either as such or in a lactol form. The occurrence of 6-keto-$PGF_{1\alpha}$ in other biological systems had been reported by several groups (24–26). PGI_2 biosynthesis has now been demonstrated in many tissues: vessel walls, stomach, heart, seminal vesicles, lung, etc. As its actions are opposite those of TXA_2, the balance between these two pathways of endoperoxide metabolism is likely to be of great importance in the body. An example of this is the control of platelet aggregability *in vivo* (27).

Novel pathways in arachidonic acid metabolism have recently been detected in leukocytes (28–34). Instead of the 12-hydroxyeicosatetraenoic acid, which is the major monohydroxy fatty acid formed by platelets, leukocytes synthesize 5-hydroxy- and 15-hydroxyeicosatetraenoic acids from arachidonic acid (28,30). Several other hydroxylated products were later identified (29,31). The first of these was identified as 5*S*,12*R*-dihydroxy-6,8,10,14-eicosatetraenoic acid (29), a compound carrying a conjugated triene structure containing two trans and one cis double bond. The stereochemistry of this triene was later identified as 6-*cis*, 8,10-*trans* (35). Four other dihydroxytetraenoic acids were also identified in this biological system (31), all of them hydroxylated at C-5 and constituting two pairs of diastereoisomers. Two compounds were isomers of the previously identified 5,12-dihydroxy acid, but with an all trans configuration of the triene double bonds, which were epimeric at C-12. The last two compounds were 5,6-dihydroxy acids, epimeric at C-6.

The mechanism of the formation of these dihydroxy acid metabolites has been studied (32). An unstable intermediate was discovered which gave rise to all the five dihydroxy acids. This intermediate was assigned the structure 5,6-oxido-7,9,11,14-eicosatetraenoic acid. This assumption was based on isotope-labeling experiments using $^{18}O_2$, on trapping experiments using methanol, ethanol, or ethylene glycol, and other studies (32). The intermediate is subsequently hydrolyzed enzymatically into 5*S*,12*R*-dihydroxy-6-*cis*, 8,10-*trans*, 14-*cis*-eicosatetraenoic acid. All the other four dihydroxy acids are nonenzymatically formed hydrolysis products of the labile intermediate. It was postulated that the 5,6-oxido intermediate was formed from a 5-hydroperoxy acid, which could also be partially reduced to a 5-hydroxy acid. This product was consistently seen in these experiments (32).

After these initial discoveries, this branch of arachidonic acid metabolism has expanded rapidly. It was discovered that the unstable 5,6-oxido intermediate is also the precursor of a novel category of metabolites of polyunsaturated fatty acids.

During the studies of arachidonic acid conversion in leukocytes it was found that the ionophore A23187 greatly stimulated the release of the novel dihydroxy acids as well as 5-hydroxyeicosatetraenoic acid (30). This was of considerable interest, because this ionophore was also known to stimulate the formation of slow reacting substance (SRS) from leukocytes. The ultraviolet absorption spectra of the dihydroxy acids also displayed a characteristic pattern, caused by the conjugated triene structure, which had strong similarities to the spectrum of purified SRS (31,36). These findings as well as others led to the concept that SRS might be an arachidonate metabolite, formed by reactions related to the 5-hydroxylated compounds.

The first SRS to be identified was isolated from murine mastocytoma cells and was found to be 5(*S*)-hydroxy-6(*R*)-*S*-glutathionyl-7,9-*trans*-11,14-*cis*-eicosatetraenoic acid (33,37). With the identification of this compound, a novel nomenclature was introduced for compounds formed via this pathway. The first compound having the conjugated triene structure is the previously mentioned unstable 5,6-oxido intermediate: this compound was given the designation leukotriene A (LTA) (later leukotriene A_4, see below) (38). The enzymatically formed hydrolysis product, 5(*S*),12(*R*)-dihydroxy-6,8,10,14-eicosatetraenoic acid, is called leukotriene B_4 (LTB_4), and the glutathionyl derivative LTC_4. The subscript was introduced as an indication of the number of double bonds (39) after the corresponding metabolites had also been identified from 5,8,11-eicosatrienoic acid and 5,8,11,14,17-eicosapentaenoic acid (40,41). The 4 series thus originates in arachidonic acid, whereas the other two precursor acids give rise to leukotrienes of the 3 and 5 series, respectively.

Subsequent studies showed that LTC_4 could be metabolized by the action of γ-glutamyltranspeptidase into the corresponding cysteinylglycine derivative (LTD_4) (42). Both these biologically very potent substances have subsequently been identified from other sources, such as rat basophilic leukemia cells (42,43), rat monocytes (44–46), and human and guinea pig lung (46,47).

Degradation of the peptide side chain may proceed even further. A compound with only a cysteinyl residue has also been identified (37,48,49) and is called LTE_4.

The entity known as slow reacting substance of anaphylaxis (SRS-A), which can originate from many different sources (50,51), has now in several cases been shown to consist of mixtures of leukotrienes of the C, D, and E type.

This area of metabolism of polyunsaturated fatty acids is still in a phase of very rapid expansion. Recently it was demonstrated that leukotriene formation was not restricted to transformation involving oxygenation at C-5 as the first step. A novel family of leukotrienes can also be formed in biological systems by initial oxygenation at C-15 (52,53). Furthermore, biologically active compounds that are either metabolites of leukotrienes or of structurally related dihydroxy acids have been identified, such as trihydroxy acids or dicarboxylic acids (54,55) (Fig. 3).

The biological activities of various slow-reacting substances have been studied for a long time. After the discoveries of the multitude of biologically active substances belonging to the leukotrienes or their related substances, interest in this area has been even further stimulated, and a great amount of new biological data have accumulated by research in recent years.

It is well known that SRS-A, besides being a potent smooth muscle stimulator and presumed mediator of antigen-induced bronchoconstriction in allergic asthma, causes microvascular leakage in turn leading to tissue edema. The first notion in this respect appears to be a statement by Brocklehurst (56) that a purified preparation of SRS-A, obtained from perfusates of sensitized guinea pig lung after antigen challenge, caused increased vascular permeability when injected intradermally. This has since been repeatedly confirmed by others, also using murine and feline SRS-A and studying the effect on other animal species, such as the rat and monkey (57,58). In particular the monkey appears to be very sensitive, and intradermal injections of SRS-A may result not only in increased vascular permeability but also in hemorrhagic necrosis of the skin. The permeability-promoting effect of SRS-A has long been regarded as an important criterion for distinguishing this compound from other presumed mediators of hypersensitivity reactions. Furthermore, these observations imply that SRS-A may have a mediator role in forms of immunological tissue injury other than antigen-induced bronchoconstriction.

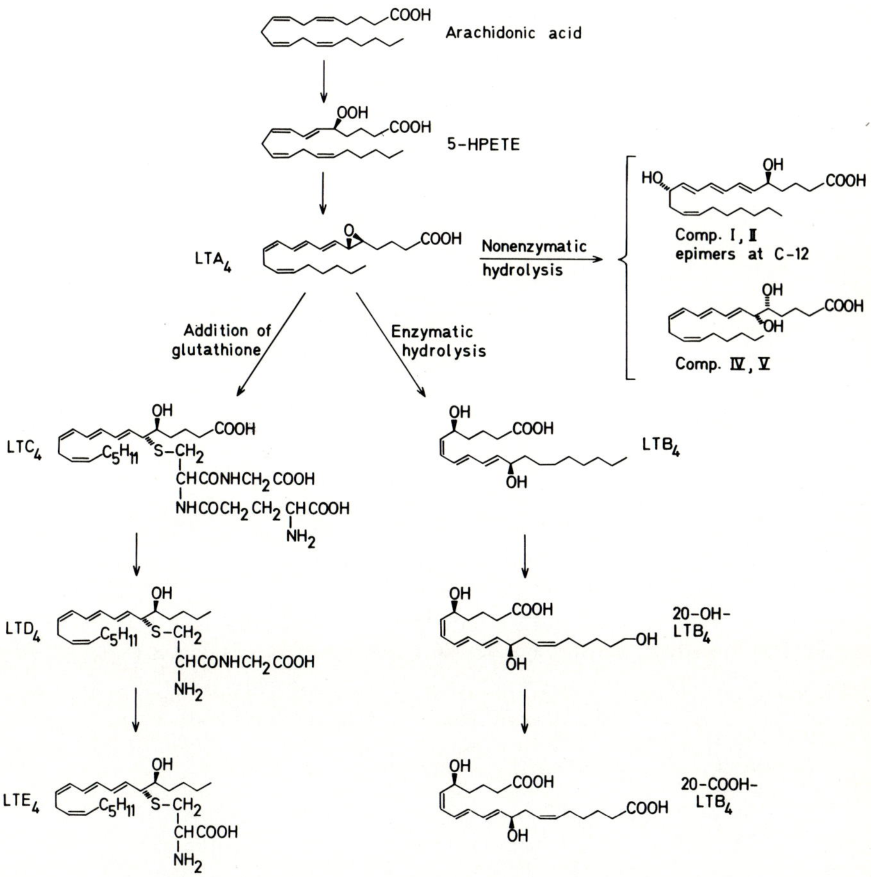

Fig. 3. Leukotriene biosynthesis from the C-5 lipoxygenase pathway.

As already pointed out in this chapter, the leukotrienes comprise a group of closely related substances biosynthesized from polyunsaturated fatty acids (mainly arachidonic acid), as outlined in Fig. 3. It is now evident that immunologically as well as nonimmunologically generated SRS-A can be identified as an entity consisting of several leukotrienes, principally the cysteinyl-containing leukotrienes of the C, D, and E series (42–46,59). Pharmacological analysis of the cysteinyl-containing leukotrienes has also confirmed that they have a spectrum of biological actions closely resembling that previously reported for SRS-A (60–62).

It is of considerable interest that the leukotrienes C_4 and D_4 not only mimic the promoting action of SRS-A on vascular permeability but seem to be remarkably potent in this respect (60,61). When injected intradermally into guinea pigs, as little as 0.1 pmol of either compound is sufficient for inducing significant extravasation of Evans blue. Higher doses produce dose-dependent and similar increases in spot diameter, the only difference relating to the appearance of the blue spot, being homogeneous with LTD_4 but having a central blanching with LTC_4. Further experiments have shown that LTE_4 is largely equipotent with LTC_4 and D_4 in causing macromolecular extravasation, whereas the two other leukotrienes, LTA_4 and LTB_4, seem to be at least 100 times less active in this respect (63).

Seemingly, the lower potency of LTB_4 on vascular leakage relative to cysteinyl-containing leukotrienes, which coincides with an equally lower potency on bronchial smooth muscle (55), might make it a second-rate leukotriene. However, it is becoming increasingly evident that LTB_4 may fulfill another important role: that of being a potent chemoattractant for leukocytes. According to *in vitro* evidence, LTB_4 induces increased undirectional and directional locomotion of human, rat, and rabbit polymorphonuclear leukocytes, as well as dose-related increases in the chemokinesis of human monocytes and rat macrophages (64–66). Furthermore, when injected intraperitoneally into guinea pigs, LTB_4 elicits an increase in white blood cell counts in the peritoneal fluid, possibly indicating that LTB_4 causes migration of leukocytes also under *in vivo* conditions (66).

The hamster cheek pouch may be used as a highly sophisticated model for microscopic *in vivo* inspection of a terminal vascular bed. When prepared according to Svensjö *et al.* (67), the experimental design permits continuous observation of different microcirculatory events, such as vessel diameter, blood flow, macromolecular permeability, and blood cell–endothelium interactions. Recently, leukotrienes were found to have important and distinct microvascular effects in this *in vivo* preparation (68). Leukotrienes C_4 and D_4, when applied topically to the cheek pouch for a 3-min period in concentrations ranging between 0.3 and 20 nM, elicited a prompt and dose-dependent constriction of arterioles—in particular, terminal arterioles. The vasoconstriction ceased rapidly after the 3-min period, and it was consistently followed by a dose-dependent and reversible leakage of macromolecules, as indicated by extravasation of fluorescein isothiocyanate-conjugated dextran (FITC-dextran). Notably, plasma leakage induced by the leukotrienes was confined to postcapillary venules, in accordance with the concept that this vessel segment is the specific target of action for agents eliciting reversible changes in vascular permeability.

The vascular effects of LTC_4 and LTD_4 were further compared with those of two other vasoactive substances, angiotensin II and histamine. Angiotensin was only slightly more potent than the leukotrienes in causing arteriolar constriction, and it did not induce any macromolecular leakage. Histamine, on the other hand, elicited a dose-dependent extravasation, but it did not result in any vasoconstriction. This would seem to indicate that vasoconstriction is not a prerequisite for leukotrienes causing leakage of plasma. In line with this contention, low concentrations of LTC_4 (1 n*M* or less) superfusing the cheek pouch for 20 min may induce near maximal leakage (300–400 leakage sites/cm^2) without causing but a slight and pulsatory vasoconstriction. According to dose–response curves assessed noncumulatively, LTC_4 and LTD_4 induced a significant increment of vascular permeability at much lower concentrations than did histamine. Leukotriene C_4 was most potent in this respect, being approximately 5000 times more potent than histamine and about five times more potent than LTD_4.

Leukotriene B_4, when administered in the same dose range as LTC_4 and LTD_4, did not elicit any vasoconstriction and had no effect on plasma leakage. However, in contrast to the cysteinyl-containing leukotrienes it caused polymorphonuclear leukocytes moving in the periphery of the axial stream to adhere to the venule wall, in particular in postcapillary venules of a diameter less than 15 μm. This effect, being dose-dependent, was visible within 1 min after application of LTB_4, reached its maximum after 6–8 min, and remained at this level until LTB_4 was withdrawn, whereafter it gradually subsided to reach the control value in 10–15 min.

Cysteinyl-containing leukotrienes have been implicated as having a mediator role in immediate hypersensitivity reactions, such as allergic asthma, principally because of their outstanding potency as bronchoconstrictors (60,62). This concept is further stressed by their conspicuous activity in causing increased vascular permeability and tissue edema, another important manifestation of asthma, being at least 1000 times more potent than histamine in this respect. However, the multifaceted actions of different leukotrienes in the hamster cheek pouch imply that they also may be involved in vascular events of more general importance. Knowing that leukotrienes may be nonimmunologically formed in various blood-borne and tissue-bound phagocyting cells, it may be more than incidental that the triad of vascular effects induced by LTB_4 together with LTC_4 or LTD_4 in the hamster cheek pouch closely mimics the early events in acute inflammatory reactions.

ACKNOWLEDGMENTS

This study was supported by grants from the Swedish Medical Research Council (projects: 03P-5804, 04X-4342, 03X-217, 03X-5914), the Swedish National Association Against Chest and Heart Diseases, and Karolinska Institutet.

REFERENCES

1. Samuelsson, B., Granström, E., Gréen, K., Hamberg, M., and Hammarström, S. (1975) *Annu. Rev. Biochem.* **44,** 669.
2. Samuelsson, B., Goldyne, M., Granström, E., Hamberg, M., Hammarström, S., and Malmsten, C. (1978) *Annu. Rev. Biochem.* **47,** 997.
3. von Euler, U. S. (1935) *Klin. Wochenschr.* **14,** 1182.
4. Flower, R. J. (1981) *In* "The Prostaglandin System" (F. Berti and G. P. Velo, eds.), p. 27. Plenum, New York.
5. Samuelsson, B. (1965) *J. Am. Chem. Soc.* **87,** 3011.
6. Hamberg, M., and Samuelsson, B. (1973) *Proc. Natl. Acad. Sci. U.S.A.* **70,** 899.
7. Nugteren, D. H., and Hazelhof, E. (1973) *Biochim. Biophys. Acta* **326,** 448.
8. Hamberg, M., Svensson, J., Wakabayashi, T., and Samuelsson, B. (1974) *Proc. Natl. Acad. Sci. U.S.A.* **71,** 345.
9. Hamberg, M., and Samuelsson, B. (1974) *Proc. Natl. Acad. Sci. U.S.A.* **71,** 3400.
10. Hamberg, M., and Samuelsson, B. (1974) *Biochem. Biophys. Res. Commun.* **61,** 942.
11. Hamberg, M., Svensson, J., and Samuelsson, B. (1974) *Proc. Natl. Acad. Sci. U.S.A.* **71,** 3824.
12. Hamberg, M., Svensson, J., and Samuelsson, B. (1975) *Proc. Natl. Acad. Sci. U.S.A.* **72,** 2994.
13. Samuelsson, B., Hamberg, M., Roberts, L. J., Oates, J. A., and Nelson, N. A. (1978) *Prostaglandins* **16,** 857.
14. Svensson, J., Strandberg, K., Tuvemo, T., and Hamberg, M. (1977) *Prostaglandins* **14,** 425.
15. Svensson, J., Hamberg, M., and Samuelsson, B. (1976) *Acta Physiol. Scand.* **98,** 285.
16. Piper, P. J., and Vane, J. R. (1969) *Nature (London)* **223,** 29.
17. Svensson, J., Hamberg, M., and Samuelsson, B. (1975) *Acta Physiol. Scand.* **94,** 222.
18. Moncada, S., Gryglewski, R., Bunting, S., and Vane, J. R. (1976) *Nature (London)* **263,** 663.
19. Gryglewski, R., Bunting, S., Moncada, S., Flower, R. J., and Vane, J. R. (1976) *Prostaglandins* **12,** 685.
20. Moncada, S., Gryglewski, R., Bunting, S., and Vane, J. R. (1976) *Prostaglandins* **12,** 715.
21. Bunting, S., Gryglewski, R., Moncada, S., and Vane, J. R. (1976) *Prostaglandins* **12,** 897.
22. Johnson, R., Morton, D., Kinner, J., Gorman, R., McGuire, J., Sun, F., Whittaker, N., Bunting, S., Salmon, J., Moncada, S., and Vane, J. R. (1976) *Prostaglandins* **12,** 915.
23. Johnson, R. A., Lincoln, F. H., Thompson, J. L., Nidy, E. G., Mizsak, S. A., and Axén, U. (1977) *J. Am. Chem. Soc.* **99,** 4182.
24. Pace-Asciak, C. (1976) *J. Am. Chem. Soc.* **98,** 2348.
25. Dawson, W., Boot, J. R., Cockerill, A. F., Mallen, D. N. B., and Osborne, D. J. (1976) *Nature (London)* **262,** 699.

26. Chang, W.-C., Murota, S., Matsuo, M., and Tsurufuji, S. (1976) *Biochem. Biophys. Res. Commun.* **72,** 1259.
27. Moncada, S., and Vane, J. R. (1978) *Br. Med. Bull.* **34,** 129.
28. Borgeat, P., Hamberg, M., and Samuelsson, B. (1976) *J. Biol. Chem.* **251,** 7816; correction **252,** 8772 (1977).
29. Borgeat, P., and Samuelsson, B. (1979) *J. Biol. Chem.* **254,** 2643.
30. Borgeat, P., and Samuelsson, B. (1979) *Proc. Natl. Acad. Sci. U.S.A.* **76,** 2148.
31. Borgeat, P., and Samuelsson, B. (1979) *J. Biol. Chem.* **254,** 7865.
32. Borgeat, P., and Samuelsson, B. (1979) *Proc. Natl. Acad. Sci. U.S.A.* **76,** 3213.
33. Murphy, R. C., Hammarström, S., and Samuelsson, B. (1979) *Proc. Natl. Acad. Sci. U.S.A.* **76,** 4275.
34. Samuelsson, B., Borgeat, P., Hammarström, S., and Murphy, R. C. (1979) *Prostaglandins* **17,** 785.
35. Corey, E. J., Marfat, A., Goto, G., and Brion, F. (1981) *J. Am. Chem. Soc.* **102,** 7984.
36. Morris, H. R., and Taylor, G. W. (1978) *FEBS Lett.* **87,** 203.
37. Hammarström, S., Murphy, R. C., Samuelsson, B., Clark, D. A., Mioskowski, C., and Corey, E. J. (1979) *Biochem. Biophys. Res. Commun.* **91,** 1266.
38. Samuelsson, B., Borgeat, P., Hammarström, S., and Murphy, R. C. (1979) *Prostaglandins* **17,** 785.
39. Samuelsson, B., and Hammarström, S. (1980) *Prostaglandins* **19,** 645.
40. Hammarström, S. (1981). *J. Biol. Chem.* **256,** 2275.
41. Hammarström, S. (1980) *J. Biol. Chem.* **255,** 7093.
42. Örning, L., Hammarström, S., and Samuelsson, B. (1980) *Proc. Natl. Acad. Sci. U.S.A.* **77,** 2014.
43. Morris, H. R., Taylor, G. W., Piper, P. J., Samhoun, M. N., and Tippins, J. R. (1980) *Prostaglandins* **19,** 185.
44. Bach, M. K., Brashler, J. R., Hammarström, S., and Samuelsson, B. (1980) *J. Immunol.* **125,** 115.
45. Bach, M. I., Brashler, J. R., Hammarström, S., and Samuelsson, B. (1980) *Biochem. Biophys. Res. Commun.* **93,** 1121.
46. Lewis, R. A., Austen, K. F., Drazen, J. M., Clark, D. A., Marfat, A., and Corey, E. J. (1980) *Proc. Natl. Acad. Sci. U.S.A.* **77,** 3710.
47. Morris, H. R., Taylor, G. W., Piper, P. J., and Tippins, J. R. (1980) *Nature (London)* **285,** 104.
48. Lewis, R. A., Drazen, J. M., Austen, K. F., Clark, D. A., and Corey, E. J. (1980) *Biochem. Biophys. Res. Commun.* **96,** 271.
49. Bernström, K., and Hammarström, S. (1981). *J. Biol. Chem.* **256,** 9579.
50. Austen, K. F. (1978) *J. Immunol.* **121,** 793.
51. Borgeat, P. (1981) *J. Med. Chem.* **24,** 121.
52. Lundberg, U., Rådmark, O., Malmsten, C., and Samuelsson, B. (1981) *FEBS Lett.* **126,** 127.
53. Jubiz, W., Rådmark, O., Lindgren, J. Å., Malmsten, C., and Samuelsson, B. (1981) *Biochem. Biophys. Res. Commun.* **99,** 976.
54. Lindgren, J. Å., Hansson, G., and Samuelsson, B. (1981) *FEBS Lett.* **128,** 329.
55. Hansson, G., Lindgren, J. Å., Dahlén, S. E., Hedqvist, P., and Samuelsson, B. (1981). *FEBS Lett.* **130,** 107.
56. Brocklehurst, W. E. (1967) *In* "International Symposium on Vaso-Active Peptides: Bradykinins and Related Kinins," p. 189. Pergamon, Oxford.
57. Orange, R. P., and Austen, K. F. (1969) *Adv. Immunol.* **10,** 105.
58. Strandberg, K., and Uvnäs, B. (1971) *Acta Physiol. Scand.* **82,** 358.

59. Samuelsson, B., Hammarström, S., Murphy, R. C., and Borgeat, P. (1980) *Allergy* **35,** 375.
60. Hedqvist, P., Dahlén, S. E., Gustafsson, L., Hammarström, S., and Samuelsson, B. (1980) *Acta Physiol Scand.* **110,** 331.
61. Drazen, J. M., Austen, F. K., Lewis, R. A., Clark, D. A., Goto, G., Marfat, A., and Corey, E. J. (1980) *Proc. Natl. Acad. Sci. U.S.A.* **77,** 4354.
62. Dahlén, S. E., Hedqvist, P., Hammarström, S., and Samuelsson, B. (1980) *Nature (London)* **288,** 484.
63. Hedqvist, P., Dahlén, S. E., and Björk, T. (1982) Adv. Prostaglandin Thromboxane Leukotriene Res. (eds. B. Samuelsson, P. Ramwell and R. Paoletti). Raven Press, New York. Vol. 9, in press.
64. Malmsten, C., Palmblad, J., Udén, A., Rådmark, O., Enstedt, L., and Samuelsson, B. (1981) *Acta Physiol. Scand.* **110,** 449.
65. Palmer, R. M. J., Stephney, R. J., Higgs, J. A., and Eakins, K. E. (1980). *Prostaglandins* **20,** 411.
66. Smith, M. H. J., Ford-Hutchinson, A. W., and Bray, M. A. (1980) *J. Pharm. Pharmacol.* **32,** 517.
67. Svensjö, E., Arfors, K. E., Artursson, G., and Kutili, G. (1978). *Upsala J. Med. Sci.* **83,** 71.
68. Dahlén, S. E., Björk, J., Hedqvist, P., Arfors, K. E., Hammarström, S., Lindgren, J. Å., and Samuelsson, B. (1981). *Proc. Natl. Acad. Sci. U.S.A.* **78,** 3887.

Synthesis of Prostacyclin by Cultured Endothelial Cells

ERIC A. JAFFE, RICHARD I. LEVIN,
BABETTE B. WEKSLER, AND AARON J. MARCUS
Division of Hematology-Oncology
Department of Medicine, Cornell University Medical College
New York, New York

INTRODUCTION

Prostacyclin (PGI_2) is a labile prostaglandin that inhibits platelet aggregation and serotonin release and causes vasodilation (1). PGI_2 was originally discovered when two independent groups of investigators noted that pieces of blood vessel walls incubated with arachidonic acid produced a material that was a potent vasodilator and inhibitor of platelet aggregation (2–6). This substance, originally called PGX (2), was subsequently identified as PGI_2 (7,8). PGI_2 inhibits platelet aggregation by stimulating the synthesis of cyclic AMP (9,10) though the mechanism by which it causes vasodilation is unclear. Subsequent investigations showed that the intima produced more PGI_2 than did either the media or adventitia (11). At the same time, we demonstrated that cultured human and bovine endothelial cells could synthesize PGI_2 (12) and that PGI_2 was the major metabolite of arachidonic acid produced by endothelial cells (12,13).

These findings led to the hypothesis that platelets and endothelial cells exert opposite effects on the state of vascular tone and platelet function. PGI_2 secreted by endothelial cells inhibits platelet function and causes vasodilation, whereas thromboxane A_2 (TXA_2) secreted by platelets stimulates platelet aggregation and induces vasoconstriction (2–6, 9–17). Because of the interest in this hypothesis, this chapter will review what is known about the regulation of endothelial cell PGI_2 synthesis.

PATHOBIOLOGY OF THE ENDOTHELIAL CELL

ISBN 0-12-521980-6

INDUCTION OF PGI_2 SYNTHESIS

PROTEOLYTIC ENZYMES

Thrombin was the first enzyme (and physiological agent) shown to induce PGI_2 production in cultured human endothelial cells (18–20). It was suggested that this pathway might represent a physiological control mechanism that would limit thrombus formation at a site of vascular injury. However, since repeated stimulation of endothelial cells with thrombin makes the cells temporarily refractory to thrombin (18,21), this physiological control mechanism may only function intermittently. The ability of thrombin to induce PGI_2 synthesis is species specific since bovine endothelial cells do not respond to either bovine or human thrombin (20,22). While stimulation of PGI_2 production by thrombin requires the presence of an intact active site (18,23), the locus of thrombin's presumably proteolytic action is unclear (23). Trypsin also induces endothelial cell PGI_2 synthesis (18,20) though it is not as active as thrombin on a molar basis.

VASOACTIVE AGENTS

Bradykinin (10 ng/ml) stimulates PGI_2 synthesis in cultured human umbilical vein and pig and calf aortic endothelial cells (20). While bradykinin also causes PGI_2 release *in vivo* and *ex vivo* (22,24,25), only a portion of the vasodilation caused by bradykinin appears due to PGI_2 production since some vasodilation still occurs *in vivo* in the presence of indomethacin (25,26). Histamine (27) and angiotensin II (25,28) also induce vascular and endothelial cell PGI_2 production, whereas epinephrine (18,25), norepinephrine (25), serotonin (25,92), acetylcholine (25,29), and ADP (18) do not.

HORMONES

Calcitonin has been shown to increase PGI_2 by porcine aortic strips (30). When injected intramuscularly in man, it increases both plasma PGI_2 levels and PGI_2 secreted by pieces of forearm veins obtained at biopsy (31,32).

PLATELET–ENDOTHELIAL CELL INTERACTIONS

Platelets release cyclic endoperoxides when they aggregate. Recently endothelial cells incubated with platelets have been shown to be able to convert these exogenous endoperoxides into PGI_2, which then blocks platelet aggregation (33). Previous studies that were unable to demonstrate this phenomenon probably failed because they used too high a ratio of platelets to endothelial cells. It is likely that this process (i.e., "endoperoxide steal") occurs most often in the capillary bed where the ratio of platelets to endothelial cells approaches 1 : 1. Platelets also release a variety of proteins when they aggregate; the α-granule protein platelet derived growth factor (PDGF) has recently been shown to increase PGI_2 synthesis as much as 100 times in cultured bovine aortic endothelium (34). Since stimulation occurs at PDGF levels below those measured *in vivo* in plasma, this effect may represent another feedback loop controlling platelet aggregation. In contrast, β-thromboglobulin and PF-4 have no effect on endothelial PGI_2 production (35).

DRUGS

Several drugs used in treating cardiovascular disorders induce endothelial PGI_2 production. We have recently shown that nitroglycerin at concentrations clinically attained in man causes a dose-related increase in the production of PGI_2 by endothelial cells (36). This has been confirmed by others in bovine coronary arteries (37). Other vasodilators which induce PGI_2 synthesis are dipyridamole (38), pentoxifylline (39, 40), a coumarin derivative called AD6 (41), and lidocaine (42). Dipyridamole also enhances the effectiveness of PGI_2 as an inhibitor of platelet aggregation because it inhibits platelet phosphodiesterase and thus tends to elevate platelet cyclic AMP levels (43). In contrast, nitroprusside (44) and captopril (45) do not induce PGI_2 synthesis. The experimental drug BAY g6575, when given to volunteers, caused their plasma to stimulate PGI_2 release from slices of rat aorta (46). Since BAY g6575 has antithrombotic properties in animals, this work suggests that the drug acts to stimulate PGI_2 release from the vessel wall (46).

ARTERIES VERSUS VEINS

Within a species, endothelial cells from most arteries produce more PGI_2 than cells from veins (47,48).

INHIBITORS OF PGI_2 SYNTHESIS

ANTIINFLAMMATORY AGENTS

Aspirin has been shown to block prostaglandin synthesis (49) by acetylating the enzyme cyclooxygenase (50) at an internal serine residue (51). In some studies, prostaglandin production *in vitro* in endothelial cells (52,53) was as sensitive to aspirin as it was in platelets (54); in cultured human endothelial cells the ID_{50} for aspirin equals 2 μM (52). In other studies, endothelial cells or aortic microsomes appeared less sensitive to aspirin than did platelets (54,55). Similarly, some *in vivo* studies in man have shown that production of PGI_2 by venous endothelium is inhibited up to 80% by daily doses of aspirin as low as 40 mg (56,57). Additionally, intravenous aspirin in rabbits at doses of 0.01 to 1 mg/kg inhibits PGI_2 and TXA_2 production 90–95% (58). However, other *in vivo* studies in man and rabbits have shown that platelets are more sensitive to aspirin than endothelial cells (59,60). Lastly, in rats aspirin inhibits prostaglandin production by platelets (ID_{50} = 3.6 mg/kg) and venous tissue (ID_{50} = 2.3 mg/kg) almost equally, though arterial tissues are less sensitive (ID_{50} = 25 mg/kg) (61). The ability of the endothelial cell to synthesize PGI_2 returns to normal (by synthesis of new cyclooxygenase) within 24–48 hr (52,53,59–61). Platelets, on the other hand, cannot regenerate their cyclooxygenase so their ability to synthesize TXA_2 returns only as the aspirin-inhibited platelets are replaced with new platelets. This process begins 2–3 days after aspirin administration and is complete by day 10 (62).

The basis for the discrepancies in the aspirin sensitivity studies of endothelial cells both *in vitro* and *in vivo* is presently unclear. However, recent work by several groups has shown that sodium salicylate, the breakdown product of aspirin, blocks and reverses in a dose-related fashion the cyclooxygenase–inactivating activity of aspirin in both platelets and the vascular wall (63–66). Possibly the different doses of aspirin and times of incubation used in the discrepant experiments gave rise to varying amounts of salicylate and thus caused differing degrees of inhibition of the aspirin used in the studies. Since salicylate is stable and is not rapidly excreted by the body, these findings also suggest that aspirin given in clinical studies to prevent stroke and myocardial infarction has, in the past, been administered too frequently and in too high a dose (66).

Other antiinflammatory agents also inhibit PGI_2 production. Indomethacin (18) and ibuprofen (67) both block PGI_2 production by cultured endothelial cells by inhibiting cyclooxygenase. These nonsteroidal drugs are thought to act at multiple sites on cyclooxygenase, and occupation of a site by a very weak inhibitor such as salicylate markedly inhibits the action of a much stronger agent such as indomethacin (68). This may explain salicylate's ability to inhibit the action of aspirin.

Antiinflammatory steroids also inhibit prostaglandin synthesis (69–71) and thus prevent the synthesis of PGI_2 (72). Steroids, however, unlike the agents mentioned earlier, inhibit phospholipase A_2 (73) probably by stimulating synthesis of lipomodulin, a specific inhibitor of phospholipase A_2 (74,75).

LIPIDS AND LIPOPROTEINS

Supplementation of endothelial cell culture medium with either dihomo-γ-linolenic (76) or linoleic acid (77) markedly diminished PGI_2 production by cultured endothelial cells. Low-density lipoproteins (LDL) also depressed PGI_2 synthesis (78) and this effect was blocked by high-density lipoprotein (HDL) (78), which, by itself, stimulates endothelial PGI_2 production (79). These observations are interesting since LDL has been shown to injure endothelial cells and HDL to protect endothelial cells against the effects of LDL (80,81). Further, HDL at physiological concentrations is a potent endothelial cell mitogen (82).

CYCLIC NUCLEOTIDES

Several recent studies have shown that PGI_2, thrombin, and β-adrenergic catecholamines activate endothelial cell adenylate cyclase and cause intraendothelial cell cAMP to rise (83–85). This elevation in intracellular cAMP in turn inhibits arachidonate release (86) and blocks PGI_2 production (84,86,87). This system may act as another feedback loop to control the production of PGI_2; however, it is presently not clear what role this system plays *in vivo* since a phosphodiesterase inhibitor must be used in experiments that demonstrate these effects (83–87).

NICOTINE AND VITAMIN E

Nicotine is thought to play a role in the increased incidence of cardiovascular disease seen in smokers. Consistent with this observation, nicotine inhibits PGI_2 synthesis in endothelial cells (88) and umbilical arteries (89) and also decreases PGI_2 release induced in rabbit hearts by hypoxia (90).

Vitamin E quinone, a contaminant of vitamin E that is hypothesized to be the actual active agent in vitamin E, depresses basal PGI_2 synthesis modestly and thrombin stimulated PGI_2 production markedly (91).

ACKNOWLEDGMENT

E. A. J. is a recipient of a NIH Research Career Development Award (5K04 HL00237). R. I. L. is the recipient of an NIH Clinical Investigator Award (5K08 HL00748). This research was supported by grants from the NIH (HL18828), the American Heart Association (78-1184), Roche Laboratories, the Alfred Jurzykowski Foundation, and the Arnold R. Krakower Foundation.

REFERENCES

1. Moncada, S., and Vane, J. R. (1978) *Br. Med. Bull.* **34,** 129–135.
2. Moncada, S., Gryglewski, R., Bunting, S., and Vane, J. R. (1976) *Nature (London)* **263,** 663–665.
3. Gryglewski, R. J., Bunting, S., Moncada, S., Flower, R. J., and Vane, J. R. (1976) *Prostaglandins* **12,** 685–714.
4. Bunting, S., Gryglewski, R. J., Moncada, S., and Vane, J. R. (1976) *Prostaglandins* **12,** 897–913.
5. Kulkarni, P. S., Roberts, R., and Needleman, P. (1976) *Prostaglandins* **12,** 337–353.
6. Raz, A., Isakson, P. C., Minkes, M. S., and Needleman, P. (1977) *J. Biol. Chem.* **252,** 1123–1126.
7. Johnson, R. A., Morton, D. R., Kinner, J. H., Gorman, R. R., McGuire, J. R., Sun, F. F., Whitaker, N., Bunting, S., Salmon, J. A., Moncada, S., and Vane, J. R. (1976) *Prostaglandins* **12,** 915–928.
8. Dusting, G. J., Moncada, S., and Vane, J. R. (1977) *Prostaglandins* **13,** 3–16.
9. Gorman, R. R., Bunting, S., and Miller, O. V. (1977) *Prostaglandins* **13,** 377–388.
10. Tateson, J. E., Moncada, S., and Vane, J. R. (1977) *Prostaglandins* **13,** 389–397.
11. Moncada, S., Herman, A. G., Higgs, E. A., and Vane, J. R. (1977) *Thromb. Res.* **11,** 323–344.
12. Weksler, B. B., Marcus, A. J., and Jaffe, E. A. (1977) *Proc. Natl. Acad. Sci. U.S.A.* **74,** 3922–3926.
13. Marcus, A. J., Weksler, B. B., and Jaffe, E. A. (1978) *J. Biol. Chem.* **253,** 7138–7141.
14. Hamberg, M., and Samuelson, B. (1974) *Proc. Natl. Acad. Sci. U.S.A.* **71,** 3400–3404.
15. Hamberg, M., Svensson, J., and Samuelsson, B. (1975) *Proc. Natl. Acad. Sci. U.S.A.* **72,** 2994–98.
16. Svensson, J., Hamberg, M., and Samuelsson, B. (1976) *Acta Physiol. Scand.* **98,** 285–294.
17. Dusting, G. J., Moncada, S., Mullane, K. M., and Vane, J. R. (1978) *Clin. Sci. Mol. Med.* **55,** 195s–198s.
18. Weksler, B. B., Ley, C. W., and Jaffe, E. A. (1978) *J. Clin. Invest.* **62,** 923–930.
19. Czervionke, R. L., Smith, J. B., Hoak, J. C., Fry, G. L., and Haycraft, D. L. (1979) *Thromb. Res.* **14,** 781–786.
20. Hong, S. L. (1980) *Thromb. Res.* **18,** 787–795.
21. Hoak, J. C., Czervionke, R. L., Fry, G. L., and Smith, B. J. (1980) *Fed. Proc., Fed. Am. Soc. Exp. Biol.* **39,** 2606–2609.
22. Goldsmith, J. C., and Jafvert, C. T. (1981) *Thromb. Haemostasis* **46,** 19.
23. Lollar, P., and Owen, W. G. (1980) *J. Biol. Chem.* **255,** 8031–8034.

24. Needleman, P., Bronson, S. D., Wyche, A., and Sivakoff, M. (1978) *J. Clin. Invest.* **61,** 839–849.
25. Mullane, K. M., and Moncada, S. (1980) *Prostaglandins* **20,** 25–49.
26. Schror, K., Metz, U., and Krevs, R. (1979) *Naunyn-Schmiedeberg's Arch. Pharmacol.* **307,** 213–221.
27. Baenziger, N. L., Force, L. E., and Becherer, P. R. (1980) *Biochem. Biophys. Res. Commun.* **92,** 1435–1440.
28. Lonchampt, M. D., Romquin, N., Bonne, C., and Regnault, F. (1981) *Thromb. Haemostasis* **46,** 39.
29. Furchgott, R. F., and Zawadzki, J. V. (1980) *Nature (London)* **288,** 373.
30. Clopath, P., and Sinzinger, H. (1980) *Prostaglandins* **19,** 1.
31. Sinzinger, H., Peskar, B. A., Clopath, P., Kovarik, J., Burghuber, O., Silberbauer, K., Leithner, C., and Woloszczuk, W. (1980) *Prostaglandins* **20,** 611–612.
32. Sinzinger, H., Clopath, P., Staehlin, A., Silberbauer, K., and Kovarik, J. (1981) *Thromb. Haemostasis* **46,** 179.
33. Marcus, A. J., Weksler, B. B., Jaffe, E. A., and Broekman, M. J. (1980) *J. Clin. Invest.* **66,** 979–986.
34. Coughlin, S. R., Maskowitz, M. A., Zetter, B. R., Antoniades, H. N., and Levine, L. (1980) *Nature (London)* **288,** 600–603.
35. Poggi, A., Niewiarowski, S., Holt, J. C., Dall'Olio, A., Stewart, G., and Smith, J. B. (1981) *Thromb. Haemostasis* **46,** 37.
36. Levin, R. I., Jaffe, E. A., Weksler, B. B., and Tack-Goldman, K. (1981) *J. Clin. Invest.* **67,** 762–769.
37. Schror, K., Grodzinska, L., and Darius, H. (1981) *Thromb. Res.* **23,** 59–67.
38. Blass, K. E., Block, H. V., Forster, W., and Ponich, K. (1980) *Br. J. Pharmacol.* **68,** 72–73.
39. Weithmann, K. U. (1981). *Thromb. Hemostasis* **46,** 129.
40. Schror, K., Matzky, R., Kahler, T., and Darius, H. (1981). *Thromb. Hemostasis* **46,** 272.
41. Dejana, E., de Castellarnau, C., Rotilio, D., and de Gaetano, G. (1981). *Thromb. Hemostasis* **46,** 273.
42. Casey, L. C., Armstrong, M. C., Fletcher, J. R., and Ramwell, P. W. (1980) *Prostaglandins* **19,** 977–984.
43. Moncada, S., and Korbut, K. (1978) *Lancet* **1,** 1286–1287.
44. Levin, R. I., Jaffe, E. A., Weksler, B. B., and Tack-Goldman, K. (1980) *Circulation* **62,** Suppl. III, 309.
45. Swartz, S. L., Williams, G. H., Hollenberg, N. W., Levine, L., Dluhy, R. G., and Moore, T. J. (1980) *J. Clin. Invest.* **65,** 1257–1264.
46. Vermylen, J., Chamone, D. A. F., and Verstraete, M. (1979) *Lancet* **1,** 518–520.
47. Skidgel, R. A., and Printz, M. P. (1978). *Prostaglandins* **16,** 1–16.
48. Johnson, A. R. (1980) *J. Clin. Invest.* **65,** 841–850.
49. Vane, J. R. (1971). *Nature New Biol.* **231,** 237–239.
50. Roth, G. J., Stanford, N., and Majerus, P. W. (1975) *Proc. Natl. Acad. Sci. U.S.A.* **72,** 3073–3076.
51. Van der Ouderaa, J., Buytenhek, M., Nugteren, D. H., and Van Dorp, D. A. (1980) *Eur. J. Biochem.* **109,** 1–8.
52. Jaffe, E. A., and Weksler, B. B. (1979) *J. Clin. Invest.* **63,** 532–535.
53. Gordon, J. L., and Pearson, J. D. (1978) *Br. J. Pharmacol.* **64,** 481–483.
54. Burch, J. W., Baenziger, N. L., Stanford, N., and Majerus, P. W. (1978) *Proc. Natl. Acad. Sci. U.S.A.* **75,** 5181–5184.

55. Baenziger, N. L., Becherer, P. R., and Majerus, P. W. (1979) *Cell* **16**, 967–974.
56. Preston, F. E., Whipps, S., Jackson, C. A., French, A. J., Wyld, P. J., and Stoddard, C. J. (1981) *N. Engl. J. Med.* **304**, 76–79.
57. Preston, F. E., Greaves, M., and Jackson, C. A. (1981) *Lancet* **1**, 1211–1212.
58. McDonald, J. W. D., and Ali, M. (1980) *Clin. Res.* **28**, 684A.
59. Hanley, S. P., Bevan, J., Cockbill, S. R., and Heptinstall, S. (1981) *Lancet* **1**, 969–971.
60. Ellis, E. F., Wright, K. F., Shearin Jones, P., Richardson, D. W., and Ellis, C. K. (1980) *J. Cardiovasc. Pharmacol.* **2**, 387–397.
61. Villa, S., Livio, M., and de Gaetano, G. (1979) *Br. J. Haematol.* **42**, 425–431.
62. Burch, J. W., Standford, N., and Majerus, P. W. (1978) *J. Clin. Invest.* **61**, 314–319.
63. Vargaftig, B. B. (1978) *Eur. J. Pharmacol.* **50**, 231–241.
64. Merino, J., Livio, M., Rajtar, G., and de Gaetano, G. (1980) *Biochem. Pharmacol.* **29**, 1093–1096.
65. Ali, M., and McDonald, J. W. D. (1979) *Prostaglandins Med.* **3**, 327–332.
66. de Castellanau, C., Cerletti, C., Dejana, E., Galletti, F., Latini, R., Livio, M., and de Gaetano, G. (1981) *Thromb. Haemostasis* **45**, 294.
67. Parks, W. M., Hoak, J. C., Czervionke, R. L., and Haycraft, D. L. (1981) *Clin. Res.* **29**, 343A.
68. Humes, J. L., Winter, C. A., Sadowski, S. J., and Kuehl, F. A. (1981) *Proc. Natl. Acad. Sci. U.S.A.* **78**, 2053–2056.
69. Gryglewski, R. J., Panczenko, B., Korbut, R., Grodzinska, A., and Ocetkiewicz, A. (1975) *Prostaglandins* **10**, 343–355.
70. Tashjian, A. H., Voelkel, E. F., and McDonough, J. (1975) *Nature (London)* **258**, 739–741.
71. Nijkamp, F. P., Flower, R. J., Moncada, S., and Vane, J. R. (1976) *Nature (London)* **263**, 479–482.
72. Blajchman, M. A., Senyi, A. F., Hirsh, J., Surya, Y., Buchanan, M., and Mustard, J. F. (1979) *J. Clin. Invest.* **63**, 1026–1035.
73. Hong, S. L., and Levine, L. (1976) *Proc. Natl. Acad. Sci. U.S.A.* **73**, 1730–1734.
74. Flower, R. J., and Blackwell, G. J. (1979) *Nature (London)* **278**, 456–459.
75. Hirata, F. (1981) *J. Biol. Chem.* **256**, 7730–7733.
76. Nordoy, A., Svensson, B., and Hoak, J. C. (1979) *Eur. J. Clin. Invest.* **9**, 5–10.
77. Spector, A. A., Hoak, J. C., Fry, G. L., Denning, G. M., Stoll, L. L., and Smith, J. B. (1980) *J. Clin. Invest.* **65**, 1003–1012.
78. Nordoy, A., Svensson, B., Wiebe, D., and Hoak, J. C. (1978) *Circ. Res.* **43**, 527–534.
79. Fleisher, L. N., Tall, A. R., Witte, L. D., Miller, R. W., and Cannon, P. J. (1981) Arteriosclerosis **1**, 359a.
80. Henriksen, T., Evensen, S. A., and Carlander, B. (1979) *Scand. J. Clin. Lab. Invest.* **39**, 361–368.
81. Henriksen, T., Evensen, S. A., and Carlander, B. (1979) *Scand. J. Clin. Lab. Invest.* **39**, 369–375.
82. Tauber, J.-P., Cheng, J., and Gospodarowicz, D. (1980) *J. Clin. Invest.* **66**, 696–708.
83. Schafer, A. I., Gimbrone, M. A., and Handin, R. I. (1980) *Biochem. Biophys. Res. Commun.* **96**, 1640–1647.
84. Hopkins, N. K., and Gorman, R. R. (1981) *J. Clin. Invest.* **67**, 540–546.
85. Dembinska-Kiec, A., Rucker, W., and Schonhofer, P. S. (1980) *Naunyn-Schmiedeberg's Arch. Pharmacol.* **311**, 67–70.
86. Adler, B., Gimbrone, M. A., Jr., Schafer, A. I., and Handin, R. I. (1981) *Blood* **58**, 514–517.

87. Brotherton, A. F. A., and Hoak, J. C. (1980) *Circulation* **62**, 111–165.
88. Weksler, B. B., Reinus, J., and Eldor, A. (1981) *In* "Prostaglandins and Cardiovascular Disease" (R. J. Hegyeli, ed.), pp. 125–137. Raven, New York.
89. Stoel, I., van Giessen, W. J., Zwolsman, E., Quadt, J. F. A., ten Hon Verheught, F. W. A., and Hugenholtz, P. G. (1980) *Circulation* **62**, Suppl. III, 97.
90. Wennmalm, A. (1980) *Br. J. Pharmacol.* **69**, 545–549.
91. Cox, A. C., Rao, G. H. R., Gerrard, J. M., and White, J. G. (1980) *Blood* **55**, 907–914.
92. Coughlin, S. R., Moskowitz, M. A., Antoniades, H. N., and Levine, L. (1981) *Proc. Natl. Acad. Sci. (U.S.A.)* **78**, 7134–7138.

Essential Fatty Acids in Platelets and Endothelial Cells

A. J. MARCUS, M. J. BROEKMAN, L. B. SAFIER,
H. L. ULLMAN, B. B. WEKSLER, E. A. JAFFE,
K. L. KAPLAN, AND K. T. H. WONG

Department of Medicine,
Division of Hematology–Oncology
New York Veterans Administration Medical Center,
New York, New York
and
Department of Medicine
Cornell University Medical College, New York, New York
and
College of Physicians and Surgeons, Columbia University, New York, New York

INTRODUCTION

Current studies of lipids in platelets and endothelial cells are concerned mainly with arachidonic acid (AA, or 20 : 4) and its metabolism. Arachidonic is a polyunsaturated fatty acid synthesized by mammals from linoleic acid, which must be obtained from the diet. Linoleic, linolenic, and arachidonic acids are known as essential fatty acids (EFA). The EFA cannot be synthesized *de novo* in cells of higher mammalian species. They must be obtained from external dietary sources. In experimental animals and in rare clinical situations, absence of EFA from the diet results in a complicated systemic metabolic disturbance known as essential fatty acid deficiency (1–3). The biosynthetic pathways for desaturation of fatty acids are oxygen-dependent, as is the synthesis of prostaglandins, thromboxanes, prostacyclin, and hydroxy fatty acids. It would appear that polyunsaturated fatty acids played a critical role in the evolutionary development of complex cell–membrane systems. For example, the major fatty acid of glycolipids in chloroplast membranes is linolenic. Linolenic acid may be essential for the structural integrity and efficiency of the photosynthetic process. In mammals, arachidonic acid is presumably important for the function of plasma and mitochondrial membranes.

PATHOBIOLOGY OF THE ENDOTHELIAL CELL

ISBN 0-12-521980-6

Among the most important developments in lipid biochemistry in the last 25 years is the elucidation of the structure of prostaglandins and the discovery that they are derived from arachidonic acid in mammalian cells.

ARACHIDONIC ACID METABOLITES AS AUTACOIDS

There is a tendency in the literature to classify prostanoids as hormones. If we think of hormones as chemical substances produced by organs or cells and transported in the bloodstream to a target tissue on which their effect is produced, then prostanoids do not quite fit into this category. On the other hand, autacoids are compounds formed in a specific organ or tissue and seem to modulate one or more functions of that tissue, or perhaps of neighboring tissues (4). Thus prostanoids appear to fit the definition of autacoids more closely than that of "hormones" or "neurotransmitters." Other chemical substances that fall into the category of autacoids are biogenic amines such as histamine and serotonin (5-HT), and polypeptides such as angiotensin and bradykinin.

No specific function can be assigned to the prostanoid family of compounds. For some cells the same prostanoid is inhibitory, while for others it is stimulatory. Currently postulated functions would include regulation of tissue blood flow and oxygen supply, carbon dioxide removal, contraction or relaxation of muscular tissues, platelet stimulation and/or inhibition, control of mineral metabolism in bone, modulation of processes such as fertility and labor, and possible involvement in the control of metastatic disease.

ARACHIDONIC ACID METABOLISM IN HUMAN PLATELETS

About 30% of the platelet phospholipid fatty acids is arachidonate. This important compound is present in both platelet membrane and granule fractions and is esterified exclusively to platelet phospholipid components (5). Arachidonate cannot be metabolized unless it is initially hydrolyzed and thus made available in its free form. It is of interest that following platelet stimulation much more AA is available than is required for metabolic processing. Platelet cyclooxygenase catalyzes addition of 2 mol of molecular oxygen and subsequent rearrangement of AA into endoperoxides, and thence mainly to thromboxane A2. Arachidonate is the preferred and most important substrate for cyclooxygenase. Lands (6) has proposed that hydroperoxides are required for optimal cyclooxygenase activation. In addition to its availability to cyclooxygenase, free 20 : 4 is distributed in several ways. Some is released into the surrounding medium of platelets, where it becomes bound to albumin if the latter is present *in vitro*, as it most certainly is *in vivo*. There are also proteins in the platelet cytoplasm that are capable of binding fatty acids such as AA. Arachidonate may also become reacylated to phospholipids such as those from which it was hydrolyzed. The platelet lipoxygenase acts on free 20 : 4 at a rate that lags behind that of cyclooxygenase. Thus, at the time when thromboxane A2 is fully formed and appears in the surrounding medium of the platelet, only relatively small quantities of lipoxygenase products—HPETE and HETE—are present in the external milieu. It is known that HETE is chemotactic for leukocytes and, therefore, this property may explain why leukocytes appear in hemostatic platelet plugs or thrombi at a later time than do platelets. It is quite difficult to study overall metabolism of arachidonate in platelets because of the multiple pathways it may take following hydrolysis. Further studies on the subcellular localization of platelet phospholipases and cyclooxygenase will be of interest.

CURRENT CONCEPTS OF ARACHIDONIC ACID MOBILIZATION

The mechanisms of hydrolysis of arachidonate from the 2-position of platelet phospholipids are under intensive investigation in several laboratories. Evidence for the involvement of several pathways has accumulated, but an unequivocal source of 20 : 4 for prostanoid formation has yet to be established.

Historically, studies involving radiolabeling indicated that a phospholipase A_2 acting on phosphatidylcholine [and possibly phosphatidylinositol (PI)] was an important mechanism (7). On the other hand, in isolated platelet membranes a phospholipase A_2 acting on diacyl phosphatidylethanolamine (PE) was demonstrated (8). Following the description of a PI-specific phospholipase C, which generated diglyceride from PI upon platelet stimulation (9), it was proposed that arachidonate was hydrolyzed by a diglyceride lipase, which therefore was indirectly utilizing PI as the source of arachidonate (10).

Subsequently Broekman *et al.* (11) carried out time course studies of changes in endogenous platelet phospholipids following stimulation. These investigators proposed that both PI (possibly via a diglyceride) and diacyl PE were sources of free platelet arachidonate. More recently, a phosphatidic acid (PA)-specific phospholipase A_2 was described by Billah and associates (12). This enzyme would utilize PA derived by phosphorylation of diglyceride that had formed from PI by a PI-specific phospholipase C. Billah *et al.* suggested this route since the phosphorylation mechanism occurs very rapidly, as evidenced by accumulation of PA (endogenous or radiolabeled), and since little diglyceride was present (12). Broekman *et al.* (13) have determined that the arachidonate content of PI and PA taken together remained constant during the first 10 sec following stimulation. Therefore arachidonate that is oxygenated to prostaglandin endoperoxides in less than 10 sec (11,14) could not have been derived from PI or PA, and probably not from diglyceride either. Thus Broekman and colleagues have suggested that arachidonate for cyclooxygenase was probably derived from another source, such as diacyl PE (13). Phospholipase A_2 activity with PI as substrate was recently suggested by Rittenhouse, also via indirect evidence (15).

Thus at the present time several mechanisms for arachidonate liberation are under study, including phospholipase A_2 acting on PE and/or PC; a PI-specific phospholipase A_2; and two pathways basically derived from the PI cycle, diglyceride lipase and PA-specific phospholipase A_2.

METABOLISM OF ARACHIDONIC ACID AND PROSTAGLANDIN ENDOPEROXIDE PGH_2 IN CULTURED HUMAN ENDOTHELIAL CELLS

In 1977 we demonstrated that cultured endothelial cells from both human umbilical veins and bovine aorta produced an inhibitor of platelet aggregation (16). To confirm the possibility that this was prostacyclin (PGI_2) we incubated microsomes from these endothelial cells with [^{3}H]arachidonic acid. This resulted in synthesis of prostacyclin (PGI_2) as demonstrated by thin-layer radiochromatography (16). These studies were subsequently extended with the use of biosynthetically prepared [^{14}C]PGH_2, which was added to intact, unstimulated endothelial cell monolayers (17).

It was demonstrated that endothelial cells were capable of synthesizing PGI_2 from exogenous PGH_2 (17). Prostacyclin and other prostaglandins formed were only present in supernatant fluids of the cultures. When [^{14}C]arachidonate was added to the monolayers, cell-associated counts were only found in the phospholipid fraction, but the prostacyclin formed from arachidonate was only present in the surrounding medium of the monolayers.

In the endoperoxide experiments the endothelial cells were pretreated with acetylsalicylic acid (ASA) in order to inhibit endogenous prostaglandin synthesis. Thus 5 min after exposure of intact ASA-treated endothelial cells to radiolabeled PGH_2, 6-keto-$PGF_{1\alpha}$ (the stable end-product of prostacyclin synthesis) and other prostaglandins were identified in the supernatant. Control incubations of PGH_2 in the absence of cells resulted in production of $PGF_{2\alpha}$, PGE_2, PGD_2, and HHT. The latter result indicated that prostacyclin synthesis was essentially enzymatic, but that transformation to other prostaglandins could also occur nonenzymatically.

From the preceding experiments it was concluded that production of PGI_2 by resting endothelial cell monolayers could be explained in two ways: (1) [^{14}C]PGH_2 entered the endothelial cell, was converted to prostacyclin and other prostaglandins, and then released; or (2) enzymatic and nonenzymatic transformation of PGH_2 actually occurred at or near the cell's outer surface. If PGH_2 could indeed be processed at the outer cell surface, it would suggest that a "natural" source of endoperoxide, i.e., stimulated platelets, might provide additional substrate for prostacyclin synthesis by vascular endothelium.

SYNTHESIS OF PROSTACYCLIN FROM PLATELET-DERIVED PGH_2 BY CULTURED HUMAN ENDOTHELIAL CELLS

We attempted to ascertain whether aspirin-treated endothelial cells that were incapable of synthesizing PGI_2 from endogenous sources would be capable of producing PGI_2 from endoperoxides released by stimulated platelets (18). [^{3}H]Arachidonic acid-prelabeled platelets were treated in aggregometer cuvettes with the calcium ionophore A23187, thrombin, or collagen in the presence of aspirin-treated endothelial cell suspensions. This procedure permitted us to carry out thin-layer radiochromatographic quantitation of [^{3}H]PGI_2 as [^{3}H]6-keto-$PGF_{1\alpha}$ and [^{3}H]thromboxane A_2 as [^{3}H]TXB_2. We were also able to analyze platelet aggregation responses in the identical sample, prior to extraction of tritium-labeled products.

In the presence of aspirin-treated endothelial cells, platelet aggregation in response to all three stimuli was inhibited. In each case [^{3}H]6-keto-$PGF_{1\alpha}$ was recovered from the supernates of the combined cell suspensions after stimulation. The order of PGI_2 production initiated by the stimuli was ionophore $>$ thrombin $>$ collagen.

It was of interest that the quantities of platelet [^{3}H]TXB_2 recovered were markedly reduced in the presence of aspirin-treated endothelial cells. In separate experiments 6-keto-$PGF_{1\alpha}$ and TXB_2 were quantitated by radioimmunoassay. The results paralled those obtained with the use of radiolabeling. The quantity of 6-keto-$PGF_{1\alpha}$ measured by radioimmunoassay represented amounts of PGI_2 sufficient to inhibit platelet aggregation (18).

The ratio of platelets to endothelial cells was an important factor in obtaining the preceding results. Thus, when 200,000 platelets per microliter were combined with 3,000–6,000 aspirin-treated endothelial cells per microliter 6-keto-$PGF_{1\alpha}$ production was measurable. At higher platelet levels, the proportion of TXB_2 to 6-keto-$PGF_{1\alpha}$ increased and platelet aggregation occurred. The effect of platelet concentration on production of PGI_2 and TXA_2 by combined suspensions of platelets and aspirin-treated endothelial cells is summarized in Table I (18).

Control studies carried out over a 90-min period indicated that the aspirin-treated endothelial cells could not synthesize PGI_2 from exogenous radioactive or endogenous arachidonate when stimulated with thrombin. Therefore the endothelial cell suspensions used in these experiments could only have employed endoperoxides from stimulated platelets.

Several specific experimental conditions were necessary in order to demonstrate production by endothelial cells of PGI_2 from en-

TABLE I

TLC–Radiochromatographic Analysis of Effect of Platelet Concentration on Production of PGI_2 and TXA_2 by Combined Suspensions of Platelets and ASA-Treated Endothelial Cells

Stimulus[a]	Platelet concentration/μl	6-keto-$PGF_{1\alpha}$		TXB_2[f]		TXB_2/6-keto-$PGF_{1\alpha}$
		cpm[d]	%[e]	cpm	%	
Ionophore	2×10^5 [b]	2,023	(8.6)	9,958	(42.3)	4.9
	31×10^5 [c]	5,586	(2.2)	193,272	(74.5)	34.6
Thrombin	2×10^5	561	(6.1)	4,724	(51.1)	8.4
	31×10^5	3,167	(2.0)	114,311	(72.5)	36.1
Collagen	2×10^5	79	(5.6)	422	(29.9)	5.3
	31×10^5	125	(1.6)	4,576	(61.3)	36.6

[a] Ionophore (1 μ*M*), thrombin (10 U/ml), collagen (30 μg/ml), Ca^{2+} (3 m*M*). Total volume was 0.5 ml.

[b] Total of 1×10^8 platelets (623,107 cpm); total of 1.45×10^6 endothelial cells.

[c] Total of 15.5×10^8 platelets (9,346,605 cpm); total of 1.45×10^6 endothelial cells.

[d] Total counts in supernate.

[e] Percentage of recovered counts minus phospholipid and solvent front.

[f] Includes a small quantity of PGE_2 that overlapped with TXB_2 on these TLC plates.

doperoxides derived from activated platelets. These include (1) enhanced platelet–endothelial cell proximity as attainable in stirred cell suspensions in aggregometer cuvettes; (2) use of increased endothelial cell–platelet ratios; (3) utilization of arachidonate of high specific activity in the radiolabeling experiments. It was also of interest that when a mixture of platelets and endothelial cells that were not treated with aspirin was stimulated with thrombin, radioimmunoassay showed generation of more than twice as much 6-keto-$PGF_{1\alpha}$ than when endothelial cells were stimulated alone. The results indicated that endothelial cells will utilize platelet endoperoxides for PGI_2 formation whenever such endoperoxides are present in the immediate vicinity of the cell. Results of the experiments described above are summarized on Table II (18).

Experiments were also carried out in the presence of two thromboxane synthetase inhibitors: imidazole and U54701. Inclusion of these compounds in endothelial cell suspension or monolayer experiments resulted in an increase in PGI_2 production, as reported by other investigators (19–21). In each instance endoperoxides synthesized by platelets accumulated in the medium as a consequence of the inhibition of thromboxane synthetase and were utilized by the endothelial cells. These results suggested to us that utilization of platelet en-

TABLE II
TLC–Radiochromatographic Analysis of Products in the Supernate of Combined Suspensions of Radiolabeled Platelets and ASA-Treated Endothelial Cells after Stimulation with Ionophore, Thrombin, and Collagen[a]

	Ionophore		Thrombin		Collagen	
Product	cpm[b]	%[c]	cpm	%	cpm	%
6-Keto-$PGF_{1\alpha}$	1,261	(5.4)	444	(4.1)	60	(3.0)
$PGF_{2\alpha}$	895	(3.9)	356	(3.3)	73	(3.7)
TXB_2	12,997	(56.1)	6,624	(61.5)	1,008	(51.0)
PGE_2	845	(3.6)	247	(2.3)	58	(2.9)
PGD_2	372	(1.6)	111	(1.0)	58	(2.9)
Hydroxy acids	4,209	(18.2)	1,713	(15.9)	451	(22.8)
Arachidonic acid	142	(0.6)	253	(2.3)	126	(6.4)

[a] Total of 1×10^8 platelets (783,678 cpm) = 200,000/μl. Total of 3.1×10^6 endothelial cells = 6,238/μl. Total volume was 0.5 ml. Ionophore (1 μM), thrombin (5 U/ml), collagen (30 μg/ml), Ca^{2+} (3 mM).

[b] Total counts in supernate.

[c] Percentage of recovered counts minus phospholipid and solvent front. Radioactive areas on TLC plates that did not correspond to standards are not included. In all samples counts obtained in the 6-keto-$PGF_{1\alpha}$ areas represented a peak of radioactivity.

doperoxides for PGI_2 synthesis by aspirin-treated endothelial cells would be possible under appropriate experimental conditions, even in the absence of thromboxane synthetase inhibitors. Those conditions have been described earlier.

It should be mentioned that in the presence of albumin the quantities of PGI_2 formed were lower than in its absence. In addition, there was an increase in PGD_2 production when albumin was present. This phenomenon has been reported before and is due to enhanced isomerization of endoperoxides by albumin (20,22). To maximize the sensitivity of the assay system, albumin was not included in the incubation media. However, the role of albumin in experiments involving unesterified fatty acids is a consideration that merits attention in all *in vitro* experimental systems (6,23,24).

STUDIES WITH ACETYL GLYCERYL ETHER PHOSPHORYLCHOLINE (AGEPC), FORMERLY KNOWN AS PLATELET ACTIVATING FACTOR (PAF)

Acetyl glyceryl ether phosphorylcholine (AGEPC, 1-*O*-hexadecyl/octadecyl-2-acetyl-*sn*-glyceryl-3-phosphorylcholine) is a unique phospholipid that mediates allergic and inflammatory reactions. In addition, AGEPC can induce aggregation and release in rabbit platelets (25–28). AGEPC has also been shown to possess antihypertensive properties (29). AGEPC was originally known as platelet activating factor (PAF), but now that its structure has been established (27,29,30), the new terminology is more appropriate. Dr. Donald J. Hanahan generously provided us with chemically pure AGEPC (31). This was free of lyso-AGEPC and other contaminants as shown by thin-layer chromatography. We studied AGEPC in two ways. First, we tested its ability to stimulate human platelets, and, second, we attempted to discern whether stimulated human platelets produced AGEPC using Dr. Hanahan's compound as a standard (32).

When AGEPC was added to platelet-rich plasma (PRP), a concentration-dependent aggregation response occurred which was preceded by a normal shape change. Concentrations of $7.6 \times 10^{-7}\ M$ AGEPC (or higher) induced irreversible aggregation in PRP from 10 of 14 donors. In the remaining four, aggregation took place but was reversible. In some cases irreversible aggregation to AGEPC took place at levels of $3.8 \times 10^{-8}\ M$, attesting to the potency of this stimulus. At higher levels of AGEPC, individual variations in dose–response curves in different donors were noted. Optimal responses to AGEPC were obtained only in the presence of calcium (3 mM). We also observed a synergistic effect between AGEPC and adenosine diphosphate (ADP). Thus when suboptimal quantities of ADP (0.5–2 μM) were added to PRP simultaneously with 1.9×10^{-8} M AGEPC, an aggregation response occurred that was greater than the sum of individual responses to each stimulus as observed by maximal amplitude of the trace (32).

When extracellular ADP was removed from the system with enzymes such as apyrase or creatine phosphate/creatine phosphokinase (CP/CPK), aggregation to AGEPC in PRP was greatly reduced and completely reversible. Shape change and aggregation were entirely abolished by incubation of PRP with prostaglandin E_1 (10 μM). In 6 of 10 donors AGEPC-induced aggregation was completely reversed by preincubation with 1 mM acetylsalicylic acid (ASA). In other donors there was no apparent inhibition by ASA or partial reversal occurred.

In three donors, 3.8×10^{-5} M AGEPC induced 31, 27, and 20% serotonin secretion, respectively. This is to be compared with release obtained with collagen (71%) or sodium arachidonate (64%). Serotonin release was markedly inhibited in the presence of ASA. Release of the platelet secretory proteins, β-thromboglobulin and platelet factor 4, paralleled serotonin release in that it was concentration-dependent and inhibited by ASA. The results we obtained in PRP were similar to those previously reported by McManus and associates (33).

AGEPC-stimulated PRP did not demonstrate a "burst" of oxygen consumption or malondialdehyde formation. This indicated that AGEPC probably did not induce significant arachidonic acid oxygenation. The experiments were extended with the use of radioimmunoassay and radio-thin-layer chromatography. The latter assays indicated that AGEPC was a relatively weak agonist for thromboxane formation in human PRP. In this respect AGEPC was similar to ADP (8 μM), which also does not induce a "burst" of oxygen consumption or malondialdehyde formation.

We added AGEPC to washed platelets in the presence of fibrinogen and calcium, but did not observe an aggregation response. This platelet preparation did aggregate to ADP under the same conditions. It was of interest, however, that when ADP and AGEPC were added simultaneously, an aggregation response took place which was greater than that elicited by ADP alone. Thus, in contrast to rabbit platelets, the washed human platelet system responded minimally, if at all, to AGEPC (32).

We tested stimulated human platelets for their capacity to generate AGEPC in the following manner. Washed platelets were prepared from 8 units of freshly collected PRP. The material was pooled and stimulated with the calcium ionophore A23187 (2.5 μM) in the presence of calcium (1.3 μM). The platelets aggregated, and the supernatant was lipid-extracted and purified by thin-layer chromatography. Material on the plates with an Rf value corresponding to that of AGEPC did induce a definite shape change and aggregation response in PRP. The amplitude of the aggregation response corresponded to that which we obtained with $3.8 \times 10^{-8} M$ AGEPC. We calculated that it represented production of approximately 12.5 ng AGEPC by the eight units of platelets. We also used thrombin as an agonist for up to 10 units of platelets in four experiments. In this case, no aggregation response was elicited by eluates from thin-layer plates that had originated from either the supernatant or pellet of thrombin-stimulated platelets. Thus only small quantities of AGEPC, if any, were generated by stimulated human platelets (32).

In comparison to arachidonic acid, collagen, and ionophore, AGEPC would appear to be a strong stimulus for platelet aggregation, a moderate agonist for release, and a relatively weak inducer of thromboxane production. By the latter criteria AGEPC more closely resembles ADP. We think the initial phase of AGEPC-induced aggregation is independent of ADP release and thromboxane formation since it is not inhibited by aspirin, apyrase, or CP/CPK. Irreversible aggregation to AGEPC always requires ADP release, and thromboxane formation is not essential in each instance. Thus, although human platelets in PRP aggregate in response to exogenous AGEPC, the washed human platelet is not a significant source of this interesting lipid mediator of cellular responses. Nevertheless, AGEPC released from activated leukocytes is potentially capable of interacting with platelets *in vitro* or *in vivo*.

ACKNOWLEDGMENTS

We acknowledge the invaluable assistance of Karen Tack-Goldman and Jean W. Ward. This work was supported by grants from the Veterans Administration, the National Institutes of Health (HL-18828 06 SCOR, HL-15486 and 21006), the New York Heart Association, and the A. R. Krakower Foundation. MJB and KLK are Senior Investigators of the New York Heart Association.

REFERENCES

1. Alfin-Slater, R. B., and Aftergood, L. (1968) *Physiol. Rev.* **48**, 758–784.
2. Guarnieri, M., and Johnson, R. M. (1970) *Adv. Lipid Res.* **8**, 115–174.

3. Mead, J. F., and Fulco, A. J. (1976) *In* "The Unsaturated and Polyunsaturated Fatty Acids in Health and Disease" (I. N. Kugelmass, ed.), pp. 1–191. Thomas, Springfield, Illinois.
4. Marcus, A. J. (1979) *Prog. Hematol.* **11,** 147–171.
5. Marcus, A. J., Ullman, H. L., and Safier, L. B. (1969) *J. Lipid Res.* **10,** 108–114.
6. Lands, W. E. M. (1979) *Annu. Rev. Physiol.* **41,** 633–652.
7. Bills, T. K., Smith, J. B., and Silver, M. J. (1977) *J. Clin. Invest.* **60,** 1–6.
8. Jesse, R. L., and Cohen, P. (1976) *Biochem. J.* **158,** 283–287.
9. Rittenhouse-Simmons, S. (1979) *J. Clin. Invest.* **63,** 580–587.
10. Bell, R. L., Kennerly, D. A., Stanford, N., and Majerus, P. W. (1979) *Proc. Natl. Acad. Sci. U.S.A.* **76,** 3238–3241.
11. Broekman, M. J., Ward, J. W., and Marcus, A. J. (1980) *J. Clin. Invest.* **66,** 275–283.
12. Billah, M. M., Lapetina, E. G., and Cuatrecasas, P. (1981) *J. Biol. Chem.* **256,** 5399–5403.
13. Broekman, M. J., Ward, J. W., and Marcus, A. J. (1981) *J. Biol. Chem.* **256,** 8271–8274.
14. Bressler, N. M., Broekman, M. J., and Marcus, A. J. (1979) *Blood* **53,** 167–178.
15. Rittenhouse-Simmons, S. (1981) *J. Biol. Chem.* **256,** 4153–4155.
16. Weksler, B. B., Marcus, A. J., and Jaffe, E. A. (1977) *Proc. Natl. Acad. Sci. U.S.A.* **74,** 3922–3926.
17. Marcus, A. J., Weksler, B. B., and Jaffe, E. A. (1978) *J. Biol. Chem.* **253,** 7138–7141.
18. Marcus, A. J., Weksler, B. B., Jaffe, E. A., and Broekman, M. J. (1980) *J. Clin. Invest.* **66,** 979–986.
19. Moncada, S., Bunting, S., Mullane, K., Thorogood, P., Vane, J. R., Raz, A., and Needleman, P. (1977) *Prostaglandins* **13,** 611–618.
20. Needleman, P., Wyche, A., and Raz, A. (1979) *J. Clin. Invest.* **63,** 345–349.
21. Baenziger, N. L., Becherer, P. R., and Majerus, P. W. (1979) *Cell* **16,** 967–974.
22. Hamberg, M., and Fredholm, B. B. (1976) *Biochim. Biophys. Acta* **431,** 189–193.
23. Spector, A. A. (1975) *J. Lipid Res.* **16,** 165–179.
24. Nordøy, A., Svensson, B., and Hoak, J. C. (1979) *Eur. J. Clin. Invest.* **9,** 5–10.
25. Henson, P. M., and Pinckard, R. N. (1977) *Monogr. Allergy* **12,** 13–26.
26. Benveniste, J., Camussi, J., and Polonsky, J. (1977) *Monogr. Allergy* **12,** 138–142.
27. Hanahan, D. J., Demopoulos, C. A., Liehr, J., and Pinckard, R. N. (1980) *J. Biol. Chem.* **255,** 5514–5516.
28. Pinckard, R. N., McManus, L. M., Demopoulos, C. A., Halonen, M., Clark, P. O., Shaw, J. O., Kniker, W. T., and Hanahan, D. J. (1980) *J. Reticuloendothel. Soc.* **28,** 95s–103s.
29. Blank, M. L., Snyder, F., Byers, L. W., Brooks, B., and Muirhead, E. E. (1979) *Biochem. Biophys. Res. Commun.* **90,** 1194–1200.
30. Benveniste, J., Tencé, M., Varenne, P., Bidault, J., Boullet, C., and Polonsky, J. (1979) *C. R. Hebd. Seances Acad. Sci.* **289,** 1037–1040.
31. Demopoulos, C. A., Pinckard, R. N., and Hanahan, D. J. (1979) *J. Biol. Chem.* **254,** 9355–9358.
32. Marcus, A. J., Safier, L. B., Ullman, H. L., Wong, K. T. H., Broekman, M. J., Weksler, B. B., and Kaplan, K. L. (1981) *Blood* **58,** 1027–1031.
33. McManus, L. M., Hanahan, D. J., and Pinckard, R. N. (1981) *J. Clin. Invest.* **67,** 903–906.

Role of Injury in Arterial Lipid Metabolism

C. RICHARD MINICK, DAVID P. HAJJAR, AND DOMENICK J. FALCONE
Department of Pathology
The New York Hospital–Cornell Medical Center
New York, New York

The hallmark of atherosclerosis is the fibrous plaque. These intimal thickenings or plaques are characterized by fibrocellular caps overlying deposits of extracellular and intracellular lipid. Such fibrous plaques are believed to result from proliferation and migration of medial smooth muscle cells into the intima, accumulation of connective tissue, and deposition of lipids primarily derived from the blood. Despite many years of research, the mechanism(s) involved in the formation of these plaques is unclear. Attempts to explain the pathogenesis of atheromatous plaques have resulted in several theories or hypotheses. These include the lipid filtration hypothesis, the thrombogenic theory, and the arterial injury and monoclonal hypotheses.

Epidemiologic, clinicopathologic, and experimental evidence that increased plasma lipid concentrations are important in the pathogenesis of atherosclerosis has led to the lipid filtration hypothesis, which attributes lipid accumulation to a gradient between increased plasma lipoproteins and those in the arterial wall (1). According to this theory, lipid filtration exceeds efflux and metabolism, leading initially to lipid accumulation and secondarily to smooth muscle proliferation and connective tissue deposition. The thrombogenic theory, a modification of the encrustation hypothesis of von Rokitansky, suggests that fibrous plaques arise from organized thrombi that form at sites of vascular injury (2). Others believe that various stimuli of a diverse nature may lead to arterial injury and that this injury and local reactive changes may stimulate intimal thickening and predispose to lipid accumulation in the arterial wall (3). Finally, the recent finding of Benditt and Benditt (4) that human fibrous plaques are monotypic with

PATHOBIOLOGY OF THE ENDOTHELIAL CELL

ISBN 0-12-521980-6

respect to isozymes of X-linked glucose 6-phosphate dehydrogenase led these investigators to suggest that these lesions are monoclonal and analogous to benign neoplasms.

Although several lines of evidence indicate the importance of increased plasma lipid concentrations in the pathogenesis of atherosclerosis and support the lipid filtration hypothesis, this theory does not explain the increased involvement of particular portions of the arterial system or the predisposition to atherosclerosis in specific segments of these arteries. The thrombogenic theory can account for the segmental occurrence of fibrous plaques. Also, some areas of intimal thickening and lipid deposition in man and experimental animals may arise in association with thrombi. However, the evidence that a significant number of plaques evolve from thrombi is not compelling. Also, the type of lipid accumulating in plaques is like that in plasma and is different than that which would be anticipated to arise from degenerating thrombi. Even though the possibility that cells within plaques represent transformed cells is intriguing, especially with regard to environmental considerations, the significance of this hypothesis to human and experimentally induced atherosclerosis remains to be elucidated.

The concept that arterial injury may act synergistically with hyperlipemia and lead to atherosclerosis is consistent with much of the experimental and clinicopathologic data (5). Further, this hypothesis may incorporate features of the other theories. For example, there is experimental evidence to indicate that hyperlipemia may be in itself a source of arterial injury, and the intrinsic pathway of coagulation as evidenced by adherence of blood platelets is believed to be important in the response of arteries to injury (3,6) leading to subsequent intimal thickening. Finally, it is conceivable that the monotypic atheromata described by Benditt and others could arise from discrete foci of injury as a result of a selective advantage of certain proliferating cells.

The morphological features of the fibrous plaque suggest that the pathogenesis of arteriosclerosis involves at least two major processes at the level of the arterial wall. One is the proliferation and migration of smooth muscle cells that leads to intimal thickening and deposition of connective tissue, and the other is the increased influx of plasma lipids at the site of injury believed to result in lipid accumulation. According to the injury hypothesis, as recently restated by Ross and colleagues, endothelium is the initial or primary site of damage in arterial injury resulting from immunologic, chemical, or physical stimuli (3). Absence of endothelium will lead to disruption of two of its important functions, i.e., nonthrombogenicity of the vascular surface and loss of the endothelial barrier function (7). Thus, if sufficiently severe or prolonged, endothelial injury may lead to endothelial denudation resulting in the adherence of platelets and blood leukocytes to the surface of the denuded arterial wall. Mitogens contained in platelets (3), monocytes (8), regenerating endothelial cells (9), low-density lipoproteins (10), and other blood components may then result in proliferation of medial smooth muscle cells and formation of connective tissue leading to intimal thickening. The importance of platelets in the resulting intimal proliferation is supported by the finding that intimal proliferation following mechanical arterial injury is abolished or significantly reduced by thrombocytopenia (11). In conclusion, results of experiments *in vivo* and *in vitro* indicate that arterial injury and the reaction of various blood components with the arterial wall can account for the proliferative component of the fibrous plaque.

Clinicopathologic observations in humans and results of numerous experiments in animals indicate that arterial injury may favor lipid deposition in the injured arterial wall (5). Even in the absence of experimental evidence directly bearing on this observation, it is widely believed that this lipid accretion is a result of disruption of the barrier function of the endothelium allowing increased transport of lipid and lipoprotein into the arterial wall. This hypothesis may not account for many instances of increased lipid accumulation, e.g., increased accumulation in segments of intimal thickening far removed in time from the initial injurious stimulus (12). Further, this hypothesis gives little consideration to the possible contribution of morphological and metabolic perturbations in the neointima of the injured arterial wall.

Having previously shown that one form of arterial injury, immunologic injury, is associated with endothelial injury and can lead to lipid accumulation (13), we next designed experiments to further study the mechanism of lipid accumulation in these damaged arteries. In particular, we tested the hypothesis that persistent absence of endothelium leads to enhanced lipid accumulation as a result of increased infiltration of lipoprotein into deendothelialized arterial wall. In these experiments we used a modification of the technique for inducing mechanical injury by the balloon catheter as described by Baumgartner (14). Specifically, we designed experiments to determine whether or not lipid accumulation would be enhanced in areas of neointima lacking endothelium as compared to adjacent reendothelialized arterial lesions (Figs. 1 and 8).

Results of morphometric and chemical analyses of arterial lesions in both normocholesterolemic and hypercholesterolemic rabbits indicated (1) that arterial injury led to enhanced accumulation of intimal

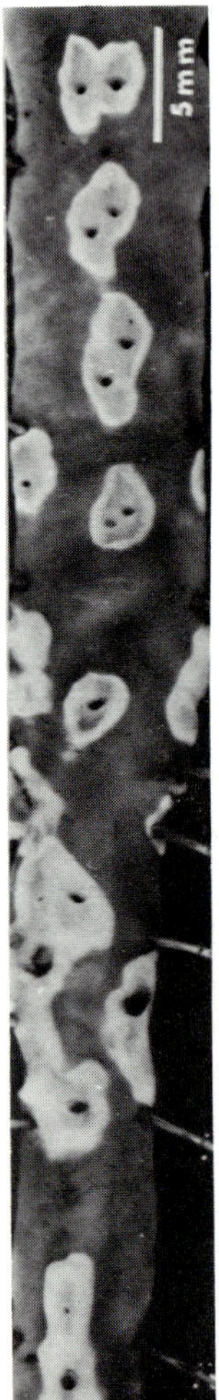
5 mm

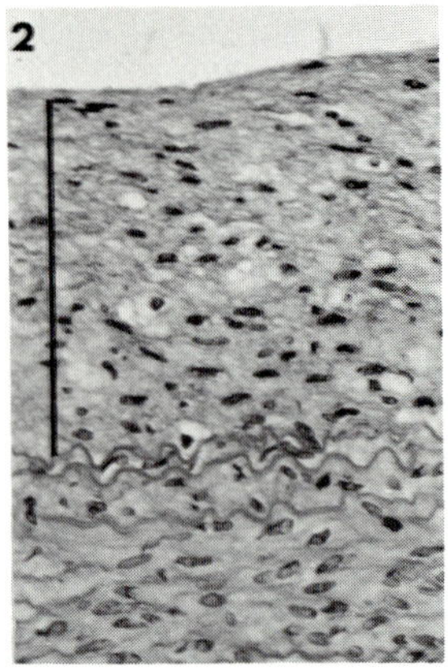
2

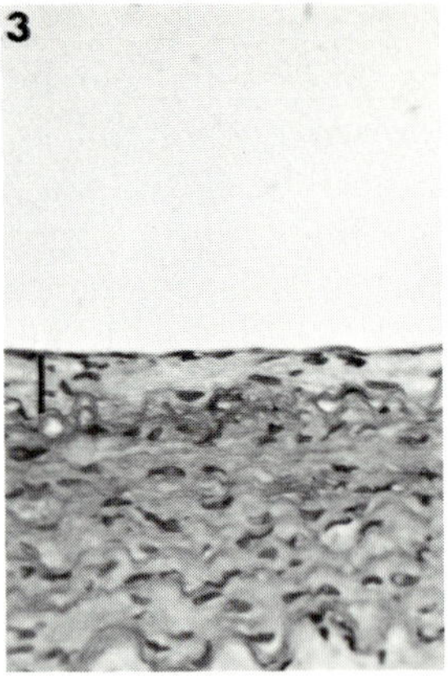
3

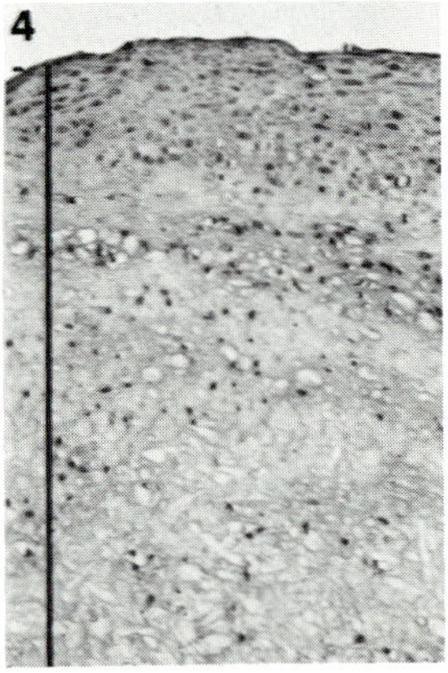
4

5

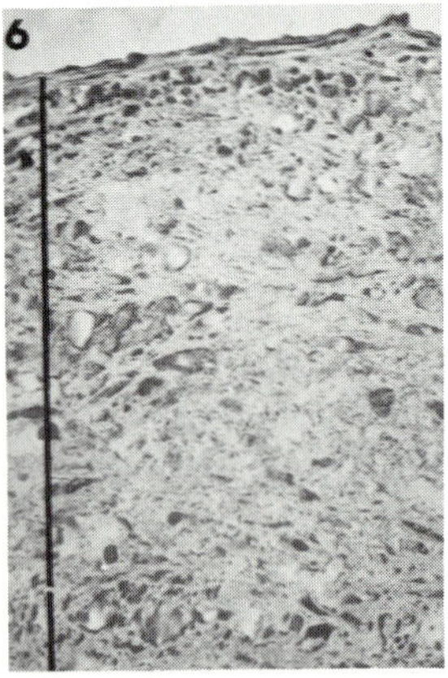
6

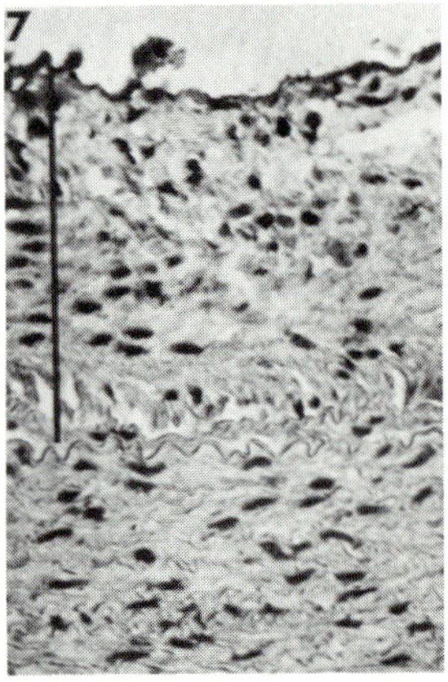
7

lipid, in particular cholesterol and cholesteryl ester and (2) that lipid accumulation was accentuated in the intima underlying regenerated endothelium and not in the deendothelialized neointima as we had anticipated (Tables I and II), (Figs. 2–7) (5,15). Results also indicated that the intima of the injured aorta was significantly thicker in the reendothelialized than in the deendothelialized areas and that hypercholesterolemia enhanced the intimal thickening in the reendothelialized but not in the deendothelialized aortas (Table III). Finally, the findings indicated that in hypercholesterolemic rabbits the total aorta cholesterol content correlated with the serum cholesterol concentration in the reendothelialized areas of the injured aorta but

---

Fig. 1. Luminal surface of thoracic and abdominal segments of rabbit aorta deendothelialized 76 days prior to killing. Evans blue injected 40 min prior to killing. Deendothelialized aorta, stained blue, surrounds unstained islands of reendothelialized aorta. Each endothelial island surrounds the ostium of a branch and consists of a central gray zone surrounded by an intensely white ringlike zone (×2.9). From Minick *et al.* (5) with permission of the *American Journal of Pathology.*

Fig. 2. Intima and media of reendothelialized island from abdominal segment of rabbit aorta deendothelialized 15 weeks prior to killing. Rabbit was fed a diet low in lipid and had normal serum cholesterol concentration. Note thickened intima with occasional fat-containing cells (vertical bar indicates thickness of intima). Compare intimal thickness and lipid accumulation with adjacent deendothelialized area illustrated in Fig. 3. (Hematoxylin and eosin, ×160). From Minick *et al.* (5) with permission of the *American Journal of Pathology.*

Fig. 3. Intima and media of deendothelialized area of aorta adjacent to reendothelialized island illustrated in Fig. 2 (vertical bar indicates thickness of intima). Note that intima of reendothelialized area (Fig. 2) is approximately four times thicker than the intima of deendothelialized area and contains appreciably more vacuolated fat-containing cells (H&E, ×160). From Minick *et al.* (5) with permission of the *American Journal of Pathology.*

Fig. 4, 5. Reendothelialized area (Fig. 4) and deendothelialized area (Fig. 5) of rabbit fed cholesterol-supplemented diet. Note thicker intima and atheromatous change in intima of reendothelialized areas compared with adjacent de-endothelialized area (vertical bars indicate thickness of intima) (H&E, Fig. 4: ×80; Figure 6: ×180). From Minick *et al.* (5) with permission of the *American Journal of Pathology.*

Figs. 6, 7. Reendothelialized area (Fig. 6) and adjacent deendothelialized area (Fig. 7) from aorta of rabbit fed cholesterol-supplemented diet. Note large quantities of lipid, staining black, in intima of reendothelialized area compared with adjacent deendothelialized area (vertical bars indicate thickness of intima) (oil red 0; Fig. 6: ×165, Fig. 7: ×182). From Minick *et al.* (5) with permission of the *American Journal of Pathology.*

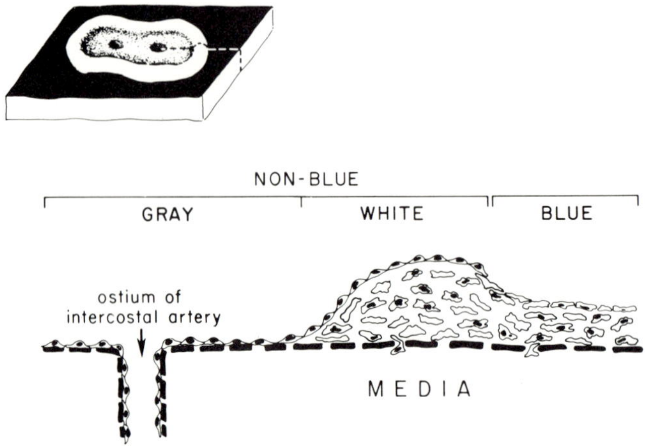

Fig. 8. Schematic representation of a cross section of a nonblue reendothelialized island and surrounding deendothelialized blue area in thoracic segment of aorta. Each reendothelialized island consists of a central gray zone exhibiting little intimal thickening and a more peripheral intensely white zone with thickened intima. The nonblue island is surrounded by a blue area characterized by thickened intima without an endothelial lining. From Falcone *et al.* (15) with permission of the *American Journal of Pathology*.

not in the deendothelialized areas over a range of serum cholesterol concentrations of approximately 900 mg/dl (Table IV). Thus, under the conditions of these experiments, atherosclerosis as defined by extent of intimal thickening and arterial lipid accumulation is most extensive in the reendothelialized aorta and not in the deendothelialized aorta, as

TABLE I
Comparison of Intimal Lipid Accumulation in Deendothelialized (DeA) and Reendothelialized (ReA) Aorta of Rabbits Fed a Lipid-Supplemented Diet*,†,‡

Area	Distribution of grades of lipid accumulation [no. (%)]			
	None	Slight	Moderate	Marked
DeA	8(18%)	17(39%)	19(43%)	0(0%)
ReA	0(0%)	0(0%)	18(41%)	26(59%)

* $\chi^2_3 = 51.03$; $p < .001$.

† For each area, $N = 44$. This includes 22 thoracic segments and 22 abdominal segments.

‡ Modified from Minick *et al.* (5) with permission of the *American Journal of Pathology*.

TABLE II
Lipid Content of Injured and Uninjured Aortas*

	Injured aortas			
	Cholesterol		Cholesteryl esters	
Diet	ReA†	DeA	ReA	DeA
Lipid-poor	0.73 ± 0.20‡	0.65 ± 0.18	1.02 ± 0.30^{af}	0.43 ± 0.11^{a}
Lipid-supplemented	3.04 ± 0.62^{bd}	1.32 ± 0.20^{b}	5.43 ± 1.20^{cg}	2.25 ± 0.57^{c}

	Uninjured aortas	
Diet	Cholesterol	Cholesteryl esters
Lipid-poor	0.47 ± 0.12^{e}	0.12 ± 0.05^{f}
Lipid-supplemented	1.20 ± 0.19^{de}	0.40 ± 0.09^{g}

* Modified from Falcone *et al.* (15) with permission of the *American Journal of Pathology*.

† ReA, reendothelialized aortas; DeA, deendothelialized aortas.

‡ Mean ± SE; μg lipid/mg wet tissue weight. Nine separate replicates were performed in each group analyzed. Values with the same corresponding superscript letter are significantly different ($p < .05$).

TABLE III
Thickness of Intima in Deendothelialized (DeA) and Reendothelialized (ReA) Aorta*

	Group I Lipid-poor diet (av. serum cholesterol = 42 mg/dl) (N = 12)		Group II Cholesterol-supplemented diet (av. serum cholesterol = 721 mg/dl) (N = 22)	
	DeA†	ReA	DeA	ReA
Thoracic segment				
Thickness of intima (μm)	61.2 ± 10.1‡^{a}	106.3 ± 12.0^{aeg}	75.1 ± 12.9^{c}	198.9 ± 29.2^{cfg}
Intimal/medial ratio	0.33 ± 0.06§^{i}	0.65 ± 0.10^{im}	0.37 ± 0.06^{k}	1.1 ± 0.20^{km}
Abdominal segment				
Thickness of intima (μm)	73.2 ± 9.8^{b}	169.9 ± 12.2^{beh}	89.5 ± 8.4^{d}	264.7 ± 19.9^{dfh}
Intimal/medial ratio	0.61 ± 0.12^{j}	1.29 ± 0.19^{jn}	0.92 ± 0.12^{l}	2.90 ± 0.51^{ln}

* Modified from Minick *et al.* (5) with permission of the *American Journal of Pathology*.

† DeA, deendothelialized aortas; ReA, reendothelialized aortas.

‡ Mean maximal thickness ± SEM.

§ Mean intimal/medial ratio ± SEM. Values with the same corresponding superscript letter are significantly different ($p < .05$).

TABLE IV
Correlation of Aortic Cholesterol with Serum Cholesterol Concentration in Injured Aortas of Rabbits Fed Lipid-Supplemented Diet*

	Aortic segment	
Area	Thoracic	Abdominal
ReA†	0.62	0.71
	$0.01 < p < .025$	$p < .005$
DeA	0.05	0.22
	n.s.	n.s.

* Modified from Falcone *et al.* (15) with permission of the American Journal of Pathology.
† ReA, re-endothelialized aortas; DeA = de-endothelialized aortas; n.s., not significant.

we had anticipated. These findings suggest that arterial lipid accumulation is not simply a result of increased filtration of lipoprotein across a deendothelialized intimal surface.

We questioned whether the accentuated lipid accumulation in reendothelialized aorta could result from (1) increased lipoprotein transport by regenerated endothelium; (2) alteration of the metabolism and the structure of the extracellular matrix, resulting in increased accumulation of extracellular lipid; or (3) alterations of cell metabolism, resulting in intracellular lipid accumulation.

Although an interesting theory proposed by Zilversmit (16) has suggested the possibility that endothelium may lead to increased formation of chylomicron remnants and perhaps increased transport of these remnants to underlying smooth muscle cells, we know of no direct evidence to indicate that endothelium enhances the transport of lipid and/or lipoprotein. In recent experiments, we assessed the accumulation of radiolabeled apo-B-lipoproteins in the reendothelialized and deendothelialized areas of the injured aortic wall over time (17). Initially, the concentration of radiolabeled lipoprotein was increased in the deendothelialized as compared to the reendothelialized aorta. However, in contrast to the situation in the deendothelialized aorta, where the concentration of radiolabeled apo-B-lipoprotein was seen to diminish with decreasing plasma concentration of labeled lipoprotein, the concentration in the reendothelialized aortic wall continued to increase. We interpret these results to indicate that apo-B-lipoproteins are retained preferentially in the reendothelialized areas and are no longer in equilibrium with the plasma. Support for this interpretation

TABLE V
Activity of Cell Marker Enzymes in Aortic Homogenates*

Enzyme	Specific activity (mU/mg DNA)					
	Injured aortas				Uninjured aortas	
	Lipid-poor diet		Lipid-supplemented diet		Lipid-poor diet	Lipid-supplemented diet
	DeA†	ReA	DeA	ReA		
N-Acetyl-β-glucosaminidase	72.4 ± 9.9‡[a]	55.3 ± 6.8[a]	67.0 ± 6.1[b]	53.6 ± 4.1[b]	61.7 ± 4.9	47.5 ± 5.2
β-Galactosidase	43.5 ± 6.9[gcd]	27.0 ± 5.1[c]	35.7 ± 5.0[e]	34.3 ± 4.2	21.7 ± 2.3[d]	18.6 ± 2.0[e]
Neutral-α-glucosidase	36.8 ± 7.4[f]	25.2 ± 4.3[f]	30.6 ± 1.9	26.4 ± 5.6	29.6 ± 2.0	24.5 ± 1.7
Cytochrome oxidase	473.7 ± 34.9[g]	313.2 ± 18.7[g]	616.2 ± 75.8[gh]	462.5 ± 34.0[i]	202.9 ± 9.7[g]	258.3 ± 23.9[ghi]
Catalase	227.6 ± 94.5[j]	163.7 ± 75.6[jk]	156.5 ± 22.1[l]	91.5 ± 8.4[l]	188.9 ± 21.5[km]	118.0 ± 15.5[m]
5′-Nucleotidase	8369 ± 1305[n]	5659 ± 897[no]	7425 ± 2203	5016 ± 1409[p]	9832 ± 1504[o]	10289 ± 1394[p]

* Modified from Hajjar *et al.* (21) with permission of the American Journal of Pathology.

† DeA, deendothelialized aortas; ReA, reendothelialized aortas.

‡ Data are presented as means ± SE. Ten separate replicates were performed in each group analyzed. Values with the same corresponding superscript letter are significantly different ($p < .05$).

comes from experiments in which we demonstrated by both morphometric and chemical analysis that reendothelialized neointima contains increased quantities of certain glycosaminoglycans (18,19). This latter finding may explain the enhanced lipoprotein accumulation in reendothelialized areas, since glycosaminoglycans have been shown to bind to low-density lipoproteins (20). Taken together, we interpret results of these experiments to indicate that endothelial loss or damage may initially lead to increased filtration of lipoprotein in the arterial wall. However, sustained increases in extracellular lipoprotein appear to result from alterations in matrix proteins, proteoglycans, even in the presence of an intact endothelium.

In addition to extracellular lipid, the quantities of intracellular lipid are also enhanced in the reendothelialized injured aortas. In order to investigate the pathogenesis of this intracellular lipid accumulation, we have recently investigated the effect of injury on cellular metabolism in the arterial wall and the effect of endothelium on the alterations in metabolism induced by injury. We assessed and compared the activities of several marker enzymes for cell organelles in homogenates

TABLE VI
Cholesteryl Ester Metabolism in Injured and Uninjured Aortas*

	Injured aortas			
	Acid cholesteryl esterase activity		Acyl-CoA : cholesterol acyltransferase activity	
Diet	ReA†	DeA	ReA	DeA
Lipid-poor	6.7 ± 1.1‡^{ad}	13.5 ± 1.6^{ac}	1.1 ± 0.1^{h}	1.1 ± 0.2^{g}
Lipid-supplemented	3.1 ± 0.5^{bd}	6.3 ± 1.2^{bce}	1.0 ± 0.2	1.1 ± 0.1^{i}

	Uninjured aortas	
Diet	Acid cholesteryl esterase activity	Acyl : CoA : cholesterol acyltransferase activity
Lipid-poor	0.5 ± 0.04^{af}	0.3 ± 0.02^{ghj}
Lipid-supplemented	2.0 ± 0.2^{ef}	0.7 ± 0.1^{ij}

* Modified from Hajjar *et al.* (21) with permission of the American Journal of Pathology.

† ReA, reendothelialized aortas; DeA, deendothelialized aortas.

‡ Specific activity (mU Act./mg DNA) ± SEM. Ten separate replicates were performed in each group analyzed. Values with the same corresponding superscript letter are significantly different ($p < .05$).

of reendothelialized and deendothelialized areas of injured aortas of normocholesterolemic and hypercholesterolemic rabbits as well as in uninjured aortas of similar rabbits (21). Enzyme activities analyzed included the major microsomal enzyme responsible for cholesteryl ester synthesis, acyl-CoA: cholesterol acyltransferase and a lysosomal enzyme responsible for cholesteryl ester degradation, acid cholesteryl ester hydrolase. Results indicate that arterial injury leads to alteration in activities of cellular enzymes in the aortic wall, particularly enzymes responsible for synthesis and degradation of cholesteryl esters, and that endothelium modulates these activities (Tables V and VI).

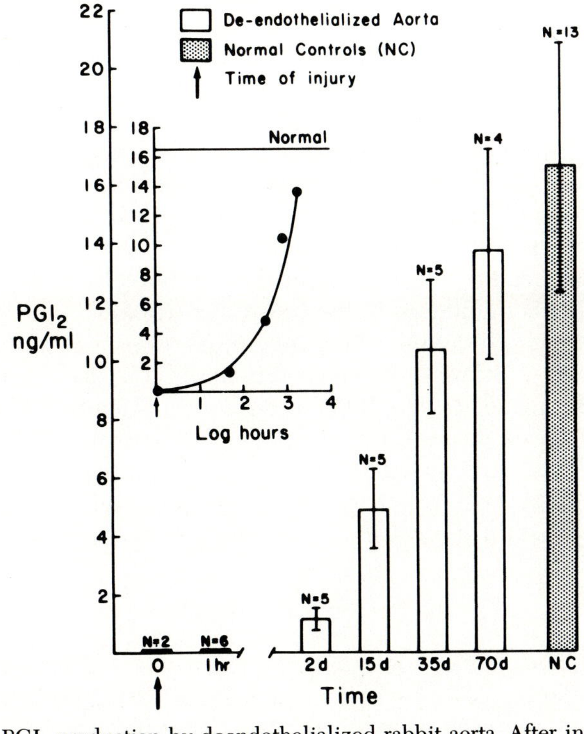

Fig. 9. PGI_2 production by deendothelialized rabbit aorta. After injury, PGI_2 synthesis is markedly decreased as compared with normal aorta. The deendothelialized aortas synthesize increased quantities of PGI_2 with time after injury. PGI_2 production expressed on a continuous time scale (inset) demonstrates an exponential relationship. At 35 and 70 days after injury, smooth muscle cells in the neointima synthesize amounts of PGI_2 comparable to normal artery. For deendothelialized aorta (open bar) at time 0, $N = 2$; 1 hr, $N = 6$; 2 days, $N = 5$; 4 days, $N = 4$; 15 days, $N = 5$; 35 days, $N = 5$; 7 days, $N = 4$; and for normal controls (NC, shaded bar) $N = 13$. ↑, time of injury. From Eldor *et al.* (22) with permission of the *Journal of Clinical Investigation.*

Especially pertinent was a relative decrease in reendothelialized areas as compared to deendothelialized areas of acid cholesteryl ester hydrolase activity. Thus, reendothelialized aorta has a decreased ability to degrade cholesteryl ester even in the presence of increased accumulation of cholesteryl ester-rich lipoprotein. We believe that these findings may at least in part account for the increased lipid accumulation in the reendothelialized areas.

Finally, stimulated by findings that neointimal smooth muscle cells in the deendothelialized aorta can be stimulated to synthesize increased quantities of prostacyclin (Fig. 9) (22), we reasoned that increased prostacyclin or its end product may have effects on acid cholesteryl ester hydrolase activity. In order to test this hypothesis, we investigated the effect of prostacyclin on lysosomal cholesteryl ester hydrolase activity and cholesterol content of cultured rabbit aortic smooth muscle cells. Results of these latter experiments indicate that nanomolar concentrations of prostacyclin and 6-keto-$PGF_{1\alpha}$ and 6-keto-PGE_1 result in a fourfold increase in the activity of cholesteryl ester hydrolase and lead to a decrease in cholesterol content of cultured aortic smooth muscle cells (23). Further, results of these experiments suggest that these increases may be mediated by cyclic AMP.

In summary, results of experiments designed to elucidate the mechanism of lipid accumulation in the injured aortic wall indicate (1) that arterial lipid accumulation following injury does not simply result from increased infiltration of lipoprotein across the deendothelialized surface of an injured artery, (2) that arterial injury leads to alteration in the metabolism of the arterial wall and endothelium modifies these changes; (3) that lipid accumulation in the injured aortic wall is a complex interaction between filtration of lipoprotein and metabolic perturbations within the wall.

ACKNOWLEDGMENTS

Research support was provided by Grants HL-18828 and HL-07423 from the National Heart, Lung and Blood Institute of the National Institutes of Health, a grant from The Cross Foundation, and an Investigatorship for Dr. Hajjar from the New York Heart Association. Dr. Hajjar is also the recipient of a Teacher-Scientist Award from the Andrew W. Mellon Foundation. We gratefully acknowledge the assistance of Ms. Carol Ibsen and Ms. Joanne Gessula in the preparation of this manuscript.

REFERENCES

1. Anitschkow, N. N. (1933) *In* "Arteriosclerosis: A Survey of the Problem" (E. V. Cowdry, ed.), pp. 271–322. Macmillan, New York.

2. Duguid J. B. (1946) *J. Pathol. Bacteriol.* **58,** 207–212.
3. Ross, R., and Glomset, J. A. (1976) *N. Engl. J. Med.* **295,** 369–377, 420–427.
4. Benditt, E. P., and Benditt, J. M. (1973) *Proc. Natl. Acad. Sci. U.S.A.* **70,** 1753–1756.
5. Minick, C. R., Stemerman, M. B., and Insull, W., Jr. (1979) *Am. J. Pathol.* **95,** 131–158.
6. Ross, R., and Harker, L. (1976) *Science* **193,** 1094–1100.
7. Gimbrone, M. A., Jr. (1980) *Atherosclerosis (Berlin)* **5,** 415–426.
8. Martin, B. M., Gimbrone, M. A. Jr., Unanue, E. R., and Cotran, R. S. (1981) *J. Immunol* (in press).
9. Gajdusek, C., Di Corleto, P., Ross, R., and Schwartz, S. (1980) *J. Cell. Biol.* **85,** 467–472.
10. Fisher-Dzoga, K., Chen, R., and Wissler, R. W. (1974) *Adv. Exp. Med. Biol.* **43,** 299–311.
11. Friedman, R. J., Stemerman, M. B., Wenz, B., Moore, S., Gauldie, J., Gent, M., Tiell, M. L., and Spaet, T. H. (1977) *J. Clin. Invest.* **60,** 1191–1201.
12. Hardin, N. J., Minick, C. R., and Murphy, G. E. (1973) *Am. J. Pathol.* **73,** 301–326.
13. Minick, C. R., Alonso, D. R., and Rankin, L. (1978) *Thromb. Haemostasis* **39,** 304–311.
14. Baumgartner, H. R., Stemerman, M. B., and Spaet, T. H. (1971) *Experientia* **27,** 283–285.
15. Falcone, D. J., Hajjar, D. P., and Minick, C. R. (1980) *Am. J. Pathol.* **99,** 81–104.
16. Zilversmit, D. B. (1979) *Circulation* **60,** 473–485.
17. Falcone, D. J., Hajjar, D. P., and Minick, C. R. (1981) *Fed. Proc., Fed. Am. Soc. Exp. Biol.* **40,** 328.
18. Minick, C. R., Litrenta, M. M., Alonso, D. R., Silane, M. F., and Stemerman, M. B. (1977) *Prog. Biochem. Pharmacol.* **14,** 115–122.
19. Wight, T. N., Curwen, K. D., Homan, W. P., and Minick, C. R. (1979) *Fed. Proc., Fed. Am. Soc. Exp. Biol.* **39,** 1075.
20. Iverius, P. H. (1972) *J. Biol. Chem.* **247,** 2607–2613.
21. Hajjar, D. P., Falcone, D. J., Fowler, S., and Minick, C. R. (1981) *Am. J. Pathol.* **102,** 28–39.
22. Eldor, A., Falcone, D. J., Hajjar, D. P., Minick, C. R., and Weksler, B. B. (1981) *J. Clin. Invest.* **67,** 735–741.
23. Hajjar, D. P., Weksler, B. B., Falcone, D. J., and Minick, C. R. (1981) *Fed. Proc., Fed. Am. Soc. Exp. Biol.* **40,** 351.

PART VI

ENDOTHELIAL CELL-CELL INTERACTIONS

Interactions of Platelets with Endothelial Cells

KAREN L. KAPLAN
Department of Medicine
College of Physicians and Surgeons
Columbia University
New York, New York

INTRODUCTION

In the last 20 years there has been considerable interest in the study of platelet interactions with subendothelium in the initiation of thrombosis and atherosclerosis (1–4) but there has been less interest in the study of the interactions of platelets with endothelial cells, perhaps because it was widely held that platelets do not interact with intact endothelial cells (3). There is evidence, however, that vascular integrity is better preserved in the presence of normal numbers of platelets than in thrombocytopenic states (5,6). As an example of this phenomenon, Fig. 1 (7) demonstrates normal capillary endothelium, while Fig. 2 (7) demonstrates capillaries from animals made thrombocytopenic with antiplatelet antiserum (platelet counts <30,000). These capillaries demonstrate marked thinning, with vesicles that bridge the endothelial thickness and regions of marked attenuation of the capillary wall. Similar changes were found in capillaries from animals made thrombocytopenic with Busulfan, suggesting that the endothelial changes were not due to the antiserum directly affecting the cells. In addition to these morphological changes in the endothelium with thrombocytopenia, there is evidence of altered function. Thorotrast has been reported to cross endothelium from thrombocytopenic, but not from control animals (7). Others have reported that passage of carbon particles across endothelium is dependent on thrombocytopenia (8). The mechanism of the effects of platelets on endothelial cell structure and function is unknown, but it is possible that there is a platelet product that has a trophic effect on endothelial cells. Platelets trans-

PATHOBIOLOGY OF THE ENDOTHELIAL CELL

ISBN 0-12-521980-6

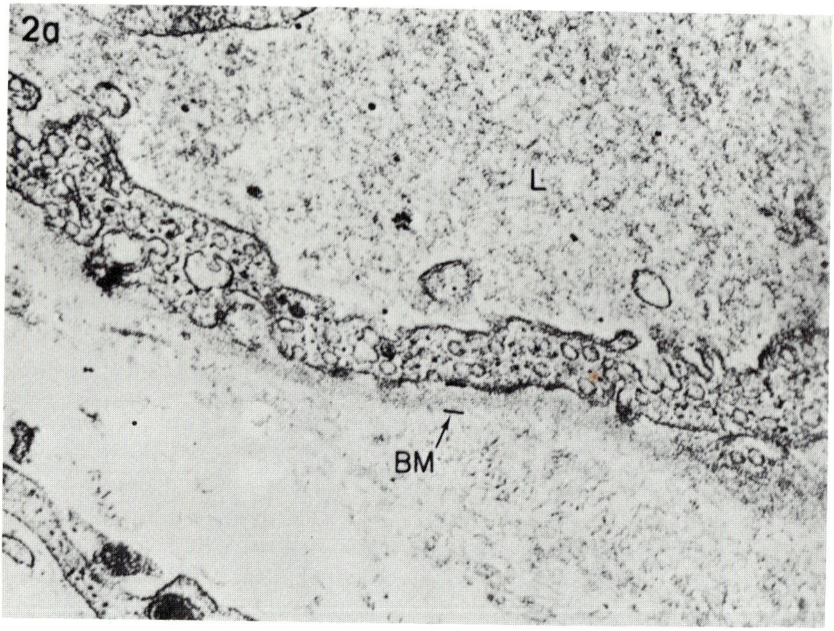

Fig. 1. Normal endothelium. Thickness 1500–2500 Å. BM = basement membrane. L = lumen (contains precipitated plasma protein). ×45000. [Reproduced by permission from Kitchens and Weiss (7).]

fused into thrombocytopenic animals have been reported to be phagocytosed rapidly by the endothelial cells (9) and restoration of a normal platelet count leads to a return of normal morphology (7,10), suggesting that a platelet product is important in restoration of endothelial integrity.

In addition to the studies in whole animals or perfused vessels cited earlier, the effects of platelets on endothelial cells in culture have been examined. Maca and colleagues (10) demonstrated that first-passage endothelial cells cultured in the presence of platelets are more dispersed and that the cells are more elongated than are cells cultured in the absence of platelets. This effect was seen with 10,000–100,000 platelets per microliter of culture medium. Migration of cells into the intercellular spaces was evident at 8 hr of culture and was maximal by 24 hr. No increase in DNA synthesis or in cell number was evident during the 24 hr of culture. Similarly, when a scratched culture was incubated with medium containing platelets the wound was closed by

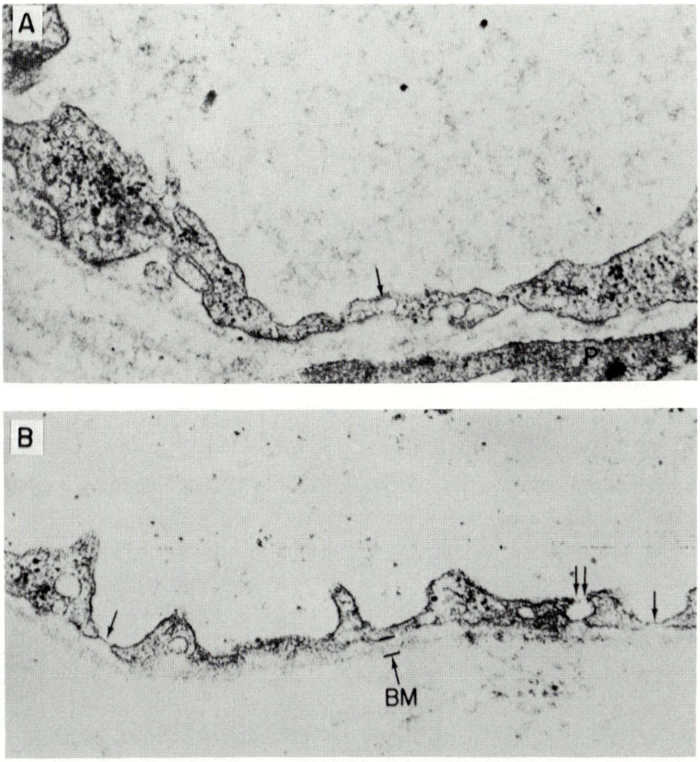

Fig. 2. Capillaries from animals made thrombocytopenic with antiplatelet antiserum. (A) Endothelium attenuated to 700 Å, allowing a vesicle (arrow) to bridge its thickness almost entirely. P = pericyte. (B) Endothelium is bridged by a vesicle (double arrow) and two fenestrations with diaphragms (arrows) are observed. Basement membrane (BM) is thinned but otherwise normal. [Reproduced by permission, from Kitchens and Weiss (7).]

migration of cells into the defect in 24–48 hours, while in the absence of platelets there was little change in the width of the wound. Leukocytes, erythrocytes, and platelets lysed by sonication or freezing were unable to substitute for intact platelets. Endothelial cells cultured with platelets for 72 hr showed an increase of 2.5- to 3.5-fold in DNA synthesis as compared with cells cultured without platelets; cell number was increased by a factor of 1.70–1.84 after 6 days of culture with platelets. Additionally, these authors noted that phagocytosis of platelets occurred after 72 hr of coculture of endothelial cells and platelets.

ADHESION OF PLATELETS TO CULTURED CELLS

Adhesion of platelets to cultured cells was studied by Wechezak *et al.* using scanning electron microscopy (11). Platelets adhered poorly to a plastic dish rinsed with platelet rich plasma, but showed moderate adhesion to a fibroblast layer. Platelets did not adhere to endothelial cells, although a single platelet adherent at an intercellular junction was seen. After retraction of the endothelial cells by versene or cold, platelets were found to adhere to exposed microfilaments. When cellular retraction was induced by trypsin, adherent platelets were not seen, apparently because of digestion of microfilaments by the trypsin. Thus platelets did not adhere to the endothelial cells themselves but to microfilaments that were not normally exposed.

In the presence of thrombin, platelets can adhere to endothelial cells in culture (12–14). Figures 3 and 4 (12) show scanning and transmission electron micrographs of thrombin-induced platelet

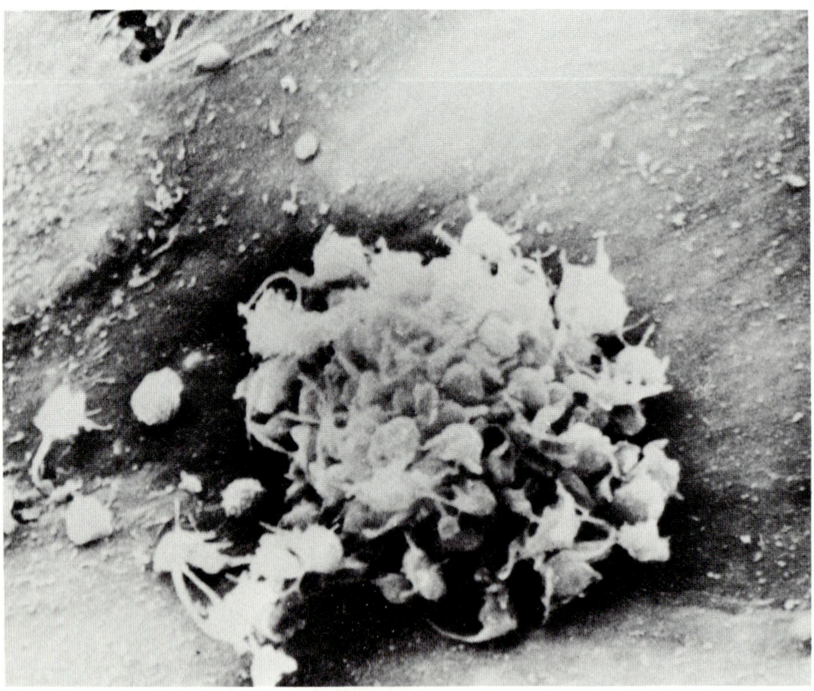

Fig. 3. Scanning electron micrograph of a thrombin-induced platelet aggregate adherent to a monolayer of cultured endothelial cells. ×2600. [Reproduced by permission from Czervionke *et al.* (12).]

aggregates adherent to the surfaces of endothelial cells. Thrombin may be interacting with endothelial cells as well as with platelets in inducing adhesion because neutralization of thrombin with hirudin prevented adhesion. Thrombin treatment of the endothelium did not reduce adhesion, but with thrombin treatment of endothelium that had been preincubated with aspirin to inhibit endothelial cell cyclooxygenase, platelet adhesion did occur (12). As discussed in an earlier paper at this meeting, thrombin binds to endothelial cells (15,16) but whether thrombin binding to the endothelial cells is related to platelet adherence to the cells is unknown.

Thrombin stimulates platelet adherence to other vessel wall cell

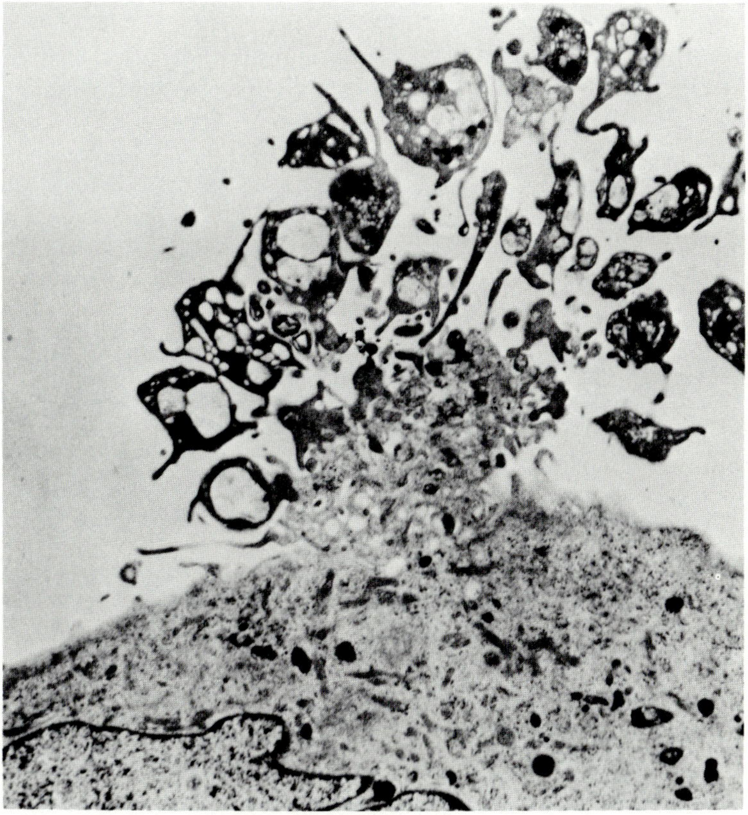

Fig. 4. Transmission electron micrograph of a thrombin-induced platelet aggregate attached to a cultured endothelial cell. ×5586. [Reproduced by permission from Czervionke *et al.* (12).]

types as well as endothelial cells, i.e., smooth muscle cells and fibroblasts (17). Thrombin-induced adhesion of platelets to endothelial cells can be stimulated by pretreatment of the endothelium with aspirin to block prostacyclin production by the endothelial cells (discussed later), and it can be completely blocked by addition of prostacyclin. Adhesion to smooth muscle cells and fibroblasts is not enhanced by treatment of the cultured cells with aspirin and is only partially inhibited by addition of prostacyclin. Figure 5 shows a dose–response curve for prostacyclin inhibition of thrombin-stimulated adhesion of platelets to various cell types, with endothelial cells showing the greatest sensitivity (17).

ENDOTHELIAL CELL PRODUCTION OF PROSTACYCLIN

Several years ago Moncada and colleagues (18) reported that vessel walls produced a substance that interfered with platelet function. This

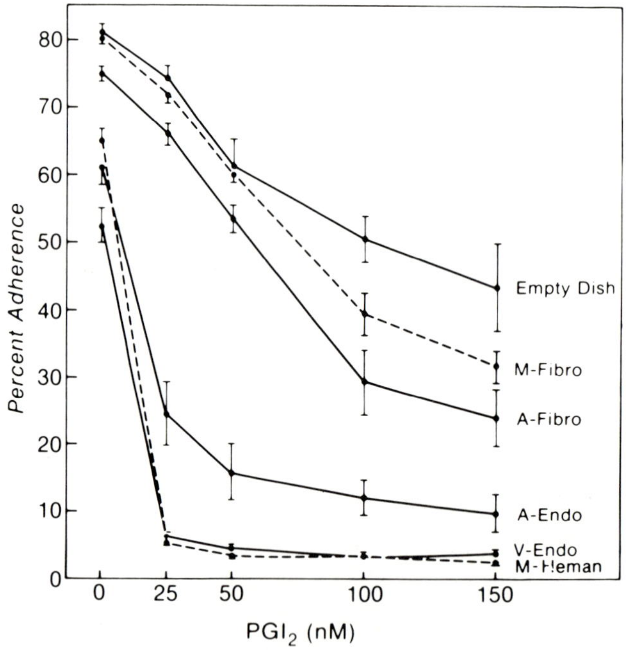

Fig. 5. Inhibition by exogenous PGI_2 of 0.5 U thrombin-induced platelet adherence to 1 m*M* aspirin-treated cell layers. PGI_2 was added just before thrombin and platelets. A-fibro, arterial fibroblast; V-endo, venous endothelium; A-endo, arterial endothelium; M-heman, mouse hemangioendothelioma; M-fibro, mouse L929 fibroblasts. [Reproduced by permission from Fry *et al.* (17).]

substance was soon identified as prostacyclin or PGI_2 (19) and was shown to be derived from prostaglandin endoperoxides (18). Weksler and colleagues (20) were the first to demonstrate production by cultured endothelial cells of a substance that inhibited platelet aggregation. The active substance from these cultures was identified as prostacyclin by thin-layer radiochromatography (20). Weksler and colleagues (21) also demonstrated that prostacyclin production by endothelial cells was stimulated by sodium arachidonate, thrombin, ionophore, and trypsin. Jaffe and Weksler (22) and Czervionke and colleagues (13) showed that aspirin blocked endothelial cell synthesis of prostacyclin but that recovery of synthetic ability was rapid when aspirin was removed from the cells. Cycloheximide blocked recovery of prostacyclin synthesis by the cells, indicating that synthesis of new cyclooxygenase was probably required for recovery. Marcus and colleagues showed that prostacyclin could be synthesized by endothelial cells from exogenous or endogenous cyclic endoperoxides (23) and also that platelets could provide endoperoxides for endothelial cell prostacyclin synthesis (24). The balance between the number of platelets and the number of endothelial cells was very important in demonstrating endothelial cell utilization of platelet endoperoxides, with inhibition of platelet aggregation seen with one platelet per endothelial cell but no inhibition with 20 platelets per endothelial cell, suggesting that the balance between platelet production of thromboxane B_2 and endothelial cell production of prostacyclin was crucial for demonstration of inhibition of aggregation by the endothelial cells.

Czervionke and colleagues (25) also studied the stimulatory effect of thrombin on production of prostacyclin by endothelial cells. Thrombin stimulation of prostacyclin production was found to be time and concentration dependent. Aspirin inhibition of prostacyclin production by endothelial cells was shown to be dependent on the aspirin concentrations (13), and with increasing inhibition of prostacyclin synthesis, there was an increase in platelet adherence to the aspirin treated cells. Thus evidence has accumulated that production of prostacyclin by endothelial cells is an important factor in the lack of adhesion of platelets to these endothelial cells in vivo or in vitro.

SECRETION OF ADP BY ENDOTHELIAL CELLS

Another way endothelial cells may influence platelet function is by secretion of ADP, a known stimulus for platelet aggregation. Pearson and Gordon labeled endothelial cells in culture with [^{3}H]adenosine for 60 minutes, then washed the cells and extracted and analyzed the

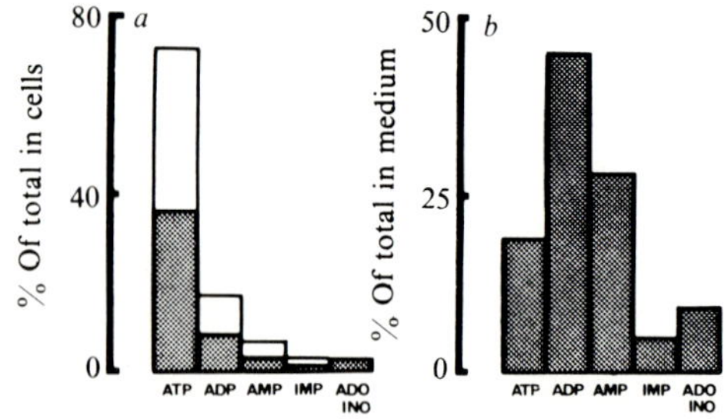

Fig. 6. Distribution of adenine nucleotides of cultured porcine aortic endothelial cells labeled with 2 μM [^{3}H]adenosine. Purines were extracted from the cells before or after treatment with 0.1% trypsin plus 0.025% EDTA or from the supernatant after treatment, were separated by thin-layer chromatography, and the radioactivity in each fraction was determined. In (a), the total height of each column represents the percentage of the total radioactivity incorporated by the cells into each compound. The shaded portion of each column represents the percentage of the total radioactivity in each compound remaining in the cells after treatment with trypsin plus EDTA. (b) The distribution of extracellular radioactivity expressed as a percentage of the total release. [Reproduced by permission from Pearson and Gordon (26).]

purine content either with or without prior treatment of the cells with trypsin (26). Figure 6 (Panel a) shows the distribution of tritium in the cells, with the total height of each bar representing the percentage of the total radioactivity incorporated into each compound and the shaded height representing the percentage of the total radioactivity retained by the cells after trypsin treatment. Panel b shows the distribution of extracellular ^{3}H from the trypsinized cells expressed as a percentage of the total released. It is evident that ADP is the predominant adenine nucleotide in the extracellular fluid. Pearson and Gordon showed that trypsinized suspensions of aspirin-treated endothelial cells (to block prostacyclin production) induced aggregation of platelets and that the aggregation was blocked by a specific antagonist of the platelet ADP receptor (26), suggesting that ADP released from the cells was the mediator of the platelet aggregation.

BINDING OF PLATELET PROTEINS TO ENDOTHELIAL CELLS

The final aspect of platelet–endothelial cell interaction to be discussed in this review is the binding of two secreted platelet proteins to endothelial cells. Platelet factor 4 has been shown by Busch and colleagues to bind to endothelial cells (27). Bound platelet factor 4 could be displaced from endothelial cells in culture by addition of various glycosaminoglycans. Heparin was the most potent agent for displacing platelet factor 4, with heparan sulfate and dermatan sulfate somewhat less potent, and chondroitin-4-sulfate having little effect. These authors also reported that treatment of endothelial cells with heparatinase, which degrades heparan sulfate, prevents binding of platelet factor 4 to the cells. Thus binding of platelet factor 4 appears to be to heparan sulfate on the cell surface. Rybak and colleagues (28) have also studied binding of platelet factor 4 to endothelial cells and have reported that binding is saturable and rapidly reversible with excess platelet factor 4 and that bound platelet factor 4 can be removed with heparin. They calculated a dissociation constant of 3.3×10^{-6} M, with 6.8×10^7 sites per cell. Although the binding constant for platelet factor 4 binding suggests a relatively low affinity binding, there is clinical evidence suggesting that platelet factor 4 binding to endothelial cells may be important physiologically. Dawes and colleagues (29) measured plasma levels of platelet factor 4 and β-thromboglobulin after infusion of heparin into normal volunteers, and, as shown in Fig. 7, there was a dramatic increase in plasma levels of platelet factor 4

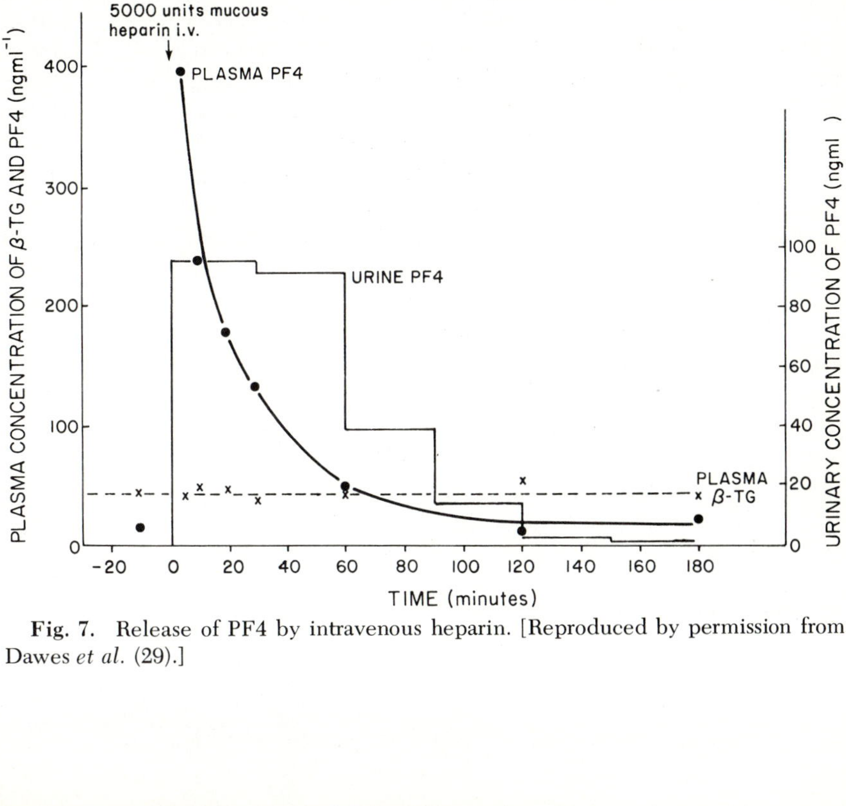

Fig. 7. Release of PF4 by intravenous heparin. [Reproduced by permission from Dawes *et al.* (29).]

with no increase in β-thromboglobulin, suggesting displacement of platelet factor 4 from a binding site. Based on the *in vitro* studies described earlier, it is likely that this binding site is the vascular endothelium. Binding of platelet factor 4 to heparan sulfate on the vascular endothelial surface may be important in modulating the interaction of thrombin and antithrombin III, which has recently been shown to be dramatically enhanced by the presence of endothelial cells (30), presumably by heparan sulfate on the cell surface. Platelet factor 4 binding to the heparan sulfate would prevent the antithrombin III interaction with the glycosaminoglycan (31) and thus would prevent enhancement of the rate of thrombin inactivation.

Binding of β-thromboglobulin to cultured bovine aortic endothelial cells in suspension was reported by Hope and colleagues (32). Scatchard analysis of the binding data showed a single class of binding sites and the calculated values for K_d and n were $2.3 \times 10^{-7}\,M$ and 2.6×10^6 sites per cell. The normal plasma concentration of β-thromboglobulin is about $0.7 \times 10^{-5}\,M$ and the serum concentration is approximately 500 times higher, in the range of the dissociation constant. Hope and colleagues suggested that at local sites of platelet thrombus formation concentrations of β-thromboglobulin similar to the K_d might be obtained with resultant binding to the endothelial cells in that region. In addition to binding to endothelial cells, these authors suggested that β-thromboglobulin inhibits endothelial cell production of prostacyclin. More endothelial cells were required to inhibit platelet aggregation by arachidonic acid when β-thromboglobulin was added to the cells than when other proteins were added. Again, the effective concentrations of β-thromboglobulin were similar to serum concentrations, so the physiological relevance of this observation is unclear. Additionally, Poggi and colleagues (33) failed to find inhibition of endothelial cell production of prostacyclin (measured by radioimmunoassay of 6-keto-$PGF_{1\alpha}$) by cultured calf aortic endothelial cells. Further studies are needed on binding of β-thromboglobulin to endothelial cells and its effect on the cells.

In addition to the interaction of platelet factor 4 with endothelial cells and the possible interaction of β-thromboglobulin with these cells, there may be other secreted platelet proteins that bind to endothelial cells with resulting effects on the functions of the cells that have not yet been investigated.

CONCLUSIONS AND QUESTIONS

Several interactions between platelets and endothelial cells have been reviewed. Platelets appear to be important for maintenance of vascular integrity *in vivo*, and they stimulate migration and proliferation of endothelial cells in tissue culture. The mechanisms of these effects are unknown and deserve further study. The requirement for intact platelets for the *in vitro* effects on cells suggests that the active agent in the platelets is not a secreted protein, but rather something that is continuously produced by the living platelets. Despite their trophic effects on endothelial cells, platelets do not adhere to normal endothelium *in vitro* and, in histological sections, have not been demonstrated to adhere to endothelial cells *in vivo*. The question of adhesion is important since platelet adhesion appears to be an early step in thrombus formation. Two factors appear to stimulate platelet adhesion to endothelial cell monolayers *in vitro*—addition of thrombin and inhibition of prostacyclin production by the cells. How these are balanced *in vivo* is unclear, since thrombin has been shown to increase prostacyclin synthesis by endothelial cells. In addition to these effects on endothelial cells, thrombin is a potent agent for activation of platelets, stimulating release, aggregation, and thromboxane A_2 production, which could induce further release and aggregation. Also, as described earlier, endothelial cells appear to be important in the inactivation of thrombin. Thus the net effect of thrombin *in vivo* on endothelial cells and platelets is likely to depend on the relative numbers of the two cell types and will require detailed study. The importance of the effect of another endothelial product on platelets, i.e. ADP, also remains to be determined because endothelial cell production of prostacyclin could well block potential effects of secreted ADP. Finally, the effects of platelet factor 4 and β-thromboglobulin on endothelial cells and their reactions also need to be better defined and other secreted platelet proteins need to be investigated for possible effects on endothelial cells.

REFERENCES

1. Mustard, J. F., Jorgensen, L., Hovig, T., Glynn, M. F., and Rowsell, H. C. (1966) *Thromb. Diath. Haemorrh., Suppl.* 21, 131–158.
2. Stemerman, M. B., and Spaet, T. H. (1972) *Bull. N. Y. Acad. Med.* 48, 289–301.
3. Stemerman, M. B. (1974) *Prog. Hemostasis Thromb.* 2, 1.
4. Baumgartner, H. R. (1977) *Thromb. Haemostasis* 37, 1–16.
5. Johnson, S. A. (1971) *In* "The Circulating Platelet" (S. A. Johnson, ed.), p. 284. Academic Press, New York.
6. Gimbrone, M. A., Jr., Aster, R. H., Cotran, R. S., Corkery, J., Jandl, J. H., and Folkman, J. (1969) *Nature (London)* 222, 33–36.
7. Kitchens, C. S., and Weiss, L. (1975) *Blood* 46, 567–578.
8. Van Horn, D. L., and Johnson, S. A. (1968) *J. Lab. Clin. Med.* 71, 301–311.
9. Stewart, B. C., Wojcik, J. D., Van Horn, D. L., and Johnson, S. A. (1970) *In* "Platelet

Adhesion and Aggregation in Thrombosis: Countermeasures" (E. F. Mammen, G. F. Anderson, and M. I. Barnhart, eds.), p. 125. Schattauer, Stuttgart.

10. Maca, R. D., Fry, G. L., Hoak, J. C., and Loh, P. T. (1977) *Thromb. Res.* **11**, 715–727.
11. Wechezak, A. R., Holbrook, K. A., Way, S. A., and Mansfield, P. B. (1979) *Blood Vessels* **16**, 35–42.
12. Czervionke, R. L., Hoak, J. C., and Fry, G. L. (1978) *J. Clin. Invest.* **62**, 847–856.
13. Czervionke, R. L., Smith, J. B., Fry, G. L., Hoak, J. C., and Haycraft, D. L. (1979) *J. Clin. Invest.* **63**, 1089–1092.
14. Hoak, J. C., Czervionke, R. L., Fry, G. L., and Smith, J. B. (1980) *Fed. Proc., Fed. Am. Soc. Exp. Biol.* **39**, 2606–2609.
15. Awbrey, B. J., Hoak, J. C., and Owen, W. G. (1979) *J. Biol. Chem.* **254**, 4092–4095.
16. Lollar, P., Hoak, J. C., and Owen, W. G. (1980) *J. Biol. Chem.* **255**, 10279–10283.
17. Fry, G. L., Czervionke, R. L., Hoak, J. C., Smith, J. B., and Haycraft, D. L. (1980) *Blood* **55**, 271–275.
18. Moncada, S., Gryglewski, R., Bunting, S., and Vane, J. R. (1976) *Nature (London)* **263**, 663–665.
19. Moncada, S., Higgs, E. A., and Vane, J. R. (1977) *Lancet* **1**, 18–20.
20. Weksler, B. B., Marcus, A. J., and Jaffe, E. A. (1977) *Proc. Natl. Acad. Sci. U.S.A.* **74**, 3922–3926.
21. Weksler, B. B., Ley, C. W., and Jaffe, E. A. (1978) *J. Clin. Invest.* **62**, 923–930.
22. Jaffe, E. A., and Weksler, B. B. (1979) *J. Clin. Invest.* **63**, 532–535.
23. Marcus, A. J., Weksler, B. B., and Jaffe, E. A. (1978) *J. Biol. Chem.* **253**, 7138–7141.
24. Marcus, A. J., Weksler, B. B., Jaffe, E. A., and Broekman, M. J. (1980) *J. Clin. Invest.* **66**, 979–986.
25. Czervionke, R. L., Smith, J. B., Hoak, J. C., Fry, G. L., and Haycraft, D. L. (1979) *Thromb. Res.* **14**, 781–786.
26. Pearson, J. D., and Gordon, J. L. (1979) *Nature (London)* **281**, 384–386.
27. Busch, C., Dawes, J., Pepper, D. S., and Wasteson, A. (1980) *Thromb. Res.* **19**, 129–137.
28. Rybak, M. E., Handin, R. I., and Gimbrone, M. A. (1980) *Circulation* **62**, Suppl. III, III-335 (abstr.).
29. Dawes, J., Smith, R. C., and Pepper, D. S. (1978) *Thromb. Res.* **12**, 851–861.
30. Busch, P. C., and Owen, W. G. (1981) *Thromb. Haemostasis* **46**, 38 (abstr.).
31. Handin, R. I., Jordan, R., and Rosenberg, R. D. (1981) *In* "The Chemistry and Biology of Heparin" (R. L. Lundblad, W. V. Brown, K. G. Mann, and H. R. Roberts, eds.), pp. 393–401. Elsevier/North-Holland Press, New York.
32. Hope, W., Martin, T. J., Chesterman, C. N., and Morgan, F. J. (1979) *Nature (London)* **282**, 210–212.
33. Poggi, A., Niewiarowski, S., Holt, J. C., Dall'Olio, A., Stewart, G. J., and Smith, J. B. (1981) *Thromb. Haemostasis* **46**, 37 (abstr.).

Secretory Products of Mononuclear Phagocytes

WILLIAM A. SCOTT AND ZANVIL A. COHN
Laboratory of Cellular Physiology and Immunology
The Rockefeller University
New York, New York

The role of macrophages as phagocytic cells in inflammatory foci and sites of tissue injury is well known. Recently it has emerged that the macrophage also acts as a secretory cell in the inflammatory response. In general, activation of these cells through exposure to lymphokines or challenge with phagocytic stimuli results in the secretion of a complex array of products (Table I). It is of note that many of these macrophage products have activities that influence the vasculature. The importance of the macrophage to the vasculature is emphasized by the presence of circulating mononuclear phagocytes, the close apposition of macrophages to capillaries, and the localization of macrophages in sinusoids.

Among the more important macrophage secretory products are the neutral proteases such as plasminogen activator, collagenase, elastase, and angiotensin convertase. These activities may provide a means for the role of the macrophage in tissue remodeling in addition to enzyme cascades with importance to the clotting reaction. The complement system illustrates the complex relations of macrophages with humoral factors. Macrophages synthesize the early components of the classical and alternative pathways and also degrade complement through the action of proteases. Two of the degradation products, the anaphylatoxins C3a and C5a, stimulate macrophage migration. On the other hand, degradation of B to Bb, a product of the alternative pathway, suppresses macrophage migration and promotes attachment and cell spreading. Macrophages contain a well-defined receptor for the C3b fragment. Activated macrophages have the ability to phagocytize particles opsonized with C3b, whereas resident cells bind, but do not ingest such particles.

PATHOBIOLOGY OF THE ENDOTHELIAL CELL

ISBN 0-12-521980-6

TABLE I
Secretory Products of Mononuclear Phagocytes

Complement Components	Hydrolases—Neutral pH optima
Classical pathway	Muramidase
C1	Neutral proteinases
C4	Plasminogen activator
C2	Collagenase
C3	Elastase
$C3_b$ Inactivation	Angiotensin convertase
B_1H	Hydrolases—Acid pH Optima
C5	Proteinases
Alternative pathway	Peptidases
Factor B	Lipases
Factor D	Phospholipases
Properdin	Ribonuclease
Coagulation Factors	Deoxyribonuclease
V	Phosphatases
VII	Glycosidases
IX	Hexosaminidases
X	Sulfatases
Thromboplastin	Arginase
Prothrombin	Lipoprotein lipase
Peptides	Binding proteins
Glutathione	Transferrin
Enzyme inhibitors	Transcobalamin II
Plasmin inhibitors	Fibronectin
α-2 Macroglobulin	Arachidonate metabolites
Nucleosides	Cyclooxygenase products
Thymidine	Prostaglandin E_2
Uridine	Prostaglandin $F_{2\alpha}$
Adenosine	Prostacyclin
Guanosine	Thromboxane
Platelet activating factor	Lipoxygenase Products
Reactive metabolites of oxygen	Leukotriene C
Superoxide anion	Leukotriene D
Hydrogen peroxide	Leukotriene E
Hydroxyl radicals	Leukotriene B?
Singlet oxygen	HETEs (to be identified)
Factors Promoting Growth of:	Factors Inhibiting Growth at:
Lymphocytes (B and T cells)	Tumor cells
Myeloid progenitors	Lymphocytes
Mononuclear phagocytes	Myeloid precursors
Erythroid progenitors	Other Hormone-like factors
Fibroblasts	Endogenous pyrogen 1 and 2
Capillary endothelium	Interferon α, β
Neutrophil chemotactic factors	Insulinlike activity
Factors regulating protein synthesis of:	
Hepatocytes	
Serum amyloid	
Haptoglobin	
Synovial cells	
Collagenase	
Plasminogen activator	

Macrophages also elaborate factors that promote the replication of lymphocytes, fibroblasts, and endothelial cells, in addition to factors that promote the growth of the microvasculature. A fourth class of secretory products are reactive metabolites of oxygen. These are produced by activated macrophages in response to pharmacological stimuli such as phorbol myristate acetate or phagocytic stimuli. Macrophages synthesize superoxide anion, which dismutates into hydrogen peroxide. Presumably, singlet oxygen and hydroxyl radical are also generated, although less is known. The tumoricidal and microbicidal activities of the activated macrophage are in part mediated by hydrogen peroxide. This is true for the destruction of intracellular parasites as well as extracellular parasites and tumor cells. Activated macrophages release quantities of hydrogen peroxide comparable to those of the neutrophil.

More recently, it has become apparent that macrophages are a potent source of arachidonic acid metabolites (1). The phospholipids of macrophages, platelets, and lymphocytes along with the macrophage progenitor, the circulating blood monocyte, are highly enriched in arachidonic acid. Approximately 25% of the macrophage phospholipid fatty acid is arachidonic acid, compared to 3–10% for most cell types, including neutrophils. When provided with a phagocytic stimulus, macrophages release 30–40 nmol of arachidonic acid/μg cell protein, which amounts to 40% of their phospholipid arachidonate content (1). These cells not only have the capacity to release a considerable percentage of the arachidonate stores but in addition quantitatively metabolize the free fatty acid to oxygenated products. This response to a phagocytic stimulus is immediate (Fig. 1). Arachidonic acid release and metabolism remain linear for 60–90 min following a zymosan challenge and closely parallel zymosan ingestion. The concentration of particles determines the quantity of products formed (Fig. 2). Provided that maximal levels of ingestion are not reached, macrophages are capable of responding to multiple challenges of a phagocytic load by initiating new rounds of arachidonic acid release and metabolism (1).

The nature of the trigger for arachidonic acid release was examined by dissecting the phagocytic process into distinct phases (2). Interiorization of the plasma membrane and its subsequent fusion with lysosomes appears not to play a role in this process. The interaction of particle-bound ligands with specific macrophage surface receptors is a sufficient criterion to initiate the release of 20:4 metabolites. Since monomeric ligands were ineffective, events in addition to ligand–receptor interaction are apparently required and may include receptor

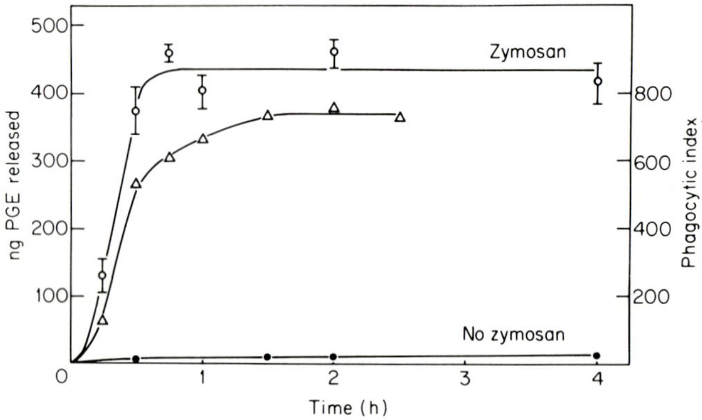

Fig. 1. Time course of PGE_2 release and zymosan ingestion by mouse peritoneal macrophages. Cultures were challenged at $t = 0$ with 160 μg of zymosan. Media from duplicate 35-mm cultures ($\sim$1–2 $\times 10^6$ macrophages) were extracted at each time point, and the PGE_2 levels were determined in duplicate by radioimmunoassay. Phagocytosis was measured in parallel cultures on glass coverslips similarly treated. Phagocytic index (△); PGE_2 release by zymosan-treated (○) and control (●) cultures. [Reproduced by permission from Scott *et al.* (1)].

clustering. These observations are important in that they suggest that the initiation of 20 : 4 metabolism is a receptor-mediated event and occurs in the absence of phagocytosis, a situation analogous to that required for mast cell degranulation.

The arachidonic acid metabolites elaborated by resident macrophages are remarkably diverse. These include cyclooxygenase and lipoxygenase products (1,3,4) which have immediate effects on the

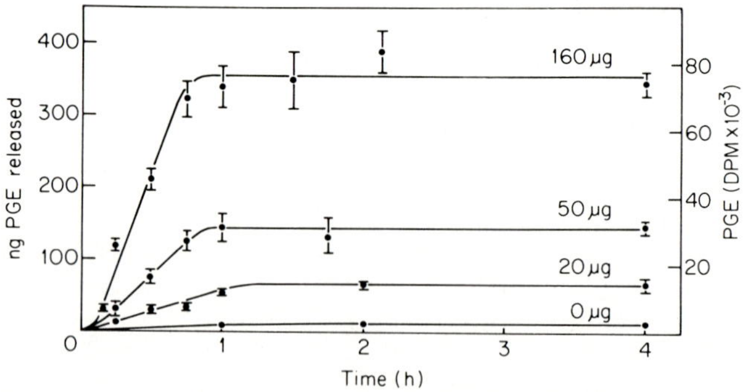

Fig. 2. Effect of zymosan concentration on PGE_2 synthesis by mouse peritoneal macrophages. Cultures in 35-mm dishes, prelabeled with tritiated arachidonic acid, were exposed to the indicated amounts of zymosan at $t = 0$. PGE_2 was measured by radioassay following extraction from culture media and isolation by silicic acid chromatography (1). [Reproduced by permission from Scott *et al.* (1)].

vasculature in addition to cellular elements of the immune system. Prostaglandins of the E and F series together with prostacyclin can result in the clinical symptoms of inflammation such as edema, erythema, pain, and heat. Approximately 50% of the macrophage products is PGE_2 and another 20% is prostacyclin. The remaining 30% is derived from the lipoxygenase pathway. HETEs (hydroxyeicosatetraenoic acids) are hydroxylated derivatives of arachidonic acid and represent 10% of the total products. The macrophage-derived HETEs are incompletely characterized. Several compounds within this group are weak chemotactic agents for neutrophils; however the function of these compounds is as yet unclear. The other major class of lipoxygenase products generated by macrophages is slow reacting substance. Leukotrienes C, D, and E have this activity. Macrophages form primarily leukotriene C, which comprises 15–20% of the arachidonic acid products (3,4). To date, the macrophage is the only normal cell population known to synthesize slow reacting substance in response to a physiological stimulus. In addition to their contracting activity on the peripheral airways of the lung, leukotrienes C and D cause intense, but short-lived, constriction of arterioles with a potency similar to angiotensin, which is followed by extravasation of macromolecules from postcapillary venules (5).

Macrophages from the peritoneum of a single mouse form 300–600 pmol of leukotriene C. Considering that 3–6 pmol is sufficient to obtain complete contraction of a guinea ileum, the standard biological assay of this material, this represents considerable quantities of SRS in biological terms. The quantities of other metabolites formed are also impressive. For example, the concentration of PGE_2 reaches the micromolar range while physiological concentrations are 10^{-7} to 10^{-11} M. Similar data for other cell types are sparse. As a result, few direct comparisons can be made. The work of Bokoch and Reed (6) indicate guinea pig neutrophils also synthesize a spectrum of cyclooxygenase and lipoxygenase products in response to the nonspecific stimulus calcium ionophore A23187; however the total quantity of products is 44 pmol/μg of cell protein or 500 to 700-fold less than the macrophage.

The activation state of macrophages is an important determinant of the ability of these cells to generate arachidonic acid metabolites. In contrast to other classes of secretory products (Table I), arachidonic acid metabolites are products of resident macrophages rather than their activated counterparts. This conclusion is based on measurements of arachidonic acid metabolism by macrophages from animals variously inoculated with the immunological microbial agents *Cornebacterium parvum* and the tuberculin bacillus BCG. For example, the capacity of these macrophage populations to synthesize prostaglandins was compared with their ability to inhibit the intracellular replication of *Toxoplasma gondii*, one measure of macrophage activation.[1] The inverse relationship between enhanced antitoxoplasma activity and reduced prostaglandin release suggests arachidonic acid metabolism is reduced in proportion to the degree of macrophage activation. Activated macrophages form thomboxane and PGE_2, but the synthesis of the vasoactive agents prostacyclin and leukotriene C is essentially eliminated. This, together with the sixfold reduction in synthetic capacity, may be a factor in limiting the immune response.

Traditionally, the role of the macrophage is thought to be in delayed-type hypersensitivity reactions. The realization that macrophages are well equipped to produce effector substances suggests a role for mononuclear phagocytes in more immediate reactions and in the expression of allergic states. This is emphasized by the fact that arachidonic acid metabolites are products of the resident tissue macrophage. Since this cell is ubiquitously distributed throughout the body, it is uniquely positioned to respond by recruitment of the host defense system. The release of arachidonic acid metabolites may be one mechanism whereby the macrophage participates and initiates the early events of the inflammatory response.

ACKNOWLEDGMENTS

Supported by U.S. Public Health Grants AI-07012 and HL27186 and by American Heart Association Grant 79-1009.

REFERENCES

1. Scott, W. A., Zrike, J. M., Hamill, A. L., Kempe, J., and Cohn, Z. A. (1980) *J. Exp. Med.* **152**, 324–335.
2. Rouzer, C. A., Scott, W. A., Kempe, J., and Cohn, Z. A. (1980) *Proc. Natl. Acad. Sci. U.S.A.* **77**, 4279–4282.
3. Rouzer, C. A., Scott, W. A., Cohn, Z. A., Blackburn, P., and Manning, J. M. (1980) *Proc. Natl. Acad. Sci. U.S.A.* **77**, 4928–4932.
4. Rouzer, C. A., Scott, W. A., Hamill, A. L., and Cohn, Z. A. (1980) *J. Exp. Med.* **152**, 1236–1247.
5. Dahlen, S.-E., Bjork, J., Hedqvist, P., Arfors, K.-E., Hammarström, S., Lindgren, J.-A., and Samuelsson, B. (1981) *Proc. Natl. Acad. Sci. U.S.A.* **78**, 3887–3891.
6. Bokoch, G. M., and Reed, P. W. (1980) *J. Biol. Chem.* **255**, 10223–10226.

[1] W. A. Scott, N. A. Pawlowski, H. W. Murray, M. Andreach, J. Zrike, and Z. A. Cohn, unpublished results.

Leukocyte-Endothelial Interactions

RICHARD L. HOOVER AND MORRIS J. KARNOVSKY
Department of Pathology
Harvard Medical School
Boston, Massachusetts

One of the initial and crucial events in the acute inflammatory response is the attachment of leukocytes to the endothelial lining of blood vessels. Normally, the polymorphonuclear leukocytes (PMN) flow freely through the vessels or, at most, adhere only momentarily to the endothelial surface (4,13,23). During acute inflammation, the PMN marginate and adhere locally along the vessel, then migrate out via the interendothelial junctions, finally accumulating at the site of injury, where they undertake their phagocytic functions. Many investigators have described this process *in vivo,* with the basis for a clear understanding of the events involved and an appreciation of their significance having been given by Julius Cohnheim in 1889. More recently, Allison *et al.* (3) accurately demonstrated experimentally with modern techniques the cellular events in the inflammatory process and provided further subtle details. What causes the PMN to adhere at a particular location along the blood vessel near an injury and how they become attached are not known, despite the fact that these initial events are decisive for the development of the acute inflammatory response.

In vivo experiments have implicated a role for a number of factors involved in the attachment and localization of the PMN: divalent cations (4,45), cell surface charge (1,2,6,48), blood flow and shear forces (5,9,35,36,50,51), cAMP levels (11), and complement/chemotactic factors (16,38). It is, however, possible to discern whether these factors are mediating these responses by affecting the endothelium or the PMN. Much work has been done on isolated PMN and what factors influence their aggregation and attachment to artificial substrates (15,17,18,21,28,30–33,39,41,52,53); however, not until recently, with

PATHOBIOLOGY OF THE ENDOTHELIAL CELL

ISBN 0-12-521980-6

the advent of methods for culturing endothelium *in vitro* (8,22,27), has the role of endothelium been studied. *In vivo*, the vessel lining is a monolayer of endothelial cells to which the PMN attach. A system can now be modeled *in vitro* using isolated PMN and cultured endothelial cells that grow into confluent monolayers, similar to the *in vivo* situation. We (25,26) as well as others (7,16,29,32,40,54) have used this *in vitro* system to examine the inflammatory process.

Basically, we have used two *in vitro* methods to investigate the adhesive interaction between the PMN and the endothelium. One method is based on the monolayer collection assay developed by Walther *et al.* (55) in which ^{51}Cr-labeled PMN are rained down on a confluent monolayer of endothelium. The cells are incubated together for a fixed period of time and rinsed to remove the unattached PMN. The remaining cells are lysed and counted on a liquid scintillation counter. Based on counts representing 100% of the PMN added, the percentage of PMN adhering can be calculated. The other method involves plating the endothelium, or any cell type to be tested, onto cover slips and allowing the cells to adhere and divide; however, the cells are not allowed to reach confluency. The cover slips are then rinsed, and the PMN are added and incubated with the endothelial cells. Under the experimental conditions, the PMN can choose to attach initially or to migrate and attach to either the cell surface or the surface of the cover slip. After the period of incubation, the cover slips are fixed and prepared for scanning electron microscopy. By counting the number of PMN attaching to each surface, one can determine which surface the PMN prefer, and by plating different cells on the cover slips, one can determine whether the PMN have an adhesive preference for cell types. We have termed this the *SEM choice assay*. The assay is based on the idea that cells will attach to the more adhesive substrate when given the choice, as suggested previously by Harris (24) and Carter (12) to explain the attachment and movement of fibroblasts to surfaces of different adhesive strengths. Using both the monolayer collection assay and the SEM choice assay, we have found that PMN prefer to adhere to endothelium than to other cell types (Table I; Figs. 1,2). For example, compare Fig. 1, which shows more PMN attaching to the endothelium than the glass, to Fig. 2, in which PMN are attaching to the glass instead of to B16F10 melanoma cells. Table I presents a list of the cell types tested in the SEM choice assay and shows that the PMN adhere better to the endothelial cells than to any other cell. The data also show that this specificity exists even when crossing species lines; i.e., human PMN prefer calf endothelium to other cells. The monolayer collection assay confirms this preference of

TABLE I
Adhesion of Calf and Human PMN after 90 min at 37°C to Different Cell Types as Monitored by Direct Cell Counts in the Scanning Electron Microscope[a]

Cell type	Calf PMN (%)	Human PMN (%)
Calf endothelium	51.6 (745)	36.2 (1586)
Human endothelium	—	36.0 (261)
Swiss 3T3	9.7 (342)	1.2 (425)
BHK	8.3 (313)	2.7 (485)
Py BHK	5.4 (56)	—
SvPy 3T3	2.5 (157)	0.2 (602)
Embryonic bovine trachea	2.5 (721)	—
B16F1	2.5 (608)	1.0 (303)
B16F10	2.5 (632)	1.0 (446)
L929	2.4 (1442)	0 (100)
Rat smooth muscle	1.3 (1185)	4.1 (967)
CHO	—	1.0 (1198)

[a] Values are expressed as percentages of PMN adhering to various cell types. Number in parentheses represents the total number of PMN counted. Standard deviation in experiments did not exceed 1.5%.

PMN for endothelium (Table II), all of which suggests that a specific relationship exists between the PMN and the endothelium. There is a difference in the percentage of PMN adhering to the endothelium in the monlayer collection assay compared to SEM choice assay (SEM choice assay < monolayer collection assay); however, this can be attributed to the design of the experiments. In the SEM choice assay the PMN can move from one type surface to another (e.g., from glass to cell), while in the monolayer collection assay the PMN encounter only the cell surface and have no choice.

We (25) as well as others (7) have shown that in the *in vitro* model, divalent cations enhance the attachment of PMN to endothelial monolayers. More specifically, we have shown (26) that Mn^{2+} causes the maximum number of PMN to attach to the endothelium, followed by $Zn^{2+} > Ni^{2+} > Mg^{2+} > Ba^{2+} > Ca^{2+}$ (Table III). This kind of specificity for divalent cations in an adhesive process has also been demonstrated by Rabinovitch and DeStefano (42,43), who found that Mn^{2+}, more so than other ions, promoted the spreading of macrophages and fibroblasts, and Martz (34), who found that Mg^{2+} and Mn^{2+}, but not Ca^{2+} and Sr^{2+}, caused optimal attachment of thymus-derived lymphocytes to specific ascites tumor target cells. Although the reason for this

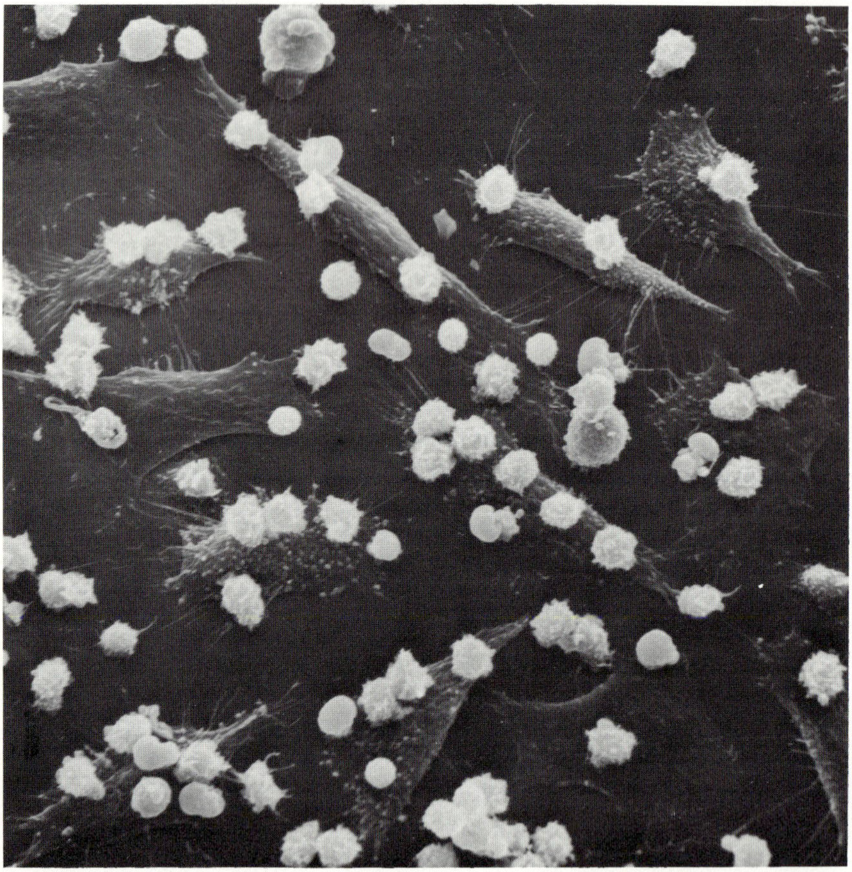

Fig. 1. Scanning electron micrograph showing the attachment of human PMN to calf endothelium. Note that most of the PMN are attached to the endothelial cells and not the glass surface. Magnification ×700.

differential effect of the divalent cations is not known, one possibility is that the net surface charge is being lowered more by Mn^{2+} than the other ions, thereby reducing to a greater extent the forces of repulsion caused by charge, which in turn would allow more cells to adhere. That charge may play a role is supported by results from our experiments in which neuraminidase treatment of either the PMN or the endothelium reduces the net surface charge and increases the number of PMN attaching to the endothelium (>50%). Upon measuring the electrokinetic properties of the PMN and endothelium treated with

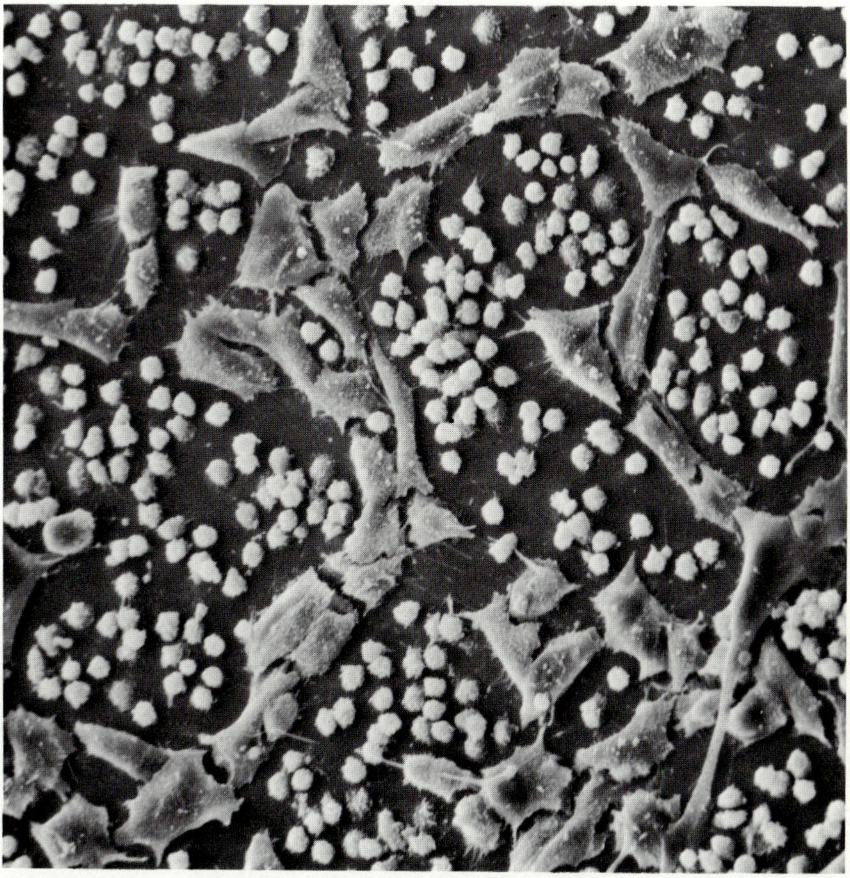

Fig. 2. Scanning electron micrograph demonstrating the interaction between human PMN and mouse B16F10 melanoma cells. In this instance, note that most of the PMN are attached to the glass and not the melanoma cells. Magnification ×500.

either Mn^{2+} or Ca^{2+}, we found that both of the ions reduce surface charge, but not to different extents, as one might expect, since Mn^{2+} increased adhesion more than Ca^{2+} (Table IV). These results do not rule out a role for net surface charge in the adherence of PMN to endothelium, but do rule out the reduction of surface charge as responsible for the differences in divalent cation effects. An alternative explanation is that the cations are acting as cofactors for certain enzyme reactions involved in adhesion. For example, Roseman (44) and Webb and Roth (56) have shown that there are glycosyl transferases, which

TABLE II
Adhesion of Human PMN to different Cells as Monitored by a Monolayer Collection Assay[a]

Monolayer	Adhesion value
Calf endothelium	1.00 (18)
Plastic	0.89 ± 0.02 (4)
CHO	0.82 ± 0.02 (4)
B16F1	0.81 ± 0.01 (4)
L929	0.75 ± 0.01 (4)
B16F10	0.69 ± 0.01 (4)
Swiss 3T3	0.57 ± 0.01 (4)

[a] Values are based on the percentages of PMN adhering to endothelium (60.6 ± 0.13).

require Mn^{2+} for optimal activation, on the surfaces of cells, and that these enzymes may be involved in cell–cell adhesion. It is also possible that the divalent cations may bind to proteins in the membrane responsible for adhesion, and in doing so cause conformational changes that affect adhesion.

Some of the most important factors that control the directional movement, attachment, and aggregation of PMN are chemotactic factors. Many investigators have shown that when PMN are treated with these substances, the adhesion of PMN to themselves (10,38,39) and to

TABLE III
The Adhesion of PMN to Endothelium and the Effect of Divalent Cations (1 m*M*)[a]

Treatment	Adhesion value
Control (no divalent ions)	1.00 ± 0.0 (10)
Mn^{2+}	1.95 ± 0.15 (6)
Zn^{2+}	1.85 ± 0.06 (2)
Ni^{2+}	1.68 ± 0.07 (3)
Mg^{2+}	1.64 ± 0.10 (5)
Ba^{2+}	1.25 ± 0.19 (2)
Ca^{2+}	1.22 ± 0.10 (5)

[a] Values are compared to controls with no divalent cations (6.2 ± 0.9%). Numbers in parentheses represent the number of experiments, each having at least two replicates. All values are significantly different from controls, $p \leq .05$.

TABLE IV
The Effects of Mn^{2+} and Ca^{2+} (1 m*M*) on the Electrophoretic Motilities (cm^2/V sec × 10^4) of PMN and Endothelium[a]

Cell type	EPM
PMN	
Mn^{2+}	1.20 ± 0.01 (4.6%)
Ca^{2+}	1.20 ± 0.01 (4.6%)
Endothelium	
Mn^{2+}	1.13 ± 0.03 (4.2%)
Ca^{2+}	1.12 ± 0.01 (4.3%)

[a] Values in parentheses represent percentage difference from controls.

artificial substrates (16,38,39,53) is increased. It has been shown that these treatments of the PMN also increase their attachment to monolayers of endothelium (16,25,26).

In the past, the endothelium was given the passive role of merely a substrate for PMN attachment; however, if one treats only the *endothelium* with factors that make the PMN more adhesive, the number of untreated PMN attaching will increase. For example, if the endothelium is treated with any of the chemotactic agents, such as C5a, zymosan-activated serum, or the synthetic tripeptide formylmethionylleucylphenylalanine (FMLP), rinsed, and then incubated with untreated PMN, the number of cells adhering is increased (Table V). Furthermore, we have recently demonstrated that endothelium, like PMN, has on its surfaces high affinity receptors ($K_D = 3 \times 10^{-10}$) for the chemotactic agent, formylmethionylleucyl-phenylalanine (Fig. 3). These results implicate a very important and active role for the endothelium in the adhesion process.

How the endothelium mediates this increase in adhesion is not known. As with the PMN, neuraminidase and divalent cations reduce the net surface charge; however, incubation with the chemotactic agent FMLP has no effect. Although it has been suggested that a reduction of net surface charge on PMN caused by chemotactic agents may be responsible for its increased adhesiveness (19,20,49), we cannot show a similar effect of net surface charge on endothelium when compared with the appropriate controls where no increase in adhesion is seen; e.g., the change in surface charge of cells treated with C5a is no different than that of cells treated with C5, which is neither chemotactic nor adhesion promoting (Table VI).

TABLE V
Adhesion of PMN to Endothelial Monolayers after Treatment with Chemotactic agents[a]

Treatment	Cells treated	
	PMN	Endothelium
Hanks balanced salt solution	1.00	1.00
Bacterial filtrate (2%)	1.18 ± 0.04[b]	1.31 ± 0.03[b]
C5 (2 U/ml)	0.98 ± 0.13	1.03 ± 0.04
C5a		
2 U/ml	1.12 ± 0.04[b]	1.22 ± 0.04[b]
10 U/ml	1.67 ± 0.10[b]	1.62 ± 0.13[b]
100 U/ml	1.80 ± 0.09[b]	1.92 ± 0.11[b]
Normal serum (10%)	1.02 ± 0.05	0.95 ± 0.12
Zymosan-activated serum (10%)	1.51 ± 0.05[b]	1.49 ± 0.09[b]
F-Met-Ala (10^{-4} *M*)	1.32 ± 0.02[b]	1.49 ± 0.06[b]
F-Met-Leu-Phe (10^{-8} *M*)	1.40 ± 0.23[b]	1.25 ± 0.18[b]

[a] Values are based on the percentage of cells attaching under control conditions in HH (11.7% ± 2.1). Each experimental condition was repeated at least three times with duplicates in each experiment.

[b] Values significantly different from controls, $p \leq .01$.

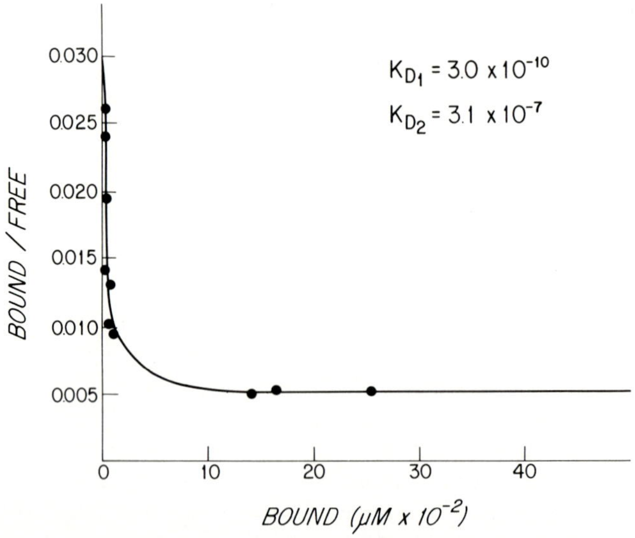

Fig. 3. A Scatchard plot of the binding of the chemotactic tripeptide [^{3}H]formylmethionylleucylphenylalanine to calf aortic endothelium.

TABLE VI
The Effects of Chemotactic Agents on the Electrophoretic Mobilities (cm^2/V sec × 10^4) of PMN and Endothelium

Treatment	PMN	Endothelium
Hanks balanced salt solution	1.54 ± 0.07	2.02 ± 0.01
C5 (2 U/ml)	1.42 ± 0.09	1.83 ± 0.03
C5a (2 U/ml)	1.40 ± 0.07	1.87 ± 0.05
Normal serum (10%)	1.39 ± 0.09	1.88 ± 0.07
Zymosan-activated serum	1.41 ± 0.07	1.94 ± 0.06
Bacterial filtrate control (2%)	1.40 ± 0.02	1.96 ± 0.02
Bacterial filtrate (2%)	1.42 ± 0.01	2.01 ± 0.03
F-Met-Leu-Phe (10^{-8} *M*)	1.52 ± 0.02	2.04 ± 0.03

It has also been shown that incubation of PMN with chemotactic agents causes the cells to produce active oxygen species (10,46). It should be considered that these oxygen-free radicals may mediate the interactions between the PMN and endothelium, or possibly damage the endothelium. Several investigations (37,46,47) have suggested that complement-activated PMN not only have an increased affinity for the endothelium, but also produce damage by releasing active oxygen species. However, in our system we cannot detect, using ^{51}Cr release as a test for viability, any damage to the endothelium in the presence of PMN with or without chemotactic factors (zymosan-activated serum) (Table VII). Furthermore, we cannot detect any superoxide production when the endothelium is treated with the chemotactic agents or when the PMN are in contact with the endothelium, even in the presence of chemotactic agents (Table VIII). It appears as if the endothelium sup-

TABLE VII
Viability of Endothelium as Monitored by ^{51}Cr Released in Supernatant[a]

	Δ%		
	15 min	30 min	120 min
Hanks Hepes (HH)	− 3%	+17%	+ 8%
HH + 10% human serum	−13%	− 3%	−29%
HH + 10% zymosan-activated serum	− 2%	+ 6%	+ 8%
HH + heat-inactivated, zymosan-activated serum	−20%	—	0

[a] Values are expressed as percentage change from monolayers of endothelium with no PMN in each experimental condition. Numbers represent the mean of three experiments.

TABLE VIII
Production of Superoxide by PMN and Endothelium[a]

	No additions	FMLP (10^{-6} *M*)
Endothelium	0.87	0
PMN on endothelium	1.03	0.21
PMN on plastic	3.61	4.28
Pretreatment of endothelium	0.87	0.56

[a] Values are expressed as nmoles/15 min/10^6 cells.

presses the PMN production of superoxide (cf. values of PMN with FMLP on glass to those on endothelium).

The most important aspect of this work is the demonstration of an active role for endothelium in acute inflammation. One can speculate that the first event to occur is the modulation of the endothelium by chemotactic agents, which results in the endothelium becoming more adhesive. As the PMN pass by this area of the blood vessel, they adhere more readily, and it is at this time that the chemotactic agents can now affect the PMN. If, as we suggest, the endothelium becomes affected first, then this would help to direct the PMN more quickly to the area of inflammation. If the chemotactic agents were going into the bloodstream and then affecting the PMN only, one would expect to see the cells adhering downstream away from the site where these factors entered the blood vessel near the injury. Furthermore, the chemotactic agents would be diluted out continuously as they moved farther downstream, away from the site of inflammation. *In vivo,* Allison *et al.* (3) have shown that at the site of inflammation, the side of the blood vessel nearest the injury is the first to be coated with PMN, suggesting that the endothelium on that side is more adhesive for PMN. If only the PMN were affected, then the cells would adhere all around the vessel wall; however, this does not occur until later. Other investigators have presented evidence supporting the idea that the substrate (in this case the endothelium) can influence adherence and migration by virtue of its adhesiveness. Carter (12) and Harris (24) have shown that fibroblasts migrate and adhere to areas that are more adhesive, and Allan and Wilkinson (2a) found that substrate modulations affect leukocyte migration. Dierich *et al.* (15) have further suggested that, *in vitro,* leukocytes migrating in tissues and on endothelium are crawling on surfaces coated and influenced by chemoattractants.

We still do not know how these agents affect the endothelium to make it more adhesive, or what role other factors not yet studied extensively, such as shear forces and prostaglandins, will play. We do know that endothelium plays a very active and important part in the adhesion process, and ultimately acute inflammation. Julius Cohnheim (14) in one of his lectures also suggested that the endothelium is an integral part of this adhesive process. In this lecture, he stated:

> . . . we hold fast the conviction which was directly forced upon us by our experiments and reflection—namely, that inflammation is the expression and consequence of a molecular alteration in the vessel walls. By it, adhesion between the vessel wall and the blood . . . is increased.

ACKNOWLEDGMENTS

We would like to acknowledge Susan Byers and Bob Rubin for their assistance in the preparation of this manuscript. The work was supported by grants HL-26191 and HL-17747 from the National Institutes of Health.

REFERENCES

1. Abramson, H. (1924) *Am. J. Med.* **167,** 702–710.
2. Abramson, H. (1927) *J. Exp. Med.* **46,** 987–1002.
2a. Allan, R., and Wilkinson, P. (1978). *Exp. Cell Res.* **111,** 191–203.
3. Allison, F., Smith, M. R., and Wood, B. (1955) *J. Exp. Med.* **102,** 655–668.
4. Atherton, A., and Born, G. V. R. (1972) *J. Physiol. (London)* **222,** 447–474.
5. Atherton, A., and Born, G. V. R. (1973) *J. Physiol. (London)* **233,** 157–165.
6. Atherton, A., and Born, G. V. R. (1973) *J. Physiol. (London)* **234,** 66p–67p.
7. Beesley, J., Pearson, J., Hutchings, A., Carleton, J. S., and Gordon, J. (1979) *J. Cell Sci.* **38,** 237–248.
8. Booyse, F. M., Sedlak, B., and Rafelson, N. (1975) *Thromb. Diath. Haemorrh.* **34,** 825–839.
9. Born, G. V. R., Planker, M., and Richardson, P. D. (1979) *J. Physiol. (London)* **289,** 76p–77p.
10. Boxer, L., Yoder, M., Bonsib, S., Schmidt, M., Ho, P., Jersild, R., and Baehner, R. (1979) *J. Lab. Clin. Med.* **93,** 506–514.
11. Buckley, R. M., Ventura, E. S., and MacGregor, R. R. (1978). *J. Clin. Invest.* **62,** 554–559.
12. Carter, S. B. (1965) *Nature (London)* **208,** 1183–1187.
13. Clark, E. R., and Clark, E. L. (1935) *Am. J. Anat.* **57,** 385–438.
14. Cohnheim, J. (1889) "Lectures on General Pathology," Sect. I. New Sydenham Society, London.
15. Dierich, M. P., Wilhelmi, D., and Till, G. (1977) *Nature (London)* **270,** 351–352.
16. Fehr, J., and Jacob, H. S. (1977) *J. Exp. Med.* **146,** 641–652.
17. Fenn, W. O. (1922–1923) *J. Gen. Physiol.* **5,** 143–167.
18. Fenn, W. O. (1922–1923) *J. Gen. Physiol.* **5,** 169–179.
19. Gallin, J. I. (1980) *J. Clin. Invest.* **65,** 298–306.
20. Gallin, J. I., Durocher, J., and Kaplan, A. (1975). *J. Clin. Invest.* **55,** 967–974.
21. Giddon, D. B., and Lindhe, J. (1972) *Am. J. Pathol.* **68,** 327–336.
22. Gimbrone, M. A., Cotran, R. S., and Folkman, J. (1974) *J. Cell Biol.* **60,** 673–684.
23. Grant, L. (1973) *In* "The Inflammatory Process" (B. W. Zweifach, L. Grant, and R. McCluskey, eds.), Vol. 2, pp. 205–249. Academic Press, New York.
24. Harris, A. K. (1973) *Exp. Cell Res.* **77,** 285–297.

25. Hoover, R. L., Briggs, R. T., and Karnovsky, M. J. (1978) *Cell* **14,** 423–428.
26. Hoover, R. L., Folger, R., Haering, W. A., Ware, B. R., and Karnovsky, M. J. (1980) *J. Cell Sci.* **45,** 73–86.
27. Jaffe, E. A., Nachman, R., Becker, C., and Minick, R. (1973) *J. Clin. Invest.* **52,** 2745–2756.
28. King, C. A., Preston, T. M., Miller, R. H., and Donovan, P. (1980) *Exp. Cell Res.* **126,** 453–458.
29. Lackie, J. M., and DeBono, D. (1977) *Microvasc. Res.* **13,** 107–112.
30. Lentnek, A., Schreiber, A., and MacGregor, R. R. (1976) *J. Clin. Invest.* **57,** 1098–1103.
31. MacGregor, R. R. (1976) *Am. J. Med.* **61,** 597–607.
32. MacGregor, R. R., Macarak, E. J., and Kefalides, N. (1978) *J. Clin. Invest.* **61,** 697–702.
33. MacGregor, R. R., Spagnuolo, P. J., and Lentnek, A. (1974) *N. Engl. J. Med.* **294,** 642–646.
34. Martz, E. (1980) *J. Cell Biol.* **84,** 584–598.
35. Mayrovitz, H., Tuma, R., and Wiedeman, M. (1980) *Microvasc. Res.* **20,** 264–274.
36. Mayrovitz, H., Wiedeman, M., and Tuma, R. (1977) *Thromb. Haemostasis* **38,** 823–830.
37. McCormick, J. R., Harkin, M., Johnson, K. J., and Ward, P. A. (1981) *Am. J. Pathol.* **102,** 55–61.
38. O'Flaherty, J. T., Craddock, P., and Jacob, H. S. (1978) *Blood* **51,** 731–739.
39. O'Flaherty, J. T., Kreutzer, L., and Ward, P. (1978) *Am. J. Pathol.* **90,** 537–550.
40. Pearson, J., Carleton, J. J., Beesley, J., Hutchings, A., and Gordon, J. (1979) *J. Cell Sci.* **38,** 225–235.
41. Penny, R., Galton, D. A. G., Scott, J. T., and Eisen, V. (1966) *Br. J. Haematol.* **12,** 623–632.
42. Rabinovitch, M., and DeStefano, M. (1973) *Exp. Cell Res.* **79,** 423–430.
43. Rabinovitch, M., and DeStefano, M. (1973) *J. Cell Biol.* **59,** 165–179.
44. Roseman, S. (1970). *Chem. Phys. Lipids* **5,** 270–297.
45. Ryan, G. B., and Majno, G. (1977) *Am. J. Pathol.* **86,** 183–276.
46. Sacks, T., Moldow, C., Craddock, P., Bowers, T., and Jacob, H. (1978) *J. Clin. Invest.* **61,** 1161–1167.
47. Sacks, T., Moldow, C. F., Craddock, P. R., and Jacob, H. S. (1977) *Clin. Res.* **25,** 347A.
48. Sawyer, P., and Pate, J. (1953) *Am. J. Physiol.* **175,** 113–117.
49. Schaack, T., Takeuchi, A., Spilberg, I., and Persillin, R. (1980) *Inflammation* **4,** 37–44.
50. Schmid-Schoenbein, G. W., Fung, V., and Zweifach, B. (1975) *Circ. Res.* **36,** 173–184.
51. Schmid-Schoenbein, G. W., Usami, S., Skalak, R., and Chien, S. (1980) *Microvasc. Res.* **19,** 45–70.
52. Smith, C. W., and Hollers, J. (1980) *J. Clin. Invest.* **65,** 804–812.
53. Smith, C. W., Hollers, J., Patrick, R. A., and Hassett, C. (1979) *J. Clin. Invest.* **63,** 221–229.
54. Taylor, R. F., Price, T. H., Schwartz, S., and Dale, D. (1981) *J. Clin. Invest.* **67,** 584–587.
55. Walther, B. T., Ohman, R., and Roseman, S. (1973) *Proc. Natl. Acad. Sci. U.S.A.* **70,** 1569–1573.
56. Webb, G., and Roth, S. (1974) *J. Cell Biol.* **63,** 796–803.

Complement-Mediated Intravascular Leukostasis and Endothelial Cell Injury

PHILIP R. CRADDOCK
Department of Medicine
University of Kentucky
Lexington, Kentucky

INTRODUCTION

The cardinal signs of acute inflammation, "rubor, et tumor cum calore et dalore," were first characterized by Cornelius Celsus in the first century A.D. (1), but it was not until the advent of the light microscope during the nineteenth century that some initial understanding of the mechanisms underlying these changes was developed. Dutrochet in 1824 was the first to describe the adherence of leukocytes to the vascular endothelium during the acute inflammatory response (2), and in his "Lectures on General Pathology" (3) Cohnheim gave a classic description of the early events of inflammation that provided considerable new insight into its mechanisms. In studies of acute inflammation in animals, he first observed vasodilatation of venules, capillaries, and arterioles accompanied by selective margination of "innumerable colorless corpuscles" on the venular and capillary walls. These events were followed by emigration of the adherent corpuscles to the exterior of the vessel and transudation of fluid through the wall, leading to infiltration and swelling of the surrounding tissues. Cohnheim contended that the critical event in acute inflammation is a "molecular change of the vessel walls" and that movement of circulating leukocytes is an entirely passive chain of secondary events. It was Metchnikoff who first suggested that additional chemotactic factors are required to draw the leukocytes from the intravascular compartment to and through the endothelium (4). Although these observations were made over a century ago, the underlying mechanisms of the early

PATHOBIOLOGY OF THE ENDOTHELIAL CELL

ISBN 0-12-521980-6

leukostatic response and the endothelial disruption that follows it are still not entirely understood. The most widely used experimental model of acute inflammation has been the Arthus phenomenon in rabbits, where vessel and tissue injury are initiated by the local conjunction of the antigen and antibody in the extravascular space (5). One of the earliest histological findings in this and other similar models is leukocyte clumping within the vessel (6,7), followed by adherence of the cellular aggregates to the vascular endothelium adjacent to the focus of tissue immune complex deposition. Simultaneous with this initial accumulation of granulocytes, endothelial swelling and increased vascular permeability are observed. Specific removal of circulating granulocytes by treatment with nitrogen mustard or heterologous anti-neutrophil antisera inhibit this vasculitic reaction in several species (8–11), and circulating granulocytes appear equally important as mediators of the systemic arteritis (but not the glomerulonephritis) associated with immune complex disease (12).

An equally important role in the acute inflammatory response is played by plasma complement. Although all factors responsible for the accumulation of granulocytes at sites of antigen–antibody deposition are not fully elucidated, much experimental attention has been focused on several complement fragments generated by activation of the classical or alternative pathways. C3a and C5a, the small cleavage fragments of C3 and C5 respectively, have anaphylotoxic properties including smooth muscle contraction, vasodilation, and histamine release from mast cells and basophils (13) and alone may account for some of the manifestations of acute inflammation. In addition, $C5a_{desarg}$, which results from rapid cleavage of the COOH-terminal arginine from C5a by a naturally occurring plasma carboxypeptidase (14), appears to be the major complement-derived chemotactic factor for granulocytes (15). Of primary importance *in vivo* in this regard are the findings that in animal models depletion of one or more components of the complement cascade leads to a diminished accumulation of granulocytes in the Arthus reaction (16). It has, however, been particularly difficult to study the mechanisms and effects of granulocyte aggregation and margination as isolated events and to define the complement fragment responsible in this system, because the initial leukostasis of the Arthus reaction is rapidly followed by chemotaxis, opsonization, and phagocytic internalization—all of which trigger several other cellular events, including oxidative metabolism, lysosomal exocytosis, and prostaglandin generation. Not only does migration of granulocytes through the vessel wall lead to mechanical disruption of the intracellular junctions between adjacent endothelial cells (17), but exposure of these migrating cells to C5a chemotaxin leads to active stimulation of lysosomal enzyme release (18), generation of toxic oxygen species such as superoxide, hydrogen peroxide, and hydroxyl radicals (19), and synthesis of several proinflammatory products of arachidonic acid catabolism, notably the thromboxanes (20) and leukotrienes (21). Subsequent opsonization and phagocytic internalization of the antigen–antibody complexes outside the vessel wall leads to further acceleration of these three complicating granulocyte responses.

GRANULOCYTE AGGREGATION AND HEMODIALYSIS LEUKOPENIA

The phenomenon of granulocyte aggregation, although first described by Metchnikoff as long ago as 1885 (4), and later by Fleck and Murczynska in 1947 (22), has received surprisingly little experimental attention. Metchnikoff noted that in septicemic frogs, "the white corpuscles, without leaving the vessel, attach themselves to the bacteria in the blood, and endeavor to surround them" and that under such circumstances where the inflammation stimulus resides primarily in the circulation, no visible extravasation of leukocytes occurs. In studies of ghetto inmates with epidemic typhus during World War II, Fleck and Murczynska found that the neutrophils in citrated blood samples from patients with active disease aggregated into tight little clusters after brief incubation *in vitro*, and that similar clumps were present in blood samples from rabbits with experimentally induced bacterial septicemia or endotoxinemia. It could be reasonably induced from first principles that the development of granulocyte aggregates around bacteria in the circulation would have a salutary effect, serving to rapidly eradicate viable organisms from the bloodstream. Indeed, subsequent studies support such a hypothesis. Following the intravenous infusion of encapsulated pneumococci into animals, polymorphonuclear leukocytes rapidly phagocytose bacteria, clearing them from the circulation and adhering them to the capillary endothelium, initially at least in the lung (23). Although this process constitutes an efficient, if not indispensable, method for the clearance of microbes from the circulation, it is of particular interest that the accumulation of the granulocytes within the pulmonary circulation causes significant lung damage manifested by increased capillary permeability and interstitial edema (24,25). Studies of the acute granulocytopenia of hemodialysis we began in 1974 produced considerable insight into the mechanism and consequences of intravascular granulocyte aggregation and pulmonary leukostasis.

Profound leukopenia, limited to granulocytes and monocytes, occurs in all patients during the first hour of hemodialysis with cellophane membrane dialyzers (26). Studies in dialyzed dogs suggested that this specific cytopenia results from reversible sequestration of these cells, not, as might be expected in the dialyzer, but in the pulmonary microvasculature (27), while Jensen *et al.* were able to induce identical leukopenia in normal humans by infusion of autologous, dialyzer-cellophane-incubated plasma (28). With the background data that certain complement fragments play an important role in stimulating granulocyte chemotaxis and phagocytosis (notably $C5a_{desarg}$ and C3b), we hypothesized that like other polysaccharides (29), dialyzer cellophane might be capable of activating the alternative pathway, and that complement fragment(s) so generated might be responsible for initiating the leukostatic response. Indeed, we were able to demonstrate that dialyzer cellophane is capable of activating the alternative pathway of the complement cascade *in vitro* and *in vivo*, and that infusion of plasma complement so activated into rabbits produced peripheral granulocytopenia and pulmonary leukostasis (Fig. 1) (30), identical to that which had been described earlier in dialyzed dogs (27). Fractionation of cellophane-incubated plasma revealed that the C fragments primarily responsible for the leukostasis were of an approximate molecular weight of 10,000–20,000—very similar to the material responsible for the granulocytopenia that results from direct intravas-

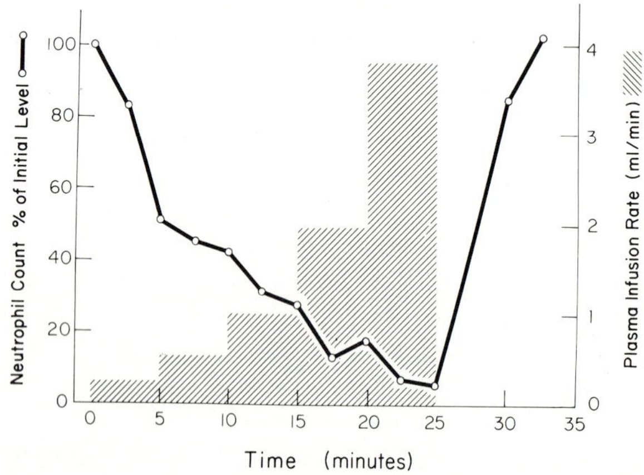

Fig. 1. The dose–response relationship between the rate at which dialyzer cellophane-incubated plasma (30) is infused into rabbits (shaded bars) and the severity of the granulocytopenia (continuous line) it produces.

cular C activation by purified cobra venom factor (31). Subsequent studies in rabbits infused with activated C (32) revealed that the site of leukostasis is primarily dependent on the site of infusion; intravenous infusion leads to pulmonary leukostasis and intraarterial infusion leads to systemic sequestration. It became evident from these studies that activated plasma complement is indeed capable of augmenting granulocyte adherence to artificial surfaces, including plastic petri dishes (30) and nylon fiber columns (32), and that there is a clear relationship between such augmented adhesiveness and C-induced granulocytopenia and pulmonary leukostasis in animals *in vivo* (32). Nevertheless, the mechanism of leukostasis was unclear and the exact nature of the complement fragment responsible was unknown.

Based on these early findings (30,32) we hypothesized that microvascular sequestration most likely resulted from C-mediated adhesion of individual granulocytes to the capillary and venular endothelium; unaware of Fleck's earlier description of granulocyte aggregation *in vivo* (22), we did not seriously consider that leukoembolism might be an important mechanism until we studied the effect of activated C on granulocytes stirring in a nephelometer (33). Contrary to our expectations, we found that the addition of activated plasma C to normal granulocytes produces a partially reversible, rapid shape change followed by a wave of aggregation (Fig. 2), and that the complement fragment responsible for this phenomenon is the chemotactic

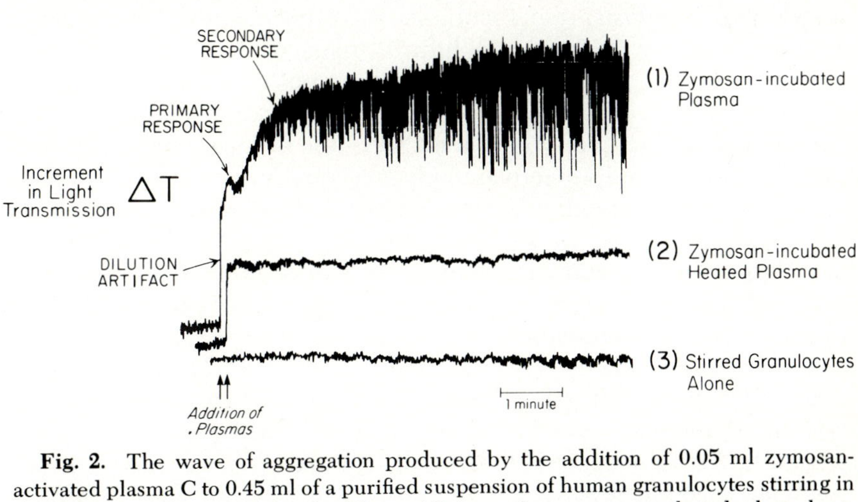

Fig. 2. The wave of aggregation produced by the addition of 0.05 ml zymosan-activated plasma C to 0.45 ml of a purified suspension of human granulocytes stirring in an aggregometer (tracing 1). The addition of zymosan-incubated, but heat-decomplemented (56°C/30 min) plasma (tracing 2) results only in a dilution artifact, and cells stirred alone (tracing 3) fail to develop any increment in light transmission or oscillation amplitude.

peptide $C5a_{desarg}$. No aggregating activity could be produced in C5-deficient plasma, aggregating activity was generated by direct tryptic digestion of pure C5, and the aggregating activity in normal zymosan- or cellophane-incubated plasma was ablated by incubation with anti-C5 but not anti-C3 antisera. That the formation of granulocyte aggregates does indeed occur *in vivo* was confirmed in studies in animals using intravital microscopy (34,35). Intraarterial infusion of autologous zymosan-activated plasma into rats produced immediate margination of fluorescein-labeled granulocytes in their mesenteric vessels, followed by the appearance of large granulocyte aggregates 30–70 μm in diameter. We concluded, therefore, that the specific granulocytopenia associated with intravascular complement activation results primarily from C5a-mediated granulocyte aggregation and embolization of these aggregates to the first large capillary bed they encounter.

ORGAN DYSFUNCTION ASSOCIATED WITH INTRAVASCULAR GRANULOCYTE AGGREGATION

It is not uncommon for patients to develop acute cardiopulmonary dysfunction (angina pectoris, dyspnea, and cardiac arrhythmias) during induction of hemodialysis (36), and we explored the possibility that pulmonary leukoembolism may play a role in such complications. We found that simultaneous with the occurrence of leukopenia during the first hour of dialysis, many patients developed arterial hypoxemia, hypocapnea, (Fig. 3), impaired diffusing capacity, and increased small airway closing volumes; one patient with severe predialysis granulocytopenia remained free of any such manifestations of lung dysfunction (37). Equally important were the findings that infusion of cellophane- or zymosan-activated plasma complement into rabbits and sheep produced equally severe lung dysfunction accompanied by readily demonstrable (in sheep) pulmonary edema and acute pulmonary hypertension, and that neutropenic animals developed no evidence of cardiopulmonary embarrassment (37). Although it is possible that such lung dysfunction might simply result from passive mechanical obstruction of the vascular bed by the leukostatic plugs, we explored the possibility that the increased vascular permeability might be due to a more active process initiated by the granulocytes; to do this we studied the effects of complement-stimulated granulocytes on cultured endothelial monolayers *in vitro* (38). The addition of human granulocytes and activated plasma C (or partially purified $C5a_{desarg}$) to such monolayers produced both functional and morphological damage

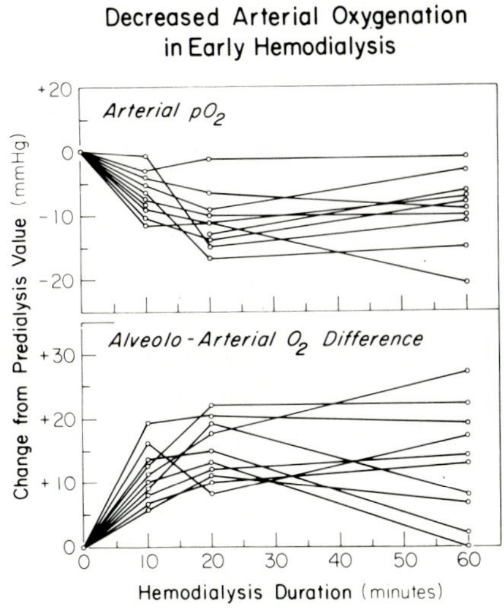

Fig. 3. The arterial hypoxemia (upper panel) and increments in alveoloarterial oxygen gradients (lower panel) that developed in 10 consecutive patients with chronic uremia during the induction of hemodialysis with cellophane membrane dialyzers.

that was dependent on close adhesion of the two cell types to each other and on the release of toxic oxygen radicals by the adherent granulocytes. Endothelial damage appeared to be independent of lysosomal exocytosis and was augmented by the addition of platelets or serotonin (39). In parallel studies we found that endotoxin will also cause granulocytes to adhere to and damage endothelium, an effect that is further potentiated by the availability of fresh plasma complement and inhibited by oxidant radical-scavengers including superoxide dismutase and catalase (40). Hoover *et al.* have subsequently reported that $C5a_{desarg}$ and other chemotaxins as well are indeed potent stimuli of granulocyte adhesion to vascular endothelium (41). We concluded therefore that some, but not all, of the lung dysfunction produced by pulmonary leukostasis is due to endothelial damage actively produced by such a mechanism; but Cooper *et al.* have recently reported that prostaglandin production may also be of importance (42). When these investigators infused autologous activated plasma complement into sheep, they detected the same pulmonary leukostasis and lung dysfunction as we had initially described, but they made the additional findings that these events were also associated with sudden and very significant increments in plasma levels of the thromboxane A_2 and PGI_2 metabolites, thromboxane B_2 and

6-keto-$PGF_{1\alpha}$, respectively, more importantly, pretreatment of the animals with indomethacin or sulfinpyrazone prevented both lung dysfunction and the increments in plasma levels of these end-products of arachidonate metabolism. Although the authors were unable to determine the cellular origin of these vasoactive prostaglandins, it is quite possible that thromboxane A_2 derived in part from leukocytes (20) may be responsible for the pulmonary hypertension they observed (43). It is unlikely that the elevated PGI_2 levels were responsible for the pulmonary hypertension in these animals because this compound generally produces vasodilation (44) and PGI_2 itself has recently been shown to ameliorate the effects of intravenous complement activation (45). Insofar as pulmonary leukostasis was equally apparent in animals treated with these prostaglandin inhibitors (42), it appears that the lung dysfunction is not simply the result of passive vascular occlusion by the granulocyte aggregates, and that a more active process such as the generation of oxidant radicals or vasoactive products of arachidonic acid are important contributing factors.

GRANULOCYTE AGGREGATION AND SHOCK LUNG

Recently it has become apparent that intravascular pulmonary leukostasis occurs in various conditions of nonimmune etiology in which activation of the plasma complement cascade was not formally considered of pathogenetic importance; one such clinical situation is the adult respiratory distress syndrome (ARDS) observed in patients as a frequent complication of sepsis, pancreatitis, and severe trauma (46,47). When lung tissue is biopsied in patients arriving in the emergency room immediately after massive trauma, the most outstanding histological finding is the presence of large numbers of polymorphonuclear leukocytes in the capillary and precapillary arterioles (46). We hypothesized on the basis of our hemodialysis studies, that this leukostasis might also be complement- and $C5a_{desarg}$-mediate, and that the granulocytes seen occluding the lung vessels might be responsible for the histological finding of endothelial cell swelling and interstitial edema seen in the systemic (48) as well as the pulmonary circulation (46) during very early ARDS. To explore the first hypothesis, we prospectively examined for $C5a_{desarg}$ plasma from patients at risk for ARDS, looking for a correlation between this, other evidence of complement activation, and the development of clinically evident respiratory insufficiency (49). Using granulocyte aggregometry as a simple and reproducible technique for the bioassay of $C5a_{desarg}$ in plasma (Fig. 4)

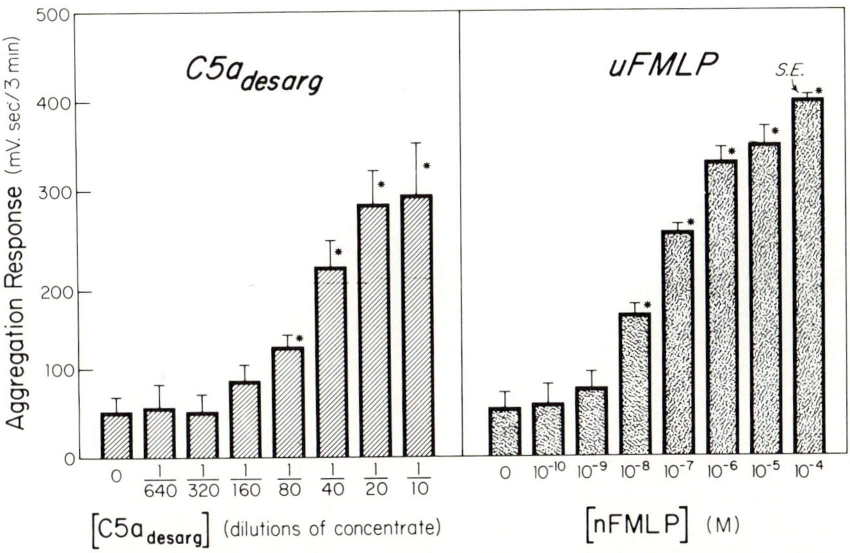

Fig. 4. Dose–response relationships between partially purified $C5a_{desarg}$ (left panel) and the synthetic chemotaxin *n*-formylmethionine leucinephenylalanine (nFMLP) (right panel), and the granulocyte aggregation responses each can produce. The aggregation responses were quantitated objectively by digital integration of the increments in light transmission (mV sec) over 3 min intervals (50).

(50,51), we found that in 61 patients at risk for ARDS as the result of trauma, sepsis or severe, prolonged hypotension, there was a strong and highly significant correlation between the presence of plasma aggregating activity and the subsequent onset of ARDS. Of the 33 patients who developed ARDS, 31 had abnormal plasma $C5a_{desarg}$ levels at diagnosis or in the previous 72 hr. Only 5 or 28 high-risk patients who did not develop ARDS had excess plasma $C5a_{desarg}$ levels. This correlation was even more striking when the 19 patients with sepsis were excluded (49). These findings supported our hypothesis not only that $C5a_{desarg}$-mediated leukostasis may also be involved in the pathogenesis of this form of severe lung injury, but also that such evidence of complement activation might be a useful prognostic factor in predicting ARDS in such patients.

Although the mechanism of its salutory effects is unknown, high doses of glucocorticoids have been advocated in the therapy of patients with ARDS and shock, regardless of etiology (52). Having shown that similar concentrations of methylprednisolone (MP) (30 mg/kg) and hydrocortisone (HC) (100 mg/kg) are capable of preventing C-mediated leukopenia and leukostasis in animals infused with purified

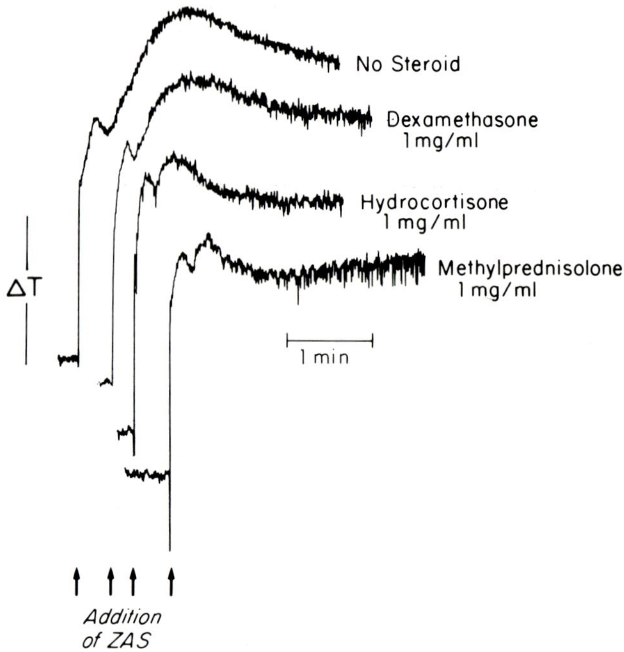

Fig. 5. The inhibition of C ($C5a_{desarg}$)-induced granulocyte aggregation responses induced by methylprednisolone at a final concentration of 1 mg/ml. Hydrocortisone and dexamethasone produced progressively less inhibitory effect. Granulocytes were exposed to the drugs for 1 min at 37°C prior to the addition of the aggregating stimulus (54).

cobra venom factor (53), we tested the capability of these two drugs and dexamethasone (DX) as well to inhibit granulocyte aggregation (Fig. 5) (54). The aggregation response engendered *in vitro* by zymosan-activated plasma ($C5a_{desarg}$) was inhibited by MP and HC at high pharmacological concentrations, whereas DX was without effect. MP and HC were also capable of retarding granulocyte aggregation induced by other stimuli, including the synthetic chemotaxin *n*-formylmethionine phenylalanine (nFMLP) and the ionophore A23187. But of even more clinical relevance was the finding that MP and HC also inhibited complement-mediated granulocyte aggregation in rats infused with zymosan-activated autologous plasma *in vivo*, and observed with the aid of fluorescence microscopy (54). To investigate the mechanism of the salutory effect of these glucocorticoids, we examined their effect on the kinetics of binding of isotopically labeled nFMLP to purified human granulocytes *in vitro* (55). Just as might have been predicted from the dampening effects of these steroids on membrane fluidity (56), both MP and HC, but not DX, produced dose-dependent inhibition of nFMLP binding, primarily by reducing the association rate constant for ligand–receptor interaction without significantly altering receptor numbers or effecting dissociation kinetics (55). We concluded that although other mechanisms, including direct inhibition of the activation of the plasma complement cascade (53), may also be operative, one of the major effects of MP and HC *in vivo* involves such interference with ligand–receptor interaction. Regardless of the exact mechanism however, it has recently been shown in sheep that premedication with MP prevents both the pulmonary hypertension and pulmonary edema usually produced by the infusion of endotoxin (57), an agent that *in vitro* at least is capable of producing complement- and granulocyte-mediated endothelial cell injury (40). MP was without effect when given after endotoxin infusion and inhibited only the high-protein pulmonary edema when it was given during infusion of endotoxin (57)—findings that are entirely consistent with the earlier observations that steroids are effective in the therapy of patients with shock only when given very early in the course of ARDS (58), but not if given 9 hr or more after it has become clinically evident (59).

THE REVERSIBILITY OF HEMODIALYSIS LEUKOPENIA

The major aspect of hemodialysis leukopenia that had not been explained was the transience of the phenomenon; granulocytes are sequestered in the lung only during the first hour of each 5 hr dialysis, after which they return to the circulation accompanied by a cohort of young cells freshly released from the bone marrow reserves (60) under the influence of another complement fragment C3e (61). When we measured the plasma $C5a_{desarg}$ generated by the dialyzer during hemodialysis with the aid of quantitative granulocyte aggregometry, we made two important observations (62). The first of these was that the $C5a_{desarg}$ present in plasma returning to the patient from the dialyzer is cleared very rapidly from the circulation and does not reappear in the arterial blood entering the dialyzer. These findings are entirely consistent with the additional data that, in rabbits at least, $C5a_{desarg}$ has a very brief intravascular half-life of only 3.0 ± 0.6 min (63). The second critical observation was that the generation of $C5a_{desarg}$ by the dialyzer continues unabated throughout dialysis, long after pulmonary leukostasis has reversed. If the primary hypothesis of $C5a_{desarg}$-mediated aggregation and leukoembolism was to remain tenable therefore, an additional explanation for this discrepancy was clearly warranted, an explanation that was first suggested by serendipitous observations in a uremic patient with disseminated cytomegalovirus infection. His predialysis plasma contained extraordinarily large quantities of $C5a_{desarg}$-related aggregating activity, and, in contrast to all other patients we have studied, he failed to develop the usual leukopenia of hemodialysis (Fig. 6). These findings raised the possibility that circulating granulocytes can become desensitized

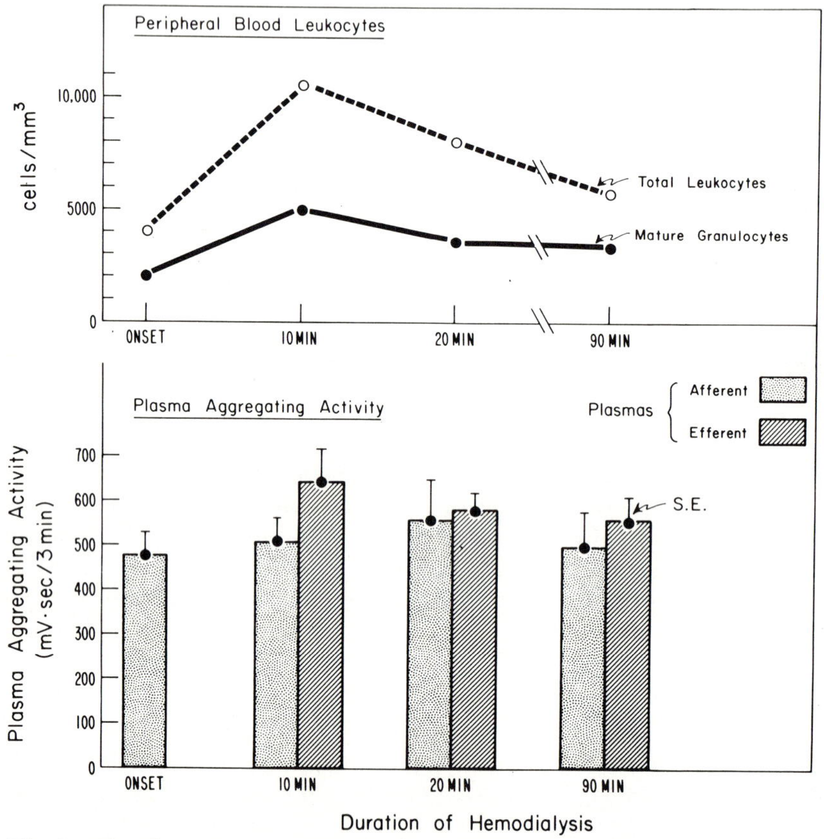

Fig. 6. The absence of the usual leukopenia and granulocytopenia (upper panel) during induction of hemodialysis in a patient with disseminated cytomegalovirus infection. When his predialysis plasma was assayed for $C5a_{desarg}$ by quantitative aggregometry (lower panel), it was found to contain very large quantities of $C5a_{desarg}$, most likely the result of the immune complexes seen in this disease. Consistently high levels of C5a were detected in his afferent (patient-to-dialyzer) plasma throughout dialysis, with relatively little contribution by the dialyzer into the efferent (dialyzer-to-patient) plasma.

to $C5a_{desarg}$ *in vivo* by prior exposure to $C5a_{desarg}$, as has already been demonstrated *in vitro* (64). Human granulocytes exposed to $C5a_{desarg}$ *in vitro* lose only their chemotactic responsiveness to $C5a_{desarg}$ itself and react normally to bacterial chemotactic factors or synthetic formylated peptides such as nFMLP; this specific deactivation is thought to result from ligand-induced and dose-related internalization of specific receptors within the plasma membrane (64,65), a process that appears to be only slowly and partially reversible. To explore this possibility we tested the aggregation responses of granulocytes from patients undergoing hemodialysis, immediately prior to and just after the occurrence of complement-mediated pulmonary leukostasis (Fig. 7). Prior to dialysis their cells responded near normally to the $C5a_{desarg}$ in efferent plasma, but after 120 min of dialysis the cells had become almost entirely unresponsive to the same aggregating stimulus. In striking contrast the same cells retained their aggregation responses to nFMLP and the higher concentrations of $C5a_{desarg}$ in zymosan-activated plasma (not shown). Chemotactic responses of these cells toward $C5a_{desarg}$

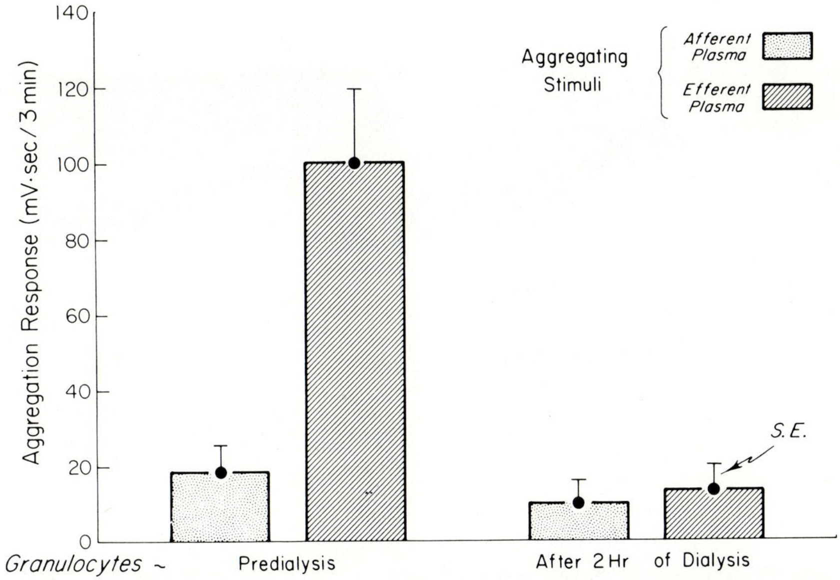

Fig. 7. The aggregation responses produced by the addition of the $C5a_{desarg}$ in efferent (dialyzer-to-patient) plasma to granulocytes drawn from five patients prior to dialysis (striped bar, left). The addition of afferent (patient-to-dialyzer) plasma, which contains very little $C5a_{desarg}$, produced no appreciable aggregation response (stippled bar, left). After 2 hr of hemodialysis, granulocytes from these five patients had become almost entirely unresponsive to the $C5a_{desarg}$ in efferent plasma (striped bar, right).

were also selectively impaired, and, regardless of the bioassay used, this selective deactivation could not be readily reversed by 2 hr incubation buffer or fresh AB-plasma. Identically selective deactivation of aggregation and chemotaxis was produced in normal granulocytes by short- or long-term exposure to $C5a_{desarg}$ *in vitro* (62). For example, cells primarily stimulated with undiluted zymosan-activated normal plasma become unresponsive when exposed to a second equipotent stimulus (Fig. 8) but retain normal responsiveness to nFMLP (not shown). On the other hand, cells exposed to a weaker primary stimulus (zymosan-activated plasma diluted to 1% in buffer) can retain some but not all of the capacity to respond to the same stimulus. We concluded, therefore, that the transience of pulmonary leukostasis of hemodialysis, and perhaps other clinical situations that result in continued intravascular generation of $C5a_{desarg}$, results primarily from this phenomenon of specific receptor down-regulation—of which hemodialysis may be the first described clinical example. The implications of these findings are twofold. The first is that some, if not necessar-

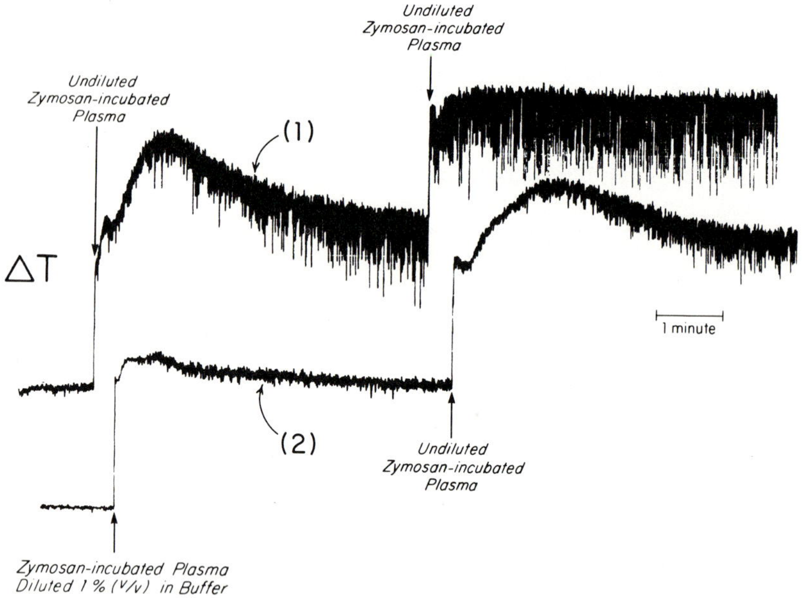

Fig. 8. Granulocytes that have undergone partially reversible aggregation in response to undiluted zymosan-activated plasma complement (tracing 1, left) developed selective refractoriness to the second stimulation with zymosan-activated plasma (tracing 1, right). Primary stimulation with zymosan-activated plasma complement diluted to 1% in buffer (tracing 2, left) results in only partial refractoriness to a second stimulation with undiluted zymosan-activated plasma complement (tracing 2, right).

ily all, of the infection propensity associated with hemodialysis (66) may result from such repetitive deactivation of granulocyte chemotactic responses. This possibility is clearly supported by the otherwise unexplained "antiinflammatory" effects of hemodialysis in patients with familial Mediterranean fever, psoriasis, and systemic lupus erythematosis (67–69). A second and perhaps more widely relevant ramification is that this down-regulation may serve to limit the deleterious effects of C-stimulated granulocytes on the integrity of the vascular endothelium of the lungs and other organs. Such a concept is quite consistent with the paradoxical frequency of autoimmune vasculitis seen in patients congenitally deficient in C5 or the earlier C components, which is not seen in patients lacking C6 through C8 (70).

CONCLUSIONS

These studies revealed therefore that the intravascular leukostasis and granulocyte aggregation seen in frogs by Metchnikoff (4) and later in humans by Fleck and Murczynska (22) may be more than just interesting trivia lacking in pathogenetic relevance. From the present studies there is little doubt that granulocyte aggregation and pulmonary leukoembolism resulting from intravascular complement activation can cause clinically evident damage to the pulmonary and perhaps systemic microvasculature, but it is becoming increasingly apparent that the same series of events may also have a salutory role in removing microorganisms from the circulation (24). Large numbers of granulocytes adherent to the endothelium of the pulmonary vascular bed clearly constitute a highly efficient filter, which can rapidly remove significant numbers of microorganisms from the bloodstream. In providing the matrix for this filtration system, however, the lung pays a price, manifest clinically as ARDS, which can itself lead to appreciable morbidity and mortality. More effective therapy for this syndrome of acute lung dysfunction will depend primarily on the development of a more complete understanding of its complex mechanism.

ACKNOWLEDGMENTS

This work was supported by a grant from the National Institutes of Health R01-CA15627, and the author has been the recipient of a Research Career Development Award HL-15627 also from the National Institutes of Health. The author is deeply indebted to his many coinvestigators and collaborators at the University of Minnesota and elsewhere, without whom these studies could not have been carried out, and is particularly grateful to Ms. Georgia Huard for her excellent technical assistance.

REFERENCES

1. Celsus, A. C. (1831) *In* "De Medicina" (translated from L. Targa's edition by A. Smith), p. 182. E. Cosc. London.
2. Dutrochet, R. J. H. (1824) "Recherches anatomiques et physiologique sur la structure intime des animaux et des vegetaux et sur leur motilité." Baillière et Fils, Paris.
3. Cohnheim, J. (1882) *In* "Lectures on General Pathology" (translated from the 2nd German edition by A. B. McKee), Vol. I. (New Sydenham Soc., London, 1889).
4. Metchnikoff, E. (1885) *J. R. Microsc. Soc.* **24,** 112–117.
5. Arthus, M. (1903) *C. R. Seances Soc. Biol. Ses. Fil.* **55,** 817–829.
6. Abell, R. G., and Schenk, H. P. (1938) *J. Immunol.* **34,** 195–213.
7. Ebert, R. H., Barclay, W. R., and Ahern, J. J. (1949) *J. Lab. Clin. Med.* **34,** 1596–1597.
8. Stetson, C. A. (1951) *J. Exp. Med.* **94,** 347–357.
9. Humphrey, J. H. (1955) *Br. J. Exp. Pathol.* **36,** 268–282.
10. Cochrane, C. G., Weigle, N. O., and Dixon, F. J. (1959) *J. Exp. Med.* **110,** 481–494.
11. Parish, W. E. (1969) *Br. J. Dermatol.* **81,** Suppl. 3, 28–35.
12. Wiggins, R. C., and Cochrane, C. G. (1981) *N. Engl. J. Med.* **304,** 518–520.
13. Cochrane, C. G., and Müller-Eberhard, H. J. (1968) *J. Exp. Med.* **127,** 371–386.
14. Bokisch, V. A., and Müller-Eberhard, H. J. (1970) *J. Clin. Invest.* **49,** 2127–2436.
15. Fernandez, H. N., Henson, P. M., Otani, A., and Hugli, T. E. (1978) *J. Immunol.* **120,** 109–115.
16. Cochrane, C. G., and Janoff, A. (1974) *In* "The Inflammatory Process" (B. W. Zweifach, L. Grant, and R. T. McCluskey, eds.), 2nd ed., Vol. 3, pp. 109–117. Academic Press, New York.
17. Marchesi, V. T., and Florey, H. W. (1960) *Q. J. Exp. Physiol. Cogn. Med. Sci.* **45,** 343–348.
18. Goldstein, I., Hoffstein, S., Gallin, J., and Weissmann, G. (1973) *Proc. Natl. Acad. Sci. U.S.A.* **70,** 2916–2920.
19. Goldstein, I. M., Roos, D., Kaplan, H. B., and Weissmann, G. (1975) *J. Clin. Invest.* **56,** 1155–1163.
20. Goldstein, I. M., Malmsten, C. L., Kindahl, H., Kaplan, H. B., Radmark, O., Samuelsson, B., and Weissmann, G. (1978) *J. Exp. Med.* **148,** 787–792.
21. Borgeat, P., Hamberg, M., and Samuelsson, B. (1976) *J. Biol. Chem.* **251,** 7816–7820.
22. Fleck, L., and Murczynska, Z. (1947) *Tex. Rep. Biol. Med.* **5,** 156–167.
23. Wood, W. B., Smith, M. R., Perry, W. D., and Berry, J. W. (1951) *J. Exp. Med.* **94,** 521–534.
24. Hosea, S., Brown, E., Hammer, C., and Frank, M. (1980) *J. Clin. Invest.* **66,** 375–382.
25. Brigham, K. L., Woolverton, W. C., Blake, L. H., and Staub, N. C. (1974) *J. Clin. Invest.* **54,** 792–804.
26. Kaplow, L. S., and Goffinet, J. A. (1968) *JAMA, J. Am. Med. Assoc.* **203,** 1135–1137.
27. Toren, M., Goffinet, J. A., and Kaplow, L. S. (1970) *Blood* **36,** 337–340.
28. Jensen, D. P., Brubaker, L. H., Nolph, K. D., Johnson, C., and Nothum, R. (1973) *Blood* **41,** 399–408.
29. Snyderman, R., and Pike, M. C. (1975) *Infect. Immunol.* **11,** 273–279.
30. Craddock, P. R., Fehr, J., Dalmasso, A. P., Brigham, K. L., and Jacob, H. S. (1977) *J. Clin. Invest.* **59,** 879–888.
31. McCall, C. E., de Chatelet, L. R., Brown, D., and Lachman, P. (1974) *Nature (London)* **249,** 841–843.

32. O'Flaherty, J. T., Craddock, P. R., and Jacob, H. S. (1978) *Blood* **51,** 731–739.
33. Craddock, P. R., Hammerschmidt, D., White, J. G., Dalmasso, A. P., and Jacob, H. S. (1977) *J. Clin. Invest.* **60,** 260–264.
34. Hammerschmidt, D. E., White, J. G., Craddock, P. R., and Jacob, H. S. (1979) *J. Clin. Invest.* **63,** 798–803.
35. Hammerschmidt, D. E., Harris, P. D., Wayland, J. H., Craddock, P. R., and Jacob, H. S. (1981) *Am. J. Pathol.* **102,** 146–150.
36. Hampers, C. L., Schupak, E., Lowrie, E. G., and Lazarus, J. J. (1973) *In* "Long Term Hemodialysis; The Management of the Patient with Chronic Renal Failure," 2nd ed., pp. 80–130. Grune & Stratton, New York.
37. Craddock, P. R., Fehr, J., Brigham, K. L., Kronenberg, R. S., and Jacob, H. S. (1977) *N. Engl. J. Med.* **296,** 769–774.
38. Sacks, T., Moldow, C. F., Craddock, P. R., Bowers, T. K., and Jacob, H. S. (1978) *J. Clin. Invest.* **61,** 1161–1167.
39. Boogaerts, M. A., Jacob, H. S., Yamada, O., Kramer, P., and Moldow, C. F. (1980) *Clin. Res.* **28,** 490a.
40. Yamada, O., Moldow, C. F., Sacks, T., Craddock, P. R., Boogaerts, M. A., and Jacob, H. S. (1981) *Inflammation* **5,** 115–126.
41. Hoover, R. L., Folger, R., Haering, W. A., Ware, B. R., and Karnovsky, M. J. (1980) *J. Cell Sci.* **45,** 73–86.
42. Cooper, J. D., McDonald, J. W. D., Ali, M., Menkes, E., Masterson, J., and Klement, P. (1980) *Surgery* **88,** 215–221.
43. Friedman, L. S., Fitzpatrick, T. M., Bloom, M. F., Ramwell, P. W., Rose, J. C., and Kot, P. A. (1979) *Circ. Res.* **44,** 748–751.
44. Hyman, P. L., Mathe, A. A., Leslie, C. A., Matthews, C. C., Bennett, J. T., Spanhake, E. W., and Kadowitz, D. J. (1978) *J. Pharmacol. Exp. Ther.* **207,** 388–401.
45. Krausz, M., Utsunomiya, T., Feverstein, G., Wolfe, J. H. N., Shepro, D., and Hechtman, H. G. (1981) *J. Clin. Invest.* **67,** 1118–1125.
46. Schlag, G., Voigt, W. H., Schnells, G., and Glatzl, A. (1976) *Anesthesist* **25,** 512–521.
47. Sandritter, W. C., Mittermayer, C., Piede, U. M., Freudenberg, N., and Grimm, H. (1978) *Beitr. Pathol.* **162,** 7–23.
48. Schlag, G., Voigt, W. H., Schnells, G., and Glatzl, A. (1977) *Anesthesist* **26,** 612–622.
49. Hammerschmidt, D. E., Weaver, L. J., Hudson, L. D., Craddock, P. R., and Jacob, H. S. (1980) *Lancet* **1,** 947–949.
50. Craddock, P. R., White, J. G., Weisdorf, D. J., and Hammerschmidt, D. E. (1980) *Inflammation* **4,** 381–395.
51. Hammerschmidt, D. E., Bowers, T. K., Lammi-Keefe, C., Jacob, H. S., and Craddock, P. R. (1980) *Blood* **55,** 898–902.
52. Sladen, A. (1976) *J. Thorac. Cardiovasc. Surg.* **21,** 800–805.
53. O'Flaherty, J. T., Craddock, P. R., and Jacob, H. S. (1977) *Proc. Soc. Exp. Biol. Med.* **154,** 206–209.
54. Hammerschmidt, D. E., White, J. G., Craddock, P. R., and Jacob, H. S. (1979) *J. Clin. Invest.* **63,** 798–803.
55. Skubitz, K. M., Craddock, P. R., Hammerschmidt, D. E., and Jacob, H. S. (1981) *J. Clin. Invest.* **67,** 1658–1667.
56. Lawrence, D. K., and Gill, E. W. (1975) *Mol. Pharmacol.* **11,** 280–286.
57. Brigham, K. L., Bowers, R. E., and McKeen, C. R. (1981) *J. Clin. Invest.* **67,** 1103–1110.
58. Schumer, W. (1976) *Ann. Surg.* **184,** 333–341.

59. Thompson, N. H., Gurley, B., Lutz, D., Jackson, L., Kuols, L., and Morris, I. (1976) *Clin. Res.* **24,** 258a.
60. Brubaker, L. H., and Nolph, K. D. (1971) *Blood* **38,** 623–631.
61. Ghebrehiwet, B., and Müller-Eberhard, H. J. (1979) *J. Immunol.* **123,** 616–621.
62. Skubitz, K. M., and Craddock, P. R. (1981) *J. Clin. Invest.* **67,** 1383–1391.
63. Weisdorf, D. J., Hammerschmidt, D. E., Jacob, H. S., and Craddock, P. R. (1981) *J. Lab. Clin. Med.* **98,** 823–830.
64. O'Flaherty, J. T., Kreutzer, D. L., Showell, H. J., Vitauskas, G., Becker, E. L., and Ward, P. A. (1979) *J. Cell Biol.* **80,** 564–572.
65. Chenoweth, D. E., and Hugli, T. E. (1982) *J. Cell Biol.* (in press).
66. Lowrie, E. G., Lazarus, J. M., Mocelin, A. J., Bailey, G. L., Hampers, C. L., Wilson, R. E., and Merrill, J. P. (1973) *N. Engl. J. Med.* **288,** 863–867.
67. Rubinger, D., Friedlander, M. M., and Popovtzer, M. M. (1979) *N. Engl. J. Med.* **301,** 142–144.
68. McEvoy, J., and Kelly, A. M. T. (1976) *Ulster Med. J.* **45,** 76–78.
69. Rao, T. K. S. (1978) *In* "Strategy in Renal Failure" (E. A. Friedman, ed.), pp. 320–343. Wiley, New York.
70. Agnello, V. (1978) *Medicine (Baltimore)* **57,** 1–23.

T-Lymphocyte Activation by Antigen Presented by HLA-DR Compatible Endothelial Cells

DENIS R. BURGER,[1,2,3] R. MARK VETTO,[1,3]
ANNE HAMBLIN,[4] AND D. C. DUMONDE[4]

[1]*Surgical Research Laboratory*
Veterans Administration Medical Center
Portland, Oregon

[2]*Department of Microbiology and Immunology*
Oregon Health Sciences University
Portland, Oregon

[3]*Department of Surgery*
Oregon Health Sciences University
Portland, Oregon

[4]*Department of Immunology*
St. Thomas' Hospital
London, England

INTRODUCTION

Endothelial cells line the vessels and lymphatics and therefore provide the barrier between circulating T cells and extravascular tissues. If endothelial cells could fulfill the dual requirement (1,2) for expression of T-cell activation, they might play a more important role in cell-mediated responses than previously recognized. Since endothelial cells express one of the two signals required for T-cell activation (Ia determinant 3,4), acquisition of the second signal (antigen) would satisfy this requirement. They could acquire the second signal when antigen is introduced into the tissue by "processing" and presenting it on the lumenal surface of capillaries. The presentation of antigen in the context of Ia determinants on intravascular capillary surfaces could provide the signal for antigen-specific T cells to localize at those sites, become activated, release inflammatory mediators, and/or egress into

PATHOBIOLOGY OF THE ENDOTHELIAL CELL

ISBN 0-12-521980-6

the tissue. This would represent a model for specific T-cell localization at the site of a cell-mediated reaction. A crucial feature for the validity of this model is the extent to which endothelial cells can signal T cells.

Recently Hirschberg *et al.* (5) have provided preliminary evidence that endothelial cells fulfill the dual requirement for expression of T-cell activation. The experiments presented here were designed to evaluate the extent to which endothelial cells fulfill this requirement and activate antigen-specific T cells. We have shown that cultured endothelial cells from umbilical veins are as effective as macrophages in signaling lymphocyte activation. Moreover, activation of T cells by antigen-pulsed endothelial cells is genetically restricted at the HLA-DR locus.

MATERIALS AND METHODS

ISOLATION OF BLOOD MONONUCLEAR CELLS (MNC)

Mononuclear cells were separated from heparinized (10 μm/ml) blood by a Percoll buoyant density technique modified after previously described procedures (6,7). Polyvinylpyrrolidone-coated colloidal silica (Percoll, Pharmacia Fine Chemicals, Uppsala, Sweden) was made isotonic by adding 1 part 10 X medium 199 (M 199) to 9 parts Percoll. This stock solution has a density of 1.1294 g/ml and is referred to as "100% Percoll." The 100% Percoll was diluted with medium RPMI 1640 (RPMI) to yield 70% (1.090 g/ml), 60% (1.077 g/ml), 50% (1.067 g/ml), and 40% (1.056 g/ml) solutions. Approximately 15 ml blood was layered on 9 ml 60% Percoll in 25 ml conical centrifuge tubes and spun at 500 g at 22–25°C for 20 min. MNC were harvested from the plasma–Percoll interface by aspiration, washed twice in RPMI, and counted.

STANDARD LYMPHOCYTE PROLIFERATION ASSAY

MNC were diluted to 1×10^6 cells/ml in RPMI supplemented with antibiotics and 30% human plasma or heat inactivated (56°C for 30 min) fetal calf serum and 100 μl (1×10^5 cells) distributed in flat bottom, 96 well Linbro (Flow Labs Ltd., Irvine, Scotland) microplates. Additional 100 μl preparations containing antigen dilutions, other cells, or media alone were added in quadruplicate to appropriate wells depending on experimental design. The microplates were incubated in a 5% CO_2, humidified atmosphere for 2–8 days, the last 24 hr with

0.2 μCi [^{3}H]thymidine (specific activity 2 Ci/mmole, The Radiochemical Centre, Amersham, England). The cultures were harvested and counted by commonly used automated techniques (8).

PRIMARY IMMUNIZATION *IN VITRO*

Modifications to techniques described by Rodey *et al.* (9) and Hensen and Elferink (10) were used. MNC were placed in primary culture in 250 ml tissue culture flasks at a cell density between 1–2 × 10^6 cells/ml (40–100 × 10^6 cells/flask) RPMI supplemented with 15% antologous human plasma or heat-inactivated (56°C for 30 min) fetal calf serum and antigen (1–100 μg/ml) for 8–22 days in a humidified 5% CO_2 atmosphere. At the end of the primary culture period, the cells were washed twice and counted and viability was assessed. Secondary (restimulation) cultures were set up in microplates in a manner identical to that described for standard lymphocyte proliferation (see above) with 1 × 10^5 viable primary cells/well.

Quadruplicate cultures were supplemented with antigen and fresh monocytes (1 × 10^4/well) or macrophages harvested from the primary culture flasks (see later). These secondary cultures were labeled with 0.2 μCi [^{3}H]thymidine (specific activity 2 Ci/mmole) for the last 24 hr of the 2–5 day secondary incubation period and harvested.

MACROPHAGE ISOLATION

Cells of the mononuclear phagocytic series (Mϕ) were isolated from MNC by adherence properties to plastic (11). Tissue culture flasks (250 ml) were pretreated (12) with 10 ml fetal calf serum at 4°C for 1–4 days before use. MNC (50–60 × 10^6 cells) were added to the pretreated, washed flasks in RPMI with 15% fetal calf serum. After 1–2 hr incubation at 37°C and 5% CO_2, the flasks were vigorously rinsed with media to remove nonadherent cells, cold RPMI with 15% fetal calf serum and 0.2% EDTA was added and the flasks were held for 20–30 min at 4°C. The macrophages were removed by shaking, washed twice, and counted.

MACROPHAGE DEPLETION

Macrophage depletion required three successive steps: adherence to plastic, passage over Sephadex G-10, and buoyant density centrifugation. MNC, routinely containing 5–15% monocytes, were first subjected to adherence on fetal calf serum-pretreated plastic flasks as described above. After 1–2 hr the nonadherent cells were washed from the flasks and resuspended in 2 ml RPMI. At this stage the MNC contain 1–2% monocytes by esterase staining (13). Sephadex G-10 (Pharmacia Fine Chemicals, Inc., Piscataway, N.J.) was resuspended in 2–3X bed volume in distilled water and autoclaved for 10 min. The sterile slurry was washed with RPMI and stored at 4°C. A column consisting of a plastic 10 ml syringe barrel was prepared and used as previously described (14). Briefly, the column was packed by gravity and washed with RPMI with 20% fetal calf serum at 37°C and 5% CO_2. Cells (2ml) were added and the column held flowless for 30 min at 37°C and 5% CO_2. The cells were eluted with RPMI with 20% fetal calf serum in 5–10 ml and pelleted (10 min at 250 g) in a 15 ml conical centrifuge tube. At this stage the cells characteristically contained 0.1–0.4% monocytes. The cell pellet is overlayered in succession with 2 ml 100%, 2 ml 70%, 2 ml 60%, 4 ml 50%, and 2 ml 40% Percoll. The discontinuous gradient was centrifuged for 10 min at 500 g at 22–25°C. T lymphocytes, free of any detectable monocytes, band at the 60–50% Percoll interface (6,7), since they cannot enter the 50% (1.067 g/ml) solution, whereas residual monocytes and B cells are found at the 50–40% interface and in the 40% Percoll solution. These T cells are devoid of detectable monocytes by latex ingestion and esterase staining (13) and will not support antigen-induced lymphocyte proliferation without Mϕ addition.

ENDOTHELIAL CELLS

Endothelial cells were isolated from the vein of umbilical cords by a modification of the technique described by Jaffe *et al.* (15). The fresh cords were thoroughly flushed with calcium–magnesium-free Puck's saline A, cannulated and filled with 0.2% collagenase (Sigma Chemical Co., No. C-0130, St. Louis, Mo.). After 10 min incubation at 37°C, the endothelial cells were flushed out with 100 ml RPMI and washed three times (10 min at 500 g). The cells were dispersed into 250 ml tissue culture flasks at a density of 500 cells/mm^2 and cultured in RPMI with 20% fetal calf serum at 37°C and 5% CO_2. Approximately 10% of the cultures grew vigorously, becoming monolayers within 4–6 days (4–5 × 10^6 endothelial cells per flask). The cells were released from the plastic by a brief treatment (3 min) with 0.2% EDTA-collagenase or 0.2% EDTA alone at room temperature and plated in flat bottom microplates at various densities. Three to 4 hr after plating the plates were X-irradiated (2000 rads, 20 Gy) or treated with 20 μg mitomycin C for 30 min and washed. The cells were confirmed as endothelial cells by light and electron microscopy, growth patterns in culture, and specific markers (factor VIII-related antigen). Smooth muscle cells and fibroblasts were grown from explants from the same umbilical cords in some experiments. These cells were identified by classic morphological criteria and were used as "control cells" in certain experiments.

HLA TYPING

Cell donors were healthy volunteers whose HLA phenotypes were established by standard serologic techniques (16) in two separate laboratories. HLA-DR phenotypes of endothelial cells were established by direct typing of the endothelial cells using a ^{51}Cr-release assay (3) and by typing the monocytes (17) from the homologous cord blood. Endothelial cells reacted poorly to polymorphic HLA-DR sera and a HLA-DR phenotype could not be established using endothelial cells alone.

ANTISERA

Polymorphic anti-HLA-DR sera were obtained from the NIH, from exchange with other investigators, and from local sources. All sera used had been extensively evaluated and tested in at least one other laboratory. The monoclonal antimacrophage sera was a gift of Howard Raff, Department of Microbiology, University of Washington, Seattle.

RESULTS AND DISCUSSION

The experiments presented here were designed to evaluate the extent to which endothelial cells could substitute for macrophages in presentation of antigen for T-cell activation. The strategy employed was first to confirm that T-cell activation in our hands was macrophage dependent, genetically restricted at the HLA-DR locus, and blocked by appropriate anti-Ia sera. After this definition of the proliferation system, antigen-pulsed endothelial cells were used to replace the macrophage requirement for T-cell proliferation. In subsequent experiments the genetic restriction of endothelial cell presentation of antigen was evaluated.

ANTIGEN PRESENTATION BY MACROPHAGES

Lymphocyte donors with marked dermal reactivity to microbial antigens (PPD, Candida, SK-SD) and hemocyanins (KLH and HCH) were selected for use in these experiments. The lymphocyte transformation assay used demonstrated dose-dependent proliferation to these antigens with typical kinetics (6-day maximal responses).

The macrophage dependence of the antigen-induced proliferation was evaluated in 13 experiments. Macrophages were removed from responding cell populations in three sequential steps: adherence to fetal calf serum-treated plastic culture vessels, absorption to Sephadex G-10, and buoyant density centrifugation in a discontinuous Percoll gradient. This combination of steps was selected based on macrophage contamination and diminution of antigen responsiveness of the resulting T cells. After depletion macrophage content was assessed by latex injection and esterase staining and was less than one positive cell in 2000 cells evaluated. Although the first two procedures were sufficient to reduce macrophage contamination to less than 0.2% and eliminate antigen-induced proliferation in some cell donors, elimination of antigen responsiveness in strongly responsive donors was not achieved until after the buoyant density step. An example of the effectiveness of this combination of steps with highly responsive cells is found in Table I (representative of 13 experiments). Response to 5 μg PPD before macrophage depletion was 27,000 cpm with a control value of 452 cpm and after depletion was reduced to 128/96 cpm (PPD/control). Extensive dose response and kinetic studies using macrophage-depleted cells eliminated the possibility that the depleted populations expressed altered response characteristics that would be unnoticed in the conventional assay (six experiments, data not shown).

To demonstrate that the depleted T cells were capable of response, macrophages were added back to the cultures. Table I shows that as little as 1% (10^3 cells) macrophage addition resulted in partial reconstitution of the strong responses. Addition of 10% (10^4 cells) macrophages routinely restored proliferation and was judged to be the most efficient concentration in each of the 13 experiments. The macrophage dependence of human T-cell activation is well documented (18–20). Addition of greater than 25% macrophages diminished the response in five of six experiments (data not shown). High concentrations of macrophages resulted in diminished responses, as has been previously reported (18,21,22).

GENETIC RESTRICTION OF ANTIGEN PRESENTATION BY MACROPHAGE

Genetic restriction was evaluated by addition of allogeneic macrophages of defined HLA-A, B, and DR specificities to cultures depleted of autologuous antigen-presenting cells. In each of nine experiments five different responding T-cell (macrophage-depleted) popula-

TABLE I
Proliferation Response of Lymphocytes Following Depletion of Antigen Presenting Cells (Macrophages)[a]

	Response of cultured lymphocytes (cpm)					
		After macrophage depletion with macrophages added (number cells/culture)				
Antigen in culture/dose (μg/ml)	Before macrophage depletion	None	10^2	10^3	3.3×10^3	10^4
None	452	96	115	250	380	441
PPD/2	14,200	121	196	6,420	9,220	16,420
PPD/5	27,000	128	318	13,813	24,500	29,400
PPD/25	22,400	78	188	10,640	21,600	26,400
KLH/2	1,300	129	—[b]	486	642	1,512
KLH/25	2,800	112	—	675	1,412	2,340
KLH/50	910	86	—	341	511	1,120
Candida/10	3,800	110	—	2,142	3,620	4,250

[a] The lymphocytes used in this experiment came from a donor who was PPD and *Candida* responsive and immunized to KLH. Standard deviations were less than 15% of the mean in all cases.

[b] (—) Indicates not done.

tions were evaluated with macrophages from four HLA-typed donors. To avoid strong MLR reactivity masking antigen responsiveness, macrophage addition at 1% concentrations ($10^3/10^5$ T cells) was employed. A correlation of responsiveness to compatibility between T cells and macrophages at the HLA-A or B locus was not evident (data not shown). However, there was a requirement for at least partial compatibility at HLA-DR in 37 of 40 combinations evaluated (Fig. 1). An example of one of the nine experiments is presented in Table II. It should be noted that even with only 1% allogeneic macrophages present some MLR reactivity was observed. Overall, when T cells and macrophages were completely disparate, 12 of 14 combinations exhibited completely negative responses. In two cases, partial responsiveness (about 50% of the response with autologous macrophages) was detected. When T cells and macrophages shared one HLA-DR determinant, 12 of 13 combinations were positive, with an average response of 51% of the autologous combination. In two cases nearly complete reconstitution of proliferation was observed. Finally, when T cells and macrophages were HLA-DR identical but allogeneic, responsiveness was significant in all 13 combinations. These combinations were as effective as autologous cells [mean response was 105% of the autologous response (Fig. 1)]. The experiments evaluating restriction using

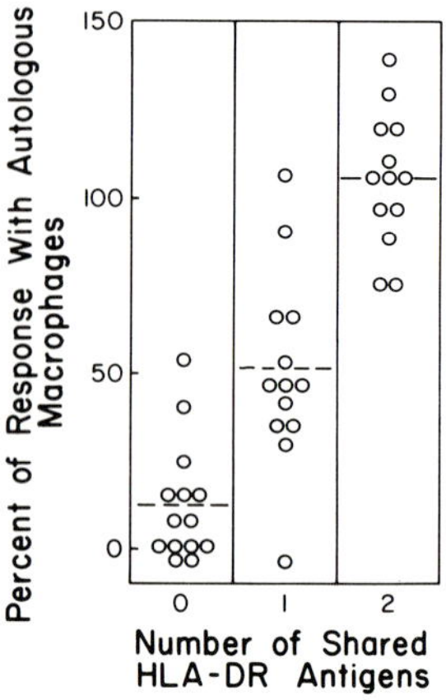

Fig. 1. Lymphocyte proliferation responses to antigen- (KLH, *Candida*, SK-SD, PPD) pulsed allogeneic macrophages are plotted as a function of disparity at the HLA-DR locus. (Dashed line represents mean response.)

macrophages generally agree with the data of others (9,23–25). Some of the disparate values in Figs. 1 and 5 could be explained by normal experimental variability or influence of HLA-DR cross-reactive groups.

TABLE II
HLA-DR Restriction of Antigen Presentation by Macrophages (Mϕ) on the Proliferation Response of Lymphocytes

Responding lymphocytes (HLA-DR)	Antigen in culture (10 μg/ml)	Response of Mϕ-depleted lymphocytes (mean cpm ± SD) with macrophages added back[a]				
		None	DR 3/5	1/5	2/2	3/6
DR 3/5	None	215 ± 31	220 ± 36	360 ± 41	620 ± 158	320 ± 22
	SK-SD	210 ± 26	2180 ± 310	1590 ± 215	1020 ± 216	1750 ± 115
DR 2/2	None	180 ± 19	620 ± 42	ND[c]	420 ± 51	1050 ± 86
	Candida	192 ± 22	710 ± 86		6200 ± 562	1100 ± 171
DR 2/4	None	250 ± 18	850 ± 111	720 ± 71	640 ± 106	1100 ± 121
	SK-SD	310 ± 6	920 ± 122	810 ± 61	2100 ± 186	960 ± 106
DR 3/6	None	310 ± 27	910 ± 116	ND	510 ± 28	420 ± 32
	KLH	210 ± 24	2250 ± 281		610 ± 59	3410 ± 301
DR 2/—[b]	None	90 ± 18	890 ± 49	ND	420 ± 31	1050 ± 88
	HCH	65 ± 6	610 ± 71		2250 ± 116	1120 ± 92

[a] The final concentration of macrophages was 1% (1×10^3 macrophages/1×10^5 lymphocytes) to avoid stimulation of strong MLR reactions by allogeneic macrophages.

[b] (—) Indicates unknown DR antigen.

[c] ND Indicates not done.

SUPPRESSION OF LYMPHOCYTE PROLIFERATION BY POLYMORPHIC ANTI-HLA-DR SERA

Since the restricted T-cell proliferation appeared to be HLA-DR dependent, antisera directed at the HLA-DR determinants of presenting cells should diminish antigen responsiveness. We evaluated this premise in eight experiments representing 35 different T-cell macrophage combinations where both HLA-DR haplotypes were shared. A representative experiment is shown in Table III. Antisera directed at HLA-DR determinants on macrophages consistently reduced T-cell proliferation. Addition of anti-DR3, w6 or anti-DR 1 + 5 reduced responsiveness to antigen-pulsed HLA-DR 3/5 macrophages by about 50%. Anti-DR 1 + 5 treatment of DR 1/5 macrophages completely diminished antigen-induced proliferation. The cumulative data from these experiments are presented in Fig. 2. It can be seen that the addition of antiserum irrelevant to the macrophage HLA-DR type had only a slight effect when compared to the response with control sera added (the mean was 88% of control). When antiserum was directed toward a single HLA-DR determinant, the reduction in responsiveness was more evident but variable (mean = 57% of control). The effect of antiserum to both DR determinants on the antigen presenting cells was pronounced and resulted in a mean reduction to 21%. These findings were in agreement with previous experiments (26–28). Previous investigations (28) have shown also that antisera against HLA-DR de-

TABLE III
Suppression of Lymphocyte Proliferation by Polymorphic Anti-HLA-DR Antisera Treatment of Macrophages

Responding lymphocytes (HLA-DR)	Antigen in culture (10 μg/ml)	Anti-HLA-DR antisera[a]	Response of macrophage-depleted lymphocytes with macrophages added back (HLA-DR)[b]		
			None	DR 3/5	DR 1/5
DR 3/5	None	None	215 ± 31	220 ± 36	360 ± 41
	SK-SD	None	210 ± 26	2180 ± 310	1590 ± 251
	SK-SD	2	185 ± 21	2040 ± 286	1440 ± 162
	SK-SD	3 + w6	170 ± 17	1180 ± 141	1510 ± 219
	SK-SD	4	245 ± 16	2140 ± 211	1610 ± 158
	SK-SD	1 + 5	165 ± 14	950 ± 108	410 ± 52

[a] Antisera titrated from 1 to 25 μl/culture and data from maximal suppression at a single antisera dilution (5 μl/200 μl culture) are shown.

[b] Final concentration of macrophages was 1% (1×10^3 macrophages/1×10^5 lymphocytes).

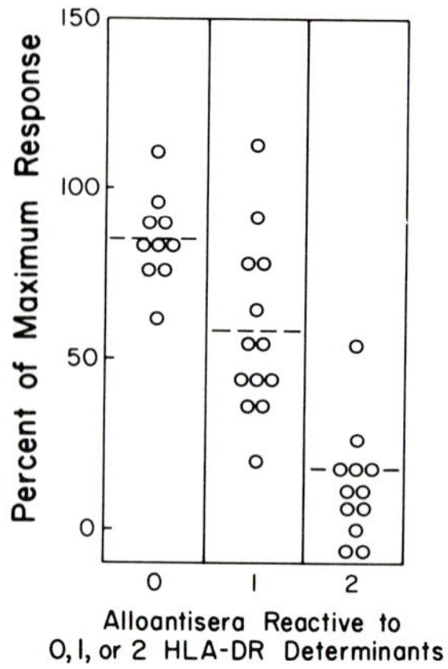

Fig. 2. Scatter diagram of pooled experiments employing allogeneic, polymorphic anti-HLA-DR antisera to inhibit antigen- (KLH, *Candida*, SK-SD, or PPD) induced lymphocyte proliferation is shown. The maximum response was the antigen-induced proliferation obtained with normal human sera added to cultures rather than the HLA-DR alloantisera.

terminants block at the level of the antigen presenting cells and do not interfere with activation at the lymphocyte level. These data, along with the graded responses observed in the restriction experiments reported in the previous section, suggest that the responsive T cells in heterozygotes represent two populations: one directed to antigen and one HLA-DR haplotype and the other directed to antigen and the second HLA-DR haplotype. Recent experiments by Hirschberg *et al.* (29) and from our laboratory (30) suggest that this is the case.

ANTIGEN PRESENTATION BY ENDOTHELIAL CELLS

Endothelial cells were cultured from umbilical veins and propagated *in vitro*. They were identified by classic morphology and production of factor VIII-related antigen (31,32). Cells from cultures in the first or second passage *in vitro* were HLA-DR typed as previously described (3). For this experiment the endothelial cells were removed from the culture vessels with EDTA, washed, and added to macrophage-depleted T cells. In all experiments a direct comparison was made between proliferation induced by antigen-pulsed macrophages (autologous) and endothelial cells (allogeneic but HLA-DR identical). One experiment representative of 10 that were done is presented in Table IV. Partial reconstitution of the T-cell response to antigen was evident with 1% endothelial cells ($10^3/10^5$ T cells) and com-

TABLE IV
Proliferation Response of Lymphocytes to Antigen Presented by Allogeneic Endothelial Cells

Antigen in culture/ dose μg/ml	Response of cultured lymphocytes (cpm)				
	Before Mϕ depletion	After Mϕ depletion with endothelial cells added back[a]			
		None	10^3	3.3×10^3	10^4
None	452	96	1,100	1,820	2,468
SK-SD/10	3,800	110	2,846	3,142	4,620
Candida/10	6,700	95	4,150	8,100	8,240
PPD/5	27,100	128	11,500	22,400	32,500
KLH/25	2,800	112	2,260	2,742	2,952

[a] Lymphocytes and endothelial cells are allogeneic and HLA-DR identical (DR 2/4). This is representative of 10 experiments that were done. Standard deviations were less than 15% of the mean in all cases.

plete reconstitution was observed with 10% addition. On a cell-for-cell basis endothelial cells and macrophages were equally effective in reconstituting the T-cell response in most experiments (Fig. 3). In two cases, endothelial cell presentation of antigen was not as effective as macrophage presentation and in one experiment endothelial cells were more effective than macrophages. Endothelial cells were pre-

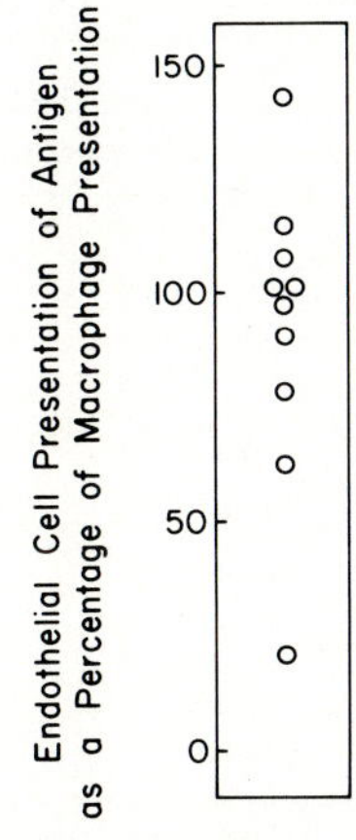

Fig. 3. Scatter diagram comparing presentation of antigen by endothelial cells to presentation of antigen by macrophages is shown. The proliferation response with endothelial cells is expressed as a percentage of the response obtained with allogeneic but HLA-DR identical macrophages.

vented from proliferating in these experiments by irradiation before addition to T cells.

Possible sources of contaminating cells in the endothelial cell cultures include fibroblasts and smooth muscle cells from subendothelial layers of the umbilical vein and blood mononuclear cells from the cord blood. Although cells other than endothelial cells were not evident in the cultures used in these experiments, additional controls were performed to reduce the possibility that a small number of contaminating cells could be responsible for the antigen presentation. To eliminate the possibility that fibroblasts or smooth muscle cells were contributing to the T-cell responses, these cells were cultured and tested directly. Table V shows data representative of two experiments that demonstrate that even 10% ($10^4/10^5$ T cells) addition of fibroblasts/smooth muscle cells could not reconstitute T-cell-depleted responses. In this example the response to 10 μg SK-SD was 5118 cpm and after depletion 314 cpm. Endothelial cells reconstituted the response to 3891 cpm, whereas addition of 10% fibroblasts/smooth muscle cells resulted in only 528 cpm.

TABLE V
Comparison of the Proliferation Response of Lymphocytes to Antigen Presented on Different Cell Types[a]

	Response of cultures lymphocytes (cpm)			
		After Mφ depletion with different cell types added (1×10^4)		
Antigen in culture/ dose (μg/ml)	Before Mφ depletion	None	Fibroblasts, smooth muscle cells	Endothelial cells
None	315	211	489	417
SK-SD/25	4296	228	511	2253
SK-SD/10	5118	314	528	3891
SK-SD/1	1856	216	454	983
KLH/25	1702	306	629	2106
KLH/10	1898	311	681	1123
KLH/1	744	281	593	527

[a] This is representative of two experiments that were done. In each case the fibroblast/smooth muscle cells were grown from the same umbilical vein as the endothelial cells. Standard deviation about the mean was less than 15% in all cases.

TABLE VI
Assessment of the Effects of Monoclonal Antimacrophage Serum on Lymphocyte Proliferation Induced by Antigen-Pulsed Macrophages

Exp.	Antigen in culture/ dose (μg/ml)	Lymphocyte response (cpm) with antimacrophage serum added[a]		
		None	10 μg/ml	100 μg/ml
1.	None	286 ± 33	254 ± 31	226 ± 27
	Candida/50	1128 ± 89	686 ± 53	328 ± 43
	Candida/10	2346 ± 236	1149 ± 127	422 ± 88
	Candida/1	1986 ± 197	875 ± 64	411 ± 58
2.	None	410 ± 30	358 ± 42	361 ± 54
	KLH/25	3256 ± 436	2131 ± 154	522 ± 36
	KLH/1	1925 ± 188	865 ± 92	527 ± 89
3.	None	316 ± 27	286 ± 92	274 ± 45
	SK-SD/50	968 ± 104	358 ± 28	258 ± 59
	SK-SD/10	1852 ± 153	762 ± 85	328 ± 33
	SK-SD/1	1122 ± 94	438 ± 36	246 ± 36

[a] Cultures without addition of monoclonal antimacrophage serum received 100 μg/ml monoclonal anti-RBC serum as control.

A more serious criticism might be that the endothelial cells could contain a small number of macrophages from the cord blood. Although none were observed, we used a monoclonal antimacrophage antibody to eliminate any responsiveness in these experiments due to macrophages. Table VI shows that in three experiments 100 μg/ml antimacrophage antibody completely reduced proliferation induced by antigen-pulsed macrophages. Table VII shows that the addition of 100 μg/ml antimacrophage had no effect on antigen-pulsed endothelial-cell-induced proliferation of T cells. As an example, maximal proliferation to 10 μg Candida (2346/286 cpm, antigen/control) was reduced significantly with antimacrophage antibody treatment of antigen presentation by macrophages (422/226 cpm, Table VI) but unchanged when this antibody was added to antigen-pulsed endothelial cells (3185/301 cpm, Table VII). We feel that this is an important control experiment that strongly indicates that the antigen presentation by the endothelial cells is not due to a small number of contaminating macrophages. In addition, this control eliminates the possible role of residual macrophages in the responding T-cell population. The monoclonal antimacrophage antibody has been previously shown to react with and block presentation by an antigen-presenting subset of macrophages (33). A summary of these experiments is presented in Fig. 4.

TABLE VII
The Effects of Monoclonal Antimacrophage Serum on Lymphocyte Proliferation Induced by Antigen-Pulsed Endothelial Cells

Exp.	Antigen in culture dose μg/ml	Proliferation responses of Mϕ depleted lymphocytes to antigen-pulsed endothelial cells (cpm) with antimacrophage serum added[a]		
		None	10 μg/ml	100 μg/ml
1.	None	292 ± 28	306 ± 22	301 ± 28
	Candida/50	1356 ± 123	1186 ± 214	1054 ± 109
	Candida/10	2786 ± 286	2906 ± 366	3185 ± 412
	Candida/1	2142 ± 242	2080 ± 280	2560 ± 376
2.	None	516 ± 53	588 ± 36	454 ± 36
	KLH/25	3654 ± 88	2988 ± 342	2914 ± 215
	KLH/1	2431 ± 316	2188 ± 317	2050 ± 311
3.	None	346 ± 29	292 ± 30	311 ± 42
	SK-SD/50	1128 ± 109	1356 ± 159	1086 ± 94
	SK-SD/10	2152 ± 241	2568 ± 310	2381 ± 192
	SK-SD/1	956 ± 102	1060 ± 105	1210 ± 138

[a] 1×10^4 Antigen-pulsed endothelial cells were added to all cultures. Cultures without addition of monoclonal antimacrophage serum received 100 μg/ml monoclonal anti-RBC as control. Responses shown are mean cpm ± SD.

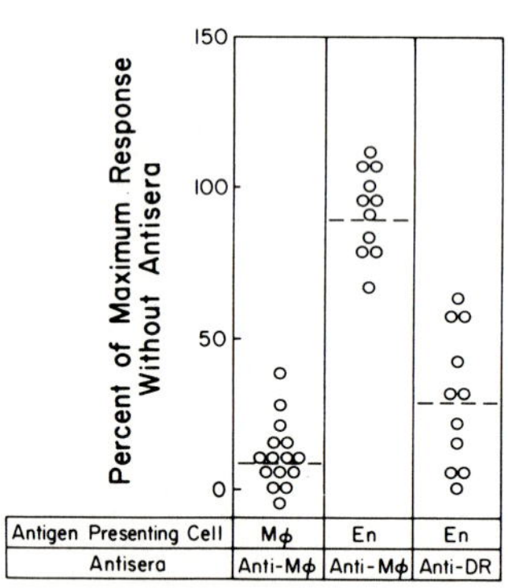

Fig. 4. Scatter diagram depicting the effects of a monoclonal antimacrophage serum or monoclonal monomorphic anti-HLA-DR serum on lymphocyte proliferation induced by antigen-pulsed macrophages (Mϕ) or endothelial cells (En) is shown. The maximum response was the proliferation obtained with antigen-pulsed macrophages or endothelial cells and normal mouse immunoglobulin added to cultures rather than the mouse monoclonal antibodies. (Dashed line represents the mean response).

GENETIC RESTRICTION OF ENDOTHELIAL CELL PRESENTATION OF ANTIGEN

In initial attempts to evaluate restriction of the endothelial cell presentation of antigen, we used conventional lymphocyte proliferation assays. HLA-A,B, and DR typed endothelial cells were pulsed with antigen and used to reconstitute macrophage-depleted T lymphocytes and the responses assessed at 6 days. Moderate-to-strong MLR reactions in most of the combinations made interpretation of results difficult (Table VIII). The antigen responses observed with HLA-DR identical but allogeneic endothelial cells appeared low because of the high control cpm (MLR reactivity). Reduction in endothelial cell concentration from 10 to 1% reduced the MLR problems but did not eliminate them. From Table VIII one can see that in each experiment, the T cells responded to each antigen when endothelial cells and T cells shared both HLA-DR specificities. However, when DR-haplotype disparate endothelial cells were employed, the increase in background (MLR reactivity) made interpretation of the antigen response difficult (S.I. less than 2). To avoid interpreting this kind of data we used cells sensitized *in vitro* to hemocyanins to evaluate restriction. The rationale was that since cells sensitized *in vitro* during 12–14 days of culture with antigen respond in secondary culture at 3 days, we could observe antigen-dependent proliferation in the absence of MLR reactivity (typically maximal at 6 days and absent at 3 days).

To prime cells *in vitro* to hemocyanins, mononuclear cells (T cells and macrophages) were cultured in 250 ml flasks with KLH or HCH

TABLE VIII
Assessment of HLA-DR Restriction of Lymphocyte Proliferation to Presentation of Antigen by Allogeneic Endothelial Cells

HLA-DR type: Lymphocytes (1×10^5)	HLA-DR type: Endothelial cells (1×10^4)	Response to antigen (cpm): *Candida* (10 μg) (+/− = S.I.)[a]	Response to antigen (cpm): SK-SD (25 μg) (+/− = S.I.)
DR 2/2	DR 2/2	6140/ 2650 = 2.2	8290/ 2650 = 3.1
	2/4	4210/ 2850 = 1.5	5410/ 2850 = 1.9
	1/5	8240/ 7240 = 1.1	7840/ 7240 = 1.1
DR 2/4	DR 2/4	8930/ 1840 = 4.9	6120/ 1840 = 3.3
	2/2	8740/ 4620 = 1.7	6240/ 4620 = 1.4
	1/5	14,100/13,200 = 1.1	13,100/13,200 = 1.0

[a] cpm with antigen divided by cpm without antigen equals Stimulation Index (S.I.).

TABLE IX
Kinetics of Secondary Culture Response of Cells Primed *in Vitro* with KLH[a]

Conditions in secondary culture		Lymphocyte proliferation (mean cpm ± SD) assessed by [³H]thymidine uptake at day:		
Antigen	Dose (μg/ml)	Day 2	Day 3	Day 4
None	—	1580 ± 114	2110 ± 185	1020 ± 161
KLH	50	2005 ± 180[b]	7050 ± 525[b]	2400 ± 286[b]
	25	1870 ± 141	6410 ± 496[b]	1850 ± 221[b]
	1	1608 ± 211	4331 ± 612[b]	1422 ± 168
HCH	50	1726 ± 183	2238 ± 321	1161 ± 88
	25	1830 ± 223	2314 ± 280	1335 ± 98
	1	1648 ± 218	1961 ± 211	1051 ± 105
SK–SD	10	1320 ± 86	2065 ± 218	1938 ± 233
PHA	10	10,300 ± 1100[b]	34,200 ± 4100[b]	24,100 ± 3100[b]

[a] Cells were primed with 25 μg/ml KLH for 12 days.
[b] Statistically significant ($p < .05$) response determined by Mann-Whitney U test.

for 12–14 days, washed, and placed in a second microculture. Antigen-pulsed macrophages were added to the second culture and the proliferation response measured at 2–4 days. Recent investigations have shown this *in vitro* sensitization to be antigen specific, macrophage dependent, and HLA-DR restricted (9,10,30,34). A typical secondary culture response of cells sensitized *in vitro* is shown in Table IX. Maximal proliferation is obtained on day 3 of secondary culture to the priming antigen. At day 3 the T-cell responsiveness to SK-SD is undetected, since the conventional antigen responses are maximal on day 6. The same rationale holds for MLR reactivity (data not shown here).

When endothelial cells are pulsed with antigen and added to T cells sensitized *in vitro*, similar antigen-specific secondary responses are observed (Table X). On day 3 of secondary culture, endothelial cells that were HLA-DR identical (but allogeneic) to T cells presented antigen for T-cell activation. When the primary cultures were sensitized with 10 μg/ml KLH, peak responsiveness in secondary cultures (19,300 cpm test to 4550 control) was observed. The same cultures could not respond to HCH (5150 cpm test to 4550 control). Likewise cultures primed with 100 μg/ml HCH only responded to an HCH challenge in secondary culture (4780 test to 1720 control).

To assess genetic restriction in this system, endothelial cells of different HLA-A, B, and DR types were pulsed with antigen and added to

TABLE X
Proliferation of Lymphocytes Primed *in Vitro* to Hemocyanins Induced by Antigen-Pulsed Endothelial Cells

Antigen/dose (μg/ml) in primary culture	Proliferation (mean cpm ± SD) in secondary culture to antigen-pulsed endothelial cells (DR identical, allogeneic)[a]		
	Not pulsed	KLH-pulsed	HCH-pulsed
	Day-2 Responses		
None	2300 ± 187	2500 ± 250	2900 ± 202
KLH/50	2400 ± 216	2510 ± 182	2300 ± 149
KLH/10	2510 ± 208	3515 ± 310	2890 ± 316
HCH/50	2300 ± 192	2240 ± 282	2270 ± 312
HCH/10	2740 ± 311	2820 ± 161	2150 ± 108
	Day-3 Responses		
None	2700 ± 192	2100 ± 203	2300 ± 310
KLH/50	4600 ± 358	4125 ± 516	5019 ± 772
KLH/10	4550 ± 413	19,300 ± 2650	5150 ± 689
HCH/50	1720 ± 162	2150 ± 148	4780 ± 315
HCH/10	2212 ± 208	2330 ± 245	3130 ± 206

[a] Lymphocytes were primed with hemocyanins for 14 days. Endothelial cells used in secondary culture were pulsed with 25 μg/ml antigen.

multiple combinations of responding T cells. An example of one of seven experiments done is shown in Table XI. In each experiment antigen in primary and secondary culture and addition of HLA-DR compatible endothelial cells was required for a response. The maximum response was obtained with HLA-DR identical endothelial cells. When DR 3/5 endothelial cells were used to present antigen to HLA-DR 3/5 lymphocytes 6860 cpm resulted (controls were 2490 cpm and 2520 cpm). When endothelial cells shared one specificity with the T cells, partial responsiveness was observed (4640 cpm in this example). Completely disparate endothelial cells did not trigger T cells for antigen responsiveness (DR 1/1 was 2830 cpm and DR 2/2 was 2560 in this example). The absence of bothersome MLR reactivity is evident from noting the similarity in control level cpm which is independent of HLA-DR type. Figure 5 represents an accumulation of the cell combinations done in this manner. When the number of shared HLA-DR specificities between responding T cells and macrophages was 2, all combinations were significantly positive compared to no antigen controls or the response produced by autologous macrophages (the maximum response). The mean was 89% of the autologous macrophage response with 10 of 10 combinations responsive. When the two cell

TABLE XI
HLA-DR Restriction of Antigen Presentation by Endothelial Cells to Lymphocytes Immunized *in Vitro*

Responding cells (HLA-DR)	Antigen in culture		Lymphocyte response to endothelial cells added to secondary cultures (HLA-DR)[a]			
	1°	2°	DR 3/5	DR 1/5	DR 1/1	DR 2/2
DR 3/5	KLH	—[b]	2490 ± 286	2280 ± 251	2450 ± 217	2180 ± 258
	—	KLH	2520 ± 258	2160 ± 186	2340 ± 189	2040 ± 201
	KLH	KLH	6860 ± 392[c]	4640 ± 342[c]	2830 ± 311	2560 ± 180
DR 2/2	KLH	—	1490 ± 138	1610 ± 162	1740 ± 128	1620 ± 111
	—	KLH	1520 ± 122	1580 ± 228	1810 ± 186	1710 ± 98
	KLH	KLH	1860 ± 159	1810 ± 196	2240 ± 202	7150 ± 1062[c]

[a] Lymphocytes were primed by 10 μg KLH for 12 days and endothelial cells pulsed with 10 μg/ml KLH.
[b] (—) Indicates nothing added.
[c] Indicates statistically significant response by the Mann-Whitney U test ($p < .05$).

types were completely disparate, eight of nine combinations were completely negative. In one case some proliferation was noted. When there was one shared specificity, the mean response was midway between combinations with two shared specificities and combinations that did not share HLA-DR specificities. There was considerable variability in some experiments as is evident from Fig. 5. There was no correlation between HLA-A or B locus determinants and antigen presenting activity.

To delineate further that the HLA-DR determinants on endothelial cells were critically involved in this genetic restriction, we used polymorphic anti-HLA-DR antisera to block antigen presentation. Eight experiments representing 30 different endothelial cell–lymphocyte combinations were evaluated by addition of anti-DR sera. When the antiserum was directed against determinants not found on the endothelial cells, consistent responses were observed (Fig. 6). The mean of eight combinations was 91% of maximal response in the presence of control sera. When the antisera was directed at both HLA-DR determinants on endothelial cells, the mean response of 12 combinations was reduced to 26% of normal. When antisera reacted to a single determinant on the endothelial cells, the responses were highly variable, with the mean of 11 combinations at 68%. In this group, when the antisera was directed against a specificity shared between the endothelial cells and T cells, the responses were lower than when a specificity was not shared (46%, six combinations versus 96%, five combinations).

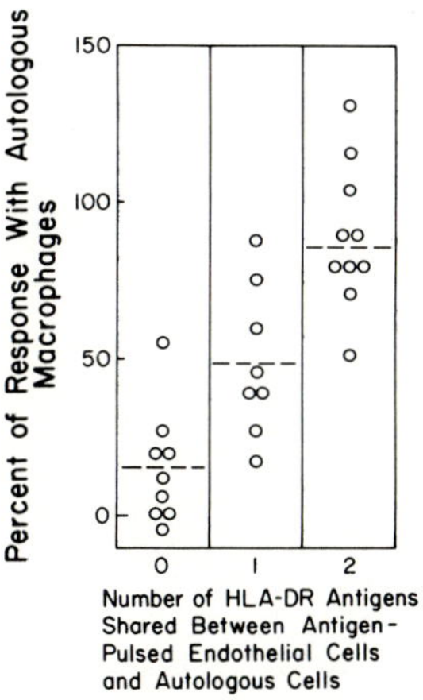

Fig. 5. Scatter diagram of lymphocyte proliferation to antigen presented by allogeneic endothelial cells. The response in secondary culture is expressed as a percentage of the response to antigen presented on autologous macrophages. The lymphocytes were primed for 14 days *in vitro* to KLH. (Dashed line represents mean response.)

Monoclonal anti-DR sera directed against a common DR determinant also resulted in inhibition of endothelial cell presentation of antigen (mean reduction to 26% of normal for 11 experiments, Fig. 4). The variability of these results could be due to difficulties in counting

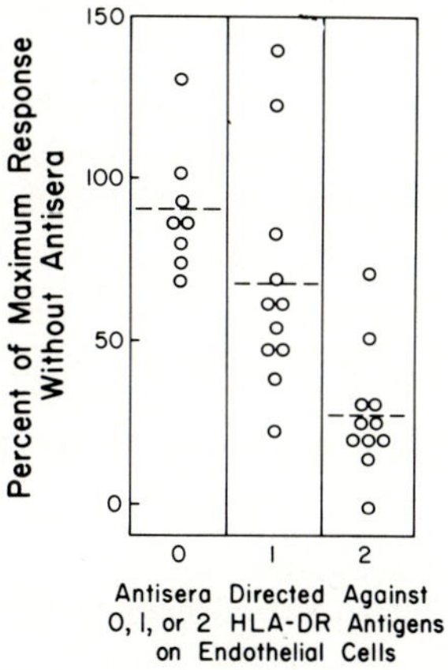

Fig. 6. Scatter diagram depicting the effects of polymorphic HLA-DR antisera on lymphocyte proliferation in secondary cultures induced by antigen-pulsed endothelial cells is shown. The maximum response was the proliferation obtained with antigen-pulsed endothelial cells and normal human sera added to cultures rather than the alloantisera. (Dashed line represents mean response.)

endothelial cells; the production of fibronectin or collagen by endothelial cells (35), which could interfere with antigen presentation; prostaglandin production (36), which has been reported to suppress T cells (37); cross-reactivity between HLA-DR antigens; or poor or incomplete HLA-DR typing or HLA-DR reagents. An additional possibility that is consistent with the difficulty we had typing endothelial cells at the DR locus is that endothelial cells display DR structure in an unconventional manner.

In conclusion, we have provided strong evidence that cultured endothelial cells present antigen to T lymphocytes *in vitro* and we suggest that vascular endothelium "processes" and presents antigen to antigen-specific T cells *in vivo*. Moreover, we have shown that this activation is genetically restricted at the HLA-DR locus. The events occurring after T-cell–endothelial interaction now become of immediate investigative interest. It remains to be shown whether this type of cell-to-cell interaction results in mediator production by either cell type and/or changes in vascular permeability. Endothelial cell–lymphocyte interaction involving antigen may be a crucial feature of T-cell-mediated immune responses.

REFERENCES

1. Rosenthal, A. S., Barcinski, M. A., and Rosenwasser, L. J. (1978) *Fed. Proc., Fed. Am. Soc. Exp. Biol.* **37**, 79.
2. Unanue, E. R. (1972) *Adv. Immunol.* **15**, 95–104.
3. Hirschberg, H., Moen, T., and Thorsby, E. (1979) *Transplant. Proc.* **11**, (1), 776–778.
4. Hirschberg, H., Moen, J., and Thorsby, E. (1979) *Transplantation* **28**, 116.
5. Hirschberg, H., Bergh, O. J., and Thorsby, E. (1980) *J. Exp. Med.* **152**, 249–255.
6. Gutierrez, C., Bernabe, R. R., Vega, J., and Kreisler, M. (1979) *J. Immunol. Methods* **29**, 57–63.
7. Kurnick, J. T., Ostberg, L., Stegagno, M., Kimura, A. K., Arn, A., and Sjöberg, O. (1979) *Scand. J. Immunol.* **10**, 563–573.
8. Uotila, K., Marnela, M., Hamblin, A., and Krohn, K. (1978) *Scand. J. Immunol.* **7**, 495–502.
9. Rodey, G. E., Leutorman, L. K., and Thomas, D. W. (1979) *J. Immunol.* **123**, 2250–2254.
10. Hensen, E. J., and Elferink, B. G. (1979) *Nature (London)* **277**, 223.
11. Mosier, D. E. (1967) *Science* **158**, 1573.
12. Kumagai, K., Itoh, D., Hinuma, S., and Tada, M. (1979) *J. Immunol. Methods* **29**, 17–25.
13. Yam, T., Lit, C. Y., and Crosby, W. H. (1971) *Am. J. Clin. Pathol.* **55**, 283.
14. Jerrells, T. R., Dean, J. H., Richardson, G. L., and Herberman, R. B. (1980) *J. Immunol. Methods* **32**, 11–29.
15. Jaffe, E. A., Nachman, R. L., Becker, C. G., and Mirick, C. R. (1973) *J. Clin. Invest.* **52**, 2745–2756.

16. Brand, D. L., Roy, J. R., Jr., Kayhoe, D. B., and McClelland, J. D. (1970) *In* "Histocompatibility Testing" (P. Terasaki, ed.), pp. 357–567. Munksgaard, Copenhagen.
17. Stux, S., Hammond, P., Fitzpatrick, D., Dubey, D., and Yunis, E. (1980) *Tissue Antigens* **15**, 152–160.
18. Bergholtz, B. O., and Thorsby, E. (1979) *Scand. J. Immunol.* **9**, 511.
19. Bergholtz, B. O., and Thorsby, E. (1977) *Scand. J. Immunol.* **6**, 779.
20. Hirschberg, H. H. (1978) *Clin. Exp. Immunol.* **34**, 46.
21. Kruiobrek, A. M. (1978) *Proc. Int. Leucocyte Cult. Conf., 12th, 1977* p. 115.
22. Unsgard, G. (1977) *Acta Pathol. Microbiol. Scand.* **85**, 381.
23. Berle, E., and Thorsby, E. (1980) *Clin. Exp. Immunol.* **3**, 668.
24. Thorsby, E., and Nousaunen, H. (1979) *Scand. J. Immunol.* **9**, 183.
25. Sonderstry Hansen, G., Rubin, B., Sorensen, S. F., and Svejgaard, A. (1978) *Eur. J. Immunol.* **8**, 520.
26. Kammer, G. M., and Unanue, E. R. (1980) *Clin. Immunol. Immunopathol.* **15**, 434–443.
27. Bergholtz, B. O., and Thorsby, E. (1978) *Scand. J. Immunol.* **8**, 63.
28. Albrechtsen, D., Solheim, B. G., and Thorsby, E. (1977) *Cell. Immunol.* **28**, 258.
29. Hirschberg, H., Bergh, O. J., and Thorsby, E. (1979) *J. Exp. Med.* **150**, 1271–1276.
30. Ford, D., Burger, D. R., and Hamblin, A. (1981) *J. Immunol.* (in press).
31. Jaffe, E. A. (1977) *N. Engl. J. Med.* **296** (7), 377–383.
32. Jaffe, E., Hoyer, L., and Nachman, R. (1973) *J. Clin. Invest.* **52**, 2757–2764.
33. Raff, H. V., Pickey, L. J., and Stobo, J. D. (1980) *J. Exp. Med.* **152**, 581.
34. Ford, D., Burger, D. R., Hamblin, A., and Dumonde, D. C. (1981) *J. Immunol.* (in press).
35. Macarab, E. J., Kirby, E., Kirk, J., and Kefalider, N. A. (1978) *Proc. Natl. Acad. Sci. U.S.A.* **75** (6), 2621–2625.
36. Gimbrone, M. A., Jr. (1975) *Science* **189**, 219–220.
37. Goodwin, J. S., Bankhust, A. D., and Messner, R. P. (1978) *J. Exp. Med.* **146**, 1719–1734.

A Recognition Function of Endothelial Cells: Directing Lymphocyte Traffic

EUGENE C. BUTCHER, GEORG KRAAL,
SUSAN K. STEVENS, AND IRVING L. WEISSMAN
Laboratory of Experimental Oncology
Department of Pathology
Stanford University School of Medicine
Stanford, California

INTRODUCTION

Lymphocytes are extraordinarily mobile cells. The average lymphocyte travels to a new lymphoid organ every 1–2 days (1). The migrating lymphocyte circulates and recirculates, following a pathway from the lymphoid organs, through the lymphatics into the venous circulation, finally leaving the systemic circulation to return to the same or another lymphoid organ. Lymphocytes are able to leave the bloodstream to enter lymphoid organs because they have the remarkable capacity to recognize the endothelium lining specialized lymphoid organ venules (see Fig. 1). Passing lymphocytes adhere to the lumenal surface of the venular endothelium and then migrate through the wall of the venule into the surrounding lymphoid parenchyma (2). These specialized lymphoid organ venules are morphologically distinct and are currently referred to as high endothelial venules (HEV) because they are lined by characteristically plump endothelial cells.

In the course of our recent studies, we have found that the migration of several major lymphocyte subpopulations is directed selectively through particular lymphoid organs and that the control of this differential migration is effected primarily by selective interaction of lym-

PATHOBIOLOGY OF THE ENDOTHELIAL CELL

ISBN 0-12-521980-6

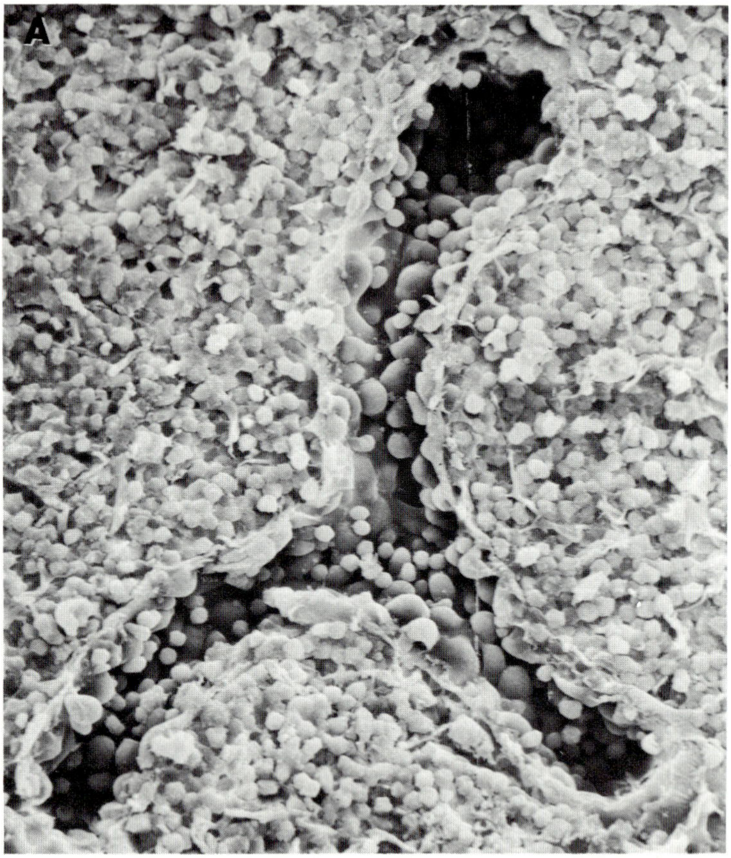

Fig. 1. (A) Scanning electron micrograph of the cut surface of a mouse lymph node showing a Y-shaped high endothelial venule (HEV) (about 200×). Lymphocytes leave the blood by binding to the high endothelium, and subsequently migrating through the vessel wall into the surrounding nodal parenchyma. [Reprinted with permission from Weissman *et al.* (1a).]

phocytes with HEV. Our findings, reviewed in this manuscript, demonstrate (1) that the endothelial cells lining HEV in peripheral lymph nodes and the gut-associated Peyer's patches present different, distinctive determinants for lymphocyte recognition and (2) that subpopulations of mature lymphocytes are in turn programmed to recognize these mucosa-associated or peripheral endothelial determinants selectively, thus directing their migration through particular regions of the body. The results thus define a remarkable and specific recognition function of endothelial cells.

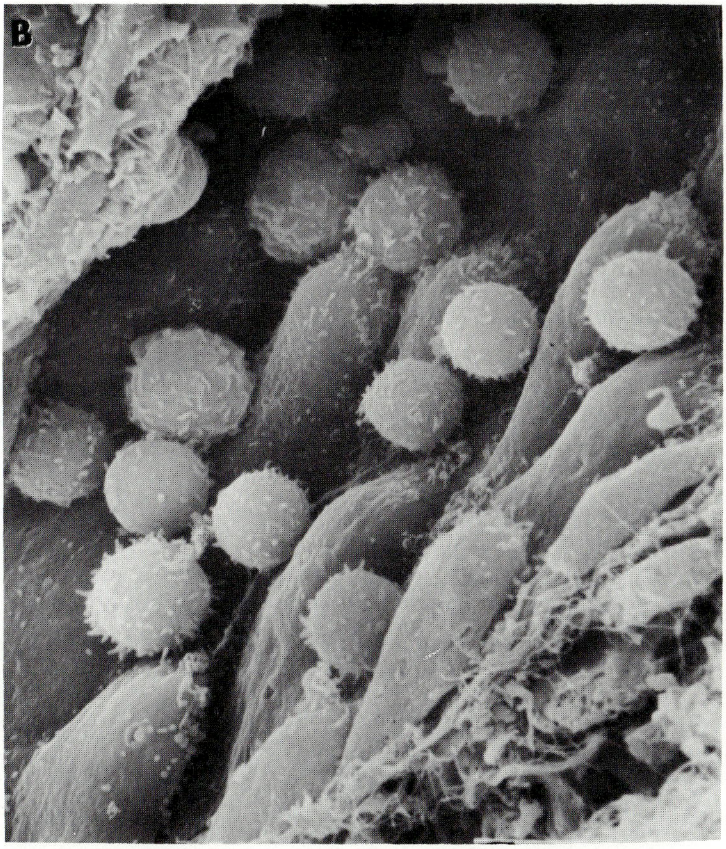

(B) Higher power showing numerous lymphocytes tightly bound to the plump endothelial cells of an HEV. Nonadherent blood elements were washed out by perfusion of the vasculature with medium prior to fixation by perfusion with glutaraldehyde in PBS.

METHODS

IN VITRO BINDING OF LYMPHOCYTES TO HEV

A previous interest in our laboratory in the phenomenon of lymphocyte–HEV interaction (3) was rekindled by the demonstration by Stamper and Woodruff (4) that lymphocytes incubated *in vitro* on frozen sections of murine lymph nodes bind selectively to the endothelial cells of HEV. This *in vitro* method not only made possible the study of an important aspect of lymphocyte migration in isolation from the complexities of cell localization *in vivo*, but also rendered the interaction of lymphocytes with HEV an attractive model system for the study of cell–cell recognition and binding interactions in general. We have modified the technique to allow quantitative comparisons of the ability of various populations to bind to HEV in mouse lymphoid organs. The detailed methodology has been described previously (5,6). Briefly, viable lymphocytes are incubated for 30 min in the cold (4–7°C) with mild agitation on freshly cut frozen sections of mouse lymph nodes or Peyer's patches. HEV-bound lymphocytes are fixed to the sections with 1% glutaraldehyde in cold PBS. In order to allow comparisons of different sample populations, an internal standard population of fluorescent lymphocytes [covalently labeled with fluorescein isothiocyanate (FITC) as previously described (5,7)] is employed. Finally, the binding ability of each sample population is related to that of a constant reference population [reference populations were either (a) mesenteric node lymphocytes or (b) peripheral lymph node T cells—the reference population employed in each experiment is identified as appropriate in figure legends]. The results are expressed as relative adherence ratios (RAR), the calculated number of sample lymphocytes binding to HEV per reference cell that would bind under the same circumstances.

We have previously shown that the results of this assay are reproducible and accurately reflect the ability of lymphocytes to bind to HEV under more physiological conditions (5). The specific binding of lymphocytes to endothelial cells in this assay system is a rapid and apparently irreversible event. Thus differences in the observed binding abilities of sample populations probably do not reflect differential affinities, but result from differences in the average probability of a successful binding event occurring when a lymphocyte comes in contact with the endothelium.

IN VIVO HOMING OR LOCALIZATION STUDIES

As a means of confirming the *in vivo* relevance of the selective recognition of HEV by lymphocytes, we have frequently compared the HEV-binding preferences of normal lymphocyte populations with their differential homing *in vivo*. In these *in vivo* studies, which will be referred to only briefly in the present manuscript, sample lymphocytes were either (1) injected into syngeneic recipients after labeling by direct fluorescent conjugation with tetramethylrhodamine isothiocyanate (TRITC) or FITC, as described previously (7,8) or (2) identified employing fluorescent antibody reagents against a donor lymphocyte cell surface alloantigen (Thy-1.1) after injection of AKR/J (Thy-1.1) lymphocytes into AKR/Cum (Thy-1.2) recipients. Recipients were sacrificed 30 min to 2 hr after injection of sample cells: These very short-term localization studies were employed to exclude significant lymphocyte recirculation or reshuffling between organs and thus to focus primarily on initial localization events, which appear to be determined largely by lymphocyte interaction with HEV. The localization of various lymphocyte populations, and their antigenically defined subpopulations, in peripheral (axillary and brachial) lymph nodes, the mesenteric lymph node, and the gut-associated Peyer's patches were compared. More detailed descriptions of methodology and examples of the data analysis, although inappropriate in this paper, have been published previously (6,8a) or will appear in subsequent manuscripts (8b).

RESULTS AND DISCUSSION

ABILITY TO RECOGNIZE HEV—A DEVELOPMENTAL PROPERTY OF MATURE, MOBILE LYMPHOCYTES

The interaction of lymphocytes with HEV is highly specific. Lymphocytes do not bind detectably to the endothelium lining other large vessels. Furthermore, of the various cell types examined (directly or indirectly) only lymphocytes during the migratory phase of their developmental life cycle appear to recognize and bind HEV. Although B and T lymphocytes from each of the secondary lymphoid organs bind well, neither their sessile precursors in the bone marrow and thymus nor germinal center cells [a sessile population of mature B cells proliferating locally within secondary lymphoid organs in response to antigenic stimulation (9)] express functional HEV receptors (see Fig. 2 and Ref. 9).

EVOLUTIONARY AND GENETIC CONSIDERATIONS

In an attempt to gain some insight into the evolution of lymphocyte–HEV recognition mechanisms and cell–cell interactions in general, we examined the species specificity of binding of lymphocytes to HEV. Lymphocytes from all vertebrate species tested, ranging from mice to chickens, are capable of recognizing and binding specifically to mouse HEV, but the binding ability declines exponentially

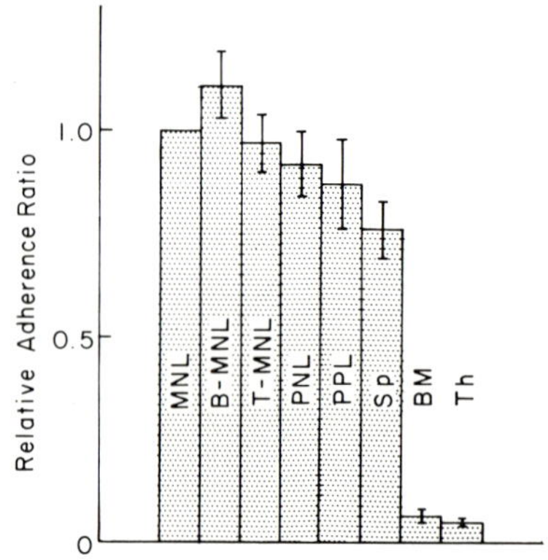

Fig. 2. *In vitro* adherence of normal lymphocyte populations to HEV in frozen sections of syngeneic mesenteric lymph nodes: The ability to bind to HEV is a developmental property of mature lymphocytes. The relative adherence ratio is the number of sample cells binding to HEV per mesenteric node lymphocyte (MNL) bound under the same conditions. B (B-MNL) and T (T-MNL) mesenteric node lymphocytes; peripheral (axillary and brachial) node lymphocytes (PNL); spleen cells (Sp); bone marrow cells (BM); and thymocytes (Th). Bars present standard errors. [Reprinted with permission from Ref. (10.)]

with increasing evolutionary distance separating the lymphocyte donor from the mouse host (Fig. 3). We have proposed the following speculative hypothesis to explain this remarkably clear-cut negative exponential relationship (10): It is well known that proteins whose function has remained constant during evolution, such as cytochrome c or hemoglobin, have nonetheless undergone a fairly uniform, linear rate of amino acid substitutions over evolutionary time. A linear rate of amino acid changes necessarily implies an exponential decline in shared amino acids between proteins in diverging lineages. Such an exponential decline in shared amino acids could possibly explain a roughly parallel decline in the compatibility of interacting proteins from different lineages, and hence in protein-mediated cell–cell recognition interactions across species barriers. Thus we propose that there has been a continuous evolutionary drift in the endothelial and lymphocyte recognition molecules involved in lymphocyte–HEV interaction. Within any lineage, selective pressure would ensure continuing interaction compatibility, but no such pressures would exist to prevent progressive development of incompatibility of recognition structures in diverging lineages. This model, if verified, implies that the preservation of cell–cell and protein–protein interactions during

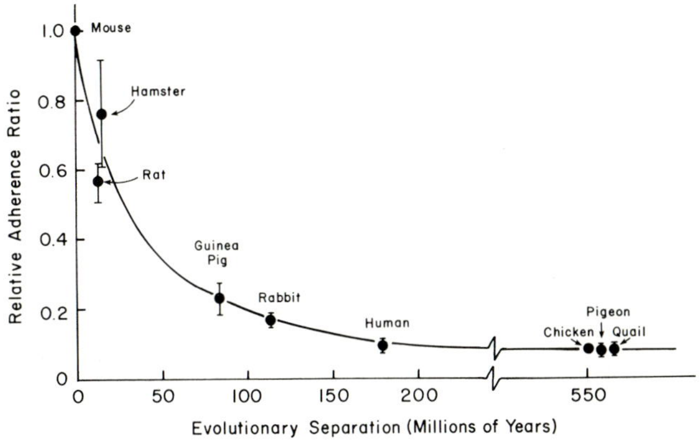

Fig. 3. The ability of mature lymphocyte populations from various species to recognize and bind to mouse HEV (in frozen sections of mouse mesenteric lymph nodes) declines exponentially with increasing evolutionary separation of the lymphocyte donor from the mouse host. [Reprinted with permission from Ref. (10.)]

evolution may not be due to stasis at the molecular level, but rather may result from a dynamic balance between natural selection and random mutation.

The observation of significant differences in lymphocytes from even closely related species suggested that small but detectable variability might exist within species as well, and we therefore examined the HEV-binding properties of lymph node lymphocytes from several common inbred strains of mice. BALB/c and C57BL lymphocytes exhibited the greatest differences: BALB/c lymphocytes were found (1) to bind better than C57BL lymphocytes to mesenteric node HEV (by about 1.5-fold) and (2) to demonstrate, relative to C57BL lymphocytes, a greater tendency to bind to peripheral node HEV as opposed to Peyer's patch HEV. These two differences co-segregated in BALB/c and C57BL recombinant inbred strains and have been mapped to a single region on chromosome 7, closely linked to histocompatibility locus 1 and the albino coat color locus c (10). A variety of mechanisms, such as regulatory controls of receptor expression, or allelic variants of the lymphocyte receptor for lymph node HEV, could explain this genetically defined strain difference. It is hoped that further studies utilizing mice congenic for this trait will yield additional insight into molecular and/or regulatory mechanisms of lymphocyte–HEV recognition.

SELECTIVE MIGRATION OF LYMPHOCYTES THROUGH MUCOSAL AND NONMUCOSAL TISSUES MEDIATED BY RECOGNITION OF ORGAN-SPECIFIC ENDOTHELIAL CELL DETERMINANTS

The mucosal and nonmucosal tissues of the body present quite different challenges to host defenses, and the immune system has evolved separate if overlapping mechanisms for providing immunity in these sites. Most notably, the humoral response in mucosal tissues is characterized by the secretion of antibodies of the IgA class, whereas IgM and IgG predominate in the extramucosal sites. One reasonable mechanism for effecting this division of immune response modalities would be the physical segregation of the responsible effector lymphocytes, perhaps by virtue of restricted migration pathways. In fact, it has been well documented that certain minor lymphocyte subpopulations migrate selectively. [For instance, immunoblasts arising from stimulation in the gut-associated lymphoid tissues localize preferentially in the gut wall and in the gut-associated Peyer's patches, whereas blast cells arising in peripheral lymph nodes tend to home to peripheral nodes or to peripheral, extramucosal sites of inflammation (11–13). Selective migration has also been shown for small recirculating Ig^- lymphocytes in sheep intestinal and peripheral lymph (14–16).] As part of a continuing effort in our laboratory to define the role and the mechanisms of lymphocyte circulation, we decide to examine the selectivity of migration of various antigenically defined lymphocyte populations and to test the possibility that lymphocyte migration could be directed by selective recognition of organ-specific endothelial cell determinants.

DIFFERENCES IN THE ENDOTHELIAL RECOGNITION SPECIFICITIES OF LYMPHOCYTE SUBPOPULATIONS

We have found significant differences in the endothelial interaction specificities of three major antigenically defined subpopulations of normal lymphocytes in the mouse: B lymphocytes, Lyt-2^-, and Lyt-2^+ lymphocytes. [The expression of Lyt-2 divides T lymphocytes roughly into two functionally distinct classes. Lyt-2 is associated with cytotoxic and suppressor activity, and possibly with a precursor population, whereas its absence is associated with helper function (17).] The greatest differences were observed in comparisons of populations of T lymphocytes (considered as a whole) versus B lymphocytes. T cells bound better than B cells to peripheral node HEV (by about 1.5-fold); both classes bound equally well to HEV in the mesenteric node; and B

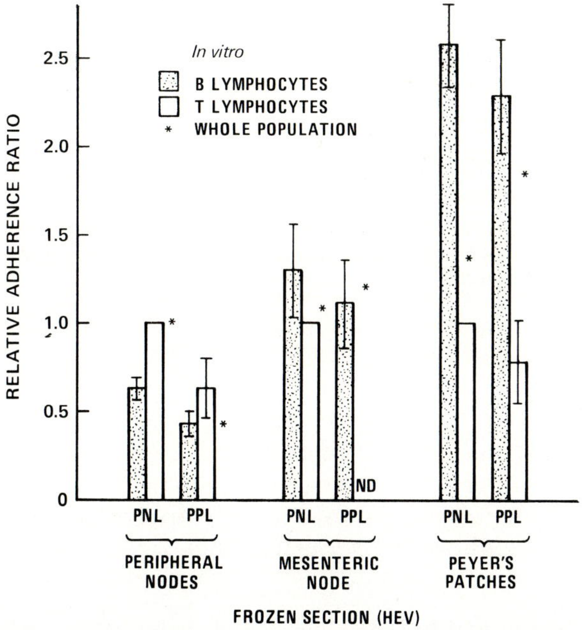

Fig. 4. Relative ability of B and T lymphocyte populations to bind *in vitro* to HEV. Adherence of peripheral node and Peyer's patch B and T cell populations was assayed on HEV in frozen sections of peripheral or mesenteric lymph nodes or Peyer's patches. The relative adherence ratio (RAR) is the calculated number of sample cells binding to HEV per peripheral node T cell bound under the same conditions. Error bars represent standard errors. Regardless of their source, B cells bind less well than T cells to peripheral node HEV and better to those in Peyer's patches. (Ref. 8a).

cells actually bound two to three times as well as T cells to Peyer's patch HEV (see Fig. 4). These differential interactions with HEV were reflected in a parallel differential distribution of B and T lymphocyte populations between these organs *in vivo* after iv injection.

Smaller but still significant differences were observed between Lyt-2^+ and Lyt-2^- subpopulations of T lymphocytes as well. Lyt-2^- and Lyt-2^+ cells bound equally well to Peyer's patch HEV, but Lyt-2^+ cells bound somewhat better (about 1.5×) to peripheral node HEV. Again, the endothelial preferences observed *in vitro* were reflected in a parallel differential distribution *in vivo*, with Lyt-2^- cells and Lyt-2^+ T cells demonstrating a slight relative preference for localizing in Peyer's patches and peripheral lymph nodes, respectively (8b).

The endothelial interaction and *in vivo* distribution patterns of these lymphocyte subpopulations are summarized in Fig. 5. It should be noted that neither the endothelial specificities nor the *in vivo* distribu-

Organ (HEV)	T cells Lyt-2$^+$		T cells Lyt-2$^-$		B cells
Peripheral Lymph Nodes	1	:	1	:	1
Peyer's Patches	1	:	1.5	:	7

Fig. 5. Summary of the *in vivo* localization and *in vitro* HEV-binding preferences of lymphocyte subclasses. Under hypothetical circumstances in which an equal number of B cells, Lyt-2$^-$ T cells, and Lyt-2$^+$ T cells bound to peripheral node HEV (or localized in peripheral nodes *in vivo*), there would be about 6–9 B cells and 1.5 Lyt-2$^-$ T cells for every Lyt-2$^+$ T cell binding to Peyer's patch HEV (or localizing in Peyer's patches *in vivo*).

tion of lymphocytes was a function of their organ source (i.e., whether the sample cells studied were from Peyer's patches, peripheral nodes, or spleen), indicating that lymphocyte class is the major determinant of the migratory properties of most small lymphocytes and suggesting that lymphocytes within these major classes are reasonably uniform.

A ROLE FOR HEV RECOGNITION IN DETERMINING THE REPRESENTATION OF LYMPHOCYTE CLASSES IN LYMPHOID ORGANS

The preceding results define a role for selective lymphocyte–endothelial cell recognition in preferentially directing the flow of B cells (as opposed to T cells) and Lyt-2$^-$ T cells (as compared with Lyt-2$^+$ T cells) through the mucosal lymphoid organs, and it is important to note that this is reflected in the differential distribution of these populations *in situ*: There are more B cells than T cells in the resident lymphocyte population of Peyer's patches (about 80% B, 10–20% T), whereas the reverse is true in peripheral nodes (25% B, 70% T). Similarly, Lyt-2$^-$ cells constitute a consistently greater proportion of the total resident T-cell population in Peyer's patches (about 75% Lyt-2$^-$, 25% Lyt-2$^+$ in AKR mice) than in peripheral nodes (65% Lyt-2$^-$, 35% Lyt-2$^+$). It is interesting to speculate that the differential migration of B cells and of Lyt-2$^-$ T cells through Peyer's patches may reflect a greater evolutionary requirement for precursors of, and help for, humoral immunity in mucosal sites. Certainly, the findings argue that because these functionally distinct lymphocyte populations are programmed to recognize particular endothelial specificites, these endothelial specificities must play an important role in determining the characteristics of local immune responses.

PERIPHERAL NODE AND PEYER'S PATCH HIGH ENDOTHELIA PRESENT UNIQUE RECOGNITION DETERMINANTS PERMITTING NEARLY ABSOLUTE SPECIFICITY OF LYMPHOCYTE INTERACTION

Although demonstrating clear preferences, each of the cell populations we have considered is nonetheless capable of recognizing and interacting with both peripheral node and Peyer's patch HEV. In order to understand the cellular basis of this *limited* specificity, it became important to determine whether even greater specificity of interaction was possible. We approached this question initially by examining the HEV-binding properties of the most accessible clonal (or presumptively clonal) lymphocyte populations available to us—namely, murine lymphomas. As shown in Figs. 6 and 7, certain lymphomas demonstrate a remarkable specificity for binding to HEV in peripheral lymph nodes, and others bind almost exclusively to HEV in Peyer's patches. Both peripheral node-specific and Peyer's patch-specific lymphomas bind fairly well to HEV in the mesenteric node (6) (data not shown). (It should be emphasized that these lymphomas were selected to illustrate this point; not all lymphomas demonstrate this degree of specificity, and some fail to bind at all.)

As previously mentioned, it is well documented that lymphoblasts (defined by their uptake of radioactive DNA precursors) can demonstrate remarkable specificity of homing to peripheral nodes or to gut-associated Peyer's patches *in vivo*. They therefore were excellent candidates for a normal counterpart of the Peyer's patch or peripheral node HEV-specific lymphomas just described. In fact, the results of preliminary studies clearly indicate that mesenteric node (gut-derived) immunoblasts bind with remarkable selectivity to Peyer's patch HEV, interacting very poorly with peripheral node HEV, whereas peripheral node blasts demonstrate a relative preference for HEV in peripheral nodes. The organ specificity of binding of these activated cells (especially of mesenteric node blasts) is much greater than that demonstrated by whole B and T lymphocyte populations. These findings suggest that differential recognition of mucosal-associated versus peripheral endothelial determinants may well be the primary controlling factor in selective lymphoblast localization (although other factors are undoubtedly involved as well).

Since these lymphoblasts presumably derive from normal small lymphocytes of more general homing ability, it is tempting to speculate that microenvironmental influences at the site of antigen-induced blastogenesis may induce the subsequent expression of homing recep-

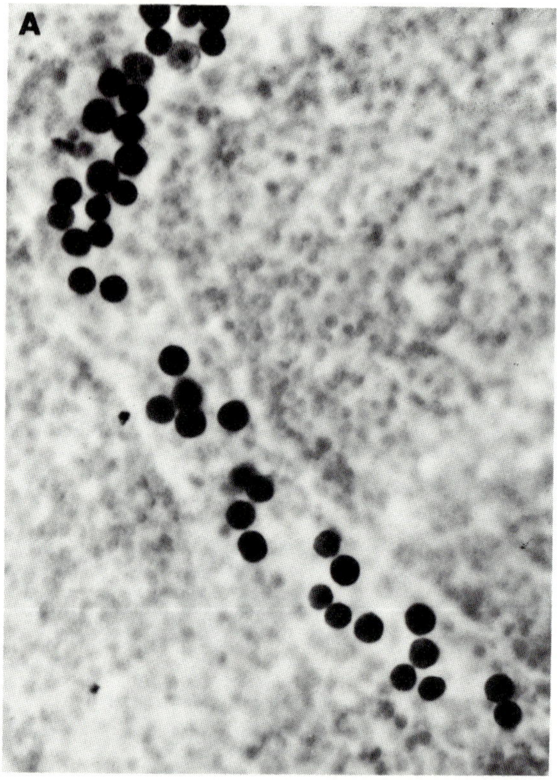

Fig. 6. Highly selective binding by a lymph node specific lymphoma. Lymphoma cells were incubated simultaneously on frozen sections of peripheral nodes and Peyer's patches. (A) Numerous cells bound to peripheral node HEV. Adherent cells are easily visualized because they are larger and stain more heavily than the underlying fragmented cells in the tissue section and lie in a higher plane of focus. *(continued)*

tors restricted for the type of tissue involved, i.e., receptors for peripheral or mucosal endothelial determinants. This mechanism may be of fundamental importance in, for instance, directing IgA plasma cell precursors stimulated by antigen in the Peyer's patches to localize specifically in the mucosal lamina propria, where they develop into the plasma cells responsible for secretory immunity.

A MODEL OF LYMPHOCYTE–ENDOTHELIAL CELL RECOGNITION

The simplest model consistent with our results is presented in Fig. 8. We propose that lymphocyte migration is directed by expression of

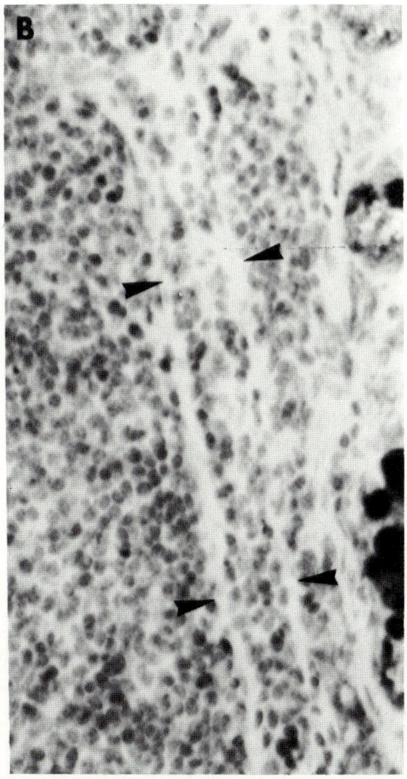

(B) No cells bound to comparable HEV (between arrows) in adjacent section of Peyer's patch.

lymphocyte surface receptors for organ specific endothelial cell determinants. At least two complementary lymphocyte–endothelial cell receptor sets exist, one mediating traffic through the peripheral nodes, and the other through Peyer's patches and possibly the gut wall. (It is possible that additional endothelial specificities, not defined by our experiments, may also exist.) As demonstrated by the nearly absolute specificity shown by certain lymphomas and lymphoblast populations, some lymphocytes may express receptors only for peripheral node or Peyer's patch HEV determinants, thus directing their migration almost exclusively through peripheral or mucosal sites (and perhaps the spleen). The *limited* organ preferences shown by most B and T cells and by T-cell subsets may be explained by their expression of different proportions of peripheral node and Peyer's patch HEV-specific recep-

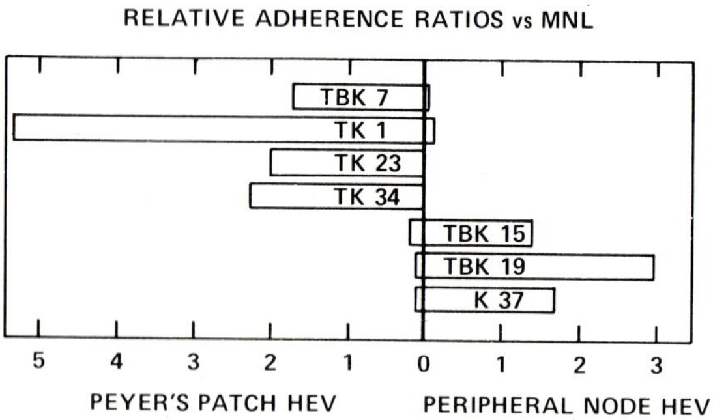

Fig. 7. HEV-binding properties of selected lymphomas from AKR (designation TK or K) or C57BL/Ka (designation TBK) mice assayed *in vitro* on frozen sections of peripheral lymph nodes or Peyer's patches. The adherence of normal mesenteric node lymphocytes (MNL) to HEV in either organ is defined as unity. The data demonstrate the existence of mechanisms permitting nearly absolute organ specificity of lymphocyte–HEV recognition. [Reprinted with permission from Butcher *et al.* (17a).]

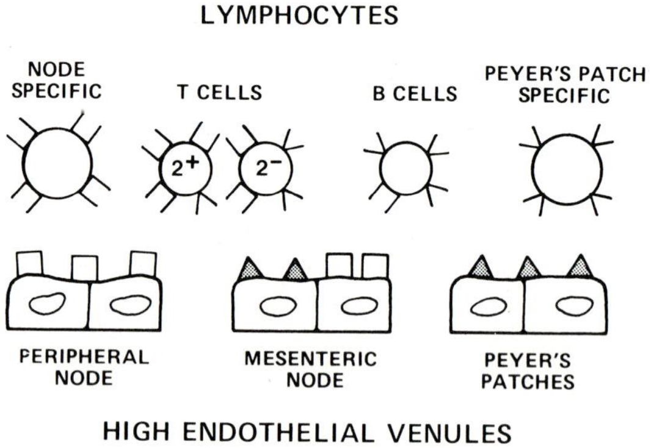

Fig. 8. Hypothetical model of the mechanism of organ-specific lymphocyte migration. HEV in Peyer's patches and nonmesenteric lymph nodes present different, distinctive surface receptors for lymphocyte recognition. Mesenteric node HEV (which bind both Peyer's patch- and node-specific cells) probably express both types of HEV determinant. The major subpopulations of small lymphocytes (B cells, and Lyt-2^- and Lyt-2^+ T cells) are capable of binding to both types of HEV, but with well-defined preferences. These small lymphocyte populations may express different proportions of receptors for peripheral node and Peyer's patch HEV, as shown, or other receptor types capable of only partial discrimination: Expression of HEV-recognition preference by most small lymphocytes is apparently a function of lymphocyte class, not organ source. On the other hand, some lymphocytes (e.g., certain immunoblasts) may express surface receptors exclusively for peripheral node or mucosal associated endothelial cell determinants, thus directing their traffic specifically through mucosal or peripheral lymphoid tissues: The exclusive expression of node or Peyer's patch HEV receptors by such activated lymphocytes is probably not class specific, but rather may be determined by microenvironmental influences at the site of antigen-induced blastogenesis.

tors (as shown in the figure) or by expression of different receptors capable of only partial discrimination. Mesenteric node HEV, which are unique in binding both peripheral node and Peyer's patch-specific cells, probably express both peripheral node and Peyer's patch HEV determinants: This finding suggests that the expression of these determinants may be controlled by local factors, perhaps related to proximity to the gut. It remains to be determined whether local endothelial specificities are immutably fixed during ontogenesis, or whether they can be altered in response to novel immune stimuli in such a way as actively to influence the flow of lymphocyte subsets through sites of inflammation. Finally, it should be noted that, although our data have clearly defined the specificities depicted, there is no direct evidence for the lock-and-key recognition mechanism shown. This may well prove to be an oversimplification of the actual molecular basis of lymphocyte–HEV recognition, as suggested by recent demonstrations of (1) a soluble factor in lymph capable of inhibiting lymphocyte–HEV adherence (18) and (2) a molecule, apparently secreted by HEV, that appears to facilitate or enhance lymphocyte binding under certain circumstances (P. Andrews and W. L. Ford, personal communication).

SUMMARY

The studies we have reviewed define a major role of endothelia in directing the traffic of lymphocytes through particular regions of the body. The specialized endothelial cells of HEV in mucosal lymphoid organs (Peyer's patches) and in peripheral lymph nodes present unique and distinct recognition determinants for migrating lymphocytes. In turn, small lymphocytes are programmed, according to their class, to exhibit limited preferences for recognizing and binding to peripheral node versus Peyer's patch HEV. Furthermore, the evidence suggests that activated lymphocytes may be induced or selected (perhaps by microenvironmental influences at the site of initial blastogenesis) to express receptors specific for the type of endothelial determinant associated with the organ where they initially encounter antigen. These endothelial and lymphocyte recognition specificities serve to direct the flow of these functionally important classes of normal and activated lymphocytes selectively through mucosal or nonmucosal organs, and thus may exert a major influence on the character and efficiency of local immune responses.

ACKNOWLEDGMENTS

We thank J. Mason and M. Beers for sedulous secretarial assistance. E. C. B., to whom correspondence should be addressed, is a Senior Fellow of the American Cancer Society, California Division. S. K. S. is a Stanford Medical School Alumni Scholar. G. K. is the recipient of a Research Fellowship Award from the Fogarty International Center.

REFERENCES

1. Gowans, J. L. (1968) *Hosp. Pract.* **3,** 34.
1a. Weissman, I. L., Butcher, E. C., Rouse, R. V., and Scollay, R. G. (1980) *In* "Strategies of Immune Regulation" (E. S. Sercarz and A. J. Cunningham, eds.), p. 77. Academic Press, New York.
2. Gowans, J. L., and Knight, E. J. (1964) *Proc. R. Soc. London Ser. B* **159,** 257.
3. Gutman, G., and Weissman, I. (1973) *Transplantation* **16,** 621.
4. Stamper, H. B., Jr., and Woodruff, J. J. (1976) *J. Exp. Med.* **144,** 828–833.
5. Butcher, E. C., Scollay, R. G., and Weissman, I. L. (1979) *J. Immunol.* **123,** 1996.
6. Butcher, E. C., Scollay, R. G., and Weissman, I. L. (1980) *Eur. J. Immunol.* **10,** 556–561.
7. Butcher, E. C., and Weissman, I. L. (1980) *J. Immunol. Methods* **37,** 97–108.
8. Butcher, E. C., Scollay, R. G., and Weissman, I. L. (1980) *J. Immunol. Methods* **37,** 109–121.
8a. Stevens, S. K., Weissman, I. L., and Butcher, E. C. (1982) *J. Immunol.* **128,** 844.
8b. Kraal, G., Weissman, I. L., and Butcher, E. C. (1982) In preparation.
9. Butcher, E. C., Reichert, R. A., Coffman, R. L., Nottenburg, C., and Weissman, I. L. *Adv. Exp. Med. Biol.* (in press).
10. Butcher, E. C. and Weissman, I. L. (1979) *Ciba Found. Symp.* [N.S.] **71,** 265–286.
11. Griscelli, C., Vassalli, P., and McCluskey, R. T. (1969) *J. Exp. Med.* **130,** 1427–1451.
12. Smith, M. E., Martin, A. F., and Ford, W. L. (1979) *Monogr. Allergy* **16,** 203.
13. Rose, M. L., Parrott, D. M. V., and Bruce, R. G. (1976) *Cell. Immunol.* **27,** 36–46.
14. Scollay, R., Hopkins, J., and Hall, J. (1976) *Nature (London)* **260,** 528–529.
15. Cahill, R. N. P., Poskitt, D. C., Grost, H., and Trnka, Z. (1977) *J. Exp. Med.* **145,** 420–428.
16. Chin, W., and Hay, J. B. (1980) *Gastroenterology* **79,** 1231–1242.
17. Cantor, H., and Weissman, I. L. (1976) *Prog. Allergy* **20,** 1.
17a. Butcher, E. C. *et al.* (1982). *In* "Recent Advances in Mucosal Immunity" (K. Sell, ed.), Raven, New York (in press).
18. Chin, Y. H., Carey, G. D., and Woodruff, J. J. (1980) *J. Immunol.* **125,** 1764.

Pulmonary Endothelial Cell Injury Induced by Antibody Fragments to Angiotensin Converting Enzyme

PETER R. B. CALDWELL,[1] H. JOACHIM WIGGER,[2]
VINCENT P. BUTLER, JR.,[1] IRMA GIGLI[3]
Departments of Medicine[1] and Pathology[2]
College of Physicians and Surgeons
Columbia University
New York, New York
Departments of Dermatology and Medicine[3]
Irvington House Institute
New York University School of Medicine
New York, New York

We have purified angiotensin converting enzyme from pulmonary membranes in the rabbit and raised antibodies to the enzyme in the goat. In studying the binding and anticatalytic properties of the antibodies we found that intravenous administration of immune serum or IgG into the rabbit resulted in lethal pulmonary edema within 5 min (1). Pathologic examination of the lungs revealed bleb formation in the capillary endothelial cells, fragmentation of endothelial membranes, and disruption of the cells. Platelets and leukocytes were sequestered in the pulmonary capillary bed. Fibrin-rich edema was present in the interstitium and alveolar spaces of the lung. Preimmune goat serum, on the other hand, did not cause this pulmonary injury.

We considered that complement activation might be involved in the pathogenesis of this injury and therefore sought to avoid it by using antibody fragments. We have prepared these by standard procedures (2,3) and obtained them free of contaminating IgG (see Fig. 1). The $F(ab')_2$ form a precipitin band against the enzyme antigen tested by immunodiffusion, whereas the Fab do not (see Fig. 2).

To determine whether complement activation might occur in the presence of binding of antibody fragments to converting enzyme we

PATHOBIOLOGY OF THE ENDOTHELIAL CELL

ISBN 0-12-521980-6

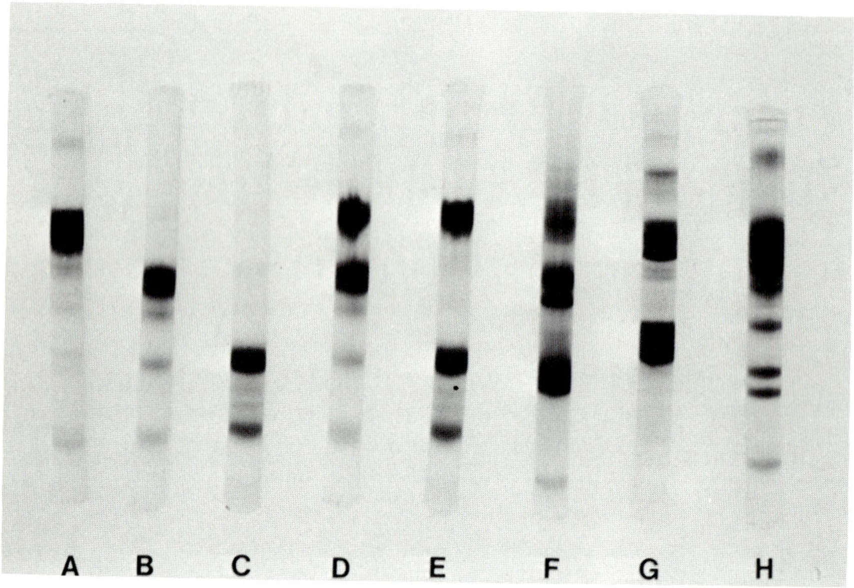

Fig. 1. 5% Polyacrylamide disk gels of immune IgG, F(ab′)$_2$, and Fab to angiotensin converting enzyme. (A) 18 μg IgG; (B) 11 μg F(ab′)$_2$; (C) 13 μg Fab; (D) same as B plus 9 μg IgG; (E) same as C plus 9 μg IgG; (F) ovalbumin MW 45,000; (G) bovine serum albumin MW 68,000; (H) bovine IgG, fraction II 160,000.

assayed for reduction in hemolytic titers of C4 and C3 *in vitro* using standard techniques (4,5). Since purified, solubilized angiotensin converting enzyme alone activates both C4 and C3 in serum, as evidenced by loss of hemolytic activity, we tested for enhancement of activation by the addition of IgG, F(ab′)$_2$, or Fab to the enzyme (6). The results are given in Table I. IgG plus enzyme enhances consumption of both C4 and C3. F(ab′)$_2$ plus enzyme shows no effect on C4 beyond that of enzyme alone, but moderately reduces C3. Fab plus enzyme shows no effect on either C4 or C3 beyond that of enzyme alone. IgG and derived fragments alone exhibit no effect on complement. These results are similar to those obtained in studies of insoluble complexes of IgG and F(ab′)$_2$ plus antigen, recently reported by Gadd and Reid using ovalbumin as antigen (7). They were not successful, however, in testing Fab in their system because of formation of soluble complexes.

When either (Fab′)$_2$ or Fab was administered intravenously in the

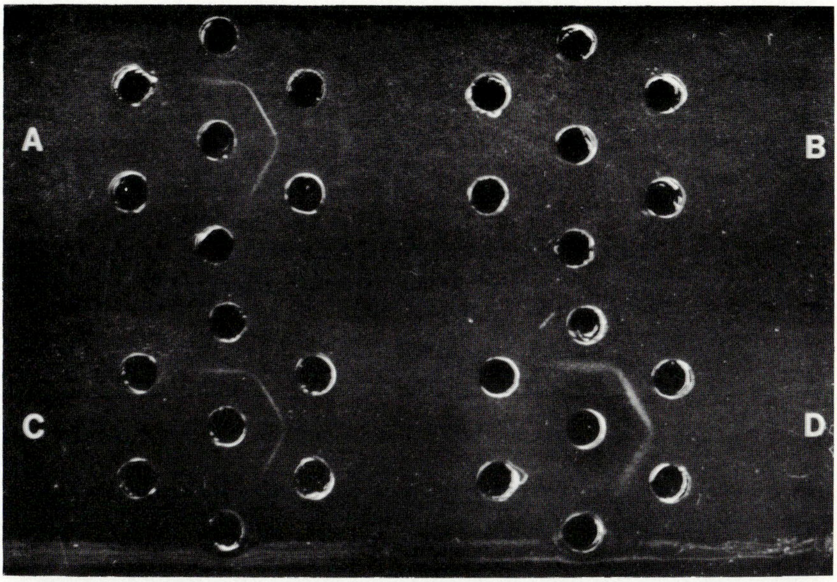

Fig. 2. Ouchterlony double diffusion of serum, IgG, $F(ab')_2$, and Fab to angiotensin converting enzyme. (A) center well, $F(ab')_2$ 40 μg; (B) center well, Fab 40 μg; (C) center well IgG 40 μg; (D) center well, immune serum 1 mg. Outer wells contain serial double dilutions of purified angiotensin converting enzyme beginning with 9 μg.

TABLE I
Percent Reduction in Hemolytic Titer of C_4 and C_3 Following Incubation with Immune Complexes[a,b]

	C4 (%)	C3 (%)
IgG + ACE	100	60
$F(ab')_2$ + ACE	60	50
Fab + ACE	60	27
ACE	60	27
IgG	0	0
$F(ab')_2$	0	0
Fab	0	0

[a] From Caldwell *et al.* (6).

[b] Angiotensin converting enzyme (ACE, 10 μg/ml), IgG (1000 μg/ml), $F(ab')_2$ (667 μg/ml), Fab (333 μg/ml) in normal human serum. From Caldwell *et al.* (6).

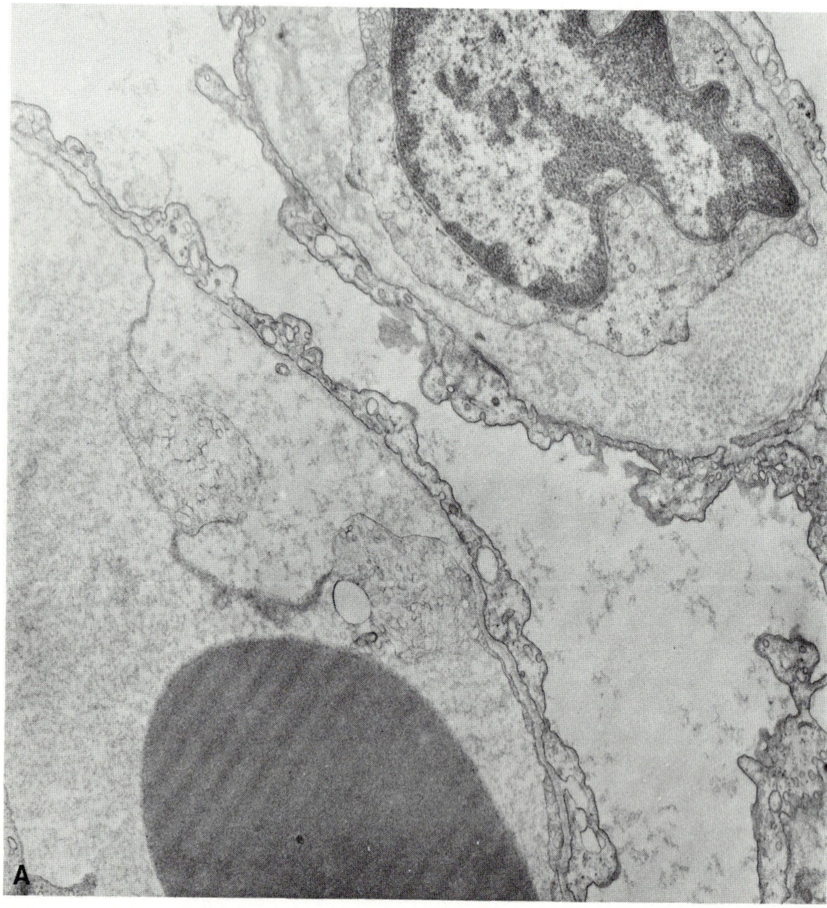

Fig. 3. Lung of rabbit following intravenous administration of 20 mg Fab to angiotensin converting enzyme. (A) Alveolar septal capillary with adjacent erythrocyte shows attenuation of endothelial cytoplasm with bleb formation and disintegration of part of the luminal cell wall.

rabbit, lethal pulmonary edema occurred. Ultrastructural examination of the lungs revealed endothelial cell injury indistinguishable from that produced by immune IgG (see Fig. 3). Thus, although Fab does not enhance complement activation by either classical or alternative pathways *in vitro*, it does cause endothelial cell injury *in vivo*. This result suggests that Fab to angiotensin converting enzyme initiates pulmonary endothelial cell injury by a mechanism independent of complement activation.

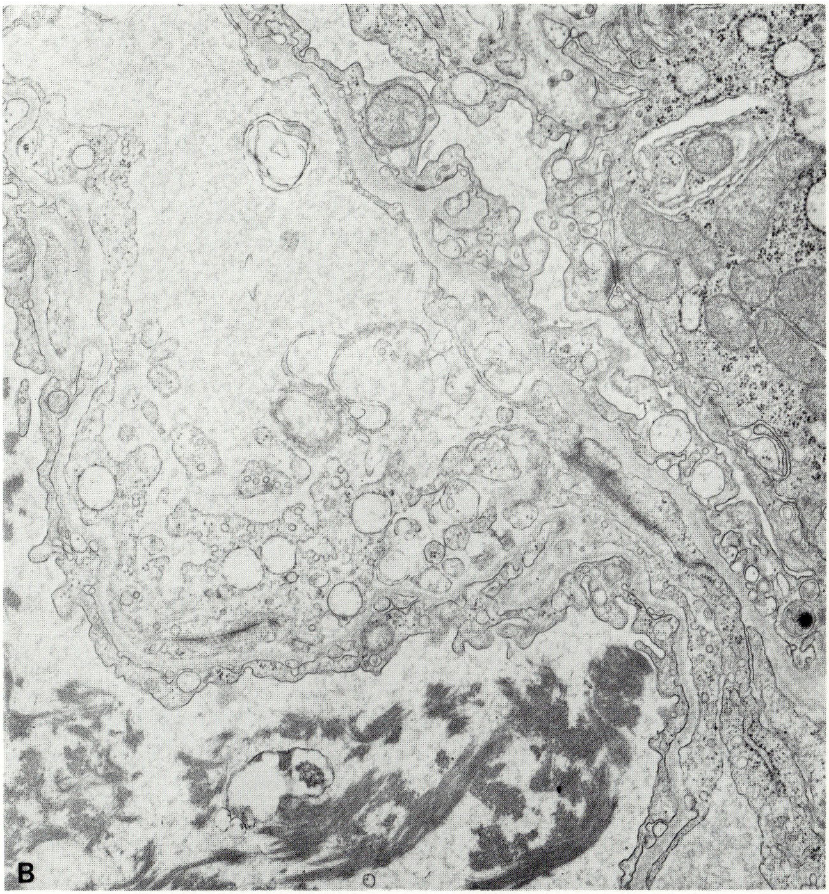

(B) Alveolar septal capillary depicting disintegration of endothelial cell with fibrin in the alveolar space. The alveolar epithelial cell appears intact. (×26,000)

ACKNOWLEDGMENT

This work was supported by NIH Grants HL-17813, HL-07018, HL-15486, HL-10608, AI-13809, a Grant-in-Aid from the New York Heart Association, and by the Joe and Emily Lowe Foundation, Inc., and the Dana Foundation. The technical assistance of Lourdes Fernandez is gratefully acknowledged.

REFERENCES

1. Caldwell, P. R. B., Wigger, H. J., Das, M., and Soffer, R. L. (1976) *FEBS Lett.* **63,** 82–84.
2. Nisonoff, A. (1964) *Methods Med. Res.* **10,** 134–141.
3. Porter, R. R. (1959) *Biochem. J.* **73,** 119–126.
4. Gigli, I., von Zabern, I., and Porter, R. R. (1977) *Biochem. J.* **165,** 439–446.
5. Nelson, R. A., Jr., Jensen, J., Gigli, I., and Tamura, N. (1966) *Immunochemistry* **3,** 111–135.
6. Caldwell, P. R. B., Wigger, H. J., Fernandez, L. T., D'Alisa, R. M., Tse-Eng, D., Butler, V. P., Jr., and Gigli, I. (1981) *Am. J. Pathol.* **105,** 54–63.
7. Gadd, K. J., and Reid, K. M. R. (1981) *Immunology* **42,** 75–82.

PART VII

VASCULAR PATHOBIOLOGY

Responses of Endothelial Cells to Injury

JOHN L. GORDON AND JEREMY D. PEARSON
A.R.C. Institute of Animal Physiology
Babraham, Cambridge, England

INTRODUCTION

There is no unanimity as to how the word *injury* should be defined and there is nothing to be gained by substituting for it the word *damage*: both are imprecise and often mean different things to different people, since we tend to see these terms in the context of our own specialized interests. Partly as a consequence of this, each specialist has his own favorite method for measuring damage—the man who works on the consequences of endothelial removal *in vivo* and on the stimuli that induce this may have little interest in those agents and processes that result only in the release of prostaglandins or lysosomal enzymes from endothelium, whereas he who works on the latter topic may hold the opposite view. Both are important in the context of endothelial injury, and we have therefore chosen to define *injury* in its widest sense: not merely in terms of physical (or other) damage that leads to progressive cellular deterioration but also in the context of any potentially damaging stimulus (or change in the pericellular environment) that provokes a response from the cells.

Much of our knowledge of endothelial biology has come from studies with cultured cells; the techniques used to culture endothelium and the characteristics of these cells *in vitro* are given elsewhere (see, for example, the papers in Part I of this volume) and will not be covered in detail in this review. The potential advantages of using cells in culture are considerable; for example, they can readily be obtained as a homogeneous population with little or no contamination by other cell types and they are maintained in a defined environment that can be readily manipulated, but it should always be recognized that the artificial conditions in which the cells are kept may alter their behavior and

PATHOBIOLOGY OF THE ENDOTHELIAL CELL

ISBN 0-12-521980-6

may, in some senses, be potentially injurious; this concept will be discussed in more detail later.

In this review we shall not provide an exhaustive catalog of cellular responses to injury; rather, we shall concentrate on selected aspects of endothelial function that are affected by environmental changes *in vivo* or *in vitro*, chosen because of their relevance to vascular physiology and hemostasis, to pathological conditions such as thrombosis and atherosclerosis, and to related subjects such as inflammation and the preservation of organs for transplantation. To set this in context, however, we should first consider how commonly encountered injurious stimuli such as ischemia or hypoxia affect general cellular functions such as membrane permeability and energy metabolism; most of the information available has come from studies with cells other than endothelium, and we shall therefore begin with a brief résumé of how membrane permeability is controlled and how it is affected by injurious stimuli, before discussing endothelial injury per se.

MEMBRANE PERMEABILITY AND THE EFFECTS OF INJURIOUS STIMULI: POSSIBLE MECHANISMS INVOLVED

Selective and/or unidirectional permeability of cell membranes to many compounds is essential for the maintenance of cellular integrity and normal function. An increase in plasma membrane permeability can be an early indication of cellular response to injury, but it should not be assumed that the processes responsible for this increased permeability merely represent overactivity of the normal mechanisms, nor is it necessarily true that a change in permeability to one molecular species means that the permeability to other types of compounds has also been altered. For convenience, we shall first consider the mechanisms controlling plasma membrane permeability to ions, small molecules and macromolecules, and then discuss the mechanisms involved in abnormal permeability that results from exposure to injurious stimuli. Most of the experimental data referred to in this section have been obtained using cell types other than endothelium (in many cases, using myocardial cells in culture), and although it is reasonable to assume that the mechanisms regulating membrane permeability in endothelium are broadly similar to those described for other cell types, there is relatively little information on this subject as yet. More work is needed to clarify this and to delineate specific features of the endothelial response to injury that may differ from those in other cells.

NORMAL PERMEABILITY TO IONS, SMALL MOLECULES, AND MACROMOLECULES

Ions, and molecules of molecular weight up to about 1000, such as amino acids, nucleosides, fatty acids, and sugars, can cross cell membranes by simple diffusion or, in certain cases, by the action of specific carriers in the membrane. Carrier-mediated systems exhibit a high degree of structural specificity for the ions or molecules transported; additionally, several of these systems are pumps that act against substantial transmembrane concentration gradients (e.g., the Na^+/K^+ exchange pump) through the expenditure of intracellular energy—usually, the hydrolysis of ATP. One consequence of the energy dependence of these membrane pumps is that if in the early stages of injury there is a deficiency in energy supply, the transmembrane gradients are not maintained; thus an early manifestation of damage is leakage of potassium ions from the cell. For many small hydrophilic species the rate of diffusion across cell membranes is very low, and for high-molecular-weight compounds it is often immeasurably slow.

Physiological processes exist for translocation of macromolecules across cell membranes, but (except for the special case of the direct interaction between intracellular membranes and nascent macromolecules destined to be integral membrane components or to be exported from the cell) these involve exocytosis or endocytosis rather than direct penetration of the cell membrane (1). From the levels of enzymes and other proteins that can be measured in blood under normal circumstances, it is clear that some release of macromolecules occurs from healthy cells all the time. Although some of this may represent controlled processes of cell turnover, translocation of intracellular enzymes to the extracellular space is apparently a normal physiological phenomenon in at least some cells (2).

ABNORMAL PERMEABILITY

A critical level of ATP production is necessary for the maintenance of a wide range of cellular functions, including that of normal membrane permeability. The failure of energy-dependent membrane pumps to maintain ion gradients when the supply of ATP is reduced is one of the earliest manifestations of cellular response to injury (3). Another early consequence of exposure to an injurious environment that results in a decrease in ATP production is a concomitant increase in anaerobic metabolism; cytoplasmic pyruvate is converted to lactate rather than to acetyl coenzyme A (which is normally removed by incorporation into the Krebs cycle), and the leakage of lactate from the cell provides another indicator of the early events in cell damage (4). The accumulation of lactate produces intracellular acidosis that may inhibit glycolysis (thus decreasing the cell's energy supply even further) and activate lysosomal hydrolases; all these events may cause conformational changes in the plasma membrane, promote further increases in membrane permeability, and induce even more cell injury (5). In addition, failure of the Na^+/K^+ pump can lead to an increased level of intracellular Ca^{2+}, because this is regulated in part by an Na^+/Ca^{2+} exchange mechanism (6), and to a general increase in the total cellular ion concentration. This is accompanied by entry of water to regulate osmolality, which causes cell swelling. This swelling stretches the plasma membrane, which increases membrane permeability and leads to the loss of small molecules such as ATP (7). Other detrimental changes following a reduction in mitochondrial activity can result from a decreased β-oxidation of fatty acids. If this occurs without any decrease in fatty acid uptake by the cells, there is an accumulation of fatty acyl coenzyme A molecules that can exert detergent effects on the plasma membrane and promote further increases in membrane permeability (8).

It is important to emphasize that the early changes in membrane permeability discussed earlier are reversible (9) but that the process of cell damage is progressive and eventually reaches a point (as yet not well characterized) where it is irreversible; for example, experimental studies on myocardial ischemia and hypoxia have shown that a large amount of enzyme leakage from the myocardium is associated with irreversible damage, whereas a small amount can be tolerated and cellular function is completely restored when conditions are normalized (10). It has been recognized for many years that leakage of macromolecules (particularly proteins) can be used to assess the severity of cell damage; the blood levels of intracellular enzymes from the heart provide a means of estimating the severity of myocardial infarction (11), and leakage of enzymes from cells is widely used as a measure of damage in both clinical and experimental situations (12,13). When making such measurements *in vivo* (or *in vitro* in experiments with perfused organs), we are handicapped by not knowing which cells are responsible for the relatively low levels of enzymes and other proteins measured in the extracellular fluid and precisely which mechanisms are involved; likewise, we cannot be sure whether the same population of cells is responsible for the increased levels found after injury and what additional pathological mechanisms are involved in this increased permeability to macromolecules. Small proteins may leak from cells simply because of the increased membrane permeability caused by cell swelling and/or the other processes discussed earlier. In addition, another mechanism by which proteins could leak out of cells is suggested by the observations made on erythrocytes by Seeman (14), who noted that slits 20–100 nm in length can open in the erythrocyte membrane for a few seconds. Membrane integrity was normally rapidly restored by a resealing process, but this resealing was impaired by an increased concentration of ionized calcium or by acid pH (15). It is possible that the onset of irreversible damage may represent a loss of this resealing capability, and in this context it is worth noting that a decrease in the production of ATP will eventually affect the cell's capacity for membrane biosynthesis, and this will interfere with membrane repair processes.

Two other phenomena may help us to understand some of the mechanisms involved in the development of abnormal membrane permeability; these are known as the oxygen paradox and the calcium paradox. Several observations made on isolated perfused hearts have shown that if the calcium concentration or the oxygen content of the perfusing fluid is reduced, subsequent restoration to normal levels causes a rapid and massive increase in the amount of enzymes leaking from the heart into the perfusate (16,17). The calcium paradox is probably the result of an increase in plasma membrane permeability to calcium when the extracellular calcium concentration is reduced. The subsequent restoration of normal extracellular calcium levels causes a massive inflow of calcium into the cell before normal calcium permeability is restored, resulting in activation of the contractile apparatus and in cell swelling caused by uptake of water. These processes, operating simultaneously, induce stretching of the plasma membrane with the consequences noted earlier, namely, increased membrane permeability and leakage of proteins from the cytoplasm. A redistribution of cellular calcium may also be involved in the oxygen paradox (18); an increase in the cytoplasmic concentration of unbound ionized calcium during anoxia results from dechelation of calcium previously bound to ATP or citrate, and also (as noted earlier) from an increased calcium influx across the plasma membrane. Subsequent oxygenation reactivates the mitochondrial electron transport chain, which can promote a rapid exchange of this cytoplasmic calcium for protons across the mitochondrial membrane; the resulting cytoplasmic acidosis then affects membrane permeability as described earlier. The primary lesion in both paradoxes may therefore be a redistribution of cytoplasmic ionized calcium, whereas the event responsible for the acute damage is a rapid calcium flux. It may well be that the intracellular redistribution of ionized calcium is an event of critical importance in all circumstances where membrane permeability is increased.

In summary, the process of cellular injury is highly heterogeneous, with constituents as small as univalent ions and as large as enzymes leaking across the plasma membrane under different circumstances. Accordingly, a single measure of cellular injury is insufficient and several indices of damage should be used in conjunction. It may not even be sufficient to choose one example as representative of a particular class of substances (e.g., enzymes); for example, although most enzymes that leak from damaged cells are cytoplasmic, some are from specific subcellular structures such as mitochondria or lysosomes (19) and therefore differential patterns of enzyme leakage may be used to assess the nature of the damage and not just the size of "holes" that may have appeared in the plasma membrane. It should also be remembered that environmental factors such as temperature can dramatically affect the leakage of constituents from cells; this may be partly because a reduction in temperature decreases the energy demands made by the cell and therefore an increased proportion of the cell's limited energy supply can be used for "protection"; additionally, a decreased temperature will reduce the rate of detrimental cellular reactions and affect the fluidity of the plasma membrane. Likewise, the metabolic and other activities of a cell under normal conditions will affect the rate at which constituents are released; for example, there is a direct relationship between the extent of contractile activity and enzyme leakage in the normal heart (2).

ENDOTHELIAL INJURY *IN VIVO:* CONSEQUENCES OF ENDOTHELIAL REMOVAL

Removal of the endothelial monolayer can be regarded as the most extreme case of endothelial damage *in vivo* and the consequences can be profound because endothelium normally forms the interface between the blood and the tissues, thus regulating the flow of fluid into and out of the blood and the interaction between the blood cells and the vascular wall; in addition, endothelial cells have metabolic and other functions that are not shared by the subjacent cells of the vessel or by the surrounding extravascular tissue (for review, see 20). Endothelial removal can be induced either by physical damage (e.g., by passing a balloon catheter up the lumen of a blood vessel) or by chemical damage (e.g., in homocysteinemia or hypercholesterolemia); examples of both may be encountered in clinical or in experimental situations (21,22).

Endothelial removal *in vivo* occurs physiologically as part of the normal process of cell turnover, but the turnover rate for vascular endothelium is normally very low (23), although it is somewhat greater at points of hemodynamic stress such as bifurcations and the ostia of branches (24). Several aspects of the consequences of endothelial removal will be discussed in detail elsewhere in this volume, and therefore this topic will be treated only very briefly here. The short summary that follows is designed to emphasize the more important characteristics of endothelial function and to indicate some of the probable consequences when the vascular wall is exposed to an injury so severe that some of the endothelial cells are removed.

Perhaps the most important characteristic of the vascular endothelium *in vivo* is that it presents a nonthrombogenic surface to the blood; that is, it discourages platelet deposition and it does not activate the coagulation system. One important factor in preventing platelet activation is the capacity of the endothelium to synthesize and release prostacyclin (PGI_2), which is discussed in more detail elsewhere in this volume. Prostacyclin is one of the most powerful inhibitors of platelet aggregation known (25); it stimulates adenylate cyclase (26), thus increasing the level of cyclic AMP in the platelets, which reduces their ability to respond to stimuli. There is some debate as to whether prostacyclin normally circulates in the bloodstream at concentrations sufficiently high to inhibit platelet aggregation; the present consensus appears to be that although prostacyclin does not circulate at biologically active concentrations, large amounts can be produced locally in response to a variety of stimuli and can thus prevent excessive platelet aggregation (27). Prostacyclin is much less effective in preventing the adhesion of platelets to cells of the vessel wall than it is at preventing platelet aggregation, i.e., platelet–platelet contact (28). Because prostacyclin has a greater inhibitory effect on aggregation than on adhesion, its apparent effect depends on whether adhesion alone or adhesion plus aggregation are measured; if no precautions are taken to prevent the formation of small platelet aggregates, the effect of prostacyclin on platelet "adhesion" may be overestimated (29). Platelets do not normally adhere to endothelial cells; this appears to be a consequence of some specific property of the external surface membrane of the endothelium (see Section IV,B). The smooth muscle cells subjacent to the endothelium in the vascular wall present a surface to the platelets that encourages platelet adhesion and aggregation, despite the fact that vascular smooth muscle cells also have the capacity to make prostacyclin. It appears that connective tissue components synthesized and secreted by vascular smooth muscle, which constitute an important part of the subendothelial basement membrane, are mainly responsible for encouraging platelet deposition when the endothelium has been removed; in particular, the fibrillar collagens types I and III are potent platelet stimulants (for review, see 30).

The fact that platelets will readily adhere to the extracellular components of vascular smooth muscle but not to endothelium should not be taken to indicate that endothelial cells present a universally "nonadhesive" surface; studies with vascular cells in culture have shown that granulocytes exhibit the opposite affinity—that is, they will readily adhere to cultured endothelium but not to vascular smooth muscle (31,32).

From the preceding discussion it is apparent that endothelial removal promotes the deposition of platelets, thus encouraging thrombus formation, and this prothrombotic tendency is further increased by the effects of endothelial removal on the coagulation system. There are binding sites (of at least two classes) for thrombin on endothelial cells but not on vascular smooth muscle cells; after binding to the endothelial surface, the thrombin is inactivated (33,34). Antithrombin III, a protein that catalyzes the inactivation of thrombin by heparin, is also present on the surface of endothelial cells (35) and presumably contributes to their anticoagulant properties; in addition, heparan and dermatan sulfates are synthesized by vascular endothelium, and these glycosaminoglycans also have anticoagulant activity (36). Thus when endothelium is removed, the anticoagulant effect of the vascular surface and its capacity to inactivate circulating thrombin are much reduced.

The final inevitable effect of endothelial removal is an increased vascular permeability to macromolecules; several experimental studies have shown that the leakage of albumin or other proteins from the plasma into the deeper layers of the vascular wall is much increased when the endothelium is physically removed (37,38).

The consequences of endothelial removal discussed earlier are all short-term effects of the initial injury, but there are long-term consequences too: the repair process begins by the remaining endothelial cells migrating into the denuded area and proliferating in an attempt to cover it as quickly as possible. If the area is sufficiently large, a "pseudoendothelium" forms, composed of cells that have migrated from the deeper layers of the vascular wall and also possibly of blood-borne cells (mononuclear phagocytes). Such newly formed endothelial layers are abnormally permeable to macromolecules (39). The final long-term effect of endothelial removal is that smooth muscle cells in the tunica media of the blood vessel proliferate, leading to thickening of the vascular wall (40). This topic is discussed in more detail in Chapter 4.

ENDOTHELIAL INJURY *IN VITRO:* OBSERVATIONS ON CULTURED CELLS

In attempting to compile a review of data that have been obtained specifically on the responses of cultured endothelial cells to injury, one is immediately struck by how little experimental work has been done on this subject. There seem to be several reasons for this:

1. The subject of endothelial injury is recognized as being important, but most previous work on this topic has concentrated on the area outlined in Section III—that is, deliberate injury to endothelium *in vivo*, resulting in denudation—and there have been few studies of the more subtle responses of endothelium to injurious stimuli that produce less than total destruction of the cells.
2. Such responses are difficult to study *in vivo* because it is not easy to determine whether the responses observed are due to endothelium alone or to another cell type; endothelial cells in culture are obviously appropriate for such studies but techniques for endothelial culture have become widely used only within the last 5 years or so (41).
3. The success of such studies depends on (among other factors) establishing the relevant normal values for endothelial cell functions in culture and on determining how these functions may have changed during culture. Unfortunately, there is relatively little information available on the normal patterns of endothelial energy metabolism, and such information is necessary before we can understand the perturbations induced by damaging stimuli. It is clear that changes in endothelial cell function do occur during culture; indeed, it is arguable that the culture environment is "injurious" to the cells, in the sense that it is not optimal. It is, therefore, important to characterize the responses of the endothelial cells to the culture conditions and to regard the changes as being possible "responses to injury," as well as to investigate the effects of specific injurious stimuli that have been deliberately introduced. Whether the changes observed in endothelium during culture represent some form of dedifferentiation or whether they represent specific responses to factors in the culture environment and are therefore reversible and/or preventable is an important ques-

tion, but the answer to this remains to be determined and may require a considerable amount of research effort. In the discussion that follows, endothelial responses to injury have been divided (somewhat arbitrarily) into three sections for reasons of convenience; these deal successively with changes in membrane permeability, surface components, and intracellular functions.

CHANGES IN MEMBRANE PERMEABILITY

Release of some intracellular constituents from endothelial cells occurs continuously; this basal rate may be increased by an injurious environment or by specific stimuli. Other constituents are not normally released and their appearance in the extracellular medium is a consequence of the cells' response to stimulation or injury. Whether the release (either basal or stimulated) of some constituents should be called "leakage" or "secretion" is not always evident; *leakage* implies a passive phenomenon, whereas *secretion* implies an active event, and *release* might apply to either.

The smallest constituents that are released from endothelial cells are ions. There is relatively little information about the basal rates of ion flux into and out of cultured endothelium or about the effects of stimuli on ion fluxes. D'Amore and Shepro (42) measured calcium influx in endothelial cultures and showed that this was transiently increased by a variety of stimuli, including thrombin, adrenaline, and histamine, but they did not attempt to compare the behavior of their cultured endothelial cells with that of other cell types or to determine whether the calcium flux was affected by the conditions of culture. Potassium flux is a valuable indicator of membrane permeability; the radioactive isotope of the potassium analogue rubidium (^{86}Rb) is widely used as a substitute for potassium in such experiments because it behaves similarly to potassium, but its half-life is much longer than that of ^{42}K or ^{43}K. There are apparently no published experiments on Rb efflux from cultured endothelium, but Mauger *et al.* (43) found that the rate of Rb efflux from a line of cultured smooth muscle cells was specifically increased by noradrenaline. Preliminary experiments from our laboratory have shown that the basal rate of Rb efflux from pig aortic endothelial cells in culture is much higher than that from the smooth muscle line used in the experiments of Mauger *et al.* or from freshly isolated arterial tissue; the efflux from cultured endothelial cells is stimulated by bradykinin but not by noradrenaline (W. Martin, unpublished; see Fig. 1).

Adenine nucleosides and nucleotides are released from cultured endothelium under different circumstances and these processes can be

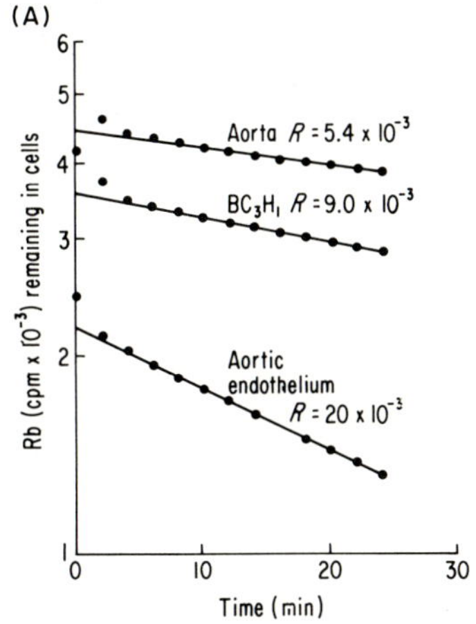

Fig. 1A. Desaturation curves for ^{86}Rb from porcine aorta, BC_3H_1 cells [data derived from Mauger *et al.* (43)], and porcine subcultured aortic endothelial cells. Tissues or cells on cover slips were loaded for 3 hr in Krebs' solution at 37°C containing 10 μCi/ml ^{86}Rb. Washout of the isotope was monitored by placing the tissue or cells in a chamber and perfusing with Krebs' solution at a rate of 5 ml/min. Fractions were collected every 2 min and radioactivity was determined in a γ counter. The desaturation curves were constructed by adding in reverse order the counts released in each fraction to the counts remaining in the tissue or cells at the end of the experiment. There is an initial rapid rate of washout from the extracellular space; then after a few minutes the efflux rate becomes constant and reflects membrane-limited efflux from the cells. The efflux rate can be calculated using the formula $R = \Delta A/\Delta t\ At$, where Δt is the time over which the fraction was collected, ΔA is the number of counts lost in that time, and At is the number of counts remaining in the tissue or cells at the end of that time.

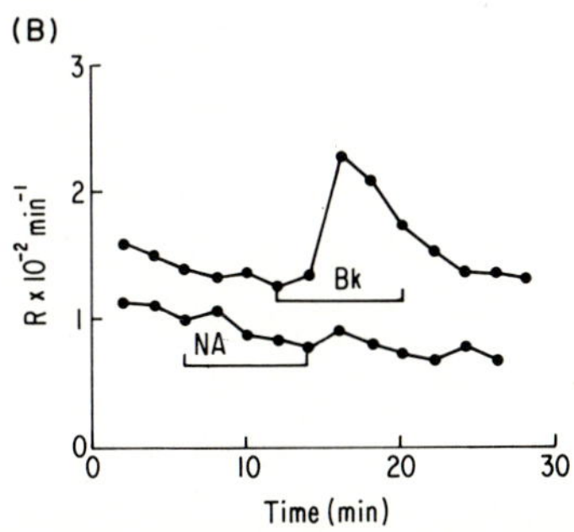

Fig. 1B. Effects of bradykinin (BK) and noradrenaline (NA) on the rate of ^{86}Rb efflux from porcine subcultured aortic endothelial cells. After loading the cells with ^{86}Rb for 3 hr at 37°C washout of the isotope was monitored at 20°C. Lowering the temperature reduces the basal rate of efflux and allows a clearer examination of the effects of vasoactive agents on efflux. The results of two separate experiments are shown. When a steady rate of efflux had been reached, bradykinin (10^{-5} M) or noradrenaline (10^{-5} *M*) was added to the perfusion fluid. The rate of efflux of ^{86}Rb from the cells was increased by bradykinin but not by noradrenaline.

conveniently monitored after labeling intracellular ATP by incubation of the cells with [^{3}H]adenine or [^{3}H]adenosine; endothelial cells have high affinity uptake processes for both of these substances and, once incorporated, they are converted mainly to ATP (44,45). In normal culture conditions, the ^{3}H released into the extracellular medium is mainly in hypoxanthine, with smaller amounts in inosine; these represent the products of intracellular ATP catabolism that diffuse from the cells down a concentration gradient. This release of radiolabeled inosine and hypoxanthine is much increased by treating the cells with cyanide plus deoxyglucose (J. D. Pearson, unpublished; see Fig. 2), but this is presumably a direct consequence of increased ATP catabolism and not due to an effect on membrane permeability. ATP itself, however, can be transiently released from endothelial cells in culture by treating the cells with agents such as thrombin or trypsin (46); subsequently, the ATP is metabolized to ADP, AMP, and

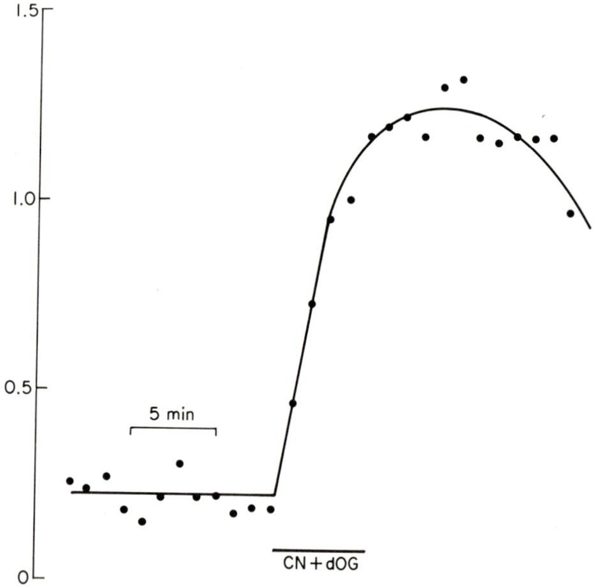

Fig. 2. Pig aortic endothelial cells cultured on Cytodex beads were incubated for 60 min with [2-^{3}H]adenosine in the absence of serum, placed in a small perfusion chamber, and perfused with minimal essential medium (gassed with 95% N_2 + 5% CO_2) at 0.5 ml/min. Subsamples of the perfusate were monitored for ^{3}H. The bar indicates the addition of 5 mM deoxyglucose + 10 mM NaCN to the perfusion medium for 5 min. Subsequent tlc analysis demonstrated that released ^{3}H comigrated with inosine and hypoxanthine. Ordinate axis: fraction of cell-associated ^{3}H [$\times 10^2$] released per minute.

adenosine by ectoenzymes on the endothelial cell surface (47). ATP release from endothelium has been further investigated by Lollar and Owen (48), who established that the dose–response curve for thrombin-induced ATP release was very similar to that for thrombin-induced arachidonic acid release (with a threshold concentration of less than 0.1 u/ml). These workers also showed that the release was linear for at least 30 min and that the high affinity sites for thrombin binding on endothelium were not involved in the release process; in addition, they showed that DIP-treated thrombin was ineffective.

ATP release is selective in that it is much more extensive than release of lactate dehydrogenase (LDH) under the same conditions. It is probably a consequence of a selective increase in membrane permeability to small molecules, but as ATP release is not inhibited in the presence of 10 m*M* extracellular ATP, and [^{3}H]ATP is not taken up by the cells under conditions in which ATP is readily released, it could be that the increase in permeability is unidirectional (46).

Preliminary experiments in our laboratory indicate that cultured endothelium releases nucleotides more readily than does fresh vascular tissue; this suggests that the plasma membranes of cultured endothelium may be more permeable than those of endothelial cells *in situ* and/or more susceptible to the effects of agents that increase permeability. This concept is important, with implications much wider than the control of nucleotide release, but further work is needed to clarify these points. One intriguing possibility is that ATP release (perhaps accompanied by the release of other low-molecular-weight constituents such as NADH or GSH) is caused by the dissociation of gap junctions between adjacent cells; these structures are known to allow the exchange of cellular constituents of molecular weight up to about 1000 (49), and agents like trypsin or thrombin could separate junctional complexes, exposing sites of increased permeability to the extracellular space until the plasma membrane reorganized.

Labeling cells with radioactive chromium (^{51}Cr) and monitoring the release of label into the medium has previously been used to assess the effects of injurious stimuli on a variety of cells. Harker and his colleagues (50) have applied this technique to cultured endothelial cells and have shown that incubating the cells with relatively high concentrations of homocysteine increases the release of ^{51}Cr. These results corroborated his previous *in vivo* studies in which homocysteinemia in baboons induced endothelial damage and desquamation (22). As a means of quantifying cell damage, ^{51}Cr release is less than ideal. It has the disadvantage that chromium is bound to numerous cellular constituents, including proteins, which may be in or on the cells. Thus,

^{51}Cr release does not monitor a single event but a variety of different processes. An additional practical disadvantage is that the basal release of ^{51}Cr can be very high, particularly if experiments necessitate long incubations. For example, Galdal and Evensen (51) used ^{51}Cr release to monitor damage to endothelial cells in culture caused by thrombin or by low density lipoproteins and the control release (without added agents) was as high as 30% over 24 hr.

Another commonly used indicator of cell damage is leakage of LDH, but this appears to be a measure of relatively severe damage; endothelial cells can be induced to release up to 50% of their ATP without detectable leakage of LDH (46). Likewise, the use of vital dyes to assess cell viability (measuring leakage in, not leakage out) is widespread, but this is another insensitive method for detecting sublethal damage. For example, in our experiments endothelial cells remained viable and did not take up trypan blue despite releasing a substantial proportion of their ATP (46), and other workers have noted that the extent of vital dye uptake does not correlate with the release of various intracellular compounds (including macromolecules) and that the relationship varies with the stimulus used (52,53). Such discrepancies could arise as a result of selective or undirectional changes in membrane permeability, and another source of variation could be variable binding of the dye to proteins, as discussed above for ^{51}Cr; trypan blue itself has a molecular weight < 1000, but binds avidly to albumin.

CHANGES IN SURFACE COMPONENTS

When cultured endothelial cells are transformed by incorporation of simian virus DNA their capacity to support platelet adhesion is much increased. This results from a change in the endothelial surface membrane and is independent of prostacyclin production by the cells; indeed, the addition of exogenous prostacyclin does not reduce platelet adhesion to normal levels (54). The main conclusion from these important observations is that although prostacyclin is an important regulator of platelet *aggregation* (see Section III), it is not the factor mainly responsible for the lack of platelet *adhesion* to normal endothelium; rather, the lack of adhesion depends on the presence of some specific component(s) on the endothelial surface.

Changes in the endothelial surface induced by culture conditions can also affect the adhesion of leukocytes to the endothelial cells. For example, de Bono (55) showed that the adhesion of lymphocytes to cultured endothelium was much greater than the adhesion to freshly isolated endothelium; he also showed that endothelial cells cultured in rapid flow conditions supported the adhesion of lymphocytes much less well than cells grown in conventional static conditions. Beesley *et al.* (31) obtained similar results when measuring the adhesion of granulocytes to flow-cultured endothelium. De Bono (56) suggested that the effects of flow on the surface properties of cultured endothelium were a consequence of a reduction in thickness of the glycocalyx on the cells, but he made no attempt to determine how the composition of the surface membrane might have changed.

Other constituents of the endothelial surface that can change during culture include 5′-nucleotidase, factor VIII antigen, and angiotensin converting enzyme (57–59; Table I), all of which may show a reduced activity in subcultures of aortic endothelial cells compared with freshly isolated cells. There is some evidence that pig aortic endothelium may be more susceptible than cells from other species or sites to the loss of such surface components (58).

The substrate on which endothelial cells are grown may be an important determinant of the amounts and distribution of surface components; the distribution of fibronectin on cultured endothelium is abnormal if the cells are grown on plastic or glass but similar to the *in vivo* distribution if the cells are grown on a natural substrate such as basement membrane (60).

Preliminary studies of surface labeling of proteins on cultured endothelium and smooth muscle cells have indicated that the endothelial surface membrane contains several proteins not present on the vascular smooth muscle cell membrane (61); further studies of this type should help to clarify the changes that occur in endothelial mem-

TABLE I
Angiotensin Converting Enzyme Activity of Porcine Endothelium in Culture[a]

	Enzyme activity (nmole substrate/10^6 cells/hr)	
Culture	Controls	+ 10 μg/ml SQ 14225
Primary	31.7 ± 7.8	2.2 ± 0.9
Fourth passage	3.4 ± 0.3	3.1 ± 0.1

[a] Values are means ± SD, n = 3. SQ14225 is a selective, potent inhibitor of angiotensin converting enzyme. Substrate used was Z-Phe-His-Leu, and enzyme activity was measured by the fluorimetric determination of the His-Leu released.

brane constituents as a consequence of environmental changes (e.g., during culture) or exposure to injurious stimuli.

CHANGES IN INTRACELLULAR FUNCTIONS

It is difficult to separate intracellular changes from other cellular responses, and it is clear that some intracellular changes may be a consequence of, or associated with, changes in surface membrane composition or function. For example, the morphological changes observed when endothelial cells in culture are exposed to a free radical generating system are profound (q.v.), but whether these changes (which include alterations in shape and structure) should be categorized as intracellular is debatable.

A method for assessing endothelial damage based on the capacity of cells to take up adenine was developed in our laboratory (44) to monitor complement-mediated or lymphocyte-mediated cytotoxicity; this proved particularly valuable when the time scale of an experiment was so long that monitoring the release of ^{51}Cr was unsatisfactory because the amount of leakage under control conditions was too great. Assessment of cell viability at the end of such experiments offered obvious advantages, and subsequent studies revealed that this terminal adenine uptake assay was more sensitive at detecting a change in cellular function or integrity than were methods based on the release of ^{51}Cr or LDH. The assay probably measures a combination of changes in intracellular adenine nucleotide metabolism and in plasma membrane adenine transport (see Section IV,A).

As mentioned briefly earlier, cultured endothelium can show profound morphological changes when exposed to various stimuli; one of the most obvious morphological changes is contraction, and we have observed endothelium in culture contracting in response to the addition of KCl or of ionophore A23187 (unpublished observations). Such contractile responses are not, however, uniformly reproducible; some cultures do not respond, and in other cultures only some of the cells are observed to contract. It is not clear whether this variable responsiveness is because some of the cells have lost the appropriate contractile machinery or because the nonresponding cells are more strongly attached to their substrate and are therefore physically prevented from contracting. Experiments with cells cultured on a variety of different substrates may help to resolve this question; alternatively, measuring one of the biochemical correlates of contractility (such as the transmembrane flux of calcium or rubidium—see earlier) may give more reproducible results than measuring contractility per se. The interpretation of such studies may, however, be complicated if the cultured cells exhibit an elevated basal rate of ion flux (see Section IV,A), as this may mean that the cells are already responding to a potentially injurious environment.

Another biochemical correlate of the contractile state of the cells is the intracellular level of cyclic AMP; increases in cyclic AMP are often associated with relaxation. Again, as with the transmembrane ion flux, it is desirable to establish that the basal level of cyclic AMP in the cultured cells is comparable to that measured in the same cells *in situ*, but the results from some of our recent experiments with endothelium and vascular smooth muscle cells in culture suggest that this may not be the case. It is difficult to compare subcultures of endothelium with freshly isolated or *in situ* endothelium, because most of the blood vessel wall is composed of smooth muscle cells and to isolate endothelium from the blood vessel may result in cellular functions being affected by the isolation procedure. However, comparisons between subcultured and freshly isolated vascular smooth muscle cells can be made, and cyclic AMP levels in the subcultured cells are substantially higher. The cyclic AMP levels in subcultured endothelium are also higher than those in fresh vascular tissue, which suggests that endothelial cells in culture, like smooth muscle cells, may have elevated levels of cyclic AMP (see Table II). Vascular cells in culture proliferate more rapidly than the same cells *in situ*, and this may be connected with the difference in cyclic AMP levels; there is evidence that cyclic AMP concentrations are related to the proliferation rate in at least some cell types (62), and we found that the cyclic AMP levels in the aortas of pigs decreased with age, from newborn animals up to 1-year-old animals, in parallel with the expected rates of proliferation (W. Martin, A. Ager, and J. L. Gordon, unpublished).

Another intracellular change that occurred as the cells were cultured was in the pattern of amine metabolism (63). Metabolism of phenylethylamine or serotonin was higher in freshly isolated endothe-

TABLE II
Cyclic AMP in Porcine Vascular Cells[a]

Cells	Cyclic AMP (pmole/10^6 cells)
Smooth muscle cells in tunica media of coronary artery	2.48 ± 0.4 (14)
Coronary smooth muscle cells in culture	15.6 ± 1.3 (7)
Aortic endothelial cells in culture	23.6 ± 2.3 (44)

[a] Values are means ± SEM (number of observations).

lial cells than in subcultured cells, suggesting that the levels of monoamine oxidase (MAO) may decrease with age in culture; a similar decline with age has been noted *in vivo* (64). Freshly isolated porcine aortic tissue contained MAO type A and type B as well as clorgyline-resistant oxidase but subcultured aortic endothelial cells apparently metabolized phenylethylamine and serotonin exclusively by MAO-A and exhibited no detectable clorgyline-resistant oxidase or MAO-B activity (63).

Endothelial cells in culture constantly release arachidonic acid from their membrane phospholipids; some of this released free fatty acid is converted to prostacyclin (PGI_2) and smaller amounts are converted to prostaglandin E_2 (65). These prostaglandins, together with some free arachidonic acid, are released into the extracellular medium and the release of arachidonate and prostaglandins is increased by various specific stimuli including bradykinin and thrombin (66,67). Prostaglandin release, as in many other cell types (68), is also increased when endothelial cells are physically damaged. It is noteworthy that a smaller percentage of the arachidonate released by pig aortic endothelial cells is converted to prostaglandins than is the case with other cell types such as smooth muscle; also, although subcultured endothelial cells produce significant amounts of PGE_2, there is little PGE formation by endothelium *in situ* or when freshly isolated (69), which indicates that the endothelial cells are responding to their conditions of culture by altering their pattern of prostaglandin production. This increase in the ratio of PGE/PGI produced is also seen when vascular smooth muscle cells or aortic segments are maintained in culture, and the change in ratio is detectable after only 1–2 hr (Table III).

Although many of the changes described earlier are responses of the cells to an alteration in their environment (e.g., isolating them and maintaining them in culture) rather than changes induced by a specific insult or injury, we have also performed some experiments in which

TABLE III

Production of 6-Oxo-$PGF_{1\alpha}$ and PGE by Segments of Porcine Aorta[a]

Time after isolation: of segment	1 hr	2 hr	24 hr
Ratio of 6-Oxo-$PGF_{1\alpha}$: PGE	8.8 ± 0.6	4.2 ± 0.5	2.2 ± 0.2

[a] Values are means ± SEM (n = 6). Segments were removed from the animal (time 0) and incubated at 37° in culture medium for 0–1, 1–2, and 0–24 hr. Prostaglandins were determined by radioimmunoassay of the culture medium after incubation.

dramatic intracellular changes were induced by deliberately exposing endothelial cells to a damaging stimulus. For example, exposure to a free radical generating system (such as hypoxanthine plus xanthine oxidase) not only causes morphological changes (q.v.) but also induces a burst of prostaglandin synthesis (A. Ager, unpublished). To take another example, as noted earlier, endothelial cells treated with cyanide plus deoxyglucose show a dramatic increase in the rate of ATP catabolism and a concomitant increase in the amounts of inosine and hypoxanthine released from the cells.

CONCLUSIONS

Much of the discussion in this review has followed the theme that endothelial cells (and by implication other cells) in culture are in an environment that is suboptimal or inappropriate and that the responses of the cells to this environment may be analogous to situations in which the cells are deliberately exposed to injurious stimuli. Several pieces of evidence have been introduced to show that the cells change during culture in a variety of ways. There could be many reasons for such changes; for example, the oxygen tension to which the cells are exposed in culture is much higher than that to which they are exposed *in vivo*—with the possible exception of endothelial cells in the pulmonary venous circulation. Additionally, the cells are usually grown on an artificial substratum, they may not be provided with all the growth factors that are necessary, the growth medium contains components of serum to which the cells are not normally constantly exposed *in vivo*, the rate of proliferation of the cells in culture is usually substantially greater than that *in vivo*, and because they are not grown in conditions where the extracellular medium is being replaced continuously (as it is *in vivo*) there is a continual accumulation of waste products in the culture medium. It is clearly important to determine how the cells change during culture and what the possible reasons for such changes might be in order to increase our understanding of endothelial cell biology and pathology, and, indeed, our understanding of cultured cells in general together with their associated artefacts.

As well as experiments designed to evaluate the effects of isolation and culture, more studies of the effects of deliberate injury upon endothelial cells are required; it is evident from the discussion in the preceding sections that this problem has yet to be investigated systematically. Future developments in this field could include experiments on physical injury (e.g., scraping the culture dish may be analogous to the effects of a balloon catheter *in vivo*, about which there is already a considerable amount of information). Further experiments involving immunological damage, hypoxia, or exposure to cytotoxic chemical agents (e.g., oxygen radicals) would be valuable. Whatever the nature of the injurious agent chosen for such experiments, it will be important to study the effects of injurious stimuli of different intensity and in each case to monitor several of the responses to injury that have been discussed earlier.

In conclusion, much of our present information on the responses of cells to injury, particularly where *in vitro* and *in vivo* comparisons have been made, has been obtained from studies with myocardial cells (see Section II). We need similar detailed studies with endothelial cells in culture, having established the artefacts and the possible injurious effects of the culture conditions first; the potential advantages of experimental studies with endothelial cells in culture are enormous, but only by discriminating between biologically relevant fact and artefact will it be possible to derive the maximum information from such experiments.

REFERENCES

1. Finean, J. B., Coleman, R., and Michell, R. H. (1974) "Membranes and Their Cellular Functions." Blackwell, Oxford.
2. De Leiris, J., Feuvray, D., and Coraboeuf, E. (1969) *Arch. Int. Physiol. Biochim.* **77**, 749–762.
3. Michell, R. H., and Coleman, R. (1979) *In* "Enzymes in Cardiology: Diagnosis and Research" (D. J. Hearse and J. De Leiris, eds.), pp. 503–526.
4. Opie, L. H., Owen, P., Thomas, M., and Samson, R. (1973) *Am. J. Cardiol.* **32**, 295–305.
5. Wildenthal, K. (1978) *J. Mol. Cell. Cardiol.* **10**, 595–603.
6. Blaustein, M. P. (1975) *Rev. Physiol., Biochem. Pharmacol.* **70**, 33–82.
7. Gebhard, M. M., Denkhaus, H., Sakai, K., and Spieckermann, P. G. (1977) *J. Mol. Med.* **2**, 271–283.
8. Opie, L. H. (1976) *Circ. Res.* **38**, Suppl. 1, 52–74.
9. Jennings, R. B., and Ganote, C. E. (1976) *Circ. Res.* **38**, Suppl. 1, 80–89.
10. Jynge, P., Hearse, D. J., De Leiris, J., Feuvray, D., and Brainbridge, M. V. (1978) *J. Thorac. Cardiovasc. Surg.* **76**, 2–15.
11. La Due, J. E., Wroblewski, F., and Karmen, A. (1954) *Science* **120**, 497–499.
12. Shell, W. E., Lavelle, J. F., Covell, J. W., and Sobel, B. E. (1973) *J. Clin. Invest.* **52**, 2579–2590.
13. Hearse, D. J., and De Leiris, J., eds. (1979) "Enzymes in Cardiology: Diagnosis and Research." Wiley, New York.
14. Seeman, P. (1974) *Fed. Proc., Fed. Am. Soc. Exp. Biol.* **33**, 2116–2124.
15. Billah, M. M., Finean, J. B., Coleman, R., and Michell, R. H. (1977) *Biochim. Biophys. Acta* **465**, 515–526.

16. Zimmermann, A. N. E., and Hulsmann, W. C. (1966) *Nature (London)* **211,** 646–647.
17. Ganote, C. E., Seabra-Gomes, R., Nayler, W. G., and Jennings, R. B. (1975). *Am. J. Pathol.* **80,** 419–450.
18. Hearse, D. J., Humphrey, S. M., and Bullock, G. R. (1978) *J. Mol. Cell. Cardiol.* **10,** 641–668.
19. Smith, A. F., Wong, P. C. P., and Oliver, M. F. (1977) *J. Mol. Med.* **2,** 265–269.
20. Thorgeirsson, G., and Robertson, A. L. (1978) *Am. J. Pathol.* **93,** 803–848.
21. Baumgartner, H. R., and Muggli, R. (1976) *In* "Platelets in Biology and Pathology" (J. L. Gordon, ed.), pp. 23–60. Elsevier/North-Holland, Amsterdam.
22. Harker, L. A., Ross, R., Slichter, S. J., and Scott, C. R. (1976) *J. Clin. Invest.* **58,** 731–741.
23. Wright, H. P. (1970) *Thromb. Diath. Haemorrh., Suppl.* **40,** 79–84.
24. Schwartz, S. M., and Benditt, E. P. (1976) *Proc. Natl. Acad. Sci. U.S.A.* **73,** 651–653.
25. Moncada, S., and Vane, J. R. (1979) *In* "Prostacyclin" (J. R. Vane and S. Bergström, eds.), pp. 5–16. Raven, New York.
26. Best, L. C., Martin, T. J., Russell, R. G. G., and Preston, F. E. (1977) *Nature (London)* **267,** 850–851.
27. Haslam, R. J., and McClenaghan, M. D. (1981) *Nature (London)* **292,** 364–366.
28. Higgs, E. A., Moncada, S., Vane, J. R., Caen, J. P., Michel, M., and Tobelem, G. (1978) *Prostaglandins* **16,** 17–22.
29. Czervionke, R. L., Smith, J. B., Fry, G. L., Hoak, J. C., and Haycraft, D. L. (1979) *J. Clin. Invest.* **63,** 1089–1092.
30. Santoro, S. A., and Cunningham, L. W. (1981) *In* "Platelets in Biology and Pathology 2" (J. L. Gordon, ed.), pp. 249–264. Elsevier/North-Holland, Amsterdam.
31. Beesley, J. E., Pearson, J. D., Carleton, J. S., Hutchings, A., and Gordon, J. L. (1978) *J. Cell Sci.* **33,** 85–101.
32. Pearson, J. D., Carleton, J. S., Beesley, J. E., Hutchings, A., and Gordon, J. L. (1979) *J. Cell Sci.* **38,** 225–235.
33. Awbrey, B. J., Hoak, J. C., and Owen, W. G. (1979) *J. Biol. Chem.* **254,** 4092–4095.
34. Lollar, P., and Owen, W. G. (1980) *J. Clin. Invest.* **66,** 1222–1230.
35. Chan, V., and Chan, K. (1979) *Thromb. Res.* **15,** 209–213.
36. Busch, C., Ljungman, C., Holdin, C.-M., Waskon, E., and Obrink, B. (1979) *Haemostasis* **8,** 142–148.
37. Packham, M. A., Rowsell, H. C., Jørgensen, L., and Mustard, J. F. (1967) *Exp. Mol. Pathol.* **7,** 214–232.
38. Packham, M. A., and Mustard, J. F. (1971) *In* "The Platelet" (K. M. Brinkhous, R. W. Shermer, and F. K. Kostofi, eds.), pp. 215–232. Williams & Wilkins, Baltimore, Maryland.
39. Minick, C. R., Stemerman, M. B., and Insull, W. (1977) *Proc. Natl. Acad. Sci. U.S.A.* **74,** 1724–1728.
40. Constaninides, P. (1965) "Experimental Atherosclerosis." Elsevier, Amsterdam.
41. Gimbrone, M. A. (1976) *Prog. Hemostasis Thromb.* **3,** 1–28.
42. d'Amore, P., and Shepro, D. (1977) *J. Cell. Physiol.* **92,** 177–184.
43. Mauger, J.-P., Moura, A. M., and Worcel, M. (1978) *Br. J. Pharmacol.* **64,** 29–36.
44. de Bono, D. P., MacIntyre, D. E., White, D. J. G., and Gordon, J. L. (1977) *Immunology* **32,** 221–226.
45. Pearson, J. D., Carleton, J. S., Hutchings, A., and Gordon, J. L. (1978) *Biochem. J.* **170,** 265–271.
46. Pearson, J. D., and Gordon, J. L. (1979) *Nature (London)* **281,** 384–386.
47. Pearson, J. D., Carleton, J. S., and Gordon, J. L. (1980) *Biochem. J.* **190,** 421–429.

48. Lollar, P., and Owen, W. G. (1981) *Ann. N.Y. Acad. Sci.* (in press).
49. Hooper, M. L., and Subak-Sharpe, J. H. (1981) *Int. Rev. Cytol.* **69,** 46–104.
50. Harker, L. A., and Ross, R. (1978) *In* "A New Approach to Reduction of Cardiac Death" (T. Abe and S. Sherry, eds.), pp. 59–66. Huber, Bern.
51. Galdal, K. S., and Evensen, S. A. (1981) *Thromb. Res.* **21,** 273–284.
52. Snow, C., and Allen, A. (1970) *Biochem. J.* **119,** 707–714.
53. Rajs, J., Sundberg, M., Härm, T., Grandinsson, M., and Söderlund, U. (1980).
54. Curwen, K. D., Gimbrone, M. A., and Handin, R. I. (1980) *Lab. Invest.* **42,** 366–374.
55. de Bono, D. P. (1976) *Cell. Immunol.* **26,** 77–88.
56. de Bono, D. P. (1977) *Adv. Microcirc.* **7,** 68–95.
57. Hayes, L. W., Goguen, C. A., Stevens, A. L., Magargal, W. W., and Slakey, L. L. (1979) *Proc. Natl. Acad. Sci. U.S.A.* **76,** 2532–2535.
58. Pederson, D. C., and Fass, D. N. (1980) *Circulation* **62,** Suppl. III, 169.
59. Ryan, U. S., Mortara, M., and Whitaker, C. (1980) *Tissue Cell* **12,** 619–635.
60. Gospodarowicz, D., Cheng, J., Hirabayashi, K., and Tauber, J.-P. (1981) *In* "Cellular Interactions" (J. T. Dingle and J. L. Gordon, eds.), pp. 135–165. Elsevier/North Holland, Amsterdam.
61. Santala, R. G., Dahle, C. E., Stein, M. N., and Hoak, J. C. (1980) *Circulation* **62,** Suppl. III, 273.
62. Pastan, I. H., Johnson, G. S., and Anderson, W. B. (1975) *Annu. Rev. Biochem.* **44,** 491–522.
63. Trevethick, M. A., Olverman, H. J., Pearson, J. D., Gordon, J. L., Lyles, G. A., and Callingham, B. A. (1981) *Biochem. Pharmacol.* **30,** 2209–2216.
64. Fuentes, J. A., Trepel, J. B., and Neff, N. M. (1978) *Exp. Gerontol.* **12,** 113–115.
65. Pearson, J. D., Ager, E. A., Trevethick, M. A., and Gordon, J. L. (1979) *Agents Actions Suppl.* **4,** 120–126.
66. Weksler, B. B., Ley, C. W., and Jaffe, E. A. (1978) *J. Clin. Invest.* **62,** 923–930.
67. Hong, S. L. (1980) *Thromb. Res.* **18,** 787–795.
68. Piper, P. J., and Vane, J. R. (1971) *Ann. N.Y. Acad. Sci.* **180,** 363–385.
69. Ager, A., Gordon, J. L., Moncada, S., Pearson, J. D., Salmon, J. A., and Trevethick, M. A. (1981) *J. Cell. Physiol.* **110,** 9–16.

Vital and Functional Activities of Endothelial Cells

UNA S. RYAN AND JAMES W. RYAN
Department of Medicine
University of Miami
Miami, Florida

INTRODUCTION

One can divide the metabolic activities of endothelial cells into two broad classes: those of vital importance and those of functional, but not vital, importance. The division is to an extent arbitrary, and it is not difficult to envisage functional activities that, episodically, become vital. Among the latter may be those endothelial cell activities called into play to limit blood loss (1), facilitate fibrinolysis (2), present antigen to T lymphocytes (3), or unmask latent receptors for the Fc portion of IgG or for C3b (4,5). Clearly, there are circumstances in which inhibition of a functional activity may contribute to injury or death of the organism, e.g., when an inhibitor of angiotensin converting enzyme is given to a volume-depleted subject with an absent or inhibited adrenergic nervous system, or, perhaps, when prostacyclin release is stopped in a hypercoagulable subject.

As for vital activities, it is generally recognized that endothelial cells must conduct fundamental metabolic activities such as oxidative phosphorylation. Similarly, the cells must be capable of cell division and migration and must be capable of maintaining their structural integrity. Structural integrity may be understood at two levels: first, in terms of each cell, and second, in terms of the formation and organization of endothelium as a tissue. Not only do endothelial cells interact with each other, but the cells also interact with cellular and soluble components of blood. Further, in some blood vessels (notably precapillary arterioles and postcapillary venules), the endothelium enters into a higher organization with smooth muscle cells (6) (Figs. 1 and 2).

PATHOBIOLOGY OF THE ENDOTHELIAL CELL

ISBN 0-12-521980-6

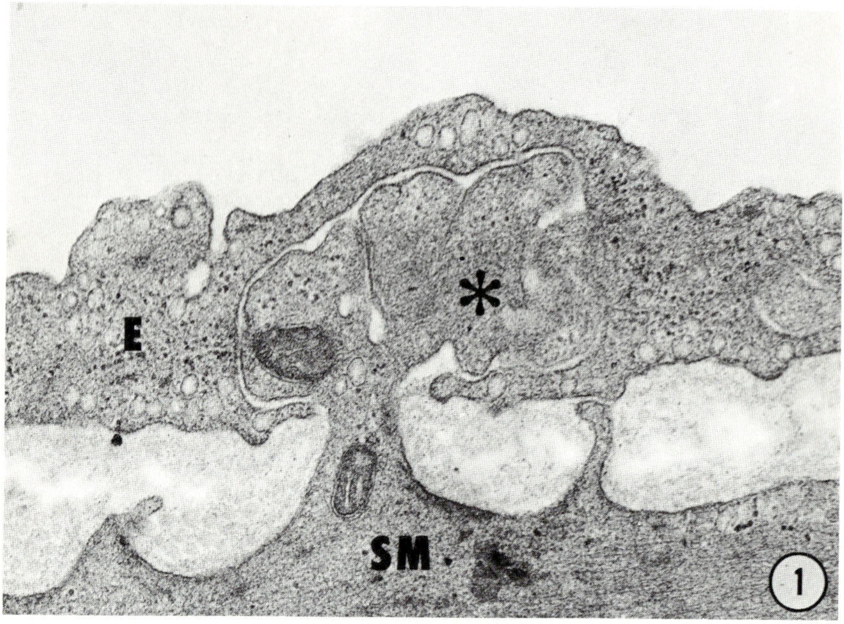

Fig. 1. Electron micrograph of a myoendothelial junction from a small pulmonary artery (approximately 210 μm in diameter). The smooth muscle cell (SM) extends a projection (*) that indents the endothelial cell (E). ×37,000.

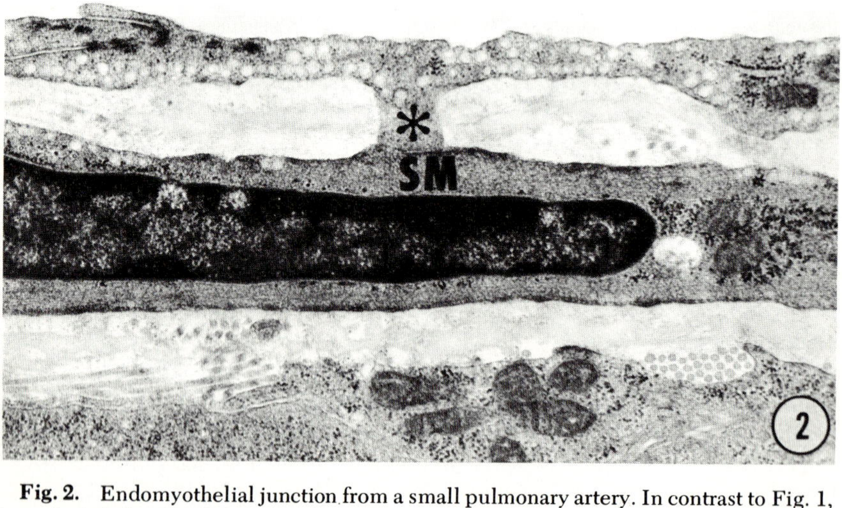

Fig. 2. Endomyothelial junction from a small pulmonary artery. In contrast to Fig. 1, an endothelial cell process (*) protrudes to make close contact with the underlying smooth muscle layer (SM). ×25,000.

Some basis for the division of activities into those of vital importance and those of functional importance can be appreciated from the effects of drugs. Drugs such as aspirin, captopril, and acetazolamide can be used chronically and in relatively large doses. In other words, cyclo-oxygenase (7), angiotensin converting enzyme (8–11), and carbonic anhydrase (12) (all enzymes possessed by endothelial cells) can be inhibited almost completely without killing endothelial cells, without damaging endothelium as a tissue, and without killing the organism. Conversely, it is not possible to use drugs that interfere with oxidative phosphorylation, cell division, cell migration, or cell structure.

Again, acknowledging that the division of endothelial cell activities into vital and functional activities is an oversimplification, we believe that the division has practical implications, especially for the development of new classes of drugs. Efforts to modulate functional activities may well lead to further therapeutic advances. Perhaps one of the most outstanding examples relates to the treatment of high blood pressure with inhibitors of angiotensin converting enzyme. Upward of 60% of all patients with high blood pressure can be so treated and in many cases the blood pressure is brought to within the normal range (13,14). The treatment appears to be without any side effects attributable to inhibition of the enzyme per se. Efforts to modulate activities later recognized to be vital will likely have the opposite effects.

The text that follows focuses on vital and functional activities of special interest to us and is not meant to be comprehensive.

VITAL ACTIVITIES

Using time-lapse cinematography, we have compared the proliferation and migration of endothelial cells in culture under two different conditions: (a) in the repair of mechanical injury to a confluent monolayer and (b) in the growth to confluence of cells seeded at low density (15).

A priori, both cell division and then migration of single cells must require separation of intercellular junctions. Further, when division is rapid (as in repair of a mechanical injury), the formation of junctions may not be complete until the new endothelium becomes confluent. Some empirical evidence can be adduced to support this concept. Endothelial cells *in situ* and in fresh isolates possess highly complex junctional structures, including gap junctions continuous with tight junctions (16). Cells in postconfluent monolayer culture are linked by junctions of the communicating type consisting of patches composed of

Fig. 3. Replica of freeze-fractured gap junction from a postconfluent culture of bovine pulmonary artery endothelial cells. The gap junction is characterized on the P face by patches composed of several thousand densely packed particles. Rows of particles frequently extend from the patches. The encircled arrow indicates the directions of shadowing. ×75,000.

several thousand particles (Fig. 3). In contrast, cells that migrate into fresh wounds or cells in preconfluent cultures possess fewer and smaller particle aggregates (5–100 particles) arrayed as patches or rows (Fig. 4).

Typically, cells seeded at low density (7.5×10^4 cells in 7 ml in a 60 mm dish), after an initial lag period, divide at a remarkably uniform rate (mean interdivision time (IDT) of 17 hr) until confluency is approached, at which time the IDT lengthens until confluency is achieved and division stops. However, those cells that participate in the repair of a mechanical injury show a very short lag phase and then a constantly accelerating division rate (IDT 20.5 hr for divisions 1–50; 9.9 hr for divisions 151–200) until confluency is reached (Fig. 5). Intercellular junctional complexes are not seen in the latter case until the wound area is clearly postconfluent.

A remarkable feature of the repair of a mechanical wound of a monolayer culture is the directionality of migration. In repopulation of

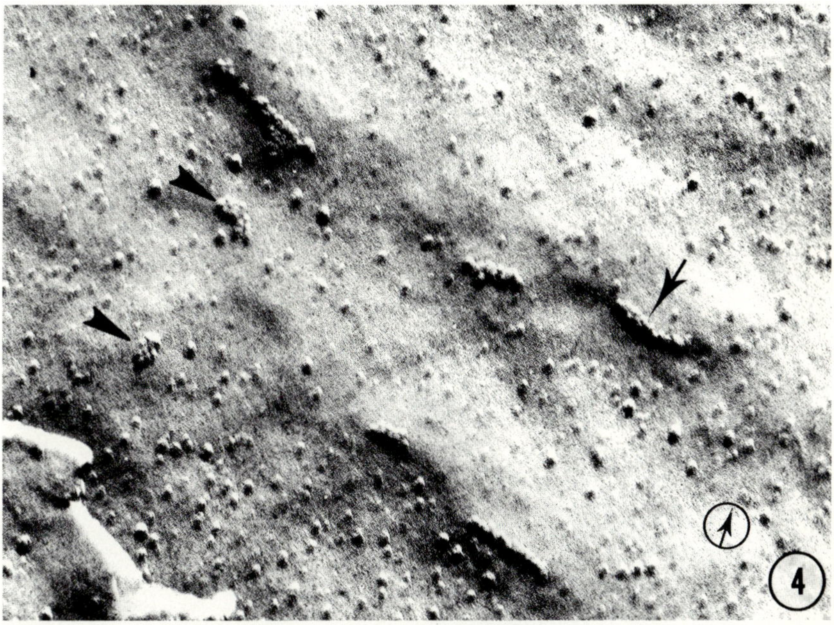

Fig. 4. Replica from a culture of bovine pulmonary artery endothelium that was in the process of repopulating a denuded area. The migrating and dividing front of cells was scraped and freeze-fractured before confluence was reached. The P face shows scattered discontinuous rows (arrows) or groups (arrowheads) of small numbers of particles. The encircled arrow indicates the direction of shadowing. ×112,000.

a defined wound area the migration of cells that detach, and of cells that detach and divide, is directly into the wound area. This is in contradistinction to the migration behavior of cells seeded into a new flask where daughter pairs of a division tend to migrate in opposite directions. The directionality of the migration of cells that repopulate the wound area would appear to be explained incompletely if at all by diffusible factors present in the culture medium.

FUNCTIONAL ACTIVITIES

It is reasonable that endothelial cells, or any other cell type, establish their ability to exist and form a coherent tissue before they express what we have referred to previously as functional activities, or activities of functional but not vital importance. Experimentally, this seems to be precisely the case. For example, a characteristic of pulmo-

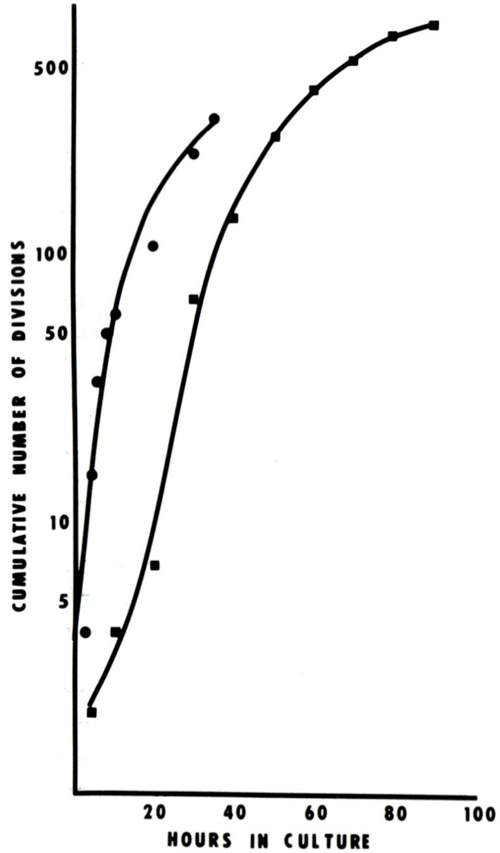

Fig. 5. Comparison of endothelial cell proliferation in wound culture ■——■ and regular culture ●——●. The cells in the wound culture began division almost immediately (58 divisions in 10 hr) and maintained a high rate of division until confluence was reached (approx 40 hr). In contrast, the regular culture showed an initial lag (four divisions in 10 hr) and a decrease in the overall proliferation as the culture approached confluence. (15)

nary endothelium *in vivo* is its ability to inactivate bradykinin and form angiotensin II, all through the action of angiotensin converting enzyme (see 17 for review). However, by using cells in culture, it is evident that angiotensin converting enzyme is not fully expressed until the cells have reached the postconfluent state (Fig. 6). This may also be true for the expression of cyclooxygenase (7), and it is reasonable to believe that processing of biogenic amines may occur at its highest rate when the cells are in the postconfluent state. To our knowledge, the latter point has not been examined experimentally.

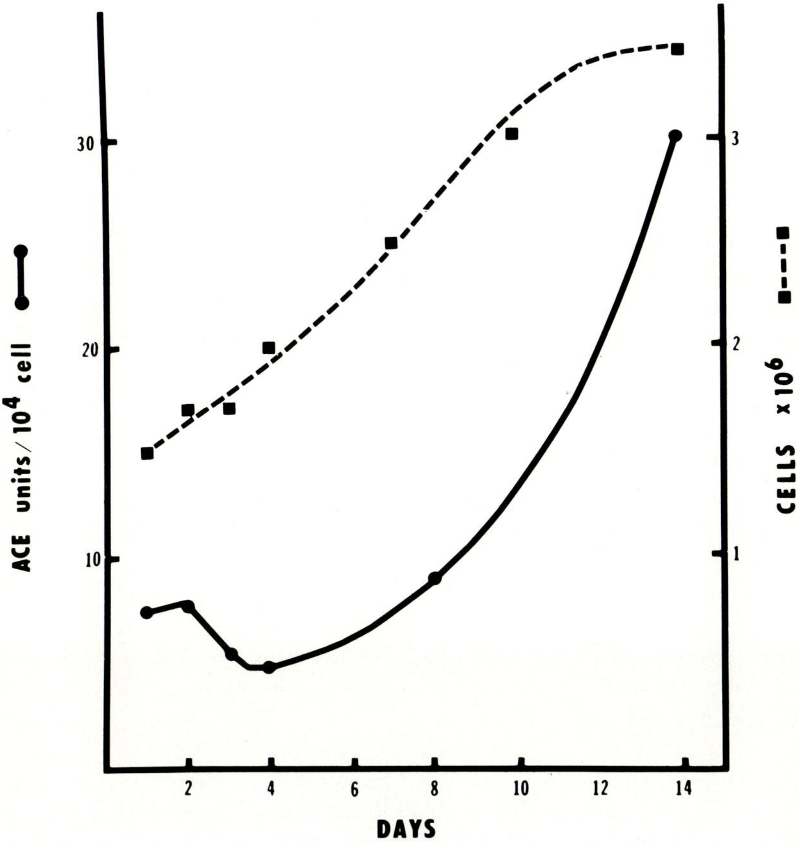

Fig. 6. Bovine pulmonary artery endothelial cells were seeded into 25 cm^2 flasks (6×10^5 cells per flask). Pairs of flasks were assayed every 2 days for angiotensin converting enzyme (ACE) activity using [^{3}H]benzoyl-Phe-Ala-Pro as substrate (16,17). Cell numbers were calculated using a hemocytometer. The solid line represents the level of ACE activity; the broken line represents the number of cells per flask. Confluence was reached on the seventh day. The culture was stationary (no mitotic figures) by the eleventh day.

Now that the culture of endothelial cells has become routine (18) there are almost unlimited possibilities for using such cultures to define both vital and functional activities. Precedent exists already for modulating some of the functional activities for therapeutic benefit. However, it is very easy to forget that endothelial cells in culture are indeed a model system and, as for any model system, can give highly misleading results. Looking back on procedures that have been used to harvest and passage endothelial cells, it is perhaps surprising that some vital activities were not lost.

We have found that functional activities of endothelial cells, such as those conducted by angiotensin converting enzyme, can be heavily damaged by the methods used commonly for cell culture. Isolation and passaging of endothelial cells by the use of protease enzymes and other commonly used agents such as EDTA may yield cells in culture that do not have all the surface enzymes and other proteins believed to exist on endothelial cells *in situ*. The factors involved in the aging process of cells in culture are not fully understood but repeated exposure to trypsin is likely to contribute to accelerated aging, loss of surface enzymes, and alterations in size, shape, and junctional contacts, together with progressively slower and more asynchronous rates of division.

Quite clearly there are serious limitations as to what can be learned from a line of cells in culture that do not retain characteristics of the same cells *in situ*. We have, therefore, begun to develop new techniques for the culture of pulmonary endothelial cells that avoid exposure to exogenous proteolytic enzymes at both the isolation step and during successive passages.

RESPONSES OF PULMONARY ENDOTHELIAL CELLS TO THE VASOACTIVE HORMONES, BRADYKININ, AND ANGIOTENSIN II

Over the past few years, we have focused almost exclusively on the ability of pulmonary endothelial cells to metabolize bradykinin and angiotensin I (19–22). However, there is growing reason to believe that bradykinin, angiotensin II, and possibly angiotensin III exert some of their effects on vascular tone via endothelial cells (6).

Generally, one regards the pulmonary vascular bed as being relatively insensitive to vasoactive hormones. Perhaps teleologically it makes sense that the pulmonary vascular bed not undergo the marked changes in tone characteristic of, e.g., the vascular beds of the kidney, skin, skeletal muscle, and small intestine. Presumably, the fundamental need for bulk exchange of gases as blood passes through the lungs would not be met efficiently if the pulmonary vascular bed were highly sensitive to vasoactive substances as they occur in central venous blood.

However, the blood vessels of the lungs do respond to bradykinin and angiotensin II (23). Typically, the response is expressed not in terms of large changes in blood pressure but in blood flow (24). As is well known, the lungs have a great capacity for accommodating changes in blood flow, say, from 6 liters/min at rest to 35–45 liters/min at maximum exercise in a well-trained athlete.

The ability of pulmonary endothelial cells to inactivate bradykinin (2) contrasted with the ability of pulmonary arteries to respond to bradykinin presents something of a paradox. The inactivation is extensive (19), and one cannot easily envision how a significant concentration of bradykinin is likely to survive passage from the vascular space, across endothelial cells, and thence on to smooth muscle cells, the latter generally being attributed with determining vascular tone.

We have postulated that passage of, e.g., bradykinin directly to smooth muscle may not occur, or may be only a small part of the overall process (6). Endothelial cells, at least in certain pulmonary arteries, form extensive junctionlike complexes with smooth muscle cells (Figs. 1 and 2). Conceivably, a vasoactive substance acting on receptors on the luminal surface of endothelial cells could initiate a series of reactions, some transmitted to smooth muscle, to affect vascular tone. The hypothesis that vasoactive polypeptides exert some of their effects on vascular tone via actions on endothelial cells requires that the cells have specific or highly selective receptors.

Endothelial cells isolated and cultured as described above are in fact responsive to angiotensins and kinins as hormones. First, it is evident that both bradykinin and angiotensin stimulate synthesis of cyclic nucleotides of endothelial cells in culture. Second, bradykinin in physiological doses stimulates the release of PGI_2 (25).

The cells are harvested by scraping the luminal surface of the pulmonary artery of a newborn calf with a scalpel. The sheet of cells thus removed can be seeded into flasks or directly onto microcarriers (Fig. 7). The cells on microcarrier beads are grown in roller bottles, a system that has numerous advantages: the surface area for attachment is enormous. The cells, on each bead, form a confluent monolayer, a feature important for assay of surface enzymes (10 and 11) (see below) while retaining the advantages of a suspension culture in terms of adequate mixing and utilization of nutrients. Perhaps the most important advantage of our microcarrier–roller culture system is that it allows the cells to be passaged without exposure to trypsin. The cell-covered bead suspension can be split into new flasks to which fresh medium and new beads are also added. The cells colonize the new beads, and the process can be repeated to provide long-term, large-scale cultures of pulmonary endothelial cells that retain the structural and functional properties of primary or early passage cultures. Using these techniques we have found that cultures in their, e.g., twenty-ninth passage show no diminution of ACE activity compared with a

Fig. 7. Microcarrier culture of bovine pulmonary endothelial cells at population doubling level of 14. The cells were cultured in roller bottles and had been isolated and passaged without the use of proteolytic enzymes. The bar represents 100 μm.

culture in its fifth passage. Studies of the functional activities of endothelial cells requiring large numbers of cells cannot be carried out on cells that have reduced functional properties. We have conducted studies of the binding of BPP_{9a} by endothelial cells. These studies suffice to underscore the need for avoiding exogenous proteolytic enzymes.

Figure 8 illustrates the binding of $[^3H]BPP_{9a}$ to bovine pulmonary endothelial cells that had been exposed to trypsin for 2 min to detach cells. The graph shows both high- and low-affinity binding sites. From the number of high-affinity binding sites (binding capacity 3×10^{-12} moles/mg protein), we calculated the number of molecules of angiotensin converting enzyme per cell to be approximately 2.5×10^5. When cells were exposed to trypsin for 10 min or longer, high-affinity binding sites were lost completely.

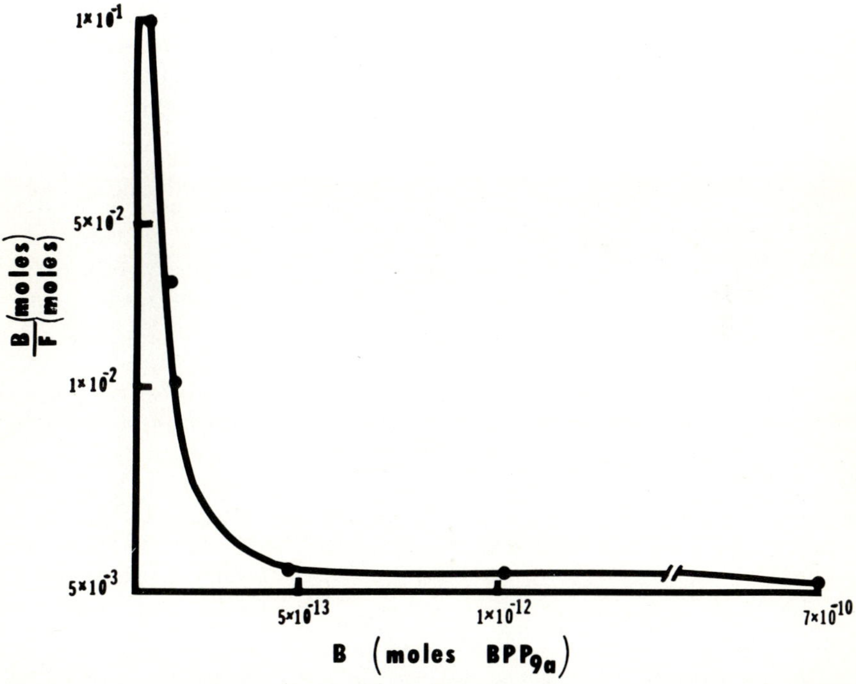

Fig. 8. The graph shows binding of $[^3H]BPP_{9a}$ to bovine pulmonary endothelial cells that had been detached from flasks by 2 min exposure to trypsin/EDTA. Cells were incubated in $[^3H]BPP_{9a}$ at concentrations of 10^{-6}–10^{-9} M in 200 μl of cold phosphate buffered saline, pH 7.4, containing sodium azide (5 mM). Thousandfold unlabeled BPP_{9a} was used to determine nonspecific binding. Both high- (binding capacity 3×10^{-12} moles/mg protein) and low-affinity binding sites are indicated.

In subsequent studies in which we assayed angiotensin converting enzyme activity of endothelial cells never exposed to exogenous proteolytic enzymes using the substrate [^{3}H]benzoyl-Phe-Ala-Pro (10,11), it became evident that each endothelial cell possesses far more than 25,000 molecules of ACE. From the specificity constant (K_{cat}/K_m) of the reaction of the substrate with human plasma ACE, each of these cells possesses approximately 1 million molecules of ACE. Thus, even brief (2 min) exposure to trypsin served to reduce severely the number of enzyme sites.

Effects of bradykinin and angiotensin II on the cyclic nucleotide content of endothelial cells are shown in Table I. In these studies, either peptide was used at a concentration higher than that likely to be obtained under physiological conditions. Nonetheless, it may be significant that bradykinin in the same and lower concentrations (as low as 10^{-10} M) stimulates the release of arachidonate and its metabolites from endothelial cells in culture, the latter primarily as 6-keto-PGF$_{1\alpha}$ (Fig. 9). It remains to be determined whether the rise in either of the cyclic nucleotides is causally related to the release of specific metabolites of arachidonic acid. However, it may be significant that Moncada and Vane (26) have reported that platelets respond to PGI$_2$ with a rise of cAMP.

Preliminary analysis of the ^{3}H-labeled material released indicates that well over 70% is apparently PGI$_2$ (prostacyclin measured as 6-keto-PGF$_{1\alpha}$). The latter is not without interest to our hypothesis in that PGI$_2$ is a potent vasodilator in its own right. Further, unlike pros-

TABLE I
Effects of Bradykinin and Angiotensin II on the Cyclic Nucleotide Contents of Bovine Pulmonary Endothelial Cells[a]

Compound	cAMP[c]	cGMP[c]	cAMP/cGMP
	±SEM		
Control[b]	6.61 ± 0.505	0.169 ± 0.025	42.95 ± 3.275
Bradykinin (10^{-8} M)[b]	10.43 ± 0.555	1.092 ± 0.234	11.02 ± 1.605
Angiotensin II(10^{-9} M)[b]	8.17 ± 0.338	0.261 ± 0.046	37.13 ± 7.054

[a] In comparison with the control, the increases in cAMP and cGMP caused by bradykinin were significant ($p < .01$ in both comparisons). In the concentration used, angiotensin II caused a significant rise in cAMP ($p < .05$). However, the rise in the latter was disproportionately high, and the ratio of cAMP/cGMP was significantly depressed ($p < .01$) in comparison with the control cells and those treated with angiotensin II.

[b] $n = 6$ for each set of experiments.

[c] pmoles/mg of protein.

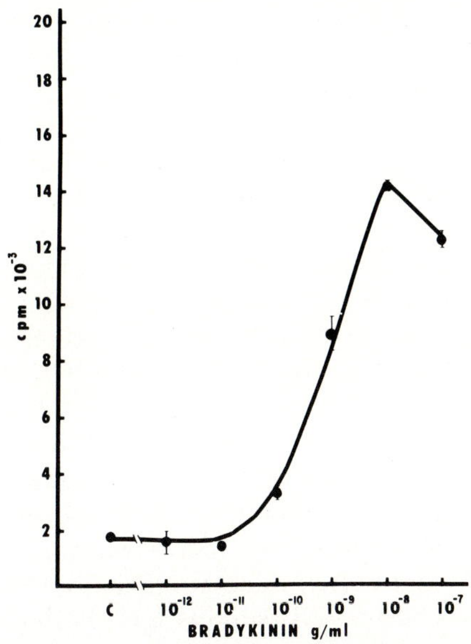

Fig. 9. Confluent monolayer cultures of bovine pulmonary endothelial cells in multiwell dishes rapidly incorporate [^{3}H]arachidonate (200 Ci/mmole; 1 μCi/well). After incorporation of the label, the medium (M199) is changed and the cells are challenged with varying concentrations of bradykinin. The linear portion of the dose–response curve (response measured as release of ^{3}H includes concentrations of bradykinin believed to occur physiologically: Up to 20% of the maximum release of ^{3}H occurs when bradykinin is at a concentration of $10^{-10}\,M$. Half-maximum release occurs with bradykinin at $5 \times 10^{-10}\,M$ and maximum release occurs with bradykinin at $10^{-9}\,M$ and higher.

taglandins of the E and F series, PGI_2 is not degraded during passage through the lungs (27) and may be capable of reaching pulmonary vascular smooth muscle (i.e., bradykinin could exert some of its effects on pulmonary vascular tone by stimulating endothelial cells to release the vasodilator, PGI_2). Curiously, thromboxane A_2 (measured as thromboxane B_2) is released in readily measurable quantities (about 5% of the total radioactivity).

It is worth noting that PGI_2 is among the most potent of the known inhibitors of platelet aggregation (cf. 27,28) and may represent a means by which bradykinin could affect, indirectly, fluidity of blood.

CONCLUSIONS

The separation of endothelial activities into vital and functional is necessarily to a degree arbitary. For example, a vital activity not considered earlier is the ability of endothelial cells to limit blood loss in the event of injury, a property perhaps explained in part by the activity of a latent tissue factor (1). There may be other episodic vital functions related to the participation of endothelial cells in, for example, acute inflammation. Previously we have shown that bovine pulmonary endothelial cells in culture do not possess binding sites for the Fc portion of IgG nor for C3b (4). More recently it has become evident that such receptors are expressed by cells exposed to lysates of polymorphonuclear neutrophils and by cells infected with influenza or cytomegalovirus (5). Not inconceivably, the expression of these receptors results in the death of designated endothelial cells, and in this sense is an episodic vital function.

We believe that the division of endothelial activities into those that are vital and those that are functional has practical implications, perhaps most notably in the development of new drugs. As is now well known, inhibition of the activity of angiotensin converting enzyme has proved to be a mild and highly effective mean of reducing high blood pressure. Upward of 60% of all patients with high blood pressure are responsive to this treatment. Many are made normotensive. It is not unreasonable to believe that efforts to modulate other functional activities may lead to further therapeutic advances. In contrast, efforts to modulate activities known, or later recognized, to be vital activities may well provide new scientific insights but are likely to be potential therapeutic disasters.

ACKNOWLEDGMENTS

It is a pleasure to thank Drs. M. Absher, D. Crutchley, and D. Lehotay for their collaboration and D. Bielefeld, G. Maxwell, M. Mortara, C. Whitaker, and L. White for technical assistance.

The work was supported by grants from the National Institutes of Health (HL-22087, HL-22896) and from the Council for Tobacco Research, U.S.A.

REFERENCES

1. Maynard, J. R., Heckman, C. A., Pitlick, F. A., and Nemerson, Y. (1975) *J. Clin. Invest.* 55, 814–824.
2. Loskutoff, D. J., this volume.
3. Berger, this volume.
4. Ryan, U. S., Schultz, D. R., Del Vecchio, P. J., and Ryan, J. W. (1980) *Science* **208**, 748–749.

5. Ryan, U. S., Schultz, D. R., and Ryan, J. W. (1981) *Science* **214**, 557–559.
6. Ryan, U. S., Ryan, J. W., and Whitaker, C. (1979) *Adv. Exp. Med. Biol.* **120A**, 357–391.
7. Ryan, J. W., Ryan, U. S., Habliston, D., and Martin, L. (1978) *Trans. Assoc. Am. Physicians* **91**, 343–350.
8. Ryan, J. W., Ryan, U. S., Schultz, D. R., Whitaker, C., Chung, A., and Dorer, F. E. (1975) *Biochem. J.* **146**, 497–499.
9. Ryan, U. S., Ryan, J. W., Whitaker, C., and Chiu, A. (1976) *Tissue Cell* **8**, 125–145.
10. Ryan, J. W., Chung, A., Martin, L. C., and Ryan, U. S. (1978) *Tissue Cell* **10**, 555–562.
11. Ryan, U. S., Clements, E., Habliston, D., and Ryan, J. W. (1978) *Tissue Cell* **10**, 535–554.
12. Ryan, U. S., Whitney, P. L., and Ryan, J. W. (1979) *Circ. Res.* **59–60**, Suppl. II, 108 (Abstr. No. 416).
13. Gavras, H., Brunner, H. R., Laragh, J. H., Sealey, J. E., Gavras, I., and Vukovich, R. A. (1974) *N. Engl. J. Med.* **291**, 817–821.
14. Gavras, H., Brunner, H. R., Turini, G. A., Kershaw, G. R., Tifft, C. P., Cuttelod, S., Gavras, I., Vukovich, R. A., and McKinstry, D. N. (1978) *N. Engl. J. Med.* **298**, 991–995.
15. Ryan, U. S., Absher, M., and Ryan, J. W. (1982) In preparation.
16. Ryan, U. S., Mortara, M., and Whitaker, C. (1980) *Tissue Cell* **12**, 619–635.
17. Ryan, J. W., and Ryan, U. S. (1980) *In* "Enzymatic Release of Vasoactive Peptides" (F. Gross and H. G. Vogel, eds.), pp. 259–274. Raven, New York.
18. Ryan, U. S. (1981) *In* "Pulmonary Toxicology" (G. E. R. Hook, ed.). Raven, New York.
19. Ryan, J. W., Roblero, J., and Stewart, J. M. (1968) *Biochem. J.* **110**, 795–797.
20. Smith, U., and Ryan, J. W. (1973) *Fed. Proc., Fed. Am. Soc. Exp. Biol.* **32**, 1957–1966.
21. Ryan, J. W., and Ryan, U. S. (1977) *Fed. Proc., Fed. Am. Soc. Exp. Biol.* **36**, 2683–2691.
22. Ryan, J. W., and Ryan, U. S. (1982) *Fed. Proc., Fed. Am. Soc. Exp. Biol.* (in press).
23. Hirshman, J., and Boucek, R. (1963) *Br. Heart J.* **25**, 375.
24. Rowe, G. G., Afonson, S., Castillo, C. A., Lioy, F., Lugo, J. E., and Crumpton, C. W. (1963) *Am. Heart J.* **65**, 656.
25. Crutchley, D. J., Ryan, J. W., Ryan, U. S., Fisher, G. H., and Paul, S. M. (1982) Agents and Actions, in press.
26. Moncada, S., and Vane, J. R. (1980) *In* "Metabolic Activities of the Lung" (R. Porter and J. Whelan, eds.), pp. 165–177. Excerpta Medica, New York.
27. Armstrong, J. M., Lattimer, N., Moncada, S., and Vane, J. R. (1978) *Br. J. Pharmacol.* **62**, 125–130.
28. Vane, J. R. (1971) *Nature (London), New Biol.* **231**, 232–235.

Endothelial Cell Function in Altered Oxygen Environments

S. ALEX STALCUP, GERARD M. TURINO, AND ROBERT B. MELLINS

Departments of Pediatrics (Pulmonary Division) and Medicine, Columbia College of Physicians and Surgeons, New York, N.Y.

The endothelial cell is the single most common cell type in the lung, comprising from between 20 to 30% of total lung cells (1). Data from several sources have accumulated to indicate that endothelial cells are critically important to the regulation of circulating levels of vasoactive substances. We define "vasoactive substances" to include circulating and tissue factors that mediate changes in blood pressure, vascular resistance, cardiac output, blood volume, and the transvascular exchange of fluid and protein (2). Endothelial cells affect the circulating concentrations of vasoactive substances either by *de novo* release, by synthesis from precursors supplied to them in blood, or by uptake from blood and subsequent catabolism. Although endothelia can apparently distinguish between vasoactive substances within a given chemical class, vasoactive substances from many different chemical classes are also handled by endothelial cells. (For example, norepinephrine is taken up and catabolized by endothelial cells while its close chemical congener epinephrine is not; see Table I.)

These several aspects of the handling of vasoactive substances by endothelial cells led John Vane to hypothesize that the lung, in addition to being the organ of gas exchange, was also an "endocrine" organ (3). By virtue of its central location, interposed between the venous and arterial circulations, and its massive surface area of endothelia exposed to blood, the lung selectively protects the arterial vascular bed from vasoactive agents with properties of local tissue release, local effect, and predominantly local catabolism. Seepage of these agents into blood, e.g., norepinephrine, serotonin, bradykinin, and the prostaglandins PGE and $PGF_{2\alpha}$, would have untoward effects if they

PATHOBIOLOGY OF THE ENDOTHELIAL CELL

ISBN 0-12-521980-6

TABLE I
Biologically Active Substances Handled in the Pulmonary Vascular Bed

Unaffected by lungs	Activated by lungs
Epinephrine	Angiotensin I
Prostaglandin A	Cyclic endoperoxides
Angiotensin II	Released by lungs
Vasopressin	Prostaglandins
Cleared by lungs	Histamine
Bradykinin	Slow reacting substance of anaphylaxis
Adenine nucleotides	Kallikreins
Norepinephrine	
Serotonin	
Prostaglandins E and F	

reached the arterial vasculature in unregulated quantities. On the other hand, the lung facilitates the function of agents that have predominantly systemic effects either by passively permitting their passage, by activating vasoactive substances from inert precursors, or by *de novo* synthesis and release. Examples of this type of lung function include the conversion of angiotensin I to angiotensin II and the release of the vasodilating and antiplatelet aggregant prostanoid, prostacyclin (PGI_2).

Thus the lung is being recognized as making an important contribution to the regulation of many different mediators, with consequent implications for the regulation of blood pressure and the transvascular exchange of fluid and protein. Is this putative lung regulatory function in turn subject to regulation from environmental influences, e.g., changes in the composition of respiratory gases or changes in the composition or flow characteristics of blood passing through the lung? Can injury to the lung alter the extent to which the lung processes vasoactive mediators? Can assessments of the various metabolic properties of endothelial cells provide an index of lung microvascular injury?

To address these questions, we chose as a representative model of the "metabolic function of the lung" the activity of angiotensin converting enzyme (ACE). The two protease pathways that share this endothelial cell-surface enzyme in common are diagrammed in Fig. 1. ACE converts angiotension I (AI) to angiotensin II (AII), one of the most potent arteriolar vasoconstrictors produced in the body. The other, and preferred, substrate for ACE is bradykinin, the nonapeptide effector peptide of the kallikrein–kinin system, which is efficiently degraded by ACE. This protease pathway is an intimate component of the coagulation and fibrinolysis pathways. Bradykinin contributes to

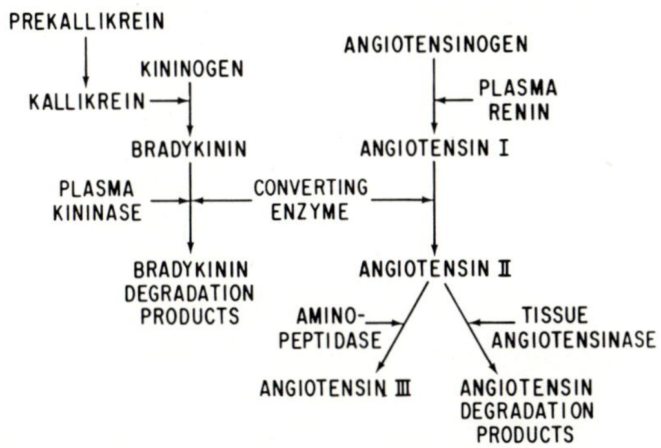

Fig. 1. Diagrammatic representation of how converting enzyme, located on the surface of endothelial cells, connects the kallikrein–kinin system with the reinin–angiotensin system. Its two substrates are bradykinin (a hypotensive and edema-producing peptide) and angiotensin I.

the generation of the inflammatory response via its two main biologic properties: bradykinin is a powerful arteriolar vasodilator and hence hypotensive agent and also increases capillary permeability to water and protein, which causes edema (4).

We examined ACE activity in a series of experiments in anesthetized dogs instrumented in a manner that permitted the calculation of angiotension conversion by a blood pressure response bioassay (Table II). Acidosis, alkalosis, and 5-hr exposures to 100% O_2 (a concentra-

TABLE II
Percent Conversion of Angiotensin I *in vivo* under Hyperoxic, Hypoxic, and Acute Acid–Base Derangements

Condition	%	*N*
100% O_2 breathing		
Controls	91 ± 7	8
Hyperoxia		
15–60 min	92 ± 8	8
90–135 min	93 ± 5	5
180–225 min	90 ± 7	2
225–300 min	92	1
Hypoxia (PO_2 = 35 mm Hg)	34 ± 15	9
Acid–base derangements		
Controls	91 ± 5	12
Respiratory acidosis	92 ± 6	6
Respiratory alkalosis	92 ± 6	6
Metabolic acidosis	94 ± 4	6
Metabolic alkalosis	94 ± 3	5

tion known to injure endothelial cells) did not affect angiotension I conversion (5). In the linear portion of the angiotensin II dose–response curve chosen for these experiments conversion was quite high—greater than 90% of injected AI was converted to AII. However, brief exposures to hypoxic gas mixtures produced striking decreases in AI conversion (6). The degree of enzyme inhibition was closely correlated to PaO_2 and the enzyme activity varied from 94% at the highest PaO_2 down to 34% at a PaO_2 of 30 mm Hg (Fig. 2). Thus the circulating concentrations of angiotensin II would appear to be critically influenced by alveolar PO_2. Direct evidence for this can be seen in the experiments of Zakheim *et al.* (7), who observed a prompt decline in radioimmunoassayable AII in rabbits breathing hypoxic gas mixtures. In similar experiments in our laboratories, we observed a decline in AII to nearly undetectable levels in dogs hypoxic to a PO_2 of 30 mm Hg, despite a progressive rise in plasma renin activity. Most importantly, with abrupt resuscitation of these dogs, extremely high AII concentrations were measured, indicating that ACE activity in the lung was restored very rapidly with return of alveolar PO_2 to normal releasing AII in a bolus fashion from AI that had accumulated during the hypoxic interval (Fig. 3) (8).

What is the effect of acute alveolar hypoxia on the metabolism of the other ACE substrate, bradykinin? We infused bradykinin into anes-

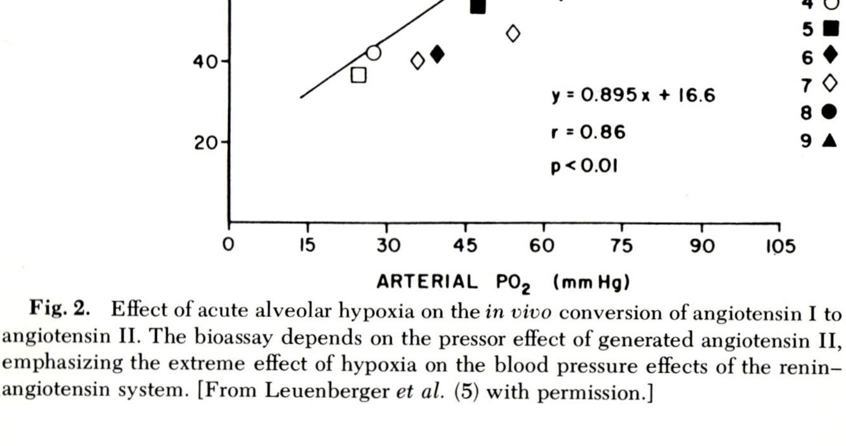

Fig. 2. Effect of acute alveolar hypoxia on the *in vivo* conversion of angiotensin I to angiotensin II. The bioassay depends on the pressor effect of generated angiotensin II, emphasizing the extreme effect of hypoxia on the blood pressure effects of the renin–angiotensin system. [From Leuenberger *et al.* (5) with permission.]

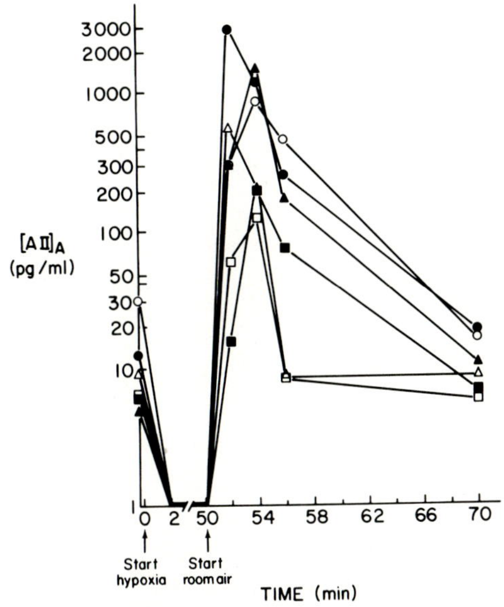

Fig. 3. Time course of changes in the arterial concentrations of angiotensin II during hypoxia and upon return to room air. Acute reoxygenation from hypoxia produces a bolus production of angiotensin II as converting enzyme activity is restored. [From Stalcup *et al.* (8) with permission.]

thetized dogs that were instrumented to permit sampling of blood from the pulmonary artery and left atrium. Using a specific radioimmunoassay we measured the bradykinin concentrations and calculated ACE activity from the transpulmonary gradient in peptide concentration. We found that the percentage of bradykinin degraded by the lung fell from greater than 95% in normoxic animals to a low of 3% in animals with a PaO_2 of 30 mm Hg. The degree of enzyme inhibition was closely correlated to PaO_2, and changes in enzyme activity occurred less than 2 min after exposure to hypoxic gas (Fig. 4) (9).

Actually, the entire range of ACE activity from low to high occurs over the range in PO_2 values observed from mixed venous to arterial blood. Indeed, except for the endothelial cells of the lung, most endothelia flourish in a PO_2 close to 40 mm Hg, a typical organ vascular PO_2. This suggests that the ACE activity of endothelia in the systemic vasculature is likely to be quite low. We examined this possibility by infusing bradykinin into the left atrium of dogs and calculating extraction gradients across the systemic vascular bed. The total converting enzyme capacity of the body (excluding the lung), as measured by the greatest arterial concentration of bradykinin that was cleared before

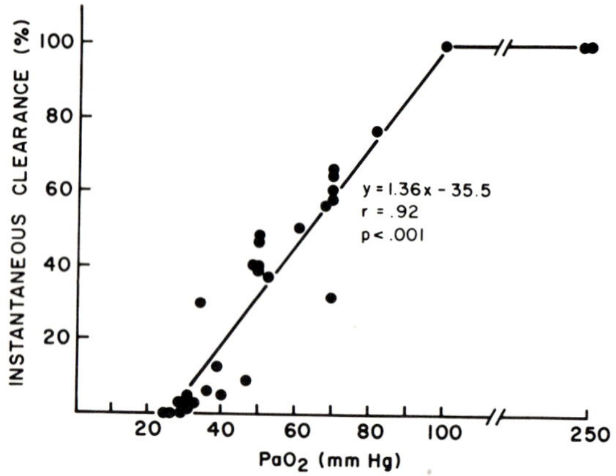

Fig. 4. Effect of acute hypoxia on pulmonary converting enzyme activity, expressed as the instantaneous clearance of bradykinin determined from the peptide concentration gradient across the lung. [From Stalcup *et al.* (9), with permission.]

venous levels rose, was 0.5 ng/ml, which is only 10% of the capacity of the lung to clear bradykinin. Reduced PaO_2 further decreased even this low level of ACE activity (9). Thus it is probably simplistic to suggest that hypoxia inhibits ACE, since ACE activity appears to be adjusted to local PO_2. Even so, hypoxia in the lung does definitely reduce ACE activity in the lung, a site in which enzyme activity is normally 10-fold higher than in the systemic vascular bed.

Although we observed rather severe hypotensive effects when we infused bradykinin into the aorta to a concentration of 0.5–1.0 ng/ml, hypoxic dogs had elevations in bradykinin concentrations to greater than 30 ng/ml, with no hypotensive effects at all (8). Simultaneous measurements of cardiac output, by the thermodilution method, provided a ready explanation for this major difference. In normoxic animals, when bradykinin was either infused directly into the aorta or infused into the venous system in sufficiently high concentrations (about 5–6 ng/ml) to exceed the barrier of the lung, prompt and severe hypotension resulted. Although hypoxia drastically reduced the pulmonary barrier, bradykinin concentrations rose at a smooth rate and instead of hypotension there was stimulation of cardiac output and blood pressure was maintained even though substantial systemic vasodilation occurred. This combination of systemic vasodilatation, augmented cardiac output, and maintained blood pressure is optimal with respect to the provision of blood and nutrients to systemic tissues during hypoxia (Fig. 5).

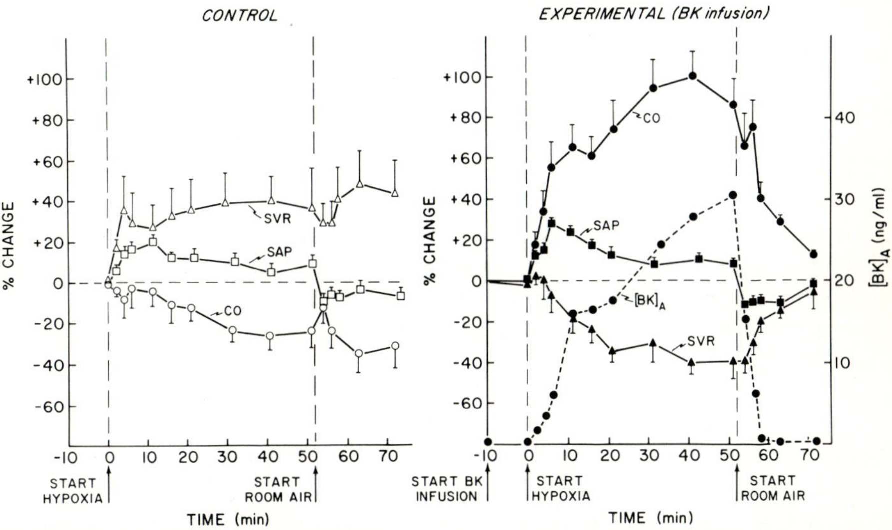

Fig. 5. Hemodynamic responses to severe hypoxia in anesthetized dogs. When bradykinin is infused, no hypotension is observed despite systemic vasodilatation because cardiac output is augmented. In normoxic animals this bradykinin concentration would produce profound hypotension. [From Stalcup *et al.* (8), with permission.]

We have come to regard the modulating effect of PO_2 on ACE activity as a physiological phenomenon, serving to redistribute and augment blood flow to vital organs during severe hypoxia. The decline in angiotensin II with the onset of hypoxia doubtless contributes to systemic vasodilatation, and thus cardiac work is lessened during a time of decreased oxygen availability. The mountain climber, exercising at high altitude, serves as an illustration of the beneficial effect of decreased ACE activity in hypoxia, in which the circulating levels of bradykinin and angiotensin II are adjusted according to PO_2 to meet the blood flow requirements of the body. In specific regional vascular beds, the local organ vascular PO_2 will adjust ACE activity and in turn the tissue concentrations of bradykinin and angiotensin II. The balance between these peptides, closely reflecting local PO_2, could provide a finely graded mechanism for regulating organ blood flow.

In another situation of major physiological stress, the time of birth, oxygen modulation of ACE activity also serves a likely physiological regulatory function. The renin–angiotensin system has an essential role at birth in defense of blood pressure, in part because the sympathetic nervous system is functioning near the top of its capacity and also because neural integrative mechanisms are functionally immature in the newborn. Bradykinin has been identified as an important contributor to the circulatory adjustment occurring during the transition from intrauterine to air-breathing life, especially the necessary decrease in pulmonary vascular resistance so that pulmonary blood flow can increase. Bradykinin also serves to initiate closure of the ductus arteriosus and increases resistance to flow of blood to the placenta at birth. The source of bradykinin at birth may be fetal granulocytes resident in the lung: the experiments of Melmon *et al.* (10) and Heymann *et al.* (11) suggest that exposure of these cells to the increased oxygen tension accompanying the first breaths leads to the explosive release of kallikrein or kallikrein activators with generation of bradykinin directly into the pulmonary vascular bed.

The rise in oxygen tension with the first breaths produces another necessary change in the control of circulating vasoactive substances—an increase in pulmonary ACE activity. The fetus lives in an oxygen milieu (PO_2 in the range of 25 to 30 mm Hg) at which ACE activity is nearly completely inhibited. Moreover, pulmonary blood flow *in utero* is less than 10% of what it will be within moments of birth. On the other hand, the oxygen tension in the placenta is higher (in the range of 35–45 mm Hg), and the placenta receives half of the cardiac output. Do placental endothelial cells respond to changes in PO_2 as do endothelia in the adult lung and in the systemic vasculature? Under controlled conditions of flow, temperature, pH, and PO_2 we perfused the intact placenta and fetus of the living guinea pig with Krebs–Henseleit solution containing either bradykinin or angiotensin I (12). Clearance of either peptide was used to calculate ACE activity. Several findings resulted from this study (Fig. 6). The fetus, at fetal PO_2, had only 20% ACE activity, while the placenta had 45% activity, indicating that the placenta was the predominant site of ACE activity in the fetal–placental unit. ACE activity on the fetal side of the placenta was sensitive to changes in the PO_2 of the maternal animal: ventilating her with hyperoxic gas increased placental ACE activity, and ventilating her with hypoxic gas drastically reduced placental ACE activity. Most importantly, increasing fetal PO_2 to neonatal PO_2 increased ACE in both the fetus and the placenta to greater than 75% in less than 5 min (Fig. 6). This means that, with the first breaths, pulmonary ACE activity will abruptly increase, adding angiotensin II to the systemic vasculature in support of blood pressure during the transition period, while limiting the period of circulating bradykinin to a narrow time window, proportional to PO_2. Successful oxygenation with birth will stimulate a sequence of interdependent biochemical events that both generate vasoactive peptides and regulate their concentration to that

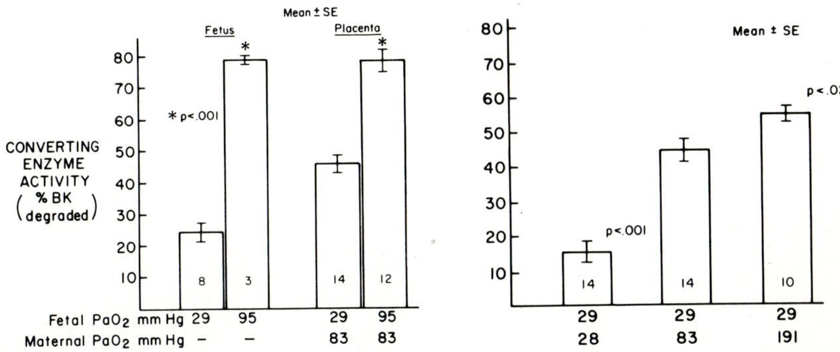

Fig. 6. (Left panel) Comparison of converting enzyme activity in the fetus and placenta ($PO_2 = 29$ mm Hg) and in the same sites 5 min after oxygenation ($PO_2 = 95$ mm Hg). The placenta is the predominant site of converting enzyme activity in the fetal–placental unit, but the fetus achieves enzyme "competence" within moments of birth because of increased PO_2. (Right panel) Effect of changing maternal PaO_2 on converting enzyme activity on the fetal side of the placenta. Both maternal hyperoxia ($PO_2 = 191$) and hypoxia ($PO_2 = 28$) affect how the fetal–placental unit metabolize vasoactive peptides. [From Davidson *et al.* (12) with permission.]

required for successful adaptation to extrauterine life. Failure of the newborn to achieve adequate oxygenation could abort this sequence, leading to impairment in angiotensin II generation and bradykinin degradation. This combination will foster hypotension and edema, which, in fact, are common clinical consequences of neonatal asphyxia (13).

What is the cellular basis for the differential expression of ACE activity in differing oxygen environments? Drawing from the methods of Ryan *et al.* (14), we harvested and propagated endothelial cells *in vitro*. By modifying the ambient gas in the tissue culture flask, acute changes in oxygen tension could be produced and the effect of these changes on the metabolism of angiotensin I and bradykinin assessed. We found that PO_2 changes from 30 to 100 mm Hg or greater increased ACE activity from close to zero to approximately 5 nm of bradykinin degraded per 10^6 cells per hour. Angiotensin I conversion increased from nearly zero to from 2.4 to 2.9 nm per 10^6 cells per hour (15). Remarkably, purified ACE was unaffected by changes in PO_2. Therefore we examined the effects of progressive membrane disruption on the degree to which ACE activity changed when PO_2 was varied. ACE activity and ACE inhibition by hypoxia was unaffected by breaking the cell by sonication, suggesting that the intact cell was not required for the expression of the O_2 phenomenon. As the membrane was progressively dissolved with detergent, however, the responsiveness

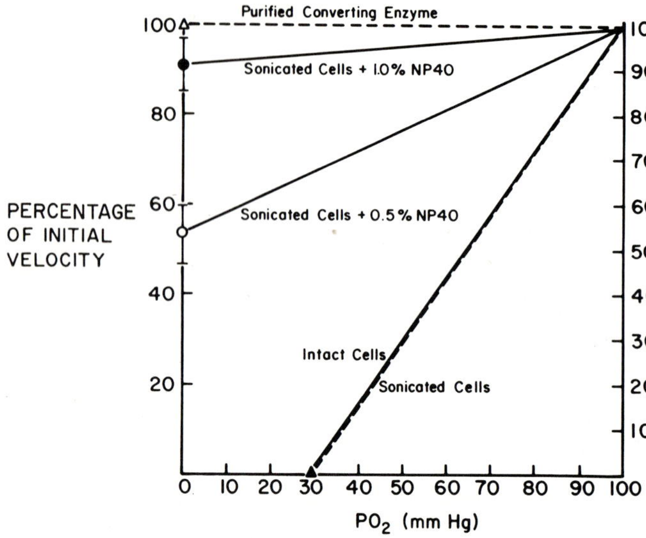

Fig. 7. Effect of progressive loss of cell membrane integrity on the inhibition of converting enzyme activity by hypoxia. As the membrane is dissolved by NP-40, a detergent, responsiveness to changes in PO_2 is lost. [From Stalcup *et al.* (15), with permission.]

of the enzyme's activity to PO_2 was lost (Fig. 7). We concluded that the modulation of ACE activity by PO_2 was a phenomenon of the intact cell membrane, not of the enzyme per se. This suggests that unusual caution must be observed in studying this phenomenon *in vitro*. Many routine tissue culture practices damage the endothelial cell membrane and decrease the cell's normoxic ACE activity. Principal among these cell insults is the harvest or serial passage of endothelia in culture using enzymatic cell dispersion methods (16). We and others (17) have observed an increase in cell size so that the confluent T-25 flask contains many fewer than the typical 4×10^6 cells and a concomitant loss of ACE activity with enzymatic (17) and other types of cell injury (18). Moreover, these studies suggest that *in vivo* injury that damages endothelial cells may be associated with impaired ACE activity and impaired O_2 responsiveness of ACE activity. The pathophysiological implications of such changes have not yet been explored.

In the analysis presented above, the differences in ACE activity seen in different oxygen environments are offered as a regulatory mechanism participating in the circulatory adjustments to hypoxic stress and at the time of birth. Hypoxia in the lung (as seen in the mountain climber) may serve to regulate the circulating levels of angiotensin II and bradykinin with consequent effects on vascular resistance, cardiac output, and the distribution of flow to vital organs. Similarly, differences in PO_2 from organ to organ (as seen in exercise) will adjust the local concentration of vasoactive peptides according to the local relationship between PO_2 and ACE. What, however, of the diseased lung, in which hypoxia develops inhomogeneously in areas of poorly distributed ventilation? Chronic lung disease has long been associated with lower than normal resting blood pressure and impaired blood pressure defenses against hypotensive insults (19). Because of the importance of the renin–angiotensin system (and probably the kallikrein–kinin system) in blood pressure regulation, we suspected that injury to the lung could interfere with ACE (and perhaps other aspects of the metabolic function of the lung). Such interference could contribute to abnormal blood pressure regulation in individuals with chronic lung disease.

To examine this hypothesis, we developed a model of pulmonary emphysema in the dog, produced by causing the animals to breathe a nebulized aerosol of elastase and papain for 6–10 weeks (20). We assessed ACE activity using both the blood pressure response bioassay of angiotensin I conversion and the bradykinin infusion technique with calculation of transpulmonary peptide concentration gradients. We also measured the capacity or efficiency of the pulmonary barrier to prevent passage of bradykinin by determining the concentration of bradykinin in the pulmonary artery sufficient to elevate the systemic arterial concentration into the detectable range of our radioimmunoassay ("spillover level") and to a concentration that produces systemic hypotension.

Angiotensin I conversion was moderately impaired in the emphysematous dogs: conversion fell from 94 to 80%. Unexpectedly the total lung clearance of bradykinin was high (92%) even though the animals were hypoxemic breathing room air, in the range of 70 mm Hg PO_2 because of their lung disease. In normal animals breathing hypoxic gas mixtures sufficient to lower their PO_2 to 70 mm Hg, conversion is of the order of 70%, so that the observed conversion was much higher than expected. Conversely, we observed an exaggerated hypotensive response to even slow infusions of bradykinin, and the measured spillover level was only one-fourth to one-half of that observed in normal animals with a comparable PO_2. Moreover, the spillover level was more typical of that seen in animals with a PO_2 below 40 mm Hg. The divergence between high clearance values and low spillover levels can be explained by the development of small foci of

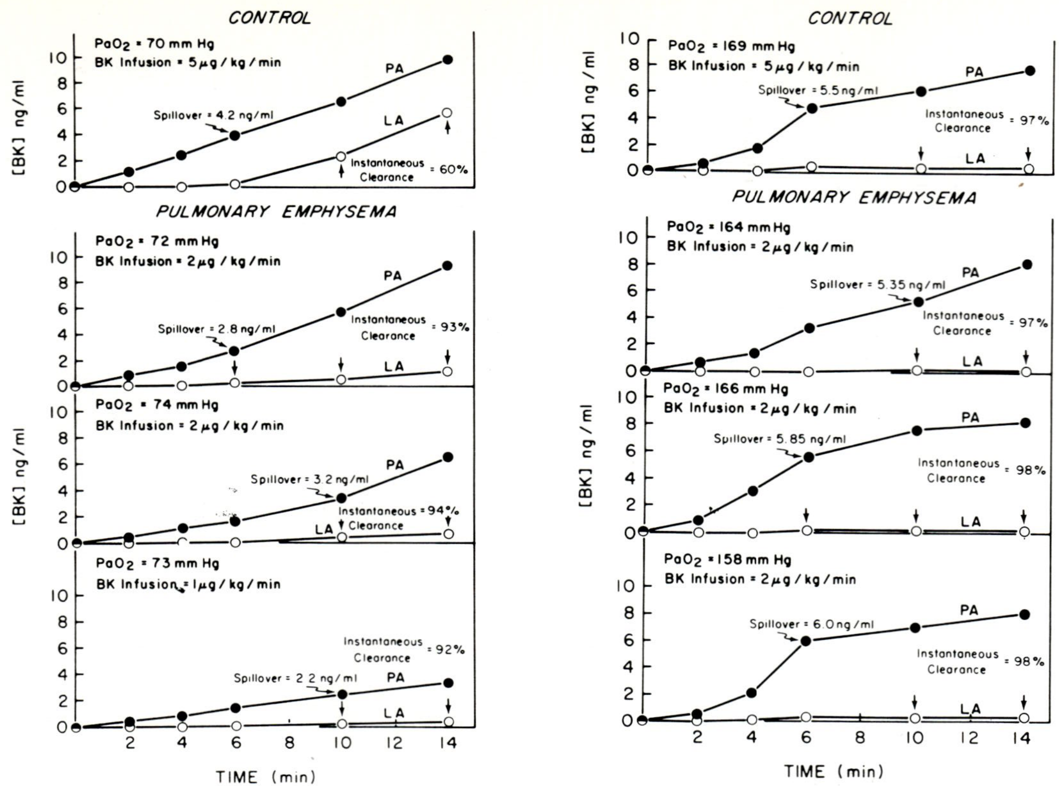

Fig. 8. (Left panel) Contrast in spillover levels and instantaneous clearance of bradykinin by the lung between three dogs with experimental pulmonary emphysema and in a control dog being ventilated with hypoxic gas sufficient to lower PO_2 to the same range as the diseased dogs. (Right panel) Effect of 40% oxygen supplementation on the same two measures of lung converting enzyme activity. Correcting regions of alveolar hypoxia corrects the defect in the pulmonary barrier to the passage of bradykinin. [From Stalcup *et al.* (20), with permission.]

regional hypoxia in the lung. This was established by providing supplemental oxygen to the dogs: with correction of their hypoxia, the lung once again became an efficient barrier to the passage of bradykinin. These experiments are illustrated in Fig. 8. Not only does development of hypoxic regions in the injured lung impair the pulmonary handling of these vasoactive peptides, assessment of this endothelial function affords a new measure of lung injury, with important implications for the control of blood pressure in stress.

Another possibly untoward consequence of the oxygen modulation of ACE activity is the development of pulmonary edema. In addition to the many different endothelial cell contributions to the regulation of coagulation, fibrinolysis, platelet function, protein synthesis and secretion, and processing of vasoactive substances, endothelial cells also serve as the first barrier to the leakage of proteins and water from capillaries. Hypoxia alone does not increase transvascular fluid exchange in the lung. However, in chronically instrumented sheep, we observed that when hypoxia occurs together with acidosis, a prompt increase in the flow of protein-rich lymph develops, indicating increased capillary permeability (21). In the same study, we measured circulating bradykinin, a well-known edema-producing substance when applied to systemic tissues. Respiratory acidosis activated kallikrein, leading to modest elevations in circulating bradykinin. In this and separate studies in which exogenous bradykinin was infused in sheep, we found no increase in capillary permeability when PO_2 was normal. However, when bradykinin endogenously produced by acidosis (21) or exogenously infused bradykinin (22) reached the hypoxic lung, permeability increased. In acidotic animals with superimposed hypoxia, ACE activity was decreased and total circulating peptide concentrations increased 25-fold (Fig. 9). Thus when hypoxia decreases ACE activity in the lung, bradykinin produces edema which would not occur when PO_2 is normal. In ischemic-acidotic systemic tissues, the extreme potency of bradykinin in increasing capillary permeability may have a physiological function. In very acidotic tissue, increasing permeability will increase the flow of buffering proteins through the interstitium, and wash out organic acids and cell wastes. This same increase in capillary permeability in the lung, however, will lead to the accumulation of interstitial fluid, which reduces lung compliance and increases resistance to air flow. Since bradykinin generation occurs in a number of clinical medical situations, e.g. sepsis, disseminated intravascular coagulation, these studies emphasize the importance of maintaining high alveolar PO_2 to prevent the complication of pulmonary edema.

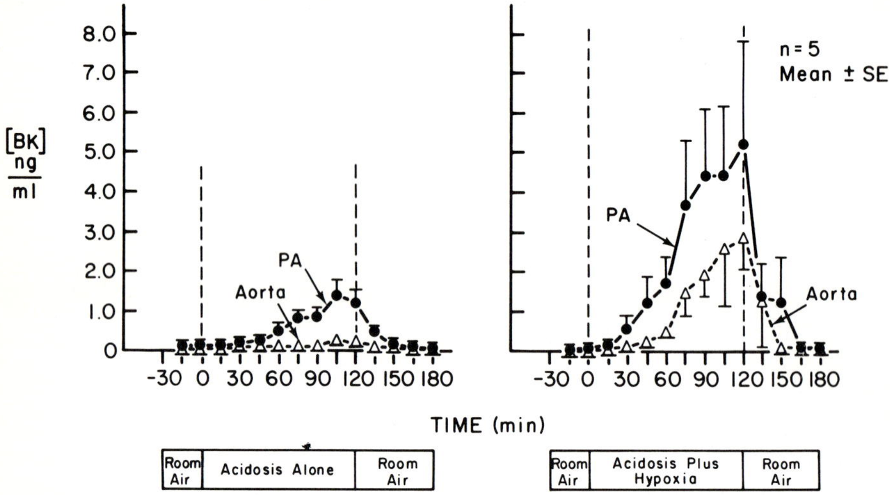

Fig. 9. Comparison of the effects of acidosis alone (pH = 7.2) versus acidosis plus hypoxia (PO_2 = 50 mm Hg) on the pulmonary artery (PA) and aortic concentrations of bradykinin. These experiments indicate that when bradykinin is generated endogenously, hypoxia inhibits ACE and causes a generalized increase in circulating bradykinin. [From O'Brodovich *et al.* (21), with permission.]

We have focused in this paper on the modulation of angiotensin converting enzyme activity by PO_2 in determining the overall effects of the renin–angiotensin and the kallikrein–kinin systems. There are, however, important interactions between these two systems and catecholamines, on the one hand, and prostanoids, on the other. Pursuit of these interrelationships under conditions of varying PO_2 is likely to deepen our understanding of the role of the metabolic activities of the lung in homeostasis as well as in pathophysiology.

ACKNOWLEDGMENT

Supported by NIH Grants 14218 (SCOR) HL07421, HL15832, and HL24997. Dr. Stalcup is an Established Investigator of the American Heart Association and receives additional support from the New York Heart Association.

REFERENCES

1. Crapo, J. D., Barry, B. E., Foscue, H. A., and Shelburne J. (1980) *Am. Rev. Resp. Dis.* **122**, 123–143.
2. Mellins, R. B., Davidson, D., and Stalcup, S. A. (1980) *In* "Metabolic Activities of

the Lung," Ciba Foundation Symposium 78 (new series), pp. 313–331. Excerpta Medica, Amsterdam.

3. Vane, J. R. (1969) *Br. J. Pharmacol.* **35,** 209–242.
4. Kellermeyer, R. W., and Graham, R. C. (1968) *N. Engl. J. Med.* **279,** 754–759, 802–807, 859–866.
5. Leuenberger, P. J., Stalcup, S. A., Greenbaum, L. M., Mellins, R. B., and Turino, G. M. (1980) *J. Appl. Physiol.* **78,** 308–312.
6. Leuenberger, P. J., Stalcup, S. A., Mellins, R. B., Greenbaum, L. M., and Turino, G. M. (1978) *Proc. Soc. Exp. Biol. Med.* **158,** 586–589.
7. Zakheim, R. M., Molteni, A., Mattioli, L., and Park, M. (1976) *J. Appl. Physiol.* **41,** 462–465.
8. Stalcup, S. A., Lipset, J. S., and Mellins, R. B. (1980) *In* "Metabolic Activities of the Lung," Ciba Foundation 78 (new series), pp. 293–311.
9. Stalcup, S. A., Lipset, J. S., Legant, P. M., Leuenberger, P. M., and Mellins, R. B. (1979) *J. Appl. Physiol.*, **46,** 227–234.
10. Melmon, K. L., Cline, J. M., Hughes, T., Niês, A. S. (1968) *J. Clin. Invest.* **47,** 1295–1302.
11. Heymann, M. A., Rudolph, A. M., Nies, A. S., and Melmon, K. L. (1969) *Circ. Res.* **25,** 521–534.
12. Davidson, D., Stalcup, S. A., and Mellins, R. B. (1981) *Circ. Res.* **48,** 286–291.
13. Stalcup, S. A., and Mellins, R. B. (1980) *In* "Pulmonary Diseases and Disorders" (A. P. Fishman, ed.), Vol. II, pp. 1653–1666. McGraw-Hill, New York.
14. Ryan, U. S., Ryan, J. W., Whitaker, C., and Chiu, A. (1976) Immunocytochemistry and Immunofluorescence. *Tissue Cell* **8,** 125–145.
15. Stalcup, S. A., Lipset, J. S., Woan, J. M., Leuenberger, P. J., and Mellins, R. B. (1979) *J. Clin. Invest.* **63,** 966–976.
16. Ryan, U. S., Mortara, M., and Whitaker, C. (1980) *Tissue Cell* **12,** 619–635.
17. Johnson, A. R., and Erdos, E. G. *J. Clin. Invest.* **59,** 684–695.
18. Block, E. R., and Stalcup, S. A. (1981) *J. Appl. Physiol.* **50,** 1212–1219.
19. Heistad, D. D., Abboud, F. M., Mark, A. L., and Schmid, P. G. (1972) *J. Clin. Invest.* **51,** 331–337.
20. Stalcup, S. A., Leuenberger, P. J., Lipset, J. S., Osman, M. M., Cerreta, J. M., Mellins, R. B., and Turino, G. M. (1981) *J. Clin. Invest.* **67,** 201–209.
21. O'Brodovich, H. M., Stalcup, S. A., Pang, L. M., Lipset, J. S., and Mellins, R. B. (1981) *J. Clin. Invest.* **67,** 514–522.
22. Pang, L. M., O'Brodovich, H. M., Mellins, R. B., and Stalcup, S. A. (1982) *J. Appl. Physiol.*, in press.

Index

F

G

H

I

L

M

N

O

P

R

S

T

U

V